T0280217

Advance Praise for *Creating Abundance*

"This is an important book! It traces the stream of biological innovation that over a period of two centuries has transformed the technical landscape of American agriculture. It demonstrates that biological innovations were essential for the movement of agriculture to new lands with more extreme climates; for maintaining and enhancing productivity in the face of evolving threats from pests, soil degradation and depletion; for creating modern livestock breeds; for enhancing feed efficiency; and for protecting animal and human health. It will be the standard against which the next generation of research in the history of agricultural technology will be evaluated."

– Vernon W. Ruttan, Regents Professor Emeritus, Department of Applied Economics and Department of Economics, University of Minnesota

"In *Creating Abundance* Alan Olmstead and Paul Rhode examine an important, but often neglected, aspect of American economic growth over the course of several centuries. Based on extensive research in a large variety of primary and secondary sources, including numerous state and national agricultural reports and related private documents, they describe in great detail the important biological developments influencing crop and livestock productivity. This is an original and superb work of scholarship. This book should be of interest to all economic historians, historians, and agricultural economists."

– Stanley L. Engerman, John H. Munro Professor of Economics and Professor of History, University of Rochester

"In an era focusing on sustainability and climate change, *Creating Abundance* should be required reading for environmentalists, economists, and policy makers, as well as for economic, environmental, and technological historians (and their students). This is so because even as virtuous cycles of biological innovation have overcome tough constraints, in agriculture biological contingencies and unanticipated consequences persist. The culmination of nearly two decades of extensive, patient, and inclusive research, *Creating Abundance* combines theoretical sophistication with a supple writing style to illuminate both a host of planning/policy advances and errors and the centrality of engaged practitioners' efforts in building flexible and responsive relationships with the natural world. A profound and compelling corrective to generations of scholarship on American agricultural development."

– Philip Scranton, Board of Governors Professor, History of Industry and Technology, Rutgers–Camden; Editor-in-Chief, *Enterprise and Society*

"Olmstead and Rhode's brilliant new book, *Creating Abundance*, is the most significant work on the economic history of American agriculture to appear in a generation. In emphasizing the early role and persistent importance of biological innovation in American agricultural development, the authors fundamentally recast the story of our farm sector. Rigorous in theoretical terms, well argued, and deeply researched, *Creating Abundance*, upon publication, becomes not merely the standard but the indispensable work in the field."

– Peter A. Coclanis, Albert R. Newsome Professor of History, UNC–Chapel Hill

"*Creating Abundance* is a panoramic survey of American agricultural development, dramatically altering the standard narrative focused exclusively on mechanization. In the process, Olmstead and Rhode relocate farming from the periphery squarely into the mainstream of American economic history. Anyone looking for the lessons of the American experience for developing economies should read this book."

– Gavin Wright, William Robertson Coe Professor of American Economic History, Stanford University

Creating Abundance

This book demonstrates that American agricultural development was far more dynamic than generally portrayed. In the two centuries before World War II, a stream of biological innovations revolutionized the crop and livestock sectors, increasing both land and labor productivity. Biological innovations were essential for the movement of agriculture onto new lands with more extreme climates, for maintaining production in the face of evolving threats from pests, and for the creation of the modern livestock sector. These innovations established the foundation for the subsequent Green and Genetic revolutions. This book challenges the misconceptions that, before the advent of hybrid corn, American farmers single-mindedly invested in laborsaving mechanical technologies and that biological technologies were static.

Alan L. Olmstead is Director of the Institute of Governmental Affairs, Distinguished Professor of Economics, and member of the Giannini Foundation of Agricultural Economics at the University of California, Davis. He is one of the six editors-in-chief of *Historical Statistics of the United States: Millennial Edition* (2006), and his writings appear regularly in leading economics and history journals.

Paul W. Rhode is McClelland Professor of Economics at the University of Arizona and Research Associate at the National Bureau of Economic Research. He is co-editor (with Gianni Toniolo) of *The Global Economy in the 1990s: A Long-Run Perspective* (2006). He is a frequent contributor to prominent economics and history journals.

The authors and publisher gratefully acknowledge publication support for this book from The Farm Foundation

and

The Giannini Foundation of Agricultural Economics, University of California

Creating Abundance

Biological Innovation and American Agricultural Development

ALAN L. OLMSTEAD
University of California, Davis

PAUL W. RHODE
University of Arizona

CAMBRIDGE
UNIVERSITY PRESS

University Printing House, Cambridge CB2 8BS, United Kingdom

One Liberty Plaza, 20th Floor, New York, NY 10006, USA

477 Williamstown Road, Port Melbourne, VIC 3207, Australia

314-321, 3rd Floor, Plot 3, Splendor Forum, Jasola District Centre, New Delhi - 110025, India

79 Anson Road, #06-04/06, Singapore 079906

Cambridge University Press is part of the University of Cambridge.

It furthers the University's mission by disseminating knowledge in the pursuit of education, learning and research at the highest international levels of excellence.

www.cambridge.org
Information on this title: www.cambridge.org/9780521673877

First published 2008
Reprinted 2018

A catalogue record for this publication is available from the British Library

Library of Congress Cataloging in Publication data
Olmstead, Alan L.
Creating abundance : biological innovation and American agricultural development /
Alan L. Olmstead, Paul W. Rhode.
p. cm.
Includes bibliographical references and index.
ISBN 978-0-521-85711-6 (hardback) – ISBN 978-0-521-67387-7 (pbk.)
1. Agricultural innovations – United States – History. 2. Agricultural biotechnology – United States – History. 3. Agricultural productivity – United States – History.
I. Rhode, Paul Webb. II. Title. III. Title: Biological innovation and American agricultural development
S494.5.I5O46 2008
630.973 – dc22 2008014226

ISBN 978-0-521-85711-6 Hardback
ISBN 978-0-521-67387-7 Paperback

To Marilyn, Dolores, Edward, and Jason

Contents

Acknowledgments

This book has been a long time in the making. Over the past decade, we have benefited immensely from the insights, advice, and assistance of many individuals. Although we had long been thinking about biological innovation, it was Julian Alston who prodded us to focus on the issue. Peter Coclanis, Peter Lindert, and Gavin Wright are probably pleased that the book is finished so we will stop bombarding them with questions. Vernon Ruttan has been particularly supportive throughout the duration of the project. Jeremy Atack, Susan Carter, Greg Clark, John Constantine, José Morilla Critz, Dana Dalrymple, Pete Daniels, Stanley Engerman, Price Fishback, Robert Gallman, Barbara Hahn, D. Gale Johnson, Bruce Johnston, Zorina Khan, Gary Libecap, Frank Lewis, Robert Margo, John Hebron Moore, Joel Moykr, Thomas Mroz, Philip Pardey, Vicente Pinilla, Wayne Rasmussen, Philip Scranton, James Simpson, Kenneth Sokoloff, Richard Steckel, Daniel Sumner, Richard Sutch, Nancy Virts, and Lorena Walsh are among a larger group of scholars that gave us comments on particular sections or chapters that helped sharpen the argument. Lee Craig, Michael Haines, John Murray, Winifred Rothenberg, and Thomas Weiss have supplied us with both data and advice that enriched our analysis.

We are economists, not agricultural scientists. The book would not have been possible without the insights of the many farmers and plant and animal scientists who helped us grapple with myriad technical issues. More than anyone, plant geneticist and wheat breeder Calvin Qualset dealt with our all-too-frequent request – "can you look at this and tell us if it makes sense?" Charles Schaller and Robert Webster also shared their knowledge of plant breeding and plant insects and diseases. Shelby

Baker, Dick Bassett, Fred Bourland, J. Jerome Boyd, Harry B. Collins, Tom Culp, Early C. Ewing, Jr., Janet Hudson, Hal Lewis, C. W. Manning, William Meredith, Larry Nelson, Gene Seigler, Macon Steele, and Henry Webb all gave us a better appreciation of the cotton plant and of cotton production. Tom Lanini informed us about weed ecology. Trish Berger, G. Eric Bradford, C. Christopher Calvert, David Hillis, David W. Hird, William H. Karasov, Kirk Klasing, Austin Lewis, and Edward Rhode responded to our questions on animal breeding and production. Roy Bainer and William Chancellor provided valuable perspectives on the workings of farm machinery.

We have been fortunate to have had the assistance of some very talented graduate students. We are indebted to Lisa Cappellari, Jeffrey Graham, and Janine L. F. Wilson. The chapter on California draws on an article coauthored with Susana Iranzo.

Throughout the project we have enjoyed working with Shelagh Matthews Mackay, whose editorial and organizational skills made our task much easier.

Our research has been facilitated by the Agricultural History Center and the Institute of Governmental Affairs at the University of California, Davis, and by grants from the International Centre for Economic Research, the Giannini Foundation of Agricultural Economics, and the National Science Foundation Collaborative Research Grants, Seeds and Slaves: Technological Change, Plantation Efficiency, and Southern Economic Development, SES-0550913 and SES-0551130. We would also like to thank the Giannini Foundation and its director, David Zilberman, and the Farm Foundation and its president, Walter J. Armbruster, for supporting the publication of the book.

1

Introduction

In the fall of 1843, John M. Harlan of Chester County, Pennsylvania, lamented that his potato crop was afflicted "by a very strange fatality which the oldest inhabitants have never before witnessed. The potatoes have been attacked by a disorder somewhat resembling the plague, generally called the rot."[1] The press and government officials tracked the devastation as the strange disease swept across the Northeast and into the Midwest between 1843 and 1846. Farmers had little recourse but to abandon potatoes and many had to cull their livestock for want of feed. This little-known American episode was a harbinger of even greater destruction; the New World disease soon jumped the Atlantic, bringing sorrow and suffering to European cultivators who were much closer to the Malthusian abyss than their American brethren. The "rot" was first reported in Belgium in June 1845 and within a few months was causing havoc in the Netherlands and Germany. The disease arrived in Ireland in September 1845, and within a few years roughly a million Irish were dead and another million had emigrated. The chain of causation was anything but straightforward. A crop disease imported to Europe from the Americas induced a massive flow of population in the opposite direction. Thus, a biological incident was a driving force in the globalization process.

The potato blight was one of a great number of diseases, insects, and weeds that significantly altered economic opportunities in the 200 years before World War II. A few prominent examples include *Phylloxera*,

[1] Peterson, Campbell, and Griffith, "James E. Teshemacher," 754–6; *Cultivator* (December 1943), 197. For other accounts in Pennsylvania see the *Pittsfield Sun* (5 October 1843), 2; *Barre Gazette* (6 October 1843), 3.

which attacked grapes; the Hessian fly and various rusts, which fed on wheat; San Jose scale, which smothered fruit trees; the boll weevil, which ravished cotton; and foot-and-mouth disease and tuberculosis, which attacked livestock. Each of these menaces threatened the economic well-being of American farmers, and some raised the specter of complete failure and crop abandonment. Fortunately, after initial periods of high losses, farmers typically learned to beat back the attacks, discovering how to coexist with the threats. In every instance, success required biological innovation.

Thousands of individual farmers contributed to crop and livestock advances, and many achieved what had been thought impossible. In 1900, two of America's preeminent plant scientists, Willet M. Hays and Andrew Boss of the Minnesota Agricultural Experiment Station, hitched up their horse and buggy and set out on a twenty-five-mile journey in search of a prize. They were drawn by the improbable report of thriving alfalfa fields far north of where the crop was thought to flourish. The reports proved accurate – an obscure farmer, Wendelin Grimm, had laboriously been acclimating alfalfa for forty years and had succeeded in developing a winter-hardy variety that would soon vastly extend the legume's domain.[2]

Agricultural production is location specific, at the mercy of conditions that likely differed from one farm to the next and certainly differed across states and regions. Settlement required biological adaptation as farmers harmonized production systems with specific local soil and climatic conditions. Learning did not end when the first settlers gained an agricultural foothold because, as areas matured, farmers generally switched to more intensive production patterns requiring new rounds of experimentation.

In addition to defending agriculture from insects, diseases, and weeds and developing better varieties and breeds for new and sometimes hostile areas, farmers, public officials, scientists, and nonfarm enterprises invested heavily in basic and applied research, often with spectacular results.[3] With rare exceptions, biological technologies in the nineteenth century were not patented, and thus the large body of research that has equated innovation with patents has concluded erroneously that there was little progress. But why would the curiosity, spirit of innovation,

[2] Russelle, "Alfalfa," 252.

[3] When discussing crops, modern agronomists have abandoned the term "variety" and adopted the term "cultivar" in its place because of the subtle distinctions as to what properly constitutes a distinct variety. Because the historical literature we cite consistently refers to "varieties," we have chosen to use the outdated terminology.

and quest for profit that drove technological change in other sectors have ceased to exist as one exited the factory, climbed out of the mine, or stepped off the train?

Everywhere we look, we find biological innovations transforming American agriculture. The varieties of wheat, corn, cotton, tobacco, and fruits grown at the dawn of the twentieth century were dramatically different from the varieties grown a hundred years earlier. Breeders also transformed farm animals – the sheep, swine, and cattle of 1940 bore little resemblance to those of 1800. These changes revolutionized farm productivity. Contemporaries understood this and in many instances attached innovators' names, dates, and places to the key changes. Joining Wendelin Grimm are Walter Burling, who gave Americans a type of cotton that, once acclimated, increased picking rates fourfold, Robert Livingston, E. I. du Pont, and others, who imported and bred early Merino sheep which, when improved by other American breeders, increased wool clip weights several-fold; Canadians Jane and David Fife, who helped open the northern plains to wheat cultivation; U.S. Department of Agriculture (USDA) wheat breeder Mark Alfred Carleton, who imported new wheat varieties from Ukraine, Russia, and central Asia and tested over a thousand varieties for rust resistance; Stephen Babcock, who devised a way to test milk for its butterfat content; William J. Spillman, who created hundreds of hybrid wheat varieties, the best of which advanced production in the Pacific Northwest; C. L. Marlatt and Charles V. Riley, whose entomological research saved countless farmers from ruin; and Eliza and Luther Tibbets, who helped introduce the navel orange tree to California, and Albert Koebele, whose beetles saved those trees from destruction.

In the nineteenth century the popular press, agricultural journals, Patent Commission reports, the U.S. Censuses, the USDA, state experiment stations, leading scientists, and farmers' journals all testified eloquently to the immense importance of biological innovation. A review of the USDA's *Yearbooks* in the second half of the nineteenth century indicates that roughly two-thirds of all articles addressed biological topics and that very few dealt with mechanical developments. Similarly, a content analysis of the leading agricultural journals published between 1860 and 1910 reveals that the space devoted to articles on new machinery and tools made up, on average, one-twelfth of the space allotted to biological topics such as animal husbandry, plant cultivation (including the use of fertilizer), pests and diseases, and water control practices.[4]

[4] Farrell, "Advice to Farmers," 209–41.

A similar picture emerges when one examines how Americans presented their agricultural sector to the world. We are familiar with the story of how Cyrus McCormick wooed the judges at the London Crystal Palace Exposition in 1851. But, before that, an 1839 London exhibition of Brother Jonathan – the mammoth, 4,000-pound American ox – drew 23,368 admirers, including "almost every branch of the Royal Family and the leading Agricultural Noblemen and Gentlemen."[5] The U.S. agricultural exhibit at the 1889 Universal Exposition in Paris (for which the Eiffel Tower served as the entryway) offers a sense of the balance between mechanical and biological displays. In addition to the wares of 32 farm equipment manufacturers, U.S. organizers exhibited 648 specimens of corn, wheat, and other small grains; 162 samples of tobacco leaf of every major type; 164 samples of cotton varieties covering the span of the cotton belt, including two complete "ready-to-pick" cotton plants; 159 examples of American wools including more than 140 fleeces; numerous examples and illustrations of citrus and deciduous fruits; 57 examples of grapes; 96 cases of assorted American wines, chiefly from California; and 349 specimens of carefully preserved insects that affected crops as well as 116 devices to combat pests and 40 different insecticides.[6] Biological topics, such as new plant varieties and animal breeds, harmful and beneficial insects, and fungal diseases, comprised the bulk of the 1,000-page report that Charles Riley wrote about agriculture at the 1889 Paris Exposition; his coverage of farm machines required less than 75 pages. This emphasis on biological innovation is not how we've been taught to view American history or the course of its agricultural development.

The Standard Account: The Primacy of Mechanization

According to the standard accounts, the period before the 1930s was primarily an epoch of mechanization – a time when machine makers such as Eli Whitney and Cyrus McCormick revolutionized American agriculture. In his influential history of American agriculture, Willard Cochrane argued that mechanization "was the principal, almost the exclusive, form of farm technological advance" between 1820 and 1920.[7] Yujiro Hayami

[5] *American Mammoth Ox.*

[6] U.S. Commission to the Paris Exposition 1889, "Number of Exhibitors in the United States Section," 342–3, and "List of the Exhibit Material," 861–76.

[7] Cochrane, *Development of American Agriculture*, 200, also 107. Griliches' treatment is less emphatic, but appears to lead to the same general conclusion. Griliches, "Agriculture," 241–5.

and Vernon Ruttan's trailblazing analysis of comparative agricultural development repeatedly echoes this general theme, as reflected in their treatment of small grains: "the advances in mechanical technology were not accompanied by parallel advances in biological technology. Nor were the advances in labor productivity accompanied by comparable advances in land productivity."[8] Other scholars have emphasized that biological innovations had to wait for relatively sophisticated advances in the plant and animal sciences. In his Richard T. Ely lecture, D. Gale Johnson succinctly captured this general view: "While American agriculture achieved very large labor savings during the last century, which made it possible to continue expanding the cultivated area with a declining share of the labor force, output per unit of land increased hardly at all.... The revolution in land productivity based on important scientific advances began very recently; its beginnings were in the 1930's with the development of hybrid corn and followed over the next several decades with equally major improvements in the yields of grain sorghum, wheat, rice, and cotton."[9] More generally, Johnson maintained that land-augmenting investments were relatively unimportant until the World War II era.

The notion that the nineteenth century was largely an era of labor-saving productivity change in agriculture is also part of the mantra of most economic historians. William Parker and Judith Klein's classic study found that, between 1840 and 1910, output per unit of labor increased more than fourfold, whereas output per unit of land only increased by about 10 percent. They attributed the vast majority of the increase in efficiency to mechanization.[10] Jeremy Atack and Fred Bateman's prominent contribution, *To Their Own Soil*, echoes this general proposition: "The great improvement in acreage yields lay almost a century into the future when chemical fertilizers, hybrid seeds, irrigation, and various scientific developments became available to farm operators. Some technological devices designed to raise labor productivity were, however, becoming available during the nineteenth century. Mechanical rather than chemical or biological, these improvements operated primarily through their effect on the usage of labor."[11] Peter McClelland's treatise on the first agricultural revolution in North America, in the period 1783–1860, devotes about 15 pages to crop rotations, fertilizers, and animal breeding (in

[8] Hayami and Ruttan, *Agricultural Development*, 209.
[9] Johnson, "Agriculture and the Wealth of Nations," 7–8.
[10] Parker and Klein, "Productivity Growth in Grain Production."
[11] Atack and Bateman, *To Their Own Soil*, 186.

appendices) and about 330 pages to farm implements.[12] Perhaps the strongest statement of the view that biological innovations were unimportant comes from the trio of eminent economic historians Atack, Bateman, and Parker: "Land abundance that encouraged extensive rather than intensive agriculture contributed to the general lack of interest in land productivity by nineteenth-century American farmers, while stimulating their interest in mechanization to substitute for labor."[13] In their view, not only did American farmers fail to invest much in biological innovation to increase land yields, *they were not even interested in doing so*. The view that mechanization was the dominant source of nineteenth-century agricultural productivity change is now a prominent fixture in standard economic history textbooks.[14] Scholars within the USDA, including Wayne D. Rasmussen, America's most eminent agricultural historian of the second half of the twentieth century, also steadfastly subscribed to the primacy of laborsaving mechanization.[15] It is little wonder that the view has permeated the broader literature on American history. Laborsaving technical change in agriculture is often equated with mechanization, as if one does not exist without other.

Induced Innovation and American Agriculture

The induced innovation hypothesis represents one of the most prominent models employed to explain technological change in American agriculture. The hypothesis, which is essentially a long-run version of the factor-substitution argument, treats the evolution of technology and institutions as endogenous responses to the forces of factor supply and product demand. In terms of its simplicity, intuitive appeal, and number of adherents, it has no close competitor. Hayami and Ruttan's book, *Agricultural Development: An International Perspective*, is the flagship of a large literature that develops, refines, and tests the induced innovation model in various national settings. The model suggests that rational farmers should have invested in saving labor because, according to the model's strongest advocates, labor was the scarce factor of production, and it was becoming

12 McClelland, *Sowing Modernity*.

13 Atack, Bateman, and Parker, "Farm, the Farmer, and the Market," 263.

14 For examples, see Walton and Rockoff, *History of the American Economy*, 334; Ratner, Soltow, and Sylla, *Evolution of the American Economy*, 264–5; Atack and Passell, *New Economic View*, 280–2; and Hughes, *American Economic History*, 275–6.

15 Rasmussen, "Impact of Technological Change," 578–91; Loomis and Barton, "Productivity of Agriculture," 6–8.

scarcer over the nineteenth century. As an example of this reasoning, "the evolution of the mechanical equipment is designed to bring about larger output per worker by increasing the land area that can be operated per worker. Furthermore, it seems apparent that the production functions which described the individual grain-harvesting technologies, from the sickle to the combine, were induced by changes in relative factor costs, reflecting the rising resource scarcity of labor relative to other inputs."[16] The model's advocates claim that the history of American agriculture conforms nicely to the model's predictions – there was a well-ordered design as farmers single-mindedly innovated to save labor. It is time to take a closer look at the induced innovation hypothesis and examine more closely the data that define the broad sweep of America's agricultural development.

The agricultural history of the United States plays a pivotal role in Hayami and Ruttan's analysis by providing a prime testing ground for the model. In their view this history, at least until about 1940, is the example *par excellence* of a high-wage economy focusing its inventive energies to save labor. A useful entry point into their analysis is the decomposition of changes in output per unit of labor into changes in output per unit of land and changes in land available per unit of labor. These two forces are viewed as "relatively independent" and are associated with two alternative development paths – one is typified by the U.S. experience, in which progress in mechanical technology "facilitated the substitution of other sources of power for human labor," and the other is typified by the Japanese experience, in which progress in biological technology increased the productivity of land. In the United States the dominant trend has been the rising land-to-labor ratio associated with the mechanical path of development.[17] Biological innovation became significant only in the 1930s when, according to Hayami and Ruttan, an increasing scarcity of land and a rapid decline in commercial fertilizer prices made such innovations profitable. This general story has been widely reproduced.

Hayami and Ruttan claim impressive empirical support for the induced innovation hypothesis, noting that it has been tested successfully against the historical records of the United States, Japan, Taiwan, Korea, the Philippines, Denmark, France, Germany, and Great Britain. On the basis of these tests, they argue that their "model provides powerful insight into

[16] Hayami and Ruttan, *Agricultural Development*, 79.

[17] Ibid., 171.

the development process in both developed and developing countries."[18] However, their brief history of the "dynamic sequences" of the evolution of key agricultural inventions rests entirely on technological change in wheat production, focusing chiefly on the diffusion of the mechanical reaper. But the wheat sector was not representative of the larger agricultural economy: between 1880 and 1930 the land-to-labor ratio in the major wheat-growing regions increased 115 percent, while over the same period the increase for the rest of the nation was only 11 percent – less than the increase in Japan. In the Pacific states the ratio actually fell. (Chapter 8 will explain why the change in the land-to-labor ratio in this progressive and highly mechanized agricultural region fell.) Thus, for the period before 1930, the key generalization about the entire country is evident only in the interior, wheat-growing states.[19]

At the very core of the test of the induced innovation story in the United States is the repeated assertion that, over the course of the nineteenth century, the price of land fell relative to the wage rate, thereby inducing laborsaving mechanization. This assertion about relative price movements has been repeated so often that it has become one of the key stylized facts of American history. Scrap one more stylized fact because, throughout most of the nineteenth century, the price of land relative to wages was *rising*, not falling. The changes were dramatic. Between 1790 and 1850 the number of days of farm labor required to purchase an acre of agricultural land more than doubled.[20] After 1850 the data become more abundant, and we can provide an index of the key relative price ratios for a 130-year period. Table 1.1 shows that the price of land relative to the wage ratio roughly tripled from 1850 to 1910. This means that real agricultural wages expressed in terms of land fell to about one-sixth of their 1790 level. Thus, the long-run relative land and labor prices were inexorably moving in the wrong direction to explain mechanization in the crucial period before 1910.

Table 1.1 also shows that the price movements after 1910 fail to support the assertion that an increasing relative scarcity of arable land and

[18] Ibid., 92.

[19] Olmstead and Rhode, "Induced Innovation (1993)," 100–18, and "Induced Innovation (1998)," 103–19.

[20] Christensen, "Land Abundance," 313; Lindert, "Long-Run Trends." Rental rates are a better measure of land scarcity than land prices, and rental rates did not rise nearly as fast as the price of land. Better accounting for this and other issues improves the performance of the induced innovation model in tracking American agricultural performance. Olmstead and Rhode, "Induced Innovation (1998)," 103–19.

TABLE 1.1. *Long-Run Trends in Factor Price Ratios in U.S. Agriculture*

1910 = 100	Land Value/ Wage Rate	Wage Rate/ Machinery Price	Land Value/ Fertilizer Price
1850	34	38	
1860	41	54	
1870	63	30	
1880	64	54	24
1890	71	70	40
1900	64	79	47
1910	100	100	100
1920	73	154	93
1930	69	128	110
1940	58	104	79
1950	43	164	123
1960	56	183	289
1970	56	210	583
1980	99	185	812

Source: Olmstead and Rhode, "Induced Innovation (1993)," 105.

the falling price of fertilizer over the previous decades induced the biological innovations of the 1930s.[21] In actuality the price of land relative to farm wages was falling between 1910 and 1950. Thus, not only were farm wages (relative to land values) falling in the early period when, according to the model, they should have been rising, in the later period they were rising when they should have been falling. The table further shows that the actual trend in the relative price of fertilizer was almost always moving in the wrong direction to support the induced innovation interpretation of American agricultural history. The upshot is that the biological revolution of the post–hybrid corn era simply could not have begun at a worse time for the adherents of the hypothesis that changing relative scarcity induced innovation. More generally, over the entire span for which we have data, the two key relative price series that represent the empirical foundation for the induced innovation model were almost always moving in the wrong direction. If one adheres strictly to the model's predictions, nineteenth- and early twentieth-century American farmers should have invested significantly in biological, land-augmenting technologies. The historical record actually supports this prediction.

[21] Hayami and Ruttan, *Agricultural Development*, 174, 177–8, 192–3.

The induced innovation model emphasizes the role of demand in determining the pace and pattern of invention and innovation, but there are also supply-side forces at work. Many argue that biological innovations were conceptually more difficult than mechanical changes and so the former had to await fundamental advances in basic science.[22] This critique applies to the modern genetic revolution, but the term *biological innovation*, when applied to other countries, such as Japan, simply refers to nonmechanical innovations – new plant varieties, fertilizers, pesticides, irrigation or drainage systems, improved cultural practices, and the like. This is also our meaning of "biological innovation." Such innovations did not require a high level of sophistication – although for their day many innovations represented significant scientific advances. It makes no more sense to assert that plant and animal scientists and farmers could not innovate because they did not understand how to splice genes than it is to question the achievements of the machinists who created industrial technologies because they knew nothing of theoretical mechanics.

Most scholars who emphasized the historic role of mechanization were impressed by crop yield data. In the United States, national average yields per harvested acre of most crops increased little, if at all, over the broad span of the nineteenth and early twentieth centuries. Yields typically shot upward sometime between 1930 and 1950, depending on the crop. But this does not imply that there were few biological innovations before 1940, nor does it warrant classifying this earlier period as primarily an era of mechanical change. Figure 1.1 offers evidence on the growth in the productivity of land and labor. Between 1910 and 1940, crude indicators show that output per unit of land grew at a rate of about 0.94 percent per year. Between 1940 and 1980, output per unit of land grew by 1.95 percent a year, more than double the previous rate. But a look at the change in the growth of labor productivity shows a striking result. Between 1910 and 1940, labor productivity in agriculture grew at a rate of 1.4 percent per year, but between 1940 and 1980 it grew at a rate of 5.5 percent per year, or more than three and one-half times as fast. Both land and labor productivity growth rates soared in the post–World War II era, but to dub this period the era of biological change is clearly misleading. It is important to emphasize that measures such as changes in output per acre might offer a crude index of the growth in land productivity, but they reveal nothing about biological innovation. Let's see why.

[22] Cochrane, *Development of American Agriculture*, 201–2.

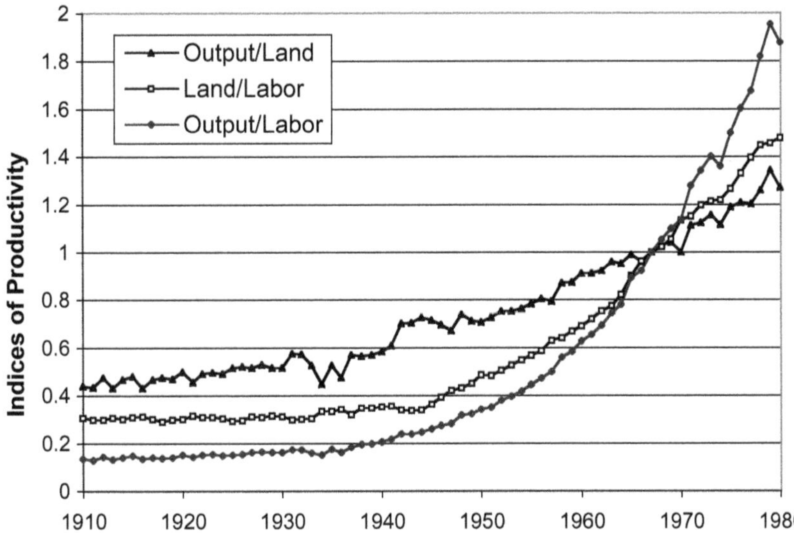

FIGURE 1.1. Indices of agricultural productivity, 1910–80. *Source*: U.S. Department of Agriculture, "Economic Indicators of the Farm Sector."

Mapping the Effects of Innovations

It has become common practice to associate mechanical innovations exclusively with saving labor and biological innovations only with saving land. For many the argument also runs in the other direction – an increase in labor productivity is a sign of mechanization, and the absence of yield improvements implies little biological innovation. These relationships have been widely asserted, more or less as an act of faith, since the general paradigm has not been put to historical test, nor have its implications been rigorously considered.

On the assumption that many readers may be skeptical of our claims, we offer some examples. The gasoline tractor was undoubtedly one of the most revolutionary technological innovations in the history of agriculture. Just to be clear, the tractor was in fact a machine, and it did save labor – lots of labor. Our estimates suggest that by 1960 it had reduced annual labor needs in agriculture by roughly 1.7 million workers. But the tractor was also one of the greatest land-saving innovations in the history of agriculture, allowing farmers to convert about 160 million acres of crop and pastureland from supporting draft animals to providing products for human consumption (we expand on this issue in Chapter 12). But there's more. By allowing farmers to perform more timely and more thorough work, the tractor had a direct impact on yields. Thus, it is very difficult to conclude whether one of the seminal mechanical inventions of all time was, on balance, land- or laborsaving.[23] Other examples abound. John Deere's plows clearly saved labor, but they also allowed farmers to do better work and to cultivate more often (because of the labor saved on any one pass over the field). Thus, like the tractor, improved implements also made land more productive.

Many biological innovations clearly saved labor. Chapter 4 analyzes how Mexican cotton, introduced in 1806, started a chain of events that more than tripled the quantity of cotton a worker could pick in a day. Thus, the greatest laborsaving innovation in the long history of the cotton industry, from the invention of the cotton gin to the diffusion of the mechanical picker roughly 150 years later, was a biological, not a mechanical, breakthrough. Other examples are easy to find. Given the terminology used in the literature, herbicides are land-saving biological inputs. But herbicides are substitutes for hand weeding, hoes, plows, harrows, and disks. What sense does it make to call an improved hoe laborsaving and an improved herbicide land-saving? Pesticides are classified as land-saving biological inputs. By reducing insect damage they increase yields. However, they also saved labor because, in the nineteenth century, workers often spent considerable time trying to control insects via very labor-intensive means – including picking them off the foliage by hand.

The conceptual problems are even more perplexing because, in many cases, there were close links and interaction effects between mechanical and biological advances. It is well known that the diffusion of a successful mechanical tomato harvester had to await the development of a new tomato designed to ripen uniformly and survive rough handling. It is less well known that a similar breeding process created cotton varieties suitable for machine picking. Examples abound. Fruit trees have been redesigned (by pruning and breeding) to make picking easier. As a rough generalization, hand harvesting rates for many crops are largely a function of area, not yields. Thus, technologies that increase yields also increase labor productivity. The most important interaction effect is both obvious and, as far as we've found, entirely ignored in the induced innovation literature. The mechanization of agriculture in the period before roughly 1910 generally entailed the substitution of machines and horse power

[23] Olmstead and Rhode, "Reshaping the Landscape," 663–98.

(and some steam engines) for less complicated tools and human power. What seems to have been ignored is that horses were a biological input. Chapter 12 will explore the implications of this observation and delve into the more general issue of draft power.

Eli Whitney's saw-toothed cotton gin is a premier example of a labor-saving innovation. Various accounts assert that, after some refinements, the gin allowed one worker to do the labor of 50 to 100 workers picking the seeds by hand. What better example of the substitution of machines for labor? However, this is only part of the story, because southern historians have long credited the gin with facilitating the spread of the cotton industry over a vast domain. At the time of the gin's invention, the commercial cotton industry was confined to a small area in the Southeast that was suitable for growing Sea Island varieties and to patches in the Gulf region that could grow a Siamese variety. These cottons could be ginned with a more primitive roller gin. By breaking the technological constraint on growing fuzzy seeded upland cotton, Whitney's contraption made it possible for landowners to convert en masse from less labor-intensive activities to cotton. Thus, the longer-run impact of the cotton gin was almost surely to increase southern labor requirements. In the terminology of economics, the output effect was large and worked counter to the substitution effect. Clearly the usefulness of the strict association of biological (or mechanical) innovations with crop yields (or labor productivity) warrants examination. There is another important reason why focusing on changes in yields per acre may fail to capture the importance of biological innovation – the Red Queen effect.

The Red Queen and Maintenance Research

In most areas of economic activity, technological advances last forever. A new idea, method, or type of machinery only becomes obsolete when it is replaced through the competitive process by an even more productive innovation. But the old idea is neither destroyed nor lost, it is simply no longer efficient. In many instances the new is built on the old, giving rise to the cumulative nature of technological development. In addition, for most sectors innovations tend to be highly footloose – an idea or a machine developed in one country is apt to work equally well in a wide variety of environments. However, agriculture is different; human intervention into biological processes predictably produces natural reactions in the form of insects and pathogens that inevitably erode the productivity of past innovations.

Agricultural production is an inherently biological process that uses organisms to transform sunlight and soil nutrients into food and fiber for humans. Agriculture is not akin to assembling parts in a factory to construct a machine. Mechanization in agriculture can reduce labor requirements, but the biological process of transforming energy and chemicals remains at the core of farming. The efficiency of the transformation process is always being shocked by evolving environmental forces such as insects, diseases, and weeds. Much like Joseph Schumpeter's description of capitalism, agriculture is, by its very nature, always in motion. Farmers could not stand still even if they wanted. To some extent the challenges to the system are endogenous because production decisions can affect the rate of decay. For example, crop rotation systems can reduce the buildup of insect populations. Conversely, vast areas of monoculture, as emerged in parts of the United States, intensified pest problems. In other instances, exogenous shocks, such as the introduction of invasive species from foreign lands, dramatically reduced yields. Introducing production to new environments required adapting organisms to meet new, often hostile, conditions. This took time and often required compromising yields. Mechanization could assist in these processes but it was always tangential to the fundamental biological endeavor.

Farmers have long understood the Red Queen's dictum – they have to run fast just to stay in one spot. For this reason a mere examination of ex post yields or some constructed measure of land productivity will most likely vastly understate the importance of biological innovations. This issue will come up repeatedly in the following chapters. We show that, before the age of modern crop and animal sciences and before the birth of the modern chemical industry, farmers and scientists closely studied the habits of insects and the nature of diseases and scored some remarkable successes in the never-ending battle against agricultural pests. Some biological innovations had dramatic spillover effects on human health. For example, steps taken to improve milk quality were, by 1940, saving well over twenty-five thousand lives a year in the United States. Such tangible savings of human lives and health are not accounted for in the standard measures of agricultural output and thus are ignored in the usual "productivity" measures. Clean milk campaigns, pure seed movements, fights against insects and diseases, and many other issues required overcoming the free rider problem, which by necessity often led to government intervention. One of the truly significant nonmechanical innovations of the nineteenth and early twentieth centuries was the creation of effective state and federal research and regulatory institutions primarily aimed at developing biological advances and controlling biological hazards.

Contours

Farmers, public officials, and scientists devoted enormous energy to promoting biological innovations, which had a huge impact on agricultural productivity. The following chapter develops two recurring themes. Without a stream of biological innovations, farmers could not have pushed wheat production into the colder and more arid west, and diseases and insects would have drastically lowered yields and labor productivity. Government research and outreach programs played a fundamental role in this process. America's rise as a major wheat exporter crucially depended on previous international transfers of biological technologies. Our analysis of corn also illustrates the importance of plant breeding because distinct northern and southern types were intermixed to increase yields and lower the barriers to production. Understanding the rise of the corn belt offers insights into the South's increased specialization in cotton production following the Civil War.

Cotton plays a major role in American historiography. Chapter 4 analyzes the creation of the upland cotton varieties, which made the United States the undisputed leader in the international cotton market and created the backbone of the antebellum plantation economy. Chapters 5 and 6 analyze how countervailing forces eroded many of the early technological advances, leading to a decline in the quality of American cottons following the Civil War. Chapter 5 deals exclusively with cotton pests, with an emphasis on the boll weevil. We offer new evidence on both the weevil's diffusion and the destruction it wrought. Chapter 6 analyses how market imperfections, together with the boll weevil, contributed to the decline in cotton quality and how the USDA and private cotton breeders contributed to the modernization of the cotton sector after World War I. This episode saw one of the largest and most successful agricultural cooperative movements in U.S. history – a movement specifically designed to promote biological innovation.

Chapter 7 explores the innovations that made tobacco the colonies' first cash crop. Tobacco is a very malleable plant, which allowed breeders to create a wide range of varieties in response to changing geographical conditions and consumer preferences. The final chapter on crops deals with the rise of the nation's leading agricultural state: California's agricultural diversity makes it an ideal historical testing ground. The transition from grains to intensive crops, the repeated attacks of insects and diseases, and the prominent role that researchers played in mitigating the effects of these onslaughts all point to the intensity of biological learning.

Our discussion of animals starts with an analysis of the characteristics of major livestock species and of the interaction of the crop and animal sectors. We show that Wendelin Grimm's innovation was part of a much larger story that saw revolutionary changes in animal feeds. Chapter 10 focuses on how farmers imported more efficient animals from around the world, and how American breeders then redesigned those horses, sheep, swine, and cattle to meet changing environmental conditions and evolving market signals. The livestock on American farms in 1940 embodied generations of biological investments. Chapter 11 tells the story of the rise of the modern dairy industry and the creation of systems and animals that allowed farmers to roughly triple milk yields and dramatically increase food safety. Chapter 12 highlights the interaction effects linking crops and animals as we trace the evolution of draft power from oxen, to horses and mules, to tractors.

In total this book portrays a far more dynamic and creative story of American agricultural development than that found in most accounts – how could it be otherwise given the widespread neglect of biological innovations. The book deals in large part with how ecology affected agricultural development and in particular how technological change focused on overcoming the barriers imposed by climate, geography, and pests. These are issues that continue to confront agriculturalists everywhere.

2

The Red Queen and the Hard Reds

Productivity Growth in American Wheat, 1800–1940

In the period before 1940, a progression of mechanical innovations – the mechanical thresher, the reaper, the binder, and the combine – transformed American wheat production. The stories of the accomplishments of leading inventors and manufacturers are a rich part of American folklore, and deservedly so. Cyrus McCormick, after all, was the man who "made bread cheap!" By allowing farmers to substitute animal power for human labor, machines vastly increased labor productivity. According to the conventional USDA estimates, producing a bushel of wheat required 140 minutes in 1840, but only 40 minutes by the late 1930s. While labor productivity was rising rapidly, the yield per acre of land harvested was almost constant. These observations have led to the erroneous conclusion that, prior to 1940, mechanization was the source of almost all productivity growth and that biological innovations were unimportant. In fact this misreading of the history of wheat production has been a major contributor to the systematic misunderstanding of the dynamics of America's agricultural past.

Contrary to the conventional wisdom, the nineteenth and early twentieth centuries witnessed a stream of biological innovations in the wheat (and small grains) sector that rivaled the importance of mechanical changes for agricultural productivity growth. These new biological technologies addressed two distinct classes of problems. First, there was a relentless campaign to discover and develop new wheat varieties and

This chapter is an expanded version of our article, "The Red Queen and the Hard Reds: Productivity Growth in American Wheat, 1800–1940." *Journal of Economic History* 62, no. 4 (2002): 929–66.

cultural methods that would allow the wheat frontier to expand into the northern prairies, the Great Plains, and the Pacific Coast states. Without these technologies, western yields would have been significantly lower, and vast areas of the United States and Canada would have been unsuitable for commercial wheat production. Second, researchers and wheat farmers made great strides in combating the growing threat of yield-sapping insects and diseases, many of which were the unintended consequences of biological globalization. With the large-scale importation of Eurasian crops to North America came hitchhikers who fed on those crops. In the absence of vigorous efforts to maintain wheat yields in the face of evolving foreign and domestic threats, land and labor productivity would have been significantly lower. In effect farmers practiced integrated pest management (IPM), in which the sensitive details of farming systems evolved in response to new threats and changing knowledge. Our analysis leads to significant revisions in the standard measures of the sources of labor productivity growth in American small grain production between 1839 and 1909, with biological innovations roughly equaling the importance of mechanical advances.

Cornerstones of the Conventional Wisdom

The prevailing dogma that biological innovations were unimportant in wheat cultivation before 1940 rests on two fundamental building blocks. The first is the time series for U.S. yields displayed in Figure 2.1. The figure shows the growth trend with a break in 1939, which maximizes the fit. These data demonstrate that output per acre harvested was nearly constant from 1866 to 1939, growing only about 0.15 percent per annum. This amounted to a meager 1.75-bushel increase over nearly three-quarters of a century. After 1939, the growth rate jumped up to 2.23 percent per annum and yields more than doubled in the course of forty years.[1]

[1] The use of average national yields to measure land productivity is subject to obvious conceptual difficulties. S. C. Salmon noted that "yields per acre are often used to measure or indicate technological improvements. They are reasonably good indices in countries in which acreage remains fairly constant or where the productivity of the new acreage does not materially differ from the old. They may be misleading, however, in a country such as the United States, where the acreage has greatly increased in areas where the conditions for growth are quite different. If an improvement reduces cost per acre, thereby permitting a larger expansion on less productive land, average over-all acre yields may actually be reduced." Although Salmon was reflecting on biological innovations, mechanization also lowered the threshold yield necessary for profitable cultivation. The expansion of the

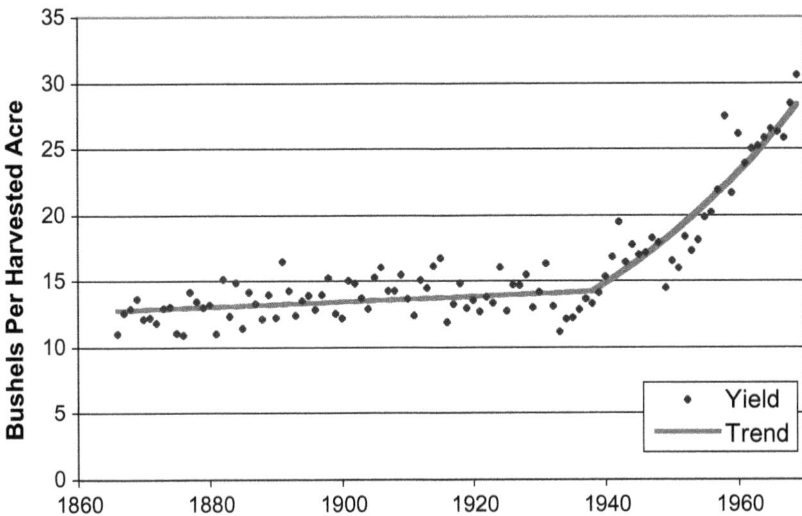

FIGURE 2.1. U.S. wheat yields, 1866–1969. *Source:* Carter et al., *Historical Statistics*, Tables Da717–729.

The second building block is the research linking labor productivity to mechanization. The most influential contribution is William Parker and Judith Klein's classic 1966 study of labor productivity growth in wheat, oats, and corn production between 1839 and1909. Their findings have become a permanent fixture in economic history textbooks and scholarly treatments of agricultural productivity growth.[2] Table 2.1 reproduces the key results of their analysis for wheat.[3] Overall, Parker and Klein

frontier of cultivation would lower measured yields, even in the absence of changes in the productivity of a specific acre of land, which in fact occurred in the United States. Salmon's views deserve our attention. He was one of America's leading agronomists and was responsible for introducing the first dwarf varieties into the United States from Japan following World War II. Salmon, Mathews, and Luekel, "Half Century of Wheat Improvement," 5. Replacing the national wheat yield data with available county-level data generates an earlier break point. See Chapter 13.

[2] 1909 is a shorthand. Parker and Klein actually used 1907–11 as the terminal years. Parker and Klein, "Productivity Growth in Grain Production," 523–82. An earlier USDA study for the interwar years reached findings similar to that of Parker and Klein about the relative importance of mechanization and yield changes for labor productivity. See Hecht and Barton, "Gains in Productivity of Farm Labor."

[3] Parker and Klein divide the United States into three major regions: the Northeast (including PA, NJ, NY, VT, MA, NH, ME, CT, RI), South (DE, MD, VA, WV, KY, TN, NC, SC, GA, FL, AL, MS, LA, AR), and West (everywhere else). In their detailed analysis, they broke the West into five regions: Corn (including OH, IN, IL, IA, MO), Dairy (MI, MN, WI), Small Grain and Western Cotton (NE, KS, SD, ND, MT, TX, OK), Range (NM,

found that wheat output per hour increased 4.17-fold over this period. In their estimation, the driving force was mechanization, which alone would have increased output per hour by 2.45-fold. The interaction of mechanization with western expansion raised this ratio to 3.77 (or about 90 percent of the total increase in labor productivity). By way of contrast, they found that biological advances played a minor role; holding all else constant, yield changes increased labor productivity by only 18 percent. Their results reinforce the general view that significant biological changes did not begin until the mid-twentieth century.[4]

A closer look at the Parker–Klein study offers insights into two other fundamental issues – changes in land productivity and the role of western settlement in the growth of total production. Parker and Klein considered output per acre only as an indirect source of labor productivity movements. With a slight change in perspective, the information in Table 2.1 also reinforces the common claim that western settlement shifted wheat cultivation onto less productive soils. In the absence of these shifts, Parker and Klein's data suggest that 1909 yields would have been 29.8 percent higher than in 1839 and 4.3 percent higher than they actually were.[5]

Over the period 1839–1909, U.S. wheat production increased almost eightfold, rising from roughly 85 million to 640 million bushels.[6] The rapid growth in output was crucially dependent on the western expansion of cultivation. These geographic shifts are illustrated in Figure 2.2, which maps the distribution of U.S. wheat output in 1839 and 1909, and in Table 2.2, which shows the changing geographic center of production

AZ, CO, UT, NV, WY), and Northwest and California (ID, OR, WA, CA). They then estimate for each region the labor required in the pre-harvest, harvest, and post-harvest operations – the direct requirements reflect the state of mechanization. The last operation is modeled to depend directly on output, whereas the first two depend directly on acreage. To determine pre-harvest and harvest labor requirements per bushel they divide by the crop yield. This is the only way that crop yields, embodying the state of biological knowledge, enter the calculation. After deriving the regional labor–output ratios, Parker and Klein use the region's weights in total production to obtain the U.S. average labor requirement per bushel of wheat. By substituting the direct labor requirements, yields, and regional weights for different periods, they decompose changes in labor productivity into the effects of (and interactions between) mechanization, biological change, and western settlement, respectively.

[4] The sum differs from 100 percent due to nonlinearities.

[5] Franklin M. Fisher and Peter Temin, "Regional Specialization," 134–49, criticized Parker and Klein for focusing exclusively on average labor productivity rather than total factor productivity.

[6] More precisely, this was a 7.54-fold (or 2.9 percent per annum) increase, which exceeded the growth in labor productivity noted previously. Thus, the wheat sector was continuing to absorb labor over this period.

TABLE 2.1. *Parker and Klein's Analysis of Labor Productivity Growth in Wheat Cultivation, 1839–1909*

	Period	United States	R1 Northeast	R2 South	R3 West	R3 Breakdown: Corn Belt	Western Dairy	Sm. Grain & W. Cotton	Range	California and Northwest
A: Pre-harvest Labor	1	13.60	19.10	11.30	12.40	12.40		nd	nd	nd
(hours per acre)	2	5.50	11.60	10.70	4.70	5.50	6.10	4.20	6.00	3.20
B: Harvest Labor	1	13.90	15.00	12.50	15.00	15.00		nd	nd	nd
(hours per acre)	2	2.40	3.00	3.00	2.30	3.00	3.00	1.80	7.50	2.00
Y: Yield per Acre	1	11.30	14.50	8.40	13.00	13.00		nd	nd	nd
(bushels per acre)	2	14.00	17.50	12.30	14.00	15.80	15.30	12.00	18.60	19.40
C: Post-harvest Labor	1	0.73	0.73	0.73	0.73	0.73		nd	nd	nd
(hours per bushel)	2	0.20	0.19	0.29	0.19	0.22	0.17	0.19	0.23	0.17
W: Acreage Shares	1	1.00	0.26	0.46	0.28	0.28		nd	nd	nd
	2	1.00	0.04	0.09	0.88	0.21	0.10	0.48	0.02	0.08
V: Output Shares	1	1.00	0.33	0.34	0.32	0.32		nd	nd	nd
	2	1.00	0.05	0.08	0.88	0.24	0.11	0.41	0.02	0.11
Total Labor Per Bushel	1	3.17	3.08	3.56	2.84	2.84		nd	nd	nd
(A + B)/Y + C	2	0.76	1.02	1.40	0.69	0.76	0.76	0.69	0.95	0.43
Mechanization Effect		1.29								
Yield Effect		2.68								
Western Settlement Effect		2.90								
Combined Mechanization and Settlement Effects		0.84								

Notes: Period 1 is 1839; period 2 is 1907–11. nd means no data. See footnote 3 for regional definitions. In 1839, the Corn Belt and Western Dairy were not segregated. "Mechanization Effect" assumes A, B, C = 2; Y, V = 1; "Yield Effect" assumes Y = 2; A, B, C, V = 1; "Western Settlement Effect" assumes V = 2; A, B, C, Y = 1; "Combined Mechanization and Settlement Effects" assumes A, B, C, V = 2; Y = 1.

Source: Parker and Klein, "Productivity Growth in Grain Production," 532–5.

TABLE 2.2. *The Changing Geographic Center of Wheat Production, 1839–1919*

	Latitude			Longitude					
	deg	min	sec	deg	min	sec	Approximate	Time Period	Miles of Movement
Mean Location									
1839	39	43	43	80	56	0	27 miles SE of Wheeling, WV	1839–49	65
1849	40	14	18	81	58	49	56 miles NE of Columbus, OH	1849–59	214
1859	39	59	59	86	1	38	17 miles NE of Indianapolis, IN	1859–69	153
1869	40	39	17	88	48	40	75 miles NE of Springfield, IL	1869–79	89
1879	40	36	14	90	30	46	75 miles NW of Springfield, IL	1879–89	157
1889	39	33	53	93	9	18	141 miles SE of Des Moines, IA	1889–99	173
1899	41	39	19	94	59	23	55 miles NE of Omaha, NE	1899–1909	76
1909	42	24	26	96	4	49	77 miles NW of Omaha, NE	1909–19	126
1919	40	36	20	95	42	39	50 miles SE of Omaha, NE	Total 1839–1909	921
Median Location									
1839	40	0	0	80	40	0	38 miles E of Wheeling, WV	1839–49	34
1849	40	23	59	81	2	22	107 miles NE of Columbus, OH	1849–59	238
1859	40	3	19	85	32	15	38 miles NE of Indianapolis, IN	1859–69	134
1869	40	36	56	87	58	36	105 miles NE of Springfield, IL	1869–79	42
1879	40	20	15	88	41	11	63 miles NE of Springfield, IL	1879–89	160
1889	40	18	15	91	43	9	132 miles SW of Des Moines, IA	1889–99	212
1899	41	10	51	95	36	36	19 miles SE of Omaha, NE	1899–1909	78
1909	41	16	0	97	7	0	182 miles SW of Omaha, NE	Total 1839–1909	988

Note: The total number of miles of movement between 1839 and 1909 is based on the starting and ending locations and, due to variations in the direction of movement, is less than the sum of decadal miles of movement.

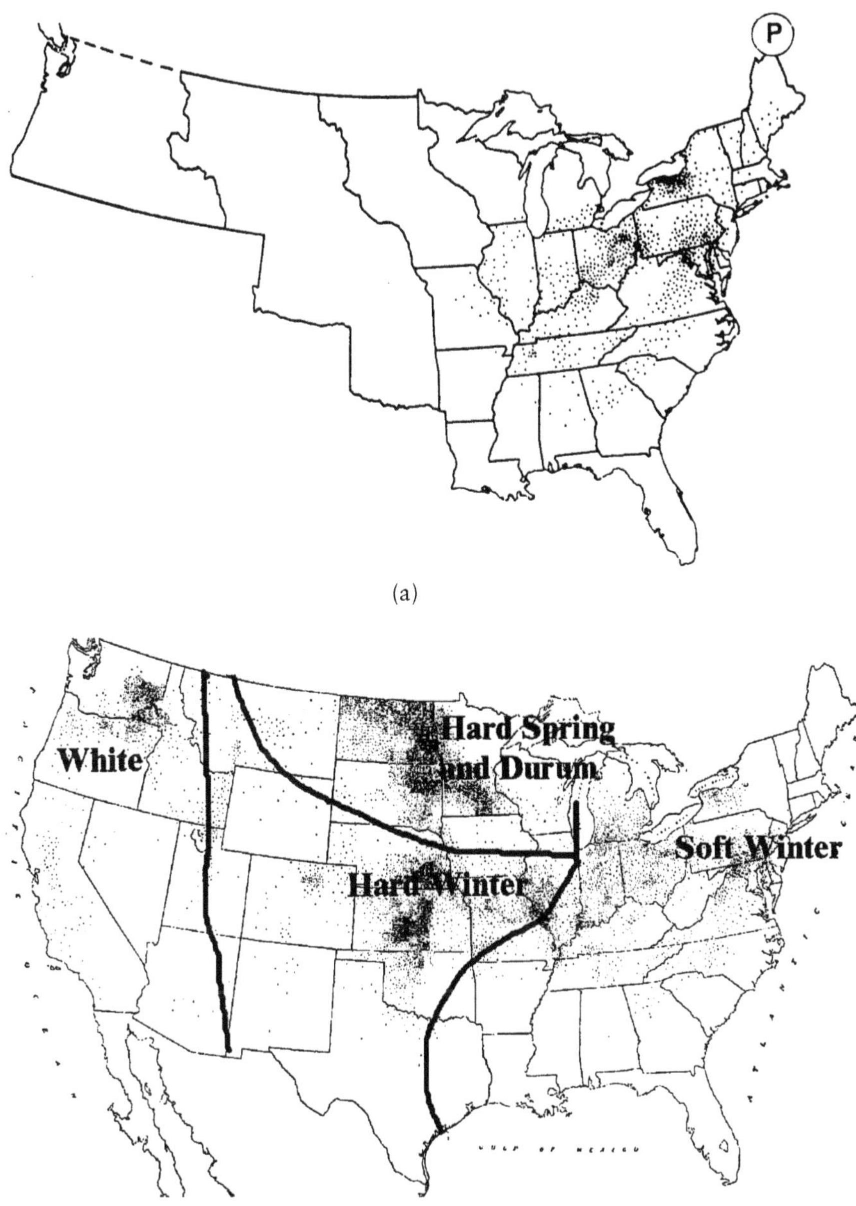

(b)

FIGURE 2.2. Wheat production in (a) 1839 and (b) 1909. *Notes:* (a) Each dot represents 100,000 bushels; (b) Each dot represents 50,000 bushels. *Sources:* (a) Paullin, *Atlas of the Historical Geography of the United States*, plate 143P. Courtesy of the Carnegie Institution of Washington; (b) U.S. Census Bureau. 13th Census 1910, v. 5, plate 3.

over the same period.[7] In 1839, the center was located east of Wheeling, (West) Virginia. Cultivation was concentrated in Ohio and upstate New York; relatively little was grown as far west as Illinois. By 1909, the center of production had moved over eight hundred miles west, to the Iowa/Nebraska borderlands. The core areas of the modern wheat belt had emerged in an area stretching from Oklahoma and Kansas in the south to the Dakotas (as well as the Canadian prairies) in the north. Another important concentration appeared in the Inland Empire of the Pacific Northwest. The western shift was so overwhelming that "new areas," not included in Parker and Klein's 1839 regions, accounted for 64 percent of 1909 output and 74 percent of the growth from 1839 to 1909. More generally, the area west of the Appalachian Mountains, which had made up less than one-half of the output in 1839, provided 92 percent of the output by 1909.

Figure 2.2 illustrates the significance of this shift in the locus of production. According to Mark Alfred Carleton, a leading USDA agronomist, these regions possessed such different geoclimatic conditions that "they are as different from each other as though they lay in different continents."[8] The key point for our reevaluation of Parker and Klein's argument is that in 1839 wheat was only extensively grown in the eastern half of one of these four regions. In addition, by 1909 the newer regions specialized in varieties – the hard reds – that were completely different from those produced in the older areas and, for the most part, had not existed in the United States in 1839.[9]

[7] We calculated the 1839 and 1909 center from Census county-level production data and the location of the county's seat. The 1839 data are from Craig, Haines, and Weiss, *U.S. Censuses of Agriculture*, and *Development, Health, Nutrition*. The 1909 data are from the U.S. Census Bureau. 13th Census 1910, v. 6–7. The information for 1849–99 and 1919 (mean only) are from the U.S. Census Bureau, *Statistical Atlas*, 22. The county seat location data are from Sechrist, *Basic Geographic and Historic Data*. The data include only U.S. production. As a result, the changes do not capture the spread of grain cultivation onto the Canadian prairies.

[8] Carleton, "Basis for the Improvement of American Wheats," 9. The four general wheat regions shown in the lower panel of Figure 2.2 represent gross demarcations because each of these areas contained important subregions.

[9] The primary distinction is between winter (-habit) and spring (-habit) wheats. ("Habit" is added because the distinction does not depend strictly on the growing season.) Winter-habit wheat requires a period of vernalization – prolonged exposure to cold temperatures – to shift into its reproductive stage. This typically involves sowing in the fall and allowing the seedlings to emerge before winter. During the cold period, the winter-habit wheat lies dormant but remains exposed to risks of winterkill. The grain is harvested in the late spring or early summer. Spring-habit wheat grows continuously without a period of vernalization. There is a third, less important category of facultative wheat that has intermediate cold tolerance but does not require vernalization to flower and develop grain.

The importance of new varieties suggests that the Parker–Klein calculations suffer from index number problems similar to the classic "new goods" issue. As Table 2.1 illustrates, data on labor requirements and yields for many of the leading wheat states in 1909 are completely lacking in 1839. In their standard approach, Parker and Klein lump together all of the states from Ohio to the Pacific Coast into the "West." To address the problem of shifts within this vast, heterogeneous region, they did explore a modified productivity calculation replacing the 1909 labor requirements and yields of their "West" with those for the five Midwestern states (their "West: Corn" region).[10] This adjustment generated minor changes in the results, but, as in the standard calculations, it missed the fundamental role that biological changes played in allowing the spread of wheat to the new lands of the West and in maintaining yields everywhere in the face of growing threats from pests and diseases.

The Introduction of New Wheat Varieties

As farmers moved wheat culture onto the northern prairies, Great Plains, and Pacific Coast, they confronted climatic conditions far different from

In Europe and North America, farmers in cold regions often sow spring-habit wheat shortly before the last freeze, harvesting the crop in mid- to late summer. But it is interesting to note that varieties with spring-habits were also used in areas with mild winters, such as the Mediterranean and California. There, the wheat was planted in the fall and grew without interruption. A longer growing season is generally associated with greater yield potential but also involves greater exposure to weather risks, diseases, and insects. Other important distinctions refer to the kernel's texture (soft, semihard, and hard) and color (white versus red). Hard wheats, which are relatively drought resistant, outperform soft wheats in the more arid areas. The rough precipitation dividing line was between 30 and 35 inches. Salmon, "Climate and Small Grain," 334–5. East of the Mississippi, soft white and red wheats were prevalent, whereas hard reds traditionally dominated in the Great Plains. Durum wheat, which became popular in selected regions of the northern Great Plains after 1900, is a species distinct from common wheat, with different flour quality and uses.

10 The latter subregion included Ohio, Indiana, Illinois, Missouri, and Iowa and encompassed most of the wheat growing areas in their 1839 "West." By this modified measure, aggregate labor productivity grew by 3.85 times, instead of the 4.17 times of their standard approach. The contribution of mechanization was lower while that of yield increases was higher. But this is not a fully satisfactory solution. Parker and Klein's modified measure retained the output weights of their standard calculation, essentially assuming that all of the wheat grown on the Great Plains, Pacific Coast, and other parts of the "West" were produced in the "West: Corn Belt." In fact, during the 1909 period, the "West: Corn Belt" accounted for only 23.5 percent of national output and 26.7 percent of the output of the "West" (which made up 87.9 percent of the national total). Parker and Klein, "Productivity Growth in Grain Production," 535–9.

TABLE 2.3. *Weather Indicators in Old and New Regions*

	Precipitation (inches)	Mean Low Temperature (degrees F)	Mean High Temperature (degrees F)	Frost-Free Days
Wooster, OH	36.2	39.1	58.5	155
Dickinson, ND	16.2	27.7	53.9	120
Ft. Hays, KS	22.8	40.1	67.3	170

Sources: Collins, *Ohio, an Atlas*, 26, 34; Goodman and Eidem, *Atlas of North Dakota*, 18; Socolofsky and Self, *Historical Atlas of Kansas*, 4; Midwestern Regional Climate Center, *Historical Climate Summaries*; High Plains Regional Climate Center, *Dickinson Exp Stn, ND*, and *Hays 1 S, KS*; Agricultural Research Center – Hays, *Weather Extremes*, and *Climatological Summary*.

those prevailing in the East.[11] Table 2.3 shows the average precipitation, the mean average high and low temperatures, and the length of the frost-free growing season at three agricultural experiment stations. These are relatively coarse indicators of the climatic conditions relevant for wheat production, but they serve to emphasize the substantial regional differences.[12] The driest year in the past 100 years at the Wooster experiment station in Ohio was wetter than the average years at the stations in Hays, Kansas, and Dickinson, North Dakota. Furthermore, the coldest year on record in Ohio was warmer than the average year in North Dakota. As a result, the pioneers suffered repeated crop failures as they attempted to grow the standard eastern varieties under the normal conditions of the Great Plains.[13]

The successful spread of wheat cultivation across the vast tracts extending from the Texas Panhandle to the Canadian prairies was dependent on the introduction of hard red winter and spring wheats that were entirely new to North America. Over the late nineteenth century, the premier hard spring wheat cultivated in North America was Red Fife (which appears identical to a variety known as Galician in Europe). According to the most widely accepted account, David and Jane Fife of Otonabee, Ontario, selected and increased the grain stock from a single wheat plant grown on their farm in 1842. The original seed was included in a sample

[11] For a classic example of the serious problems associated with finding varieties suitable for the frontier, see Murray, *Valley Comes of Age*, 37; Pritchett, *Red River Valley*, 113, 228.

[12] For a discussion of the effects of weather conditions on wheat, see Cook and Veseth, *Wheat Health*, 21–4.

[13] Clark and Martin, "Varietal Experiments," 1.

of winter wheat shipped from Danzig via Glasgow. It was not introduced into the United States until the mid-1850s. Red Fife was the first hard spring wheat grown in North America and became the basis for the spread of the wheat frontier into Wisconsin, Minnesota, the Dakotas, and Canada. It also provided much of the parent stock for later wheat innovations, including Marquis. At the time of the first reliable survey of wheat varieties in 1919, North Dakota, South Dakota, and Minnesota grew hard red spring and durum wheats to the virtual exclusion of all others.

Another notable breakthrough was the introduction of "Turkey" wheat, a hard red winter variety suited to Kansas, Nebraska, Oklahoma, and the surrounding region. The standard account credits German Mennonites, who migrated to the Great Plains from southern Russia, with the introduction of this strain in 1873.[14] James Malin's careful treatment describes the long process of adaptation and experimentation, after which the new varieties gained widespread acceptance only in the 1890s. In 1919, Turkey-type wheat made up about "83 percent of the wheat acreage in Nebraska, 82 percent in Kansas, 67 percent in Colorado, 69 percent in Oklahoma, and 34 percent in Texas. It... made up 30 percent of total wheat acreage and 99 percent of the hard winter wheat acreage in the U.S."[15] A similar story holds for the Pacific Coast. The main varieties that would gain acceptance in California and the Pacific Northwest differed in nature and origin (Chile, Spain, and Australia) from those cultivated in the humid East in 1839.

Wheat cultivation in the East was also in a constant state of flux – many varieties were tried and abandoned and others took root where they

[14] Ball, "History of American Wheat," 63. Turkey was also new to southern Russia, having been introduced by the Mennonites in 1860. Bernhard Warkentin, one of the early Mennonite settlers in Kansas, reportedly imported 25,000 bushels of seed from Russia and had as many as 300 test plots near his home in Kansas. In 1904 black rust destroyed a large part of the soft wheat, but the new Russian wheat was hardly affected. Stucky, *Century of Russian Mennonite History*, 27–30.

[15] Quisenberry and Reitz, "Turkey Wheat," 98–114. Improvements in flour-milling technologies contributed to the spread of hard red wheat, thereby creating an example of the synergism of biological and mechanical innovations. Using the traditional stone-grinding methods, millers found that hard red wheat yielded darker, less valuable flour than the softer white wheat varieties. The introduction of the middling purifier (to separate the bran from the flour) in 1870 and the new roller grinding process in 1878 allowed millers to make high-quality flour from the new varieties. Over this period, flour from hard red wheat, which had formerly sold at a substantial discount relative to that ground from white winter wheat, began to sell at a premium. Knopf, "Changes in Wheat Production," 233; Malin, *Winter Wheat*, 188–9.

proved better suited to evolving local conditions. The most notable change in the East in the mid-nineteenth century was the replacement of soft white varieties by soft reds. Leading this transition was Mediterranean, a late-sown variety introduced from Europe in 1819, which gained wide favor in the 1840s and 1850s. The field of competing varieties was large and ever changing. Clarence Danhof notes that around 1840 a survey listed 41 varieties that were being grown in New York State, "of which, nine winter wheats and nine spring wheats were most important."[16] In 1857, the Ohio State Board of Agriculture cataloged 111 varieties (96 winter and 15 spring) that had been grown locally in recent years. The report detailed the time of ripening, performance in different soils and climates, flour quality, and resistance to pests and diseases. Of the 86 varieties that we could date, 28 percent had been introduced into Ohio within the previous five years.[17]

This evidence suggests that today's rapid turnover of wheat varieties, which many view as a product of modern science, has nineteenth-century roots.[18] In the past, as today, new wheat varieties could be secured by introduction from other regions, selection of naturally occurring mutations and crosses, and deliberate hybridization. The balance across methods has shifted in modern times, but the commercial spread of wheat varieties derived from hybridization (and subsequent selection) began before 1870.[19]

The U.S. government had been active in the search for new wheat varieties since the days of Washington and Jefferson. These activities reflected Jefferson's belief, stated in 1821, that "the greatest service which can be rendered any country is to add an useful plant to its culture; especially, a bread grain."[20] The 1854 Commissioner of Patents report notes that "a considerable share of the money appropriated by Congress for Agricultural purposes has been devoted to the procurement and distribution of seeds, roots, and cuttings."[21] The report describes 14 varieties of wheat

[16] Danhof, *Changes in Agriculture*, 157.

[17] Ohio State Board of Agriculture, *Annual Report 1857*, 737–61. It is likely that some varieties were listed under different names.

[18] Johnson and Gustafson, *Grain Yields*, 119; Pardey et al., *Hidden Harvest*, 8–12; Dalrymple, "Changes in Wheat Varieties," 23–7.

[19] Large, *Advance of Fungi*, 302–4. In the United States, the first wheat variety derived from hybridization is usually traced back to 1870 when Cyrus G. Pringle marketed Champaign, but Sereno E. Todd dates American wheat hybridization to the 1840s. Ball, "History of American Wheat," 48–71.

[20] Foley, *Jefferson Cyclopedia*, 697.

[21] U.S. Patent Office, *Report of the Commissioner 1854*, v, x–xiii.

that had been recently imported from 9 countries. In 1866 the newly formed Department of Agriculture (USDA) tested 122 varieties (55 winter and 67 spring), including "nine from Glasgow, eight from the Royal Agricultural Exhibition at Vienna ... several varieties from Germany," and a number from the Mediterranean and Black seas.[22]

Private breeders were also at work. In 1862, Abraham Fultz of Mifflin County, Pennsylvania, found three spikes of bald wheat in a field of Lancaster wheat, a variant of Mediterranean. The selected seed proved hardy, ripened relatively early, and produced semihard red grain of good quality. "Fultz" was so superior that it spread quickly, with the USDA distributing the seed by 1871. Even before this date, Garrett Clawson of Seneca, New York, had selected several superior heads from a field of Fultz that yielded a good white wheat, White Clawson or Goldcoin (1865). By 1886, S. M. Schindel, a seedman in Hagerstown, Maryland, hybridized Fultz and Lancaster to produce Fulcaster, which was "considerably resistant to rust and drought." It soon competed with Fultz as the most popular soft red winter wheat. Other hybrids, such as Diehl–Mediterranean (1884) and Fultzo–Mediterranean (1898), also gained favor in the East.[23] In the northern plains, breeders such as J. B. Power, L. H. Haynes, and Willet M. Hays were also at work during this period, producing many improved varieties.

As a rule breeders and farmers were looking for varieties that improved yields, were more resistant to lodging and plant enemies, and, as the wheat belt pushed westward and northward, varieties that were more tolerant of heat and drought and less subject to winterkill.[24] Data from Canadian experiment stations show that changes in cultural methods and varieties shortened the ripening period by 12 days between 1885 and 1910. Given the region's harsh and variable climate, this was often the difference between success and failure.[25] The general progression in varieties allowed the North American wheat belt to push hundreds of

[22] U.S. Department of Agriculture [hereafter USDA], *Report of the Commissioner*, 8.

[23] Carleton, "Basis for the Improvement of American Wheats," 65, 70; Clark, Martin, and Ball, "Classification," 83–5, 135, 160; Patterson and Allan, "Soft Wheat Breeding," 36–41.

[24] Scholars often focus on yields as a summary measure of biological improvement in wheat. But breeders and farmers were also keenly interested in a number of other economically significant characteristics unrelated to yield, including milling quality, protein and gluten content, color, baking quality, and the percentage of the kernel weight that was converted to flour.

[25] Norrie, "Rate of Settlement," 410–27; Ward, "Origins of the Canadian Wheat Boom," 864–83. A. H. Reginald Buller, *Essays on Wheat*, 175–6, credits Marquis with giving

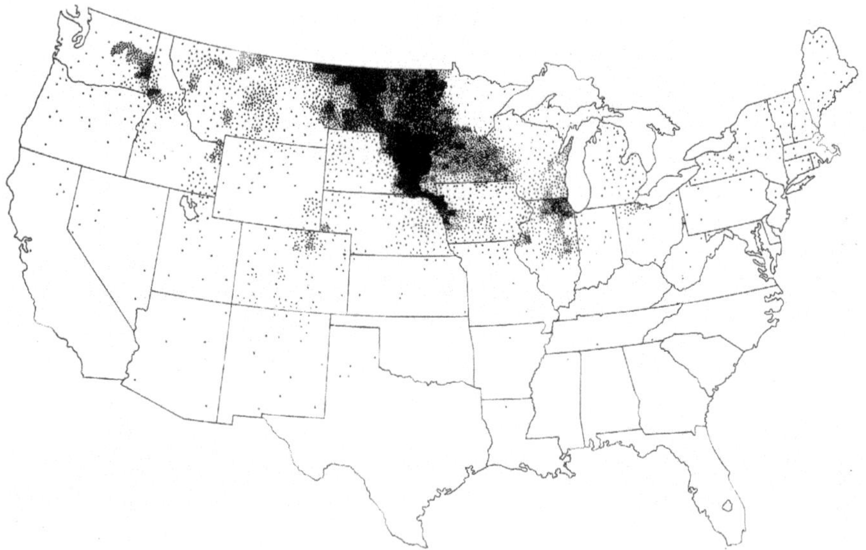

FIGURE 2.3. Marquis' range in 1919, seven years after its introduction. *Source:* USDA. Division of Cereal Crops and Diseases. Photograph Collection. No. 4176. U.S. National Agricultural Agricultural Library Manuscript Collections, Beltsville, MD. Courtesy of Special Collections, National Agricultural Library.

miles northward and westward and significantly reduced the risks of crop damage everywhere.

One of the most important of the early twentieth-century innovations was Marquis, a cross of Red Fife with Red Calcutta, bred in Canada by Charles Saunders. The USDA introduced and tested Marquis seed in the period 1912–13. By 1916, Marquis was the leading variety in the northern grain belt, and by 1919 its range stretched from Washington to northern Illinois (see Figure 2.3).[26]

The spread of Marquis was not an isolated case. Following extensive expeditions to the Russian plains, Carleton introduced Kubanka and several other durum varieties in 1900.[27] These hardy spring wheats proved to be relatively rust resistant. By 1903, durum production, which was

adopters about one extra week between harvest and freezeup (which put an end to fall plowing).

26 Clark, Martin, and Ball, "Classification," 90–1.

27 Ball and Clark, "Experiments with Durum Wheat," 3–7; Clark, Martin, and Smith, "Varietal Experiments with Spring Wheat," 8–9.

TABLE 2.4. *Diffusion of New Wheat Varieties in the Hard Spring Wheat States, 1914–21*

Shares of Production by Variety in Minnesota, North and South Dakota							
	(1) Marquis	(2) Durum	(3) Velvet Chaff	(4) Bluestem	(5) Fife	(6) Other	(7) Total New (1)+(2)
	percent	percent	percent	percent	percent	percent	percent
1914	4.1	11.6	20.9	44.2	15.6	3.7	15.7
1916	33.2	12.5	22.4	22.3	9.1	0.4	45.7
1917	45.0	16.5	18.8	14.1	5.0	0.7	61.5
1918	55.2	19.3	14.3	6.5	3.6	1.2	74.5
1919	56.9	23.6	10.7	5.3	2.7	0.7	80.4
1920	54.8	28.5	9.0	3.9	2.4	1.4	83.3
1921	49.8	37.6	5.8	2.9	2.3	1.6	87.4

Source: USDA, "Statistics of Grain Crops 1921," 530–1.

concentrated in Minnesota and the Dakotas, approached 7 million bushels. In 1904, the region's Fife and Bluestem crops succumbed to a rust epidemic with an estimated loss of 25 to 40 million bushels, but the durum crop was unaffected. By 1906, durum production soared to 50 million bushels.[28] The introduction of Marquis and various durum varieties to the United States illustrates the rapid spread of new varieties in the early twentieth century. Table 2.4 displays the transformation of the wheat stock in the northern Great Plains in the late 1910s. As the new Marquis and durum varieties took hold, the production share of the older varieties, such as Velvet Chaff, Bluestem, and Fife, fell from 84 percent in 1914 to 13 percent by 1921. These rates of diffusion are comparable with those for hybrid corn in the Midwest during the 1930s.

The national turnover of varieties is evident in USDA surveys of wheat distribution, first systematically collected in 1919 and reported thereafter roughly every five years until 1984. Using the 1919 survey together with information on the date of introduction or release of specific varieties, we can gain a clear picture of the changing composition of the wheat varieties grown in the United States.[29] In that year, roughly 24.2 percent of U.S. wheat acreage was in hard red spring wheat, 6.4 percent in durum,

[28] Carleton, "Hard Wheats," 404–8.

[29] Clark, Martin, and Ball, "Classification." A variety's "vintage" is measured since first introduction or creation, not when it became generally available following testing.

TABLE 2.5. *Vintage of U.S. Wheat Varieties in 1919*

Introduced	Percent of Acreage	Introduced	Percent
Before 1800	0.2	1860–69	6.7
1800–09	0.2	1870–79	31.6
1810–19	3.6	1880–89	9.7
1820–29	0.7	1890–99	8.7
1830–39	1.7	1900–09	9.8
1840–49	1.2	1910–19	17.0
1850–59	2.0	Unknown	6.9

Sources: Clark, Martin, and Ball, "Classification"; U.S. Agricultural Research Service. *GrainGenes*.

32.0 percent in hard red winter, 30.1 percent in soft red winter, and 7.1 percent in white. It is important to note that there was essentially no production of durum or the hard reds in 1839, but they composed 62.8 percent of the 1919 total. Table 2.5 provides further evidence of the age distribution of wheat varieties in 1919. Of the 133 varieties that could be dated, the acreage-weighted mean "vintage" was 1881, or less than 40 years old. The median was 1873, which corresponded to the introduction of Turkey. This is not surprising given that Turkey was the largest single type, making up almost 30 percent of total acreage. Note that even the soft red winter varieties experienced significant turnover. Their mean "vintage" was 1868. And of the top four soft red winter wheats in 1919 – Fultz, Fulcaster, Mediterranean, and Poole – only Mediterranean was introduced before 1839.[30] The key results are that in 1919, well before the usual dating of the onset of the biological revolution, roughly 80 percent of U.S. wheat acreage consisted of varieties that had not existed in North America before 1873, and less than 8 percent was planted in varieties dating earlier than 1840.

Farmers of the Great Plains, mountain states, and Pacific Coast showed a strong preference for varieties different from those grown in the wheat belt of 1839. But were the advantages of the new wheats large or small? The controlled settings of the experiment station variety trials provide the best information on this issue. For example, from the late 1880s onward, the stations in Minnesota and North Dakota cooperated to test hundreds of spring wheat varieties. Because the agronomists dropped unsuccessful varieties after one to three years, the eastern stocks rarely

[30] Clark and Quisenberry, "Distribution of the Varieties," 37.

even appeared in these trials. However, in the period 1892–94, China Tea, an early maturing soft spring wheat, was tested on Red River Valley plots. China Tea's average yields were about 88 percent of the leading Fife and Bluestem varieties. Moreover, China Tea was consistently classed a "reject" suitable only for animal feed and subject to almost 50 percent price discounts because of its low quality. The 1892–93 Fargo trials included Lost Nation, a soft spring wheat popular farther east in the 1870s and 1880s. It proved less reliable and only gave 80 percent of Red Fife's yield.[31] In addition, Lost Nation's lower quality resulted in roughly a 10 percent price discount relative to the price of Fife.

These experimental results left the Minnesota officials "disappointed" because they would "heartily welcome" a high-yielding soft spring variety. To provide perspective, researchers estimated that soft wheats of standard grade would have to outyield their "famous" hard wheats by five bushels per acre to overcome the quality differential.[32] The North Dakota officials left no doubt where matters stood in their state. "North Dakota is beyond the northern limit of winter wheat.... Not only has the growing of the wheat been limited to spring varieties but almost exclusively to those kinds hardest in berry and strongest in gluten. Even among these the limitations has [*sic*] included little else than Fife.... Fife ... is the great wheat of this region.... Averaging all conditions, this wheat is a superior yielder (and) ... is as hardy as any variety of spring wheat.... [T]he value of this wheat can hardly be overstated."[33] Combining the quantity and quality differences meant that the soft wheats suffered an effective yield disadvantage relative to Fife of 28 to 54 percent. This gap would have been far greater in the colder and drier expanses to the west of the Red River Valley.

The contrasts between Red Fife, and China Tea and Lost Nation, as large as they are, significantly understate the extent of technological change because, by 1909, Red Fife had been largely replaced by even more

[31] China Tea, also known as Black Tea, Siberian, Java, and Early Java, was imported to New York from Switzerland around 1837. Clark, Martin, and Ball, "Classification," 140–1. Given that it takes several years to increase the seed, the variety could not have been widely available in 1839. Thus, using China Tea as the 1839 reference variety biases the case against biological innovation.

[32] North Dakota Agricultural Experiment Station, "Grain and Forage Crops," no. 10, 5–10, and "Grain and Forage Crops," no. 11, 1–17; Minnesota Agricultural Experiment Station, *Annual Report 1894*, 253–61.

[33] North Dakota Agricultural Experiment Station, "Grain and Forage Crops," no. 10, 1, 12–13.

superior varieties, including Bluestem and Preston, and various durum wheats. As the 1914–21 production data underlying Table 2.4 reveal (consistent with earlier experiment station results), the durum yields were roughly one-third higher than that of Fife, and the newer hard spring wheats outyielded Fife by about 16 percent. The net result is that, in the northern plains, the varieties available around World War I offered a net return (combining yield and quality differences) that about doubled what could have been earned by growing the defunct varieties available in the United States and Canada in 1839.[34]

The situation was similar in the hard winter wheat belt. Early settlers in Kansas experimented with scores of soft winter varieties common to the eastern states.[35] According to the Kansas State Board of Agriculture, "as long as farming was confined to eastern Kansas these [soft] varieties did fairly well, but when settlement moved westward it was found they would not survive the cold winters and hot, dry summers of the plains."[36] The evidence on winterkill lends credence to this view. Data for four east central counties for the period 1885–90 show that more than 42 percent of the planted acres were abandoned. For the decade 1911–20, after the adoption of hard winter wheat, the winterkill rate in these counties averaged about 20 percent.[37] Mark Carleton also left his imprint on Kansas (see Figure 2.4). In 1900 he introduced Kharkof from Russia. This hard winter wheat adapted well to the cold, dry climate in western and northern Kansas, and by 1914 it accounted for about one-half of the entire Kansas crop.[38]

Drawing from decades of research, S. C. Salmon, O. R. Mathews, and R. W. Luekel noted that for Kansas "the soft winter varieties then grown yielded no more than two-thirds as much, and the spring wheat no more than one-third or one-half as much, as the Turkey wheat grown

[34] This actually understates the advantages of the new varieties because, if the older varieties had been planted continuously, they almost surely would have become highly susceptible to diseases.

[35] Malin, *Winter Wheat*, 96–101.

[36] Salmon, "Developing Better Varieties of Wheat," 210.

[37] Malin, *Winter Wheat*, 156–9. Winterkill rates for 1911–20 are calculated using data from Salmon, Mathews, and Luekel, "Half Century of Wheat Improvement," 6, 78–9. The search for varieties suitable for Kansas echoed the earlier experiences of settlers in other states. In the 1840s pioneer farmers attempted to grow winter wheat on the Wisconsin prairie. Repeated failures due to winterkill eventually forced the adoption of spring varieties. Hibbard, "History of Agriculture in Dane County," 125–6.

[38] Carleton, "Hard Wheats," 404–8.

FIGURE 2.4. Special Agent Mark Alfred Carleton crossing wheats at Garrett Park, Maryland, in 1894. *Source:* Galloway, Beverly Thomas. Papers. Series 7, Photographs. No. 34. U.S. National Agricultural Library Manuscript Collections, Beltsville, MD. Courtesy of Special Collections, National Agricultural Library.

somewhat later."[39] In 1920, Salmon concluded that, without these new varieties, "the wheat crop of Kansas today would be no more than half what it is, and the farmers of Nebraska, Montana and Iowa would have no choice but to grow spring wheat" which offered much lower yields.[40]

By the eve of World War I, Nebraska had emerged as the nation's fourth leading wheat producer. Its farmers experienced many of the same challenges as growers in Kansas and relied primarily on spring wheat until after 1900 because of the high winterkill losses suffered by the soft winter varieties. "Some measure of the benefit derived from the general culture of Turkey wheat in Nebraska after 1900 is afforded by comparing its average yield with that of spring wheat at the North Platte Station in western Nebraska. During the twenty-eight-year period ending in 1939 . . . winter wheat yielded on the average 20.6 bushels as compared with 14.3 for spring wheat, a gain of more than 44 per cent. At Lincoln, in eastern Nebraska, the corresponding gain for this 31-year period is 14.2 bushels, or 96 per cent."[41] The movement in actual statewide yields bolsters this evidence. Yields had averaged about 12.5 bushels per acre for the period 1870–1900, but jumped by about 40 percent to 17.5 bushels during 1900–09. At the time scientists attributed the vast majority of this increase to the substitution of Turkey red for spring wheats.[42]

Field tests conducted across the Great Plains and in the Pacific Northwest between 1906 and the early 1920s offer further evidence that hard winter wheat outperformed soft winter varieties in yield, days to maturity, and survival rates.[43] The summary finding was that "hard red winter wheat is now the principal crop in many sections of limited rainfall, including much of Kansas and Nebraska, western Oklahoma, northeastern Colorado, central Montana, and the drier portions of the Columbia Basin of Oregon and Washington. In these areas farming was

[39] Salmon, Mathews, and Luekel, "Half Century of Wheat Improvement," 14.

[40] Salmon, "Developing Better Varieties," 211–12.

[41] Salmon, Mathews, and Luekel, "Half Century of Wheat Improvement," 16.

[42] Montgomery, "Wheat Breeding Experiments," 4–7. Also see Kiesselbach, "Winter Wheat Investigations," 6–7, 103, 107. Several strains of Turkey Red, including Malakoff, Kharkov, Crimean, and Beloglina, were adapted for Nebraska conditions.

[43] Clark and Martin, "Varietal Experiments." The tests comparing Turkey with spring wheats and soft winter varieties significantly understate the advantage that Turkey would have had over wheats available in 1839, because many of the soft varieties tested were themselves hybrids of Turkey and other varieties recently bred for arid conditions.

not practiced or was exceedingly hazardous before this class of wheat was grown."[44]

Wherever it is feasible, farmers prefer to grow winter wheat instead of spring wheat. Winter wheat generally offers higher yields and is much less subject to damage from insects and diseases. The problem is that in colder climates winter wheat suffers high losses to winterkill. The agronomy literature commonly recognizes that the development of heartier winter varieties that could be grown in harsher climates was a great achievement. Just how much land was affected by this fundamental change in farming practices? County-level data on spring and winter wheat production found in the agricultural censuses of 1869 and 1929 allow us to map the northern shift in the "spring wheat–winter wheat" frontier in the plains and prairie states.[45] Table 2.6 reports estimates (derived from regression analysis) for each degree of longitude between 87° and 105° of the latitude where spring wheat output equaled winter wheat output in 1869 and 1929.[46] In both years, except in isolated pockets, spring wheat output exceeded winter wheat output north of the estimated frontier, and winter wheat dominated south of the frontier. In most places the break was sharp with a narrow transition zone. Farmers grew little winter wheat just 30 miles north of the demarcation line and little spring wheat 30 miles south of the line. In 1869, the frontier generally followed the 40th parallel for longitudes between 87° and 94° and then swept down to the southwest across eastern Kansas. (Given the prevailing limits of wheat cultivation, the frontier cannot be mapped for higher longitudes in 1869.) By 1929, the spring wheat frontier had shifted dramatically to the north and west. In that year, the frontier followed roughly the 43rd parallel between 87° and 100° and then took a southwest course. Thus, over this sixty-year period, the frontier crept northward across most of Kansas and Iowa as well as southern Nebraska. The area between the 1869 and 1929

44 Clark and Martin, "Varietal Experiments," 1. Besides Fife, another important hard red spring wheat grown in the northern Great Plains was Haynes Bluestem, developed by L. H. Haynes of Fargo, North Dakota, by 1885. The Minnesota experiment station further improved the variety, creating a pure line variety, Minn. No. 169, by the late 1890s. Clark, Martin, and Ball, "Classification," 124–5.

45 The 1869 data were provided by Craig, Haines, and Weiss, *Development, Health, Nutrition*. The 1929 data are from U.S. Census Bureau. 15th Census 1930, *Agriculture*.

46 To derive the estimates, we performed logit regressions for the winter wheat share of wheat output in each county in a given longitude grouping. We again used the county seat as a measure of the county's location. For each degree of longitude, we used the latitude where the winter wheat share equaled one-half.

TABLE 2.6. *The Northern Shift of the Spring Wheat–Winter Wheat Frontiers, 1869–1929*

	Latitude of Equal Output (in decimals)	
Longitude	1869	1929
87	42.63	43.39
88	40.98	43.33
89	40.55	43.38
90	40.86	43.63
91	40.86	43.62
92	40.53	43.57
93	40.53	43.74
94	39.71	44.25
95	38.27	42.96
96	33.57	43.03
97	33.64	43.07
98		42.69
99		43.38
100		43.05
101		42.64
102		41.37
103		41.72
104		40.81
105		36.19

Note: North of the frontier for any given longitude, spring wheat output exceeds winter wheat output except in isolated pockets.

spring wheat–winter wheat frontiers accounted for almost 30 percent of U.S. wheat output in 1929!

An examination of the spread of wheat culture in the Pacific states supports this general view of the crucial importance of varietal adaptation. By the end of the nineteenth century the Inland Empire, comprising parts of Idaho, Washington, and Oregon, had emerged as a major wheat producer. In 1909, combined production in this region rivaled that of Minnesota. Although the Inland Empire is often treated as a single, distinct area, it was hardly homogeneous. To the contrary, large differences in rainfall, soils, and elevation contributed to great intraregional diversity in growing conditions.

Early settlers experimented with wheat culture in the coastal regions; the first recorded wheat harvest occurred in 1825 at Fort Vancouver, Washington. Circa 1839, Club wheats were the dominant varieties, but these were inappropriate for the harsher climates in the semiarid zones to the east where, in the view of one of the region's leading agronomists, "wheat growing was formerly considered impossible."[47] Gradually, farmers accomplished the impossible by identifying a number of suitable soft spring wheat varieties. By the mid-1890s, a survey of the types of wheat grown in Washington showed the variety of choice was closely matched to the subregion's expected rainfall. In areas with greater than 20 inches, Little Club dominated; in areas with 18 inches, Red Chaff was the clear favorite; and areas with less than 17 inches grew Pacific Bluestem. This pattern reflected the proven superiority of each variety for the local microclimates.

Cultural systems also evolved to match specific local conditions. Farmers in some areas followed the California practice of planting spring wheats in the fall. This increased yields in normal years but often had disastrous results when cold weather hit. Around 1900, Turkey strains made significant inroads because they offered high yields "when the season was favorable."[48] In addition, Baart and Federation wheats, imported from Australia in 1900 and 1914, respectively, gained popularity. The addition of new wheats increased the ability of farmers to push production into even more arid zones. As an example, Baart could be produced in regions with as little as 10 inches of rain.[49] In the period 1918–19, twelve varieties accounted for 93 percent of Washington wheat output. None of these varieties had existed in the United States in 1839.[50] In part because of the need for regionally adapted varieties, Washington developed one of the most impressive wheat research programs in the world. Around 1890 the Washington state experiment station began collecting and testing hard winter wheats. All the winter wheats had serious problems because their straw was too weak to withstand the high winds, and they were prone to high shattering losses. In 1899 William J. Spillman started work crossing the best spring and winter varieties, and in 1907 the experiment station began the release of his new hybrids.[51] Between 1911 and 1926, "647 varieties and selections have been introduced from outside sources and 1240 produced by hybridization, all of which have been included in varietal tests." Out of these nearly two thousand varieties,

47 Elliott and Lawrence, "Some New Hybrid Wheats," 1.

48 Spillman, "Hybrid Wheats," 7.

49 Elliott and Lawrence, "Some New Hybrid Wheats," 1–8.

50 Schafer, Gaines, and Barbee, "Two Important Varieties of Winter Wheat," 7–9.

51 Spillman, "Hybrid Wheats," 7–8, 22–7; Salmon, Mathews, and Luekel, "Half Century of Wheat Improvement," 16–17, 84–6; Carlson, *William J. Spillman*, 20.

the experiment station saw fit to continue testing only 47 in 1926.[52] The best of Spillman's hybrids offered yield advantages of 5 to 10 bushels an acre in a wide range of general field conditions and rapidly gained favor with farmers. Similar, albeit less extensive, activities were under way in Oregon and Idaho.[53]

The California experience perhaps best exemplifies what would have happened more generally in the absence of biological innovation. After learning to cultivate Sonora and Club wheats in the 1850s to 1870s, California grain growers appear to have focused their innovative efforts almost entirely on mechanization. They pioneered the adoption of labor-saving gang plows and combined harvesters but purportedly did little to improve cultural practices, to introduce new varieties, or even to maintain the quality of their seed stock. The result was such sharply declining yields in many areas that wheat, formerly the state's leading staple, ceased to be a paying crop and was virtually abandoned. Acreage had ranged between two and three million during the 1880s and 1890s but dropped to roughly one-half million by 1910. Only after this collapse did the state's agricultural research establishment, which had focused primarily on horticultural and viticultural activities, begin to devote serious attention to biological innovations in grain culture.[54]

To this juncture we have argued that a stream of scientific advances, coupled with widespread experimentation by countless farmers, transformed wheat production in every region of the country and allowed wheat cultivation to move into vast regions that in 1839 were considered impossible to farm. Rather than merely being the primitive ancestors to the modern era, the earlier innovations were a necessary condition for the expansion of wheat culture far beyond its 1839 boundaries.

The Curse of the Red Queen

In addition to the demand to find varieties well adapted to different geoclimatic regions, the threat from pests and pathogens provided another imperative for biological innovation. The danger imposed by pests and pathogens increased over time due to invasions by new threats from foreign lands and because the vast expanses of continuously cropped

[52] Schafer, Gaines, and Barbee, "Wheat Varieties in Washington," 5.

[53] Spillman, "Hybrid Wheats," 25.

[54] See Shaw, "How to Increase the Yield of Wheat," 255–7; Blanchard, "Improvement of the Wheat Crop," 1–5; Rhode, "Learning, Capital Accumulation," 775–7, 786–7.

wheat lands created an ideal breeding ground for the enemies of wheat to multiply and evolve. Wheat farmers were cursed by the Red Queen's dictum: they had to run hard just to stay in one place.[55] Without significant investments in maintenance operations, grain yields would have plummeted as the plant's enemies evolved.

To illustrate this problem, we draw from the work of D. Gale Johnson and Robert Gustafson regarding the sources of productivity growth in U.S. grain production from 1928 and 1954. They found that yields in the West increased by about 21 percent, and the introduction of new varieties accounted for 60 percent of this increase. But these estimates understate the importance of new varieties because "in the absence of the research and the adoption of new varieties it is quite clear that the yields of the small grains would have declined over time."[56] On the northern plains, wheat yields per seeded acre shrunk from 14.5 bushels in the decade 1941–51 to 9.7 bushels in 1952, 6.2 in 1953, and 3.0 in 1954.[57] In the early 1950s a new race of stem rust, 15B, evolved to overwhelm the previously resistant durum varieties. Only the introduction of new varieties allowed yields to recover, because once a wheat variety fell victim to rust, its economic value was permanently diminished.

Farmers have long been aware that over time wheat varieties become more susceptible to diseases and pests. Rusts, which come in many varieties, typically are the most destructive diseases that affect wheat. These windblown fungi attack a plant's stems and leaves, causing lodging and shriveled grain. Stem and leaf rusts thrive in hot, humid climates and attack wheat in most grain-growing regions of North America. Stripe rust thrives in cooler climates and in most years is limited to the Mountain and Pacific regions.[58] In the span of a couple of weeks, stem rust can destroy a crop. There were two fundamental ways that a wheat variety might avoid rust damage. First, it might have genetic resistance to the rust races currently in the area. Finding such varieties was a top

[55] In the words of the Red Queen "Now, HERE, you see, it takes all the running YOU can do, to keep in the same place. If you want to get somewhere else, you must run at least twice as fast as that!" Lewis Carroll, *Through the Looking Glass*. For an application of the Red Queen problem to evolutionary theory, see Van Valen, "New Evolutionary Law," 1–30. The idea that biological innovations are not only alive but also necessarily generate destructive coevolutionary processes is not new. As pioneering plant pathologist E. C. Stakman memorably put it in his 1947 *American Scientist* article, "Plant Diseases Are Shifty Enemies," 321–50.

[56] Johnson and Gustafson, *Grain Yields*, 120.

[57] Ibid.

[58] Loegering, Johnston, and Hendrix, "Wheat Rusts," 307–35.

priority. Before the modern age, this was a haphazard process, but breeders made significant progress. Second, a variety might mature before the rust did much damage (although under more ideal conditions, early maturation often compromised quality and yield). Because winter wheats ripened much earlier than spring wheats, the former were generally less vulnerable to damage. This is one reason why the development of hardier winter wheats, which allowed farmers to shift millions of acres out of spring varieties around 1900, was such a great achievement. (See Plate 2.1 for a photo of rust spores in a grain field.)

As early as the 1660s, the Puritans were enacting a scenario that would be repeated thousands of times as farmers sought to match crops to their local conditions. Early introductions of English winter wheat failed in the harsh New England winters. After some trial and error, the Puritans succeeded in growing soft spring varieties. But in 1664 black stem rust appeared in Massachusetts, ravaging the wheat crop by 1665. Farmers attempted to substitute earlier maturing winter wheats without much success. The inability to find winter-hardy, rust-resistant varieties largely explains why New England never emerged as a serious wheat-producing region.[59] The high incidence of leaf rust in the southeastern United States is a major reason why little wheat was grown in that region despite generations of attempts. In the late nineteenth century, attacks by stem rust forced large sections of Iowa and Texas to abandon wheat production, at least temporarily.[60]

Normal stem rust losses are estimated to have been 5 to 10 percent of the wheat crop in the late nineteenth and early twentieth centuries.[61] Regional epidemics in 1878, 1904, 1914, 1916, 1923, 1925, 1935, and

[59] Carrier, *Beginnings of Agriculture in America*, 147. Clarence Albert Clay, *History of Maine Agriculture*, 38, dates the arrival of the blast in New England to 1660. Bidwell and Falconer, *History of Agriculture*, 13–14; Flint, "Progress in Agriculture," 72–3.

[60] Carleton, "Cereal Rusts," 13–19, and "Basis for the Improvement of American Wheats," 11–22.

[61] Beginning in 1918 the USDA's *Plant Disease Bulletin* (*Reporter* beginning 1923) began collecting estimates by polling plant specialists about the damage in each state or region. These estimates show that national stem rust damage averaged around 3 percent over the period 1919–39, with peak losses of 23 percent in 1935. National leaf rust damage averaged around 2 percent, with a 9.6 percent peak in 1938. Alan Roelfs, "Estimated Losses Caused by Rust in Small Grain," summarizes these results for the period 1918–76. But subsequent studies suggest that the reporting scientists did not fully recognize the damage caused by rust and tended to report only damage in excess of normal damage. As an example, K. Starr Chester, "Plant Disease Losses," 210–12, argues that the estimates of the losses to leaf rust for the years 1900–35 "must be regarded as gross

1937 pushed losses much higher. The 1916 stem rust epidemic destroyed about 200 million bushels in the United States (over 30 percent of the harvested crop) and 100 million bushels in Canada.[62] In many locales the entire crop was lost. The emergence of vast concentrations of wheat in the Great Plains increased the breeding ground for rusts (and other enemies) and, thus, the frequency and severity of rust epidemics.[63] The spread of grain cultivation to the southern plains (Oklahoma and Texas) provided warm overwintering grounds for the fungi that would blow north in the spring and summer, further increasing rust problems. The added incidence of rust is just one reason why agronomists maintain that the wheat-growing environment had seriously deteriorated by the early twentieth century.[64]

Compared to the advances after World War II, the early efforts to control rusts seem primitive. But that was not the perspective in 1940, when E. C. Large proclaimed that the "greatest single undertaking in the history of applied Plant Pathology was to be the attack on the Rust diseases of cereals."[65] What accomplishments so excited Large? A systemic analysis of rusts in the United States dates back to the contributions of Mark Carleton in the 1890s. Carleton tested over one thousand wheat varieties for yield, winter hardiness, rust and insect resistance, and other qualities. The work of numerous other American scientists, along with research in Australia, Canada, and Europe, unlocked many of the mysteries of rust diseases. Aided by the rediscovery of Mendel's laws around 1900 and the publication of Wilhelm Johannsen's pure line theory in 1901, this research accelerated the development of rust-resistant hybrids.[66]

There is indisputable evidence that farmers and wheat breeders were systematically developing and adopting more rust-resistant and earlier maturing varieties. For its day, Red Fife, which gained such favor in the northern Great Plains, had excellent rust resistance and was an early

under estimates." He claims annual losses were at least 5 percent and possibly much higher.

[62] Roelfs, "Effects of Barberry Eradication," 177–81; Miller et al., "Diseases of Durum Wheat," 75–83; Carleton, "Hard Wheats," 407–8; Dondlinger, *Book of Wheat*, 167–8.

[63] Peterson, *Wheat*, 201–4; Hamilton, "Stem Rust," 157.

[64] Salmon, Mathews, and Luekel, "Half Century of Wheat Improvement," 16.

[65] Large, *Advance of Fungi*, 292.

[66] Ibid., 292–312; Salmon, Mathews, and Luekel, "Half Century of Wheat Improvement," 113–14. For a treatment of the early history of rust research see Bushnell and Roelfs, *Cereal Rusts*, 3–38.

ripening variety.[67] Early Manitoba wheat farmers noted that Fife matured ten days earlier than the Prairie Du Chien variety that it replaced. Marquis, which followed Red Fife, further cut the ripening period by seven to ten days, thereby providing significant rust protection. Kubanka, which Carleton had introduced in 1900, proved remarkably resistant to the 1904 epidemic that hammered the Bluestem and Fife crops.[68] When rusts evolved to attack Kubanka, it was replaced by Mindan (1918), which in turn was replaced in 1943 by the Carleton and Stewart varieties. At the time of their release these two varieties were highly resistant to the prevailing stem rust races. They maintained their resistance until race 15B suddenly made them obsolete.[69] A similar progression took place in the hard winter wheat belt because the new Turkey wheats, which became the dominant variety by 1900, also had excellent rust resistance when first introduced. Subsequent releases, including Kharkof (1900), Kanred (1917), and Blackhull (1917), were chosen in large part for their rust resistance and because previously resistant varieties had come under attack.[70] The successive changes in varieties that began in the early colonial period were neither random nor haphazard. Rather the process led to a progression of varieties that were better able to cope with the evolving disease environment. By the end of the nineteenth century, researchers were playing an increasingly prominent role in the identification, creation, and diffusion of new varieties. The rapid rates of diffusion also testify to the economic value of the new releases – without this continuous process of technological replacement, wheat yields would have plummeted and remained low.

A better understanding of the stem rust life cycle allowed farmers and scientists to attack its breeding ground in barberry bushes. In 1660 farmers in Rouen, France, observed that wheat growing near barberry bushes was more apt to be damaged by stem rust, and they tore out the bushes. In the mid-eighteenth century Connecticut, Massachusetts, and Rhode Island enacted legislation against the barberry. In 1865 Anton De Bary scientifically demonstrated the role of barberry bushes as a host,

[67] Fife probably had genetic resistance, because when it was first selected "it proved at harvest to be entirely free from rust, when all wheat in the neighborhood was badly rusted." Carleton, "Hard Wheats," 393.

[68] Carleton, "Hard Wheats," 407–8.

[69] Miller et al., "Diseases of Durum Wheat," 69–92.

[70] Cox et al., "Genetic Improvement," 756–60. Kanred (1917) ranked first out of 150 varieties tested for stem and leaf rust. Salmon, "Developing Better Varieties of Wheat," 214, 228.

but this knowledge was slow to diffuse. The widespread presence of barberry bushes may have contributed to the stem rust epidemic that devastated the Minnesota wheat crop in 1878, but it was not until 1918 that Minnesota outlawed the barberry bush. This was part of a larger federal–state cooperative campaign initiated in the aftermath of the 1916 rust epidemic to eradicate barberry bushes across the 13 north central states. By 1950 about 340 million bushes had been destroyed.[71] It is unlikely that the early scientists who promoted the eradication program fully understood the barberry's role in the propagation of rusts. Not only did the barberry provide a home for the stem rust to carry over and multiply, the rust passed through its sexual recombination stage on the barberry. Thus, the bush was the breeding ground where rusts mutated and hybridized to develop new races. Alan Roelfs estimates that the eradication program delayed the disease's onset by about ten days and, by removing the site of the rust's sexual reproduction, significantly slowed the evolution of new destructive races. Although rust spores could migrate over vast areas, the eradication program meant that new races would have to migrate from Texas and Mexico, or from barberry bushes still remaining in remote regions.[72]

Taken as a whole, the century of biological changes prior to the modern biological revolution had an enormous impact on limiting the damage that rust otherwise would have caused. The introduction of new varieties of spring wheat along with better cultural methods probably shortened the ripening time by about twenty days. The destruction of barberry bushes effectively gave farmers up to another ten days of protection and reduced the rate of rust mutation. In addition, the new wheat varieties almost always had better resistance to rust than did the varieties they replaced. Parallel changes took place in the winter wheat areas, and the introduction of hard winter wheats allowed vast areas to convert from growing spring to growing winter varieties that could be harvested much earlier in the summer.

In addition to rusts, various smut fungi did great damage to wheat throughout North America. Stinking smut (or bunt) was the most destructive. "In a ripe but bunted ear of wheat the grains were swollen and

[71] Ball, "History of American Wheat," 48–71; Hamilton, "Stem Rust," 156–64; Large, *Advance of Fungi*, 121–46; Elwood et al., *Changes in Technology and Labor Requirements*, 80; Salmon, Mathews, and Luekel, "Half Century of Wheat Improvement," 123; Campbell and Long, "Campaign to Eradicate the Common Barberry," 20–31.

[72] Roelfs, "Effects of Barberry Eradication," 177–81.

black, still whole, but with all their inner substance transformed into a pulverulent mass."[73] Milder cases damaged the grain and lowered its value. In 1908, Peter Tracy Dondlinger noted that "formerly at least one-fifth of the cereal crops was [*sic*] annually destroyed by smut."[74] In addition, H. T. Gussow and L. L. Conners observed that "previous to 1900 bunt was alarmingly serious and threatened to be a limiting factor in wheat production" in southern Canada.[75] Even if Dondlinger's figure is an exaggeration, both of these accounts suggest that the damage from smut was declining by the turn of the century.[76] This was a direct result of scientific advances and farmer education. In an exhaustive series of experiments in the mid-1700s, Mathieu Tillet of France proved that smut was a seedborne disease, and he developed a number of treatments. Other researchers built on this discovery, leading to increasingly effective chemicals. In the nineteenth century, many American farmers soaked seeds in hot water to control loose smut and employed lime and copper sulfate solutions to fight stinking smut. By 1900 cheaper formaldehyde solutions became available (see Figure 2.5) and by the early 1920s mercury solutions and copper carbonate dust came on the market. In 1927 roughly 10 percent of U.S. wheat acreage "was sown with seed treated with copper-carbonate," yielding an annual savings of at least 5 million bushels. There were still losses to smut, but they were far lower than experienced previously.[77]

Another disease particularly damaging in the soft wheat belt was wheat scab (or head blight), which in some years destroyed about 5 percent of

[73] Large, *Advance of Fungi*, 70.

[74] Dondlinger, *Book of Wheat*, 162.

[75] As reported by Salmon, Mathews, and Luekel, "Half Century of Wheat Improvement," 16. They also discuss the rise of the bunt problem in the Pacific Northwest after 1900. "Nowhere in the United States and probably nowhere in the world has bunt been so serious or so difficult to control," 84.

[76] Plant scientist Charles Schaller, interviewed by the authors April 25, 2000, believed that Dondlinger's estimates were realistic because bunt losses in modern tests with untreated seeds often exceeded 20 percent.

[77] Large, *Advance of Fungi*, 70–82; Salmon, Mathews, and Luekel, "Half Century of Wheat Improvement," 125–6; Freeman and Stakman, "Smuts of Grain Crops," 33–64; Boss et al., "Seed Grain," 370–9; Powell, *Bureau of Plant Industry*, 34. These chemical breakthroughs also applied to other crops damaged by smut. As an example, in 1920 kernel smut was the most damaging disease of sorghum in the United States, causing over $2 million in losses per year in Kansas. With new chemical treatments, the Kansas experiment station estimated that this loss could be prevented with an expenditure for labor, equipment, and chemicals of less than $24. Johnston and Melchers, "Control of Sorghum Kernel Smut," 6.

FIGURE 2.5. 1918 USDA experiment on the effects of storage of formaldehyde-treated seeds on germination. *Source:* USDA. Division of Cereal Crops and Diseases. Photograph Collection. No. 3889. U.S. National Agricultural Library Manuscript Collections, Beltsville, MD. Courtesy of Special Collections, National Agricultural Library.

the national crop. The disease both harms wheat seedlings and attacks the heads, causing shrunken and aborted kernels.[78] As with many other diseases, scab increased in importance over time as its domain grew. As an example, it probably did not enter Minnesota until around 1900, when corn became an important crop. A 1919 epidemic destroyed 80 million bushels of wheat and reduced the quality of many more in Kansas, Nebraska, and the Dakotas. Among the recommended control measures were to keep the fields clean and plant when the soil was cool – plant winter wheat as late as possible and spring wheat as early as possible.[79] By the early twentieth century, scientists had determined that growing wheat in rotation with corn actually intensified scab outbreaks in wheat. The spores carried over from the wheat crop and multiplied, causing considerable damage in cornfields, and subsequently migrated back to the wheat fields. Thus, the fight against scab yielded the interesting realization

78 Salmon, Mathews, and Luekel, "Half Century of Wheat Improvement," 127–9.
79 Johnson and Dickson, "Wheat Scab and Its Control," 3.

that some crop rotations increased diseases rather than reduced them, as was commonly believed.[80]

Insects represent another arrow in the Red Queen's quiver. The Hessian fly, whose maggots sucked the sap from young plants, was the most destructive of the scores of insects that attack wheat. Its spread reduced yields and led to wholesale changes in the varieties planted and in cultural practices. Although folklore has the Hessian fly entering the United States at Long Island in 1776 in the straw of Hessian mercenaries, many scholars have placed the insect in New York and New Jersey before the revolution. In any case, it spread into Pennsylvania in 1786, arrived in upstate New York by 1789, entered Virginia by 1794, swept across the Alleghenies by 1797, hit Ohio by the mid-1820s, Michigan in 1843, Illinois by 1844, Kansas by 1871, and reached the Pacific Coast in 1884. We know the minute details of the fly's early northern advances thanks to the investigations of two presidents-to-be. "In the spring of 1791 Thomas Jefferson and James Madison set out on a journey to chart the insect's destructive path. They traveled by carriage through New York and Connecticut, 'where this animal has raged much.'"[81] The new scourge was obviously an issue of national concern. Appropriately named *Cecidomyia destructor*, the Hessian fly shifted American wheat farmers onto a significantly lower production possibility frontier. To stand still, farmers now had to run much faster. In many areas, the arrival of the fly led farmers to abandon wheat.

By carefully studying the fly's behavior, farmers learned that the pest might have several broods, and that the most damaging (for winter wheat) was the fall brood, whose maggots sucked sap from the young plants. If

80 McInnes and Fogelman, "Wheat Scab in Minnesota," 1–43; Salmon, Mathews, and Luekel, "Half Century of Wheat Improvement," 16. It was also well known that corn provided a green bridge that allowed chinch bugs to carry over. Later discoveries demonstrated other harmful effects of wheat–corn rotations. As an example, planting winter wheat downwind from mature corn significantly increased the migration of aphids carrying barley yellow dwarf virus. Cook and Veseth, *Wheat Health*, 57, 76–7.

81 Fletcher, *Pennsylvania Agriculture, 1640–1840*, 147. Seldom has an insect so occupied national leaders. When England banned American wheat imports in June 1788 due to a growing alarm over the spread of the fly, it fell to Ambassador Thomas Jefferson to denounce the policy in the Court of Saint James. Hunter, "Creative Destruction," 237, 243; Pauly, "Fighting the Hessian Fly," 377–400; Headlee and Parker, "Hessian Fly," 88; Marlatt, "Annual Losses Occasioned by Destructive Insects," 14; Webster, "Hessian Fly," 259–60. Other sources offer slightly different chronologies of the fly's spread. There are several strains of the Hessian fly and later treatments often focus on *Mayetiola destructor*. Dahms, "Insects Attacking Wheat," 428–31.

not killed, the infested plants were stunted and susceptible to lodging.[82] Gradually, farmers also learned that they could reduce the damage by sowing winter wheat late (or spring wheat early) and by better cleaning their fields to reduce the carryover of the fly population. Planting late delayed the harvest, increasing the danger from rust and frost, but many farmers were willing to take this risk. As a dramatic example of the extent of the cultural changes, one local account from Connecticut indicates that by 1811 the date of planting had shifted from the third week in August to the end of September or early October. Farther south, in New Jersey, the sowing date moved from late August to early October.[83] The fly also triggered a search for new varieties that could be sown later and which had stronger stalks to resist the maggots. Shortly after the first serious attack in 1779, Long Island farmers adopted "a yellow-bearded, southern variety of wheat, which seemed to be less affected by the attacks of the fly."[84] Among the notable proponents of the yellow-bearded wheat was George Washington, who, along with James Madison, tested varieties for their fly resistance.[85] By far the most important biological innovation was the introduction of Mediterranean wheat from Europe in 1819. This variety proved suitable for late planting and had gained wide favor by the 1840s and 1850s.[86]

Just when American farmers were learning to live with the Hessian fly, a new scourge appeared. An insect that contemporary observers called the grain midge first entered Vermont from Canada in the 1820s. This one insect had such a profound effect that the 1860 Census of Agriculture devoted more attention to it than to the mechanical reaper. The Census traced the midge's gradual spread, describing the horrible damage it wrought. It appeared in Washington County, New York, in 1830 and by 1832 "had so multiplied as to completely destroy the crop in many fields." In 1834 and 1835 the midge moved south into Rensselaer and Saratoga

82 Headlee and Parker, "Hessian Fly," 113.

83 Bidwell and Falconer, *History of Agriculture*, 96; Schmidt, *Agriculture in New Jersey*, 92.

84 Webster, "Hessian Fly," 259.

85 See Washington, *George Washington to John Beale Bordley*; Mosiman, *Philadelphia Society for Promoting Agriculture.*

86 Conflicting stories claim that this wheat came from the Mediterranean islands or from Hesse via Holland. See Klose, *America's Crop Heritage*, 66; Ohio State Board of Agriculture, *Annual Report 1857*, 700–1; Fletcher, *Pennsylvania Agriculture, 1640–1840*, 148.

counties, "devastating the wheat-fields." In 1835 and 1836, "over all the territory to which it had extended ... it was so extremely destructive that further attempts to cultivate grain were abandoned."[87] The New York State Agricultural Society estimated that in 1854 the midge caused at least $15 million in damage to the state's wheat crop. In that year the pest entered the fertile Genesee Valley. "In 1856 it destroyed from one-half to two-thirds of the crop on the uplands, and nearly all on the flats. In 1857 it was still worse, taking over two-thirds of the crop." The Census reported that the midge also caused problems in Pennsylvania but that farther to the south the insect did little damage, "owing, it is thought, to the warmer climate."[88] Between 1849 and 1859, wheat production in New York fell by 44 percent. The Census directly attributed most of this decline to the effects of the midge, as "spring crops and winter barley took the place of wheat."[89]

Initially farmers "knew little of the habits of this minute insect, and were unable to offer it any resistance."[90] But once again they adjusted their cultural practices to survive the midge. Many farmers shifted from winter to spring wheats, which helped, but resulted in significantly lower yields. Farmers faced a dilemma because the key to fighting the Hessian fly was to delay planting winter wheat, but the trick with the midge was to harvest as early as possible. All else equal, this required planting earlier. Thus, the midge further constricted the available options by narrowing the window in which planting and harvesting had to take place. In New York, the sowing date, which had been pushed from August to late September or early October because of the Hessian fly, now had to be recalibrated to the first three weeks of September because of the midge.[91] Moving to an earlier planting date might allow for a slightly earlier harvest, but it was more important to find earlier-ripening varieties.

Experience with midge infestations showed that "the injury has been almost entirely confined to the high quality 'white' varieties, the Mediterranean escaping altogether."[92] By the 1850s, Mediterranean had become the dominant variety in the United States even though its flour quality and yield (in the absence of insects) were inferior to those of many abandoned

[87] U.S. Census Bureau. 8th Census 1860, *Agriculture*, xxxi–xlv, quote from xxxiii; Hedrick, *History of Agriculture in the State of New York*, 332–5.

[88] U.S. Census Bureau. 8th Census 1860, *Agriculture*, xxxiv.

[89] Ibid., xxxv.

[90] Ibid.

[91] Ibid., xl, xxxv; Bidwell and Falconer, *History of Agriculture*, 239.

[92] U.S. Census Bureau. 8th Census 1860, *Agriculture*, xxxiv.

varieties.[93] Although the 1860 Census called the midge the "greatest of all pests which has infested the wheat-crop," adjustments in cultural practices – including plowing deep, burning the chaff from infected fields, and rotating crops – soon demoted it to a lesser status.[94]

The battle against the Hessian fly intensified as countless farmers and researchers investigated the fly's behavior and tested cultural practices and wheat varieties to limit its damage. Out of necessity farmers adopted so-called fly-safe varieties that allowed for late planting and, gradually, researchers publicized fly-safe dates for every nook and cranny that grew wheat. The recommended dates varied by about two months with latitude, longitude, elevation, soil conditions, rainfall, and wheat varieties. As noted earlier, the planting decision involved a delicate balancing of several threats, and as wheat culture moved onto the Great Plains the problem became even more difficult. Planting late to avoid the fly made the crop more susceptible to winterkill and reduced the time for the root system to develop. Delaying the harvest exposed the crop to heat, drought, and myriad insects and diseases.[95] As a 1923 Kansas report noted, "the proper time of seeding must be determined for each locality by experimental sowings extending over a period of years."[96] Preventive measures had a collective dimension because the benefits of destroying volunteer wheat and cleaning infected fields of stubble were spread throughout the neighborhood.

Despite considerable precautions, there were local fly outbreaks every year and serious regional infestations roughly every five to six years. As examples, in 1900 over one-half of the wheat acreage in Ohio and Indiana was abandoned due to fly damage, and yields on the harvested land fell by about 60 percent. The following year the fly destroyed over half of New York's wheat crop. Kansas experienced six serious outbreaks between 1884 and 1913 with losses peaking at about 27 percent of the crop.[97] Damage tended to be more serious in years with unseasonably warm falls,

93 Ohio State Board of Agriculture, *Annual Report 1857*, 685; U.S. Patent Office, *Report of the Commissioner 1847–1854*.

94 U.S. Census Bureau. 8th Census 1860, *Agriculture*, xxxiii. Much later, tests would confirm that Mediterranean in fact had fly-resistant qualities. Salmon, Mathews, and Luekel, "Half Century of Wheat Improvement," 98.

95 Cook and Veseth, *Wheat Health*, 84.

96 McColloch, "Hessian Fly in Kansas," 80.

97 Marlatt, "Annual Losses Occasioned by Destructive Insects," 461–74. For a small sample of the studies conducted and for estimates of state losses in bad years see Roberts, Slingerland, and Stone, "Hessian Fly," 113, and McColloch, "Hessian Fly in Kansas."

in wet years, and in years with large volunteer crops. Nationally, estimates of annual losses caused by the Hessian fly around 1900 hovered at 10 percent of the wheat crop.[98] In 1938, USDA entomologist J. A. Hyslop noted the "general adoption, throughout the greater part of the regions infested by the hessian fly, of the practice of planting wheat after the fly-free date has materially reduced" the losses from 6.0 percent of the crop over the 1923–27 period to about 2.2 percent over the 1928–35 period.[99]

What would have happened if farmers did not learn to adjust to the fly? Numerous accounts from the late eighteenth and early nineteenth centuries tell us that those who did not change simply lost their crops.[100] For later years, experiment station investigations repeatedly showed that planting a week or two earlier than recommended led to heavy losses.[101] One Kansas study demonstrated that in the absence of normal precautions, such as planting early and destroying volunteer wheat, yields fell by about 80 percent.[102] Studies conducted in other states also found that in most seasons early sown wheat suffered moderate to heavy fly damage, whereas wheat sown later escaped infestation. As an example, an experiment conducted at eight locations in Illinois over eight years showed that wheat sown after the fly-safe date yielded 29 percent more on average than wheat sown before the date.[103]

More recent studies by modern agronomists show similar results. As an example, in 1981 when researchers took no precautions on test plots near Colfax, Washington, the entire crop was destroyed.[104] When we asked three senior agronomists specializing in wheat what would have happened to farmers in the early twentieth century if they had not followed normal precautions, their response was uniform: "those farmers would not have

[98] Marlatt, "Principal Insect Enemies," 13. Numerous other sources place the actual losses in this general magnitude ranging between $50 and $100 million. Dondlinger, *Book of Wheat*, 172–3, 627–43, asserts that 10 percent is a lower-bound estimate. The direct estimates of losses to the fly are mostly low, because when farmers adjusted varieties and cultural practices to avoid the fly there were real costs in terms of yield losses and increased exposure to winterkill and rust. The USDA estimated in 1904 that annual wheat losses to all insects were "at least 20 percent of the crop." Marlatt, "Annual Losses Occasioned by Destructive Insects," 468.

[99] Hyslop, Losses Occasioned by Insects, 9.

[100] As an example, see Fletcher, *Pennsylvania Agriculture, 1640–1840*, 147–8.

[101] Headlee and Parker, "Hessian Fly," 115.

[102] McColloch, "Hessian Fly," 91–4.

[103] Metcalf and Flint, *Destructive and Useful Insects*, 410–11.

[104] Cook and Veseth, *Wheat Health*, 56.

had a wheat crop worth harvesting."[105] These findings lend credence to the 1909 assessment of C. L. Marlatt, a leading scientist with the U.S. Bureau of Entomology, that the "prevention of loss from the Hessian fly, due to knowledge of proper seasons for planting wheat, and other direct and cultural methods, results in the saving of from $100,000,000 to $200,000,000 annually."[106] Relative to Marlatt's reference value for the wheat crop ($500 million), the biological investments to control this one pest led to yield savings of 20 to 40 percent.

As wheat culture expanded, several other pests, including chinch bugs, grasshoppers, and greenbugs, became of increasing concern. Chinch bugs were first noticed sucking the sap of wheat plants in Orange County, North Carolina, around 1783. By 1790, the insect had spread through North Carolina and Virginia, causing substantial damage to wheat and corn. In fact the chinch bug did the most damage when these two crops were grown in close proximity with one another because the extended food supply allowed the insect to multiply. Over the next half-century, the chinch bug spread from the south Atlantic states into the vast grain fields of the Midwest. There were notable outbreaks in the Carolinas and Virginia in 1839, in Illinois in 1844–45 and 1854–58, Iowa in 1847, and Indiana in 1848 and 1854. During the major infestations of 1863–65, some accounts assert that the bug wiped out three-fourths of the wheat crop and one-half of the corn crop in Illinois and did considerable damage in surrounding states. Serious regional outbreaks occurred in 1868, 1871, 1874, 1887, and 1892–97.[107]

Taking 1887 as an example, the insect reduced the wheat crop by an estimated 50 percent in Iowa, 20 percent in Minnesota, 19 percent in Kansas, and 13 percent in Illinois. In 1901 the ever active Marlatt ranked the chinch bug as by far the most destructive insect to American crops and as more damaging to wheat than the Hessian fly. In 1913 researchers at the Kansas experiment station reached a similar conclusion, noting that

105 Interviews by the authors with Charles Schaller, 27 February 2001; plant pathologist Robert Webster, 25 April 2000; and plant scientist Calvin Qualset, 25 April 2000. The Hessian fly remains the "number one" wheat pest because the insect continues to mutate to "find a way around the plants' defenses." See Talley, "Hessian Fly Genomics Research."

106 Marlatt's estimates recognize that even with preventative measures the Hessian fly caused losses on the order of about 10 percent of the annual wheat crop early in the twentieth century. Marlatt, "Losses Due to Insects," 308, and "Annual Losses Occasioned by Destructive Insects," 463; Dondlinger, *Book of Wheat*, 174.

107 Howard, "Chinch Bug," 7–9; Hibbard, "History of Agriculture in Dane County," 131.

"the chinch bug has damaged Kansas crops to a greater extent than has any other injurious insect, for from the time the settlers began to plant the prairie to the present it has exacted [a] merciless toll."[108] According to J. R. Horton and A. F. Satterthwait, over the entire 1850–1915 period, the "average annual losses sustained by the most heavily infested States" were about 5 percent of the wheat crop (plus sizable portions of several other crops). These losses were suffered in spite of control measures that included destroying the bug's hibernating places in bunch grasses, leaves, and litter, protecting fields with barrier strips, and spraying infected areas in the late spring.[109]

Plagues of grasshoppers were all too familiar during the formative stage of agriculture on the Great Plains, and they contributed to widespread farm abandonment in the southern plains in the 1880s and 1890s. Because grasshoppers thrived in drier climates, their damage significantly increased in severity as the wheat frontier moved west. Severe grasshopper outbreaks occurred at least every decade, with each lasting from one to six years. Annual losses to all crops in the decade 1925–34 averaged about $25 million, but crude estimates for the nineteenth century suggest even higher losses. Given the location of the attacks, the destruction would have been concentrated in wheat. Grasshoppers were not effectively controlled until the 1940s, but starting in 1885 a series of increasingly useful chemical baits became available. In addition, settlement began to encroach on their breeding grounds.[110]

Greenbugs, or spring grain aphids, which first gained notice in Virginia in 1882, were yet another member of the parade of new pests. These European immigrants were considered unimportant before the 1890s, but their proclivity for sucking the sap of wheat plants and for spreading the barley yellow dwarf virus had increasingly devastating effects after the turn of the twentieth century. As with rust, the spread of grain cultivation to the southern plains worsened insect problems by creating a home to overwinter. These enemies of wheat could then fly or blow north in the spring and summer. A virus outbreak in 1907 destroyed 50 million

[108] Other estimates of losses caused by the chinch bug are not consistent with the hyperbole that this was the most destructive insect to wheat, but there is no doubt that it did considerable damage. Headlee and McColloch, "Chinch Bug," 289; Marlatt, "Principal Insect Enemies," 6–7; Shelford and Flint, "Populations of the Chinch Bug," 435–7.

[109] Horton and Satterthwait, "Chinch Bug and Its Control," 4.

[110] Riley, *Locust Plague in the United States*, 29–54; Haeussler, "Insects as Destroyers," 141–6; Salmon, Mathews, and Luekel, "Half Century of Wheat Improvement," 136–9.

bushels of wheat and oats in the southern plains and led Texas wheat growers to abandon as much as 70 percent of their acreage. Smaller outbreaks occurred in 1904, 1911, and 1916. Over this period, insecticides proved ineffectual in controlling the aphid, although natural enemies such as a wasp-like parasite did provide a check under the right climatic conditions. The recommended control measure was cultural and collective in nature – destroy volunteer grain crops during the summer and early fall to prevent survival between the harvest and next year's planting. By the mid-twentieth century, this relative newcomer was destroying more wheat than the Hessian fly.[111]

The Red Queen had yet another arrow in her quiver. Weeds became a more serious problem, in part due to new introductions from other parts of the world. Referring to the northern Great Plains, Salmon asserted that "weeds were not an important factor on the new lands until near the end of the [nineteenth] century," and for California he noted that "previous to 1900 any improvements in per acre yield resulting from a choice of better varieties and from the increasing use of fallow probably were more than offset by the increase in weeds."[112] Along with bindweed and wild oats, one of the most damaging weeds was the Russian thistle – a tumbleweed that entered the United States in the mid-1870s. The "best authorities" place and date the thistle's introduction to Scotland, South Dakota, around 1873. The weed spread to Iowa, Nebraska, and North Dakota by 1888; to Minnesota, Wisconsin, Illinois, and Indiana by 1890–91; and to Kansas, Montana, and Idaho by 1894. Adapting to the times, the thistle hitchhiked rides on the railroad, reaching as far east as New York and as far west as California by the mid-1890s. Where it became established, the weed caused crop losses estimated between 15 and 20 percent. An Illinois observer noted that "no other weed has caused such widespread discussion, or been the subject of such great fear." In the 1890s numerous states and the USDA initiated programs to destroy the weeds. The aforementioned campaign to eliminate the barberry bush was a part of this larger endeavor. In spite of these efforts, by the early twentieth century, USDA experts estimated that weeds reduced the yield of spring wheat by 12 to 15 percent and of winter wheat by 5 to 9 percent.[113] Comparison with other areas offers a hint of what might have happened without control measures. In Russia, for example, with no

[111] Walton, "Green-Bug or Spring Grain-Aphis," 7–8.

[112] Salmon, Mathews, and Luekel, "Half Century of Wheat Improvement," 16, 19.

[113] Cates, "Weed Problem in American Agriculture," 205.

similar collective efforts, "the cultivation of crops has been abandoned over large areas."[114]

A better understanding of the pest environment in northern agriculture sheds light on the causes of international yield differences. Numerous observers have noted that wheat farmers in many parts of Europe achieved significantly higher yields than their American counterparts, and the difference is attributed to the more labor-intensive methods and the greater attention given to maintaining land fertility in labor-abundant Europe. This explanation is incomplete because it fails to take into account the dramatically different threats that American farmers faced. Marlatt noted that "our system of growing the same grain crops over vast areas year after year furnishes at once the very best conditions for the multiplication of the insect enemies of such crops. In addition to this is the fact that America, with its long, hot summers, presents the most favorable conditions for the multiplication of most insects. These two reasons undoubtedly account for the far greater losses experienced in this country as compared with Europe, the summers of which are very cool and short."[115] Marlatt's observation about insects also applied to wheat diseases and to other crops. In 1889 Charles V. Riley, who led the campaign to create the U.S. Entomology Commission, noted that "the injury to agriculture occasioned by insects is more marked in the United States than in any other country in the world."[116]

The great advances in modern science since the 1940s have not entirely freed the grain farmers of today from such threats. In 1986 a new insect threat, the Russian wheat aphis, entered the United States from Mexico and by 1988 had spread north through the western grain belt to the Canadian border. Its appetite for wheat caused losses totaling $891 million over the period 1987–93.[117] In recent years, fusarium head blight (or scab) has reemerged as the major yield-sapping disease in the northern plains. Throughout the 1990s, scab destroyed over $2.5 billion in wheat (over 500 million bushels) in the United States and at least $520 million in Canada. What had been a "minor problem" has placed North America's "breadbasket under siege."[118]

[114] Clinton, "Russian Thistle," 87–97; Dondlinger, *Book of Wheat*, 151–2.

[115] Marlatt, "Principal Insect Enemies," 5, further noted that the smaller holdings and more intensive cultivation in Europe led to more careful inspection and more aggressive responses to insect outbreaks.

[116] Riley, "Injurious and Beneficial Insects," 603.

[117] Morrison and Pearis, "Response Model Concept and Economic Impact," 1–11.

[118] McMullen, Jones, and Gallenberg, "Scab of Wheat and Barley," 1340; Windels, "Economic and Social Impacts," 17–21.

Turn-of-the-century observers clearly identified the collective action problem facing individual farmers. An individual who decided to adopt rotation schemes along the lines of those practiced in parts of Europe might reap little benefit because the insects and diseases would simply migrate from nearby fields. There were also important dynamic implications to the differing pest and disease environments. All else equal, wheat grown on land with a high nitrogen content – prime bottomland, land that had been fertilized, or land that had been left fallow – took considerably longer to mature than wheat grown on lower quality land or on land that had been cropped in wheat the previous year. This is why the Census of 1860 noted that the midge caused more destruction on the "flats" than on the "uplands" in the Genesee Valley. However, the problem was not limited to New York. On the Canadian prairie, Red Fife ripened about eight to ten days earlier on stubble land than on fallowed land.[119] Thus, wheat farmers everywhere faced a conundrum. If they took obvious steps to increase yields, they greatly increased the risk that rust and insects might wipe out the crop. In the spring wheat belt, farmers also increased the risk of losing their crop to early frosts. On the Great Plains, farmers discovered that if land was plowed and then harrowed to keep down the weeds and left fallow it would produce an excellent wheat crop, even in dry years when land that had been planted continuously yielded almost nothing. But to use this technique delayed the harvest about ten days. One of the contributions of earlier ripening wheats like Marquis was that it allowed farmers in the northern plains to employ this dry farming technique with far greater assurance that they would beat the late summer or fall frosts.

Our discussion has touched on only some of the most important of the hundreds of insects, diseases, and weeds injurious to wheat. But there is a common pattern. In all cases the severity of the potential problems grew significantly between 1839 and the early twentieth century, and in all cases the actions of scientists, government agencies, and individual farmers in changing cultural practices dramatically reduced the severity of the problems.

Rethinking the Sources of Productivity Growth

This section offers revisions to the Parker–Klein estimates of the sources of nineteenth-century labor productivity growth for wheat. We shun the heroic task of modeling how diseases and pests might have evolved

[119] de Kruif, *Hunger Fighters*, 41–2.

differently and how the wheat economy might have changed if biological technologies had stagnated. Rather we simply impose our estimates of the importance of IPM systems and new varieties on top of the Parker–Klein analysis. Our counterfactual asks: What would land and labor productivity in wheat cultivation have been in 1909 if grain growers had continued using 1839 varieties and failed to invest any effort to combat the rising threats from insects, weeds, and plant diseases? Taking the 1909 distribution of wheat acreage as our starting point, we next estimate how much of this 1909 acreage would have fallen below a plausible yield threshold of commercial viability.

Table 2.7 details our estimates of what 1909 yields and output per hour of work would have been in the absence of the biological changes. This exercise is in the spirit of modern "crop loss assessment" in the agricultural subdiscipline of plant protection. Even today, one of the leading practitioners of this methodology notes that "crop loss assessment is not an exact science... the alternative would be no estimates at all."[120] This is precisely what the historical literature has done by implicitly attributing zero weights to the biological investments made to ward off yield declines. Our approach is intended to produce conservative, lower-bound estimates of the impact of biological investments. In line with the experience during the 1950s when durum yields fell by over 70 percent due to the emergence of stem rust race 15B, the agronomy literature suggests that, in the absence of biological adjustments to control damage, disease epidemics and pest problems would have soon gotten out of hand, inflicting staggering yield losses.

To capture the direct effects of varietal changes, we use Parker and Klein's 1839 yields in their Northeast, South, and West: Corn Belt regions in place of the 1909 yields. For the other regions of the West, we follow the lead of Salmon, Mathew, and Luekel, and reduce the 1909 yields by one-third. The relatively poor performance of China Tea and Lost Nation vis-à-vis Fife in the North Dakota–Minnesota trials, as well as the subsequent widespread switch from Fife to yet higher-yielding hard red spring and durum varieties by 1909, suggest that our assumed 33 percent decline in yields would be an underestimate for the northern grain belt. The same conclusion applies to the Pacific region, which between 1839 and 1909 witnessed important changes in the location of production, several wholesale turnovers in varieties, and the development of cultural methods different than those used in the East.

[120] Oerke, "Estimated Crop Losses Due to Pathogens," 72.

TABLE 2.7. *Estimated Impact of Biological Innovation on Land and Labor Productivity in 1909*

	Period	United States	R1 Northeast	R2 South	R3 West	R3 Breakdown: Corn Belt		Western Dairy	Sm. Grain & W. Cotton	Range	California and Northwest
Yields											
1 Actual	1	11.30	14.50	8.40	13.00		13.00		nd	nd	nd
2 Actual	2	14.00	17.50	12.30	14.00	15.80		15.30	12.00	18.60	19.40
3 Counterfactual	2	7.50	13.10	7.60	7.30	10.40		6.50	5.60	9.30	10.00
4 Relative to Actual		0.54	0.75	0.61	0.52	0.66		0.42	0.47	0.50	0.51
Yield Adjustments											
5 Varieties			1839 Y	1839 Y		1839 Y		2/3*1909 Y	2/3*1909 Y	2/3*1909 Y	2/3*1909 Y
6 Insects & Weeds			0.90	0.90		0.80		0.80	0.80	0.80	0.80
7 Plant Diseases			1.00	1.00		1.00		0.79	0.87	0.93	0.96
Total Labor Per Bushel											
8 Actual	1	3.167	3.082	3.563	2.838		2.838		nd	nd	nd
9 Actual	2	0.760	1.020	1.400	0.690	0.762		0.760	0.690	0.952	0.433
10 Counterfactual	2	1.246	1.309	2.102	1.132	1.041		1.570	1.262	1.682	0.686
Bushels per Hour											
11 Actual	1	0.316	0.324	0.281	0.352		0.352		nd	nd	nd
12 Actual	2	1.318	0.976	0.712	1.449	1.312		1.316	1.449	1.051	2.309
13 Counterfactual	2	0.803	0.764	0.476	0.883	0.960		0.637	0.792	0.595	1.458

Notes: The regions R1, R2, R3 (and its subregions) refer to Parker and Klein's regions. See text footnote 3 for regional definitions. In 1839 the Corn Belt and Western Dairy were not segregated. The yield adjustments are based on the assumption that (a) the Northeast, South, and West: Corn Belt regions used 1839 varieties and received 1839 yields; (b) following Salmon, Mathews, and Luekel, "Half Century of Wheat Improvement," that yields in other regions of the West are two-thirds of 1909 yields; (c) insect and weed losses are 10 percent everywhere and an additional 10 percent higher in the West region first hit by the Hessian fly, the chinch bug, and other insects after 1839; (d) the plant disease losses in the new areas of the West equal the difference between the average and the peak three years in the period 1919–39 as indicated in the *Plant Disease Bulletin* and *Plant Disease Reporter*.

To account for the increasing insect and weed problems, we reduce yields by 10 percent in all regions and by an additional 10 percent (for a total of 20 percent) in the West, which first suffered serious infestations of Hessian flies, chinch bugs, and other insects after 1839. The 20 percent figure is likely a serious underestimate of the pest control savings because it is equal to Marlatt's 1909 lower-bound estimates of the savings from Hessian fly prevention measures alone, and thus ignores the vigorous efforts directed against locust, chinch bugs, greenbugs, tumbleweeds, and hundreds of lesser animal and plant enemies of wheat.[121]

An equally important task is to quantify the effect of controls for plant diseases. We can construct lower-bound regional estimates of the magnitude of the difference between potential and actual losses by examining the excess damage reported during periods of serious disease outbreaks. Our estimates use the state-level loss estimates published in the *Plant Disease Bulletin* and *Plant Disease Reporter* over the period 1919–39 to compare the damage in the worst three years with the average damage. This results in national yield losses that average about 11.5 percent.[122] We take this estimate to represent the additional decline in yields due to disease if biological technologies had remained constant.

There is a risk of double counting – the same wheat crop cannot be killed by the Hessian fly and then be damaged again by rust or the chinch bug. (However, a crop weakened by one enemy might be more susceptible to another.) To address this problem, we have taken lower-bound loss estimates and adopted the standard practice in the crop protection literature of modeling the percentage losses as having a multiplicative rather than an additive effect on yields.

The resulting upper-bound counterfactual yield estimates, presented in the third row of Table 2.7, generate a stark picture. Without biological innovations, 1909 yields in Parker and Klein's West region (R3) would have been less than one-half of what was actually achieved. Yields would have fallen to roughly 7.3 bushels per acre, attaining low, noneconomic levels in many subregions of the West. In other regions yields would have

[121] As noted previously, Marlatt's lower-bound estimate is well below that for fly losses when recommended procedures were not followed.

[122] *Plant Disease Bulletin*, 1917–22, and *Plant Disease Reporter*, 1923–39. This is a lower-bound estimate because, in the complete absence of biological learning, diseases would have evolved to be far more devastating than they were during the "bad" years of the relatively enlightened 1919–39 period. By region, the excess losses were West: Dairy, 21 percent; Small Grain, 13 percent; Range, 7 percent; and California and the Northwest, 4 percent.

been about one-third lower than actually achieved. National yields would have been about 54 percent of those actually achieved in 1909 and about 67 percent of those prevailing in 1839.[123]

Inserting the revised yield estimates into the Parker–Klein framework offers a fresh perspective on the sources of growth in labor productivity. Parker and Klein show that, nationally, bushels per hour of labor increased from 0.316 in 1839 to 1.318 in 1909 (rows 12 and 13), meaning labor productivity increased 4.17-fold. But our estimates show that, without biological innovation, bushels per hour of labor in 1909 would have increased to only 0.803. By this reckoning, biological innovations increased the output per hour of labor by 0.515 bushels (that is, subject to rounding, 1.318–0.803), accounting for about one-half of the total increase in labor productivity.

Using our alternative yield estimates, U.S. wheat production circa 1909 would have been 46 percent lower. This calculation presumes that all land planted with wheat in 1909 remained in wheat. This is unlikely. With lower yields, substantial acreage would have dropped below the threshold for sustained commercial viability in grain production. Although commercial viability clearly depends on input and output prices, a breakpoint of 6.5 bushels per acre can serve as a rough-and-ready standard. Yields below this breakpoint were commonly considered "poor crops" or "failures" and very little wheat – less than 1 percent of 1909 output – was produced in counties with average yields of less than this level.[124] Applying our yield adjustments to the county-level wheat cultivation data from the 1909 Census offers an estimate on how much acreage would not have been viable. These calculations show that, without biological learning, over one-quarter (28 percent) of U.S. wheat land in 1909 would have fallen below our 6.5 bushel standard. Much of this acreage would presumably have remained rangeland. Of course, the reduction in production might have increased prices, leading to shifts back into wheat cultivation in the East.[125] The key point remains that without biological learning the story of American agriculture over the

[123] Our estimates are in line with Salmon, Mathews, and Luekel, "Half Century of Wheat Improvement," 110, who found that the improved varieties introduced since 1900 increased annual output by about 231.8 million bushels or roughly 21 percent of 1949 output.

[124] See, for example, Patton, "Relationship of Weather to Crops," 43.

[125] Removing the unviable acreage from the cropland base would have reduced 1909 wheat production by an additional 10 percent. It would also increase "measured" land and labor productivity relative to that reported in the counterfactual estimates.

nineteenth and early twentieth centuries would have been fundamentally different.

Conclusion

In the mid-nineteenth century John Klippart, the corresponding secretary of the Ohio State Board of Agriculture, was arguably the most informed individual on wheat culture in the United States. In 1858 he published a 700-page tome detailing much of what was then known about the wheat plant and wheat farming.[126] In his view the commercial wheat belt would be forever limited to Ohio, Pennsylvania, and western New York. The soils and climate of Illinois, Iowa, and Wisconsin would doom those states to the haphazard production of low-quality and low-yielding spring wheat. Farther west the climate and soils would make any wheat production unlikely. Klippart believed that the entire territory south of southern Indiana and southern Illinois could never yield reliable crops because of rust. As a result, unless the United States husbanded its resources it would soon be an importer of wheat.

How could Klippart have been so far off the mark? He obviously was familiar with the mechanical reaper and thresher, and he would not have been surprised by the next generation of harvesting equipment – the self-binder. These are the machines that the standard accounts assert made the settlement of the West possible. What so blurred Klippart's vision was his inability to predict the wholesale changes in the genetic makeup of the wheat varieties that would become available to North American farmers. Mechanical innovations certainly lowered the cost of growing wheat in the West, but the binding constraint was biological. Without a biological revolution (assisted by the transportation revolution), the centers of wheat production in the United States and Canada could not have assumed their late-nineteenth-century dimensions.[127]

During the nineteenth century the wheat-growing environment seriously deteriorated as diseases, insects, and weeds multiplied. If, as the literature assumes, generations of wheat farmers had simply followed

[126] Klippart, *Wheat Plant*; Ohio State Board of Agriculture, *Annual Report 1857*, 675–816.

[127] If the traditional view for the United States was correct, there should have been a similar emphasis on mechanization in the histories of other land-abundant and labor-scarce frontier economies. This is not the case. The Canadian literature emphasizes Charles Saunders' path-breaking creation of Marquis, and the Australian literature highlights William Farrar's contributions in creating Federation and other drought-tolerant varieties.

in their fathers' footsteps (apart from adopting laborsaving machinery), their crops would have been ravaged. The fact that national yields increased slightly between 1839 and 1909 is a strong testament to biological innovation. This is especially true because of the wholesale shift in production to more marginal lands. Modern agricultural scientists have long appreciated the importance of maintenance research to overcome the effects of crop depreciation. One survey of 744 researchers yielded a mean estimate that maintenance efforts constituted over 41 percent of all wheat research.[128] A significant fraction of nineteenth-century research also was needed for maintenance – it simply allowed farmers to stay in place.

Nineteenth-century biological innovations carried over into the Green Revolution era, because much of the genetic material that modern wheat breeders used to produce the first generations of post–World War II hybrids came from Turkey wheat and other varieties introduced to the United States from around the world. In 1969, 11 varieties of hard red winter wheat were grown on one million or more acres. Turkey was important in the pedigree for all of these varieties. The semidwarf characteristics that are the hallmark of Green Revolution wheat derive from a Japanese variety called Norin 10. But one of the parents of Norin 10 was Turkey, which the Japanese had imported from the United States around 1890.[129]

By allowing wheat production to move into more hostile climates, the new wheat technologies significantly contributed to the pressure on eastern farmers to abandon wheat and seek other crops and production systems. The effect was also felt in Europe; without the widespread adoption of Red Fife, Turkey, and other new varieties, the grain invasion described by Kevin O'Rourke and others would not have been possible.[130]

Mark Carleton's introductions of foreign wheat varieties and Charles Saunders' creation of Marquis are beacons of wise government investments. Cyrus McCormick has long been eulogized as the man who "made bread cheap," but he needed considerable help. It is time that we add the names of Mark Carleton, Charles Saunders, David and Jane Fife, Abraham Fultz, William J. Spillman, and the other researchers who revolutionized North American wheat production to the high pantheon of nineteenth-century inventors.

128 Adusei and Norton, "Magnitude of Agricultural Maintenance Research," 1–6.

129 Quisenberry and Reitz, "Turkey Wheat," 110.

130 O'Rourke, "European Grain Invasion," 775–801.

3

Corn

America's Crop

Corn is America's leading crop in terms of both acres harvested and value of output.[1] The common view is that the modern age of biological advances in agriculture began in the 1920s and 1930s with the development and diffusion of hybrid corn. We do not dispute that hybrid corn represented a technological milestone. However, this does not imply that there were no earlier advances of consequence. By the 1920s farmers and breeders had developed roughly one thousand varieties of corn tailored to varied geoclimatic zones. These early developments, which often involved deliberate hybridization, combined the desirable characteristics of southern Dents with those of northern Flints. The repeated crossing of two distinct races of corn represented an achievement for its age comparable in importance to the development of F1 and F2 hybrids in the 1910s. As with the diffusion of new wheat varieties, these efforts increased yields and facilitated a significant geographic shift in corn cultivation.

The history of corn has notable differences from that of wheat. Most significantly, because corn was cross-pollinated, whereas wheat was largely self-pollinated, the use of hybridization in seed production was both more feasible and of greater value. In addition, because corn was generally less affected by insects and diseases, there was less need for breeders to concentrate on developing pest-resistant varieties. Furthermore, because corn was native to the Americas, there was no worldwide search for varieties.

[1] Corn is also a leading component of the American diet via its contribution to the production of meats, oils, sweeteners, etc. Pollan, *Omnivore's Dilemma*, 15–119.

There were other significant differences. In the twentieth century, wheat straw became a nuisance, rather than a valuable by-product, due to the replacement of horses by tractors and the diffusion of combined harvesting. For corn, the spread of ensilage, which began in the last quarter of the nineteenth century, increased the usefulness of its stalks. Most wheat was destined for human consumption whereas the vast majority of corn was used for animal feed. The increased use of corn and corn silage reflected an intensification of animal feed and management systems. In addition, the pre-1940 era witnessed several chemical and processing innovations that made possible the conversion of corn into corn starch, corn sugar, corn oil, and myriad industrial products. These biological innovations expanded the market for corn.

Hybrid Corn

The standard account of the development of modern hybrid corn commonly begins in 1877 when William J. Beal, of the Michigan Agricultural College (now Michigan State University), first demonstrated the value of hybrid vigor. Agronomists did not pursue Beal's insight until the rediscovery of Mendel's laws around 1900, when George H. Shull and Edward M. East began experiments with inbred, pure line maize. Farmers had long known that inbreeding produced runty-looking ears, which offended those raised to appreciate the beauty of highly uniform "show corn." In addition, early experiments with hybrid corn had resulted in poor yields, which offended those raised to appreciate the beauty of money. To practical breeders, the theoretical exercise of developing pure lines had little immediate appeal. Indeed, Cyril G. Hopkins, head of the Illinois corn program in the early 1900s, purportedly justified his rejection of East's proposed inbreeding work by stating that, "We know what inbreeding does and I do not propose to spend people's money to learn how to reduce corn yields."[2]

Undaunted, East and Shull demonstrated by the mid-1910s that crossing two pure lines provided a systematic means of producing strains that outproduced open-pollinating varieties. Initially, the pioneers of hybridization found that F1 crosses yielded too few seeds to be feasible for commercial production. In 1917, Donald Jones developed the idea

[2] Fitzgerald, *Business of Breeding*, 22. Deborah Fitzgerald, 9–42, offers an excellent analysis of the theory and practice of corn breeding and a discussion of the backward state of corn breeding science in the Bureau of Plant Industry, 44–9. See also Crabb, *Hybrid-Corn*.

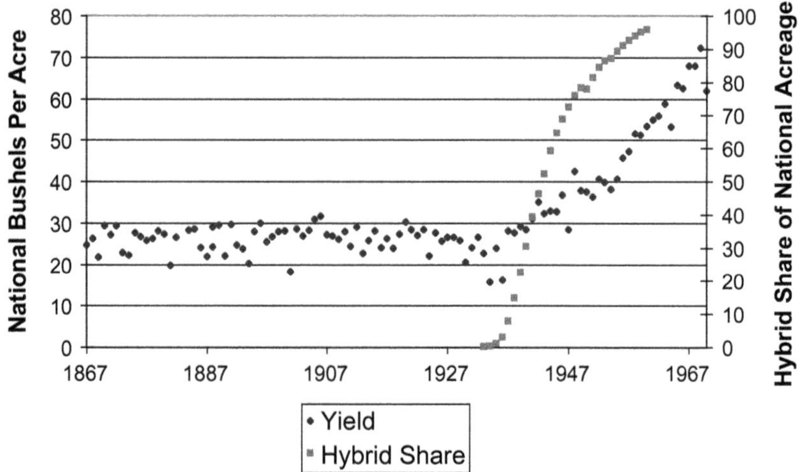

FIGURE 3.1. Corn yields and the spread of hybrids. *Sources:* Carter et al., *Historical Statistics*, Tables Da693–694; USDA, *Agricultural Statistics 1962*, 41.

of multiplying the available seed by crossing plants of the F1 generation to produce F2 seed. The "double cross" hybrids tended to be less productive than the "single cross" generation but the seeds were sufficiently numerous for commercial uses.[3]

Henry A. Wallace offered the first commercial hybrid corn, Copper Cross, to the farming public in 1923 without generating significant sales.[4] Ten years later, less than 1 percent of corn acreage in the corn belt was planted with hybrid seed; but once adoption began, it was rapid and thoroughgoing.[5] The new seed accounted for more than one-half of the region's corn acreage by 1939 and virtually all by 1947. Figure 3.1 charts the diffusion of hybrid seed and average corn yields.

The creation of modern hybrid corn represented more than the development of a single new line of high-yielding seed. Rather it was, according to Zvi Griliches, the invention of a new way of inventing. In addition, it

[3] Thus, the takeoff of hybrid corn was associated with F2 seed. Eventually, improvements in hybridization techniques made using F1 seed feasible and, after 1960, such seed accounted for most of the planting stock.

[4] Henry A. Wallace continued his association with his Hi-Bred seed company even after becoming Secretary of Agriculture. Early on, he objected to the prevailing focus on the appearance and uniformity of corn, arguing that performance and yield were of greater economic relevance. As *Wallace's Farmer* put it: "The look of an ear of corn makes no difference to a hog." Quoted in Baker, "Indian Corn," 97.

[5] For economists, the classic work on this issue is Zvi Griliches, "Hybrid Corn," 501–22.

TABLE 3.1. *Selected Statistics on the Pioneer Hi-Bred Seed Company, 1926–36*

	Hybrid Seed (lbs)		Company Finances	
Season	Produced	Sold	Capital ($)	After-tax Profit ($)
1926–27	800	605	7,000	(1,235)
1927–28	1,200	928	7,000	34
1928–29	2,146	2,055	7,000	7,270
1929–30	2,632	2,281	7,000	7,999
1930–31	3,050	1,870	14,000	2,857
1931–32	4,329	2,656	15,000	(1,508)
1932–33	5,970	3,643	30,000	4,313
1933–34	18,617	16,252	29,600	37,593
1934–35	24,971	23,771	29,600	45,768
1935–36	32,880	31,700	59,200	50,594
1936–37	41,035	40,586	103,000	125,323

Source: Pioneer Hi-Bred International, *History*.

launched a new industry of seed producers on the cutting edge of plant genetics. Table 3.1 reveals the sales and revenues of Wallace's firm, the Hi-Bred Corn Company, founded in 1926 (and rechristened Pioneer Hi-Bred in 1935). A venture that had been something of a hobby developed, in the late 1920s and early 1930s, into a struggling but exciting collaboration of enthusiastic, if amateur, entrepreneurs and scientists. By the mid-1930s the company emerged as a highly profitable enterprise. Of course, Pioneer was not alone in exploiting this new opportunity – by the late 1930s its sales represented less than one-half of the hybrid seed market and faced stiff competition from firms such as Funk, DeKalb, and Pfister. Some 400 hybrid seed companies operated in Iowa alone in 1940.[6] But Pioneer did become an industry leader and remains prominent to this day. Given this often-told story of the development of hybrid corn, we now turn to the less well-known story of earlier advances.

Technology Transfer

Native to the Americas, maize was the leading staple of the indigenous peoples on the eve of European colonization. A number of distinct races of corn had been cultivated for at least two millennia before the arrival of Europeans. Centuries of careful selection matched corn to

[6] Pioneer Hi-Bred International, "History"; Troyer, "Temperate Corn," 460.

specific locales. Over a longer period, Central and South American farmers transformed the typical ear of corn from a size smaller than a child's little finger into a product resembling modern corn. By the time of Columbus, this transformation was so extensive that corn could reproduce itself only with human intervention.[7]

Stories of the starving times in the early Jamestown and Plymouth settlements are familiar to every schoolchild. Indigenous peoples helped save the colonists by sharing their ancestral seed stocks (gourdseed corn in the Chesapeake and Flint corn in New England) and their knowledge of how to cultivate the crop.[8] A story of the founding of the Virginia colony relates how, in 1609, Captain John Smith of Jamestown captured Kemps and Tassore, who revealed the mysteries of maize. In Smith's words, these prisoners "did double taskes and tought us how to order and plant our fields."[9]

A similar story of the Plymouth colony focuses on Squanto – a native who was captured in 1614, sold into slavery in Europe, and returned to his native land in 1621. He taught the Pilgrims, in the words of William Bradford, how "to plant their corne, ... showing them both ye manner how to set it, and after how to dress and tend it."[10] Whether the local Indians commonly fertilized the corn hills with fish, as Squanto purportedly taught the newcomers, is in doubt. Initially, New England farmers followed the native polycultural practices. But over time, the settlers moved toward a more permanent system of growing monoculture fields of corn in rotation with crops such as oats and hay.[11] Corn remained the classic pioneer's crop in the forested East. After clearing trees, settlers could grow corn in hills between the decaying stumps.

The technological transfer of corn was part of a far larger story. "English expectations about the weather patterns in North America were based on the common-sense assumption that climate is constant in any latitude around the globe. Newfoundland, which is south of London, was

[7] Goodman and Brown, "Races of Corn," 59–68; Smith, *Crop Production*, 12–15. There is considerable controversy as to how early maize became an important part of the diet in North America. Galinat, "Domestication and Diffusion of Maize," 245, 269–70; Lusteck, "Migrations of Maize," 523; Clinton, "Origin and Spread of Maize," 539.

[8] This story illustrates just how slowly knowledge diffused. Columbus introduced maize into Europe in 1493. Roughly 114 years later the crop was still unknown to the Jamestown settlers.

[9] Smith, *Generall Historie of Virginia*, 86.

[10] Hardeman, *Shucks, Shocks, and Hominy Blocks*, 16–17; Mann, *1491*, 31–43.

[11] Merchant, *Ecological Revolutions*, 78, 155–6, 296; Ceci, "Fish Fertilizer," 26–30.

expected to have a moderate climate, and Virginia was expected to be like southern Spain."[12] It required decades of harsh empirical evidence before settlers gave up their dreams of growing oranges and olive trees in Virginia. Biological learning was often a slow and costly process.

Sunlight, Corn, and Settlement

The European settlers' commonsense notions about the importance of latitude proved useful for understanding the spread of corn cultivation. In a well-known article, Richard Steckel invoked the photoperiodic properties of maize to explain the east–west pattern of U.S. migration during the nineteenth century.[13] Steckel notes that corn is classified as a short-day plant. Such plants flower after the number of hours of daylight falls below a certain maximum threshold. For corn, the shortening of days in the latter part of summer triggers flowering. Long-day plants such as wheat and small grains, by way of contrast, time their flowering to occur after the number of hours of daylight increases above a certain minimum. Steckel further observes that "long-day or short-day plants that are grown outside their latitude of adaptation mature too early or too late for optimal performance."[14]

Farmers had long understood that the number of hours of sunlight was significant. Indeed they adopted various practices, such as planting by the moon, which reflected long-standing folk wisdom. A scientific understanding of photoperiod sensitivity began in 1918 with an accidental discovery by two U.S. Department of Agriculture (USDA) scientists, Harry A. Allard and W. W. Garner, working with tobacco. They noticed that one tobacco mutant, Maryland Mammoth, responded differently to daylight conditions than was typical.[15] Subsequent research revealed that the developmental sequences of many plants were influenced by pigments that measured the length of the photoperiod. Long- and short-day plants relied on the same measuring mechanism but responded differently to the signal.

[12] Kupperman, "Puzzle of the American Climate," 1262.

[13] Steckel, "Economic Foundations of East–West Migration," 14–36.

[14] Experiments in controlled environments show that photoperiod sensitivity is determined by the length of the night rather than the length of the day. Steckel, "Economic Foundations of East–West Migration," 20, does an excellent job of introducing the concept of photoperiod sensitivity to nonspecialists. However, Chapter 4 offers some qualifications to his statement that "Cotton is a day-neutral plant."

[15] Garner and Allard, "Effect of the Relative Length," 553–606.

Steckel quantified the effect of growing the "right" corn by using historical data from experiment station trials. Between 1888 and 1894 the Illinois Agricultural Experiment Station at Champaign tested a variety of corn seeds adapted "to about 80 different locations" in the Midwest and Northeast. Drawing on these field trials, Steckel's econometric analysis found that the "yields of seeds adapted 250 miles south and 250 miles north were only 62 and 72%, respectively, of the yield of seed adapted to Champaign. Yields of seeds adapted up to 250 miles east were slightly higher than those adapted to Champaign, whereas the yields of seed adopted 250 miles west was 93% of the yield of seeds adapted to Champaign."[16] Here is solid evidence of the importance of matching corn varieties to geoclimatic conditions. For corn, north–south variations mattered significantly, but east–west variations were relatively minor.

Steckel argued that pioneering farmers learned that their seed corn was adapted to the seasonal daylight conditions of their own latitude. When moving to new areas of settlement, they "probably took their own supplies of seed grain." Thus, they would be disinclined to change latitudes significantly for fear that their seeds would generate substantially lower yields. "Farmers who went too far north or south had poor yields and sent relatively unfavorable reports back to the community from which they left."[17] Although westward settlement occurred across a broad front, for many it involved movement along an east–west line. This latitudinal thinking made more sense within a limited range of east–west settlement within North America than between the Atlantic coasts of Western Europe and North America.

Steckel's findings imply that farmers had to be especially observant to choose corn varieties that fit local conditions. According to Willet M. Hays, one of America's most successful plant breeders, "a variety of wheat may dominate in several states, a variety of corn is more likely to be adapted to a group of counties."[18] Through constant experimentation

[16] Steckel, "Economic Foundations of East–West Migration," 22. Tests conducted at the Arkansas Experiment Station in 1888 and 1889 bolster Steckel's findings. Newman, *Comparative Yield of Corn*, 109–21. In 1880 Brewer found that corn farmers reported "native seed" was "the best." But they also thought higher yields resulted "from carrying seed eastward than westward" and "in numerous instances . . . corn of some excellent variety carried from New England or eastern New York to the more fertile West was inclined to run to stalk, the quality of the grain and the yield being inferior." Brewer, "Report on the Cereal Production," 99.

[17] Steckel, "Economic Foundations of East–West Migration," 23.

[18] Hays, *Breeding Animals and Plants*, 19. Hays recommended having a corn breeder in every township. See Troyer, "Persistent and Popular Germplasm," 224. This picture is

and a careful process of selection and crossbreeding, American pioneer farmers and breeders developed varieties that would yield more and survive better in the West.

Inventing the Corn Belt Dents

One important amendment to Steckel's account is that the varieties that came to predominate in the Midwest, the so-called Corn Belt Dents, were not identical to those grown along the same latitude on the Atlantic coast. Farmers in the Northeast typically grew Flint corns, a set of early maturing, slender-stalked varieties that produced one or more long, cylindrical ears with eight to ten rows of smooth, hard kernels. Farmers in the southern United States typically grew later-maturing, heavier-stalked gourdseed or Dent corns which produced several rounded, many-rowed ears with softer, dimpled kernels. The late-maturing Dents were far more productive than Flints in areas where conditions allowed both to be grown. The Flints tended to be yellow, the gourdseed white.[19] In the early period, the soft gourdseed were generally considered better for animal feed and the Flints for the family's food. Dents and Flints belonged "to races of maize so different that, were they wild grasses, they would certainly be assigned to different species and perhaps to different genera." Among the most important differences was that the southern Dents were heterogeneous and readily hybridized in the field. Unlike the early Dents, the Flints that went into making the early hybrids were much more like a pure line. "The northern flints were essentially homogeneous at the eastern end of their range in New York and New England but became increasingly variable as the Great Plains were approached."[20]

Raymond Baker, a leading breeder for Pioneer, offers an informed account that adds significantly to Steckel's story. Early nineteenth-century farmers did carry their seed corn from east to west, but, in the areas where the migration flows from New England and from Virginia and Kentucky

confirmed in reports of the state boards of agriculture and experiment stations. As an example see Stadler and Helms, "Corn in Missouri," 39, 50–1. The Missouri Experiment Station had been testing maize throughout the state since 1905. Less than one-third of the more than fifty varieties tested were recommended and none for a large area.

19 Anderson and Brown, "History," 4, quote John Lorain's 1825 posthumous memoir, *Nature and Reason Harmonized*, 203. Recent genetic research largely supports the findings of an earlier generation of scholars. Labate et al., "Molecular and Historical Aspects," 80–91.

20 Anderson and Brown, "Origin," 148; Brown and Anderson, "Northern Flint Corns," 2.

overlapped, the northern Flints hybridized with the southern Dents. Such crosses, occurring thousands of times during the settlement of the Midwest, combined the higher yields of the Dents with the earlier maturation and greater hardiness of the Flints, creating the famed Corn Belt Dents. Edgar Anderson and William Brown viewed twentieth-century efforts to hybridize corn essentially as attempts to repeat this process – to reestablish pure-line Flints and Dents and then cross them to gain the benefits from heterosis.[21] In the 1810s, John Lorain of Philpsburg, Pennsylvania, began crossbreeding Dents and Flints, and many accounts credit Christopher and Jacob Leaming of southwestern Ohio with initiating development of an important line of Flint–Dent crosses (Leaming) as early as 1826.[22] (See Plate 3.1 for a photo of a Flint–Dent cross.)

The classic "founding story" for the Corn Belt Dents highlights the development of Reid Yellow Dent. In the period 1845–46, Robert Reid moved from Brown County in southern Ohio to Tazwell County in central Illinois. In 1846 and 1847, he planted Gordon Hopkins, a late, red variety, originally from the Shenandoah Valley in Virginia, which he had carried from southern Ohio. The variety performed poorly so he replanted the "missing hills" with Little Yellow, a local Flint corn, which naturally crossed with the Gordon Hopkins. Many of these accidental crosses possessed the best qualities of both races. Reid subsequently carefully selected the best progeny to develop an early, high-yielding strain with 18 to 24 rows of relatively smooth kernels. After winning the prize for best corn at the 1893 Chicago World's Fair, Reid Yellow Dent gained wide popularity in the Midwest and served as the basis for many other improved varieties of open-pollinated corn, including Funks Yellow Dent, and Krug.[23] Reid Yellow Dent, Leaming, and their descendants provided the bulk of the germ stock used in early hybrid corn seed.

Robert Reid's achievement had as much to do with the long process of selection as with recognition of the value of his first hybrids. Raymond Baker maintained that Reid was following a common practice of planting early Flints and high-yielding gourdseed types next to each other and thus

[21] Baker, "Indian Corn," 96–9; Anderson and Brown, "History," 2, claim that "Most maize breeders have not understood that the hybrid vigor they now capitalize is largely the dispersed heterosis of the flint–dent mongrels."

[22] Troyer, "Persistent and Popular Germplasm," 155, 158, 162–93; Lloyd, *J. S. Leaming*, 3; Wallace and Bressman, *Corn and Corn Growing* (1937), 208–9, contest the 1826 date, asserting that Leaming was developed after 1856.

[23] Wallace and Bressman, *Corn and Corn Growing* (1937), 208–9.

unwittingly creating new hybrids. Anderson and Brown agreed that the "mixture of the southern dents and northern flints began so early that it has been largely forgotten. The often-told story of the origin of Reid's Yellow Dent... is merely one of thousands of such mixtures and took place relatively late in the history of corn-belt dents."[24] The 1850 U.S. Patent Office *Report* offers contemporary testimony that mixed Dent–Flint varieties were widespread by the time of Reid's discovery.[25] By 1850, the crossbreeding "process was actively under way from Pennsylvania to Iowa and south to the Gulf States. By the 70s and 80s, a new type of corn had emerged from this blending, though crossing and recrossing continued up to the advent of modern hybrid corn."[26]

We can gain a better sense of the contribution of Flint–Dent crosses by examining the results of the Illinois corn trials investigated by Steckel.[27] Flints and soft flour corns appeared in some of the early trials but were largely dropped after performing poorly. In the four years that Flints were tested (1888–90 and 1895) their yields averaged 71 percent of those for Leaming, by then a standard Corn Belt Dent. Four soft corns (tested in 1889, 1890, 1893, and 1895) produced only 64 percent of Leaming's yield. The reports disparaged both of these founding types. Flints, including those grown from Illinois seed, were described as "novelties" and "of no general practical value in this state." A well-known soft variety matured so late as to be "utterly worthless at this latitude."[28] These yield differences were on par with the extreme north–south deviations that Steckel found, and they suggest what would have happened if farmers had in fact continuously planted the varieties that they had grown in the East.

[24] Baker, "Indian Corn," 96–7; Anderson and Brown, "History," 5.

[25] U.S. Patent Office, *Report of the Commissioner. Agriculture 1850*, 232, 245, 301, 371, 400–1, 454, reports on farmers in Alabama, Illinois, Ohio, and South Carolina growing Dent–Flint crosses.

[26] Anderson and Brown, "History," 8. Studies employing isozyme electrophoresis confirm the claims of the traditional historical literature and more classical botanical studies. John Doebley et al., "Origin of Cornbelt Maize," 129, show that "Midwestern Dents arose from the hybridization of Northern Flint from New England and Southern Dent." Their findings "further indicate that Midwestern Dents contain a greater share of Southern Dent germplasm as compared to Northern Flint germplasm." This is also true of the modern hybrids derived from midwestern Dents.

[27] "Field Experiments with Corn 1888," 37–87, and "1889," 215–45, and "1890," 389–404; Morrow, Gardner, and Farrington, "Corn and Oats Experiments," 337–51; Davenport and Fraser, "Corn Experiments," 163–73.

[28] "Field Experiments with Corn, 1888," 48, and "1889," 221.

"Indian" Corn Farther West

A second important amendment to Steckel's account is that, whereas pioneering farmers did carry their seeds west, in many cases the settlers adopted varieties developed by the indigenous populations which were well suited to the local environment. The transfer of seeds from Indians to European settlers did not end at the Jamestown and Plymouth colonies. Native varieties became especially important with the extension of corn cultivation into the northern Great Plains and the arid West. It "soon became evident that eastern varieties of corn could not be depended upon to make a crop," inducing settlers to become "really interested in the hardier types of the native corn."[29] Settlers in the lower Missouri Valley initially grew Omaha blue corn and Mandan "squaw corn." These were displaced by Dents from "back home" but only after a period of adaptation, when Dents presumably mixed with the local corns. By this process the Dent corn area extended into Iowa and Nebraska. "As the frontier moved up the Missouri valley into South Dakota, however, the triumphal progress of the dent corns began to slacken. As the latitude and altitude both increased and the rainfall became less, it grew to be a much longer and more difficult task to acclimate dent varieties of corn. In this region also the flints and flour corns of the Rees and Mandans had acquired some reputation... for their extreme hardiness."[30]

The region's earliest corn growers, women of the Mandan, Arikara (Ree), and Hidasta tribes, had developed many hardy varieties of Flint, flour, and sweet corns. They "carefully cared for their seed supplies in a manner which might well put to shame the white men who have now usurped their lands; thus handing down to their white agricultural successors many ideally acclimated strains and varieties" of maize. "Fortunately for the future of agriculture in North Dakota a few of the early white settlers of the Missouri Valley realized the potential value of these long-acclimated varieties and proceeded to select and improve them."[31] The technological transfer of corn from Native Americans to white farmers was even more complicated, because the corns that many settlers brought with them very likely had only recently arrived in the East from the prairie states. Antebellum accounts tell of eastern farmers seeking higher

[29] Will and Hyde, *Corn Among the Indians*, 20.

[30] Ibid., 23–4.

[31] Olson, Walster, and Hopper, "Corn for North Dakota," 3.

yields by introducing Maha (Omaha Indian), Sioux, and Canadian Flint varieties grown by indigenous peoples in the West.[32]

Thus, it appears likely that these western corns introduced in the East mixed with eastern varieties to form new hybrids. When these hybrids were carried west, further crosses occurred. The latter stages of this process are well documented. From the 1880s on, settlers bred native Mandan, Ree, and Arikara corns to produce improved strains of mixed Flints. Varieties noted for their hardy and early ripening characteristics include Dakota White Flint (1886), Gehu Yellow Flint (ca. 1886), and Burleigh County Mixed (1887). Until the final years of the nineteenth century, "these hardy types of corn were the only ones grown in the Northwest, and many a struggling homesteader in the lean years owed his home and the foundation of his success to them."[33] Only after 1896 did an unusually hardy and early red Dent – Northwestern Dent developed by Oscar Will – begin to replace the native-type Flints.[34]

Unconstant Maize

As an indication of the dynamism of U.S. maize stocks, it is likely that "three-fourths of the present varieties of corn [as of 1913] have been developed since 1840."[35] Anderson and Brown noted that the "controlled breeding of new varieties by farmers themselves was more frequent than anyone would believe who has not looked into the record."[36] Over the late nineteenth and early twentieth centuries, maize varieties proliferated. William Brewer opined in 1883 that maize "varieties are practically numberless" and it is "highly probable that as many as 150 or 200 varieties of corn are cultivated" in the United States.[37] E. L. Strutevant's 1899 *Varieties of Corn* listed 507 separately named strains (with 163 synonyms) grown in the United States, including 323 Dent corns, 69 Flints, 63 sweet corns, 27 soft or flour corns, and 25 pop corns.[38] Edward

32 Wallace and Bressman, *Corn and Corn Growing* (1923), 4.

33 Atkinson and Wilson, "Corn in Montana"; Will and Hyde, *Corn Among the Indians*, 29.

34 Troyer, "Background of U.S. Hybrid Corn," 617–18.

35 Montgomery, *Corn Crops*, 79.

36 Anderson and Brown, "History," 7.

37 Brewer, "Report on the Cereal Production," 97.

38 Strutevant, "Varieties of Corn." Troyer, "Background of U.S. Hybrid Corn," 604, states that Strutevant "described 538 cultivars and 232 synonyms in the USA and Canada in 1899." Poneleit, "Breeding White Endosperm Corn," 241.

Montgomery estimated that, circa 1913, a complete catalog listing all locally named varieties would have over one thousand entries.[39] The number of varieties was large and expanding rapidly.

Most of the open-pollinated varieties prominent in the corn belt on the eve of the introduction of modern hybrid corn were of relatively recent origin. As examples, Lancaster Sure Crop was developed by Isaac Hershey in the mid-1910s, Silvermine dated from 1890, Funk Yellow Dent in the 1890s, Johnson County White and Minnesota No. 13 in 1893, Hilderth Yellow Dent in 1901, Commercial White from 1902, Krug in 1921, and so on. As noted earlier, Reid Yellow Dent was the product of Robert Reid's accidental cross of Hopkins and Little Yellow in 1847, but more than fifty years of continuous selection perfected the strain popular in 1900.[40] There is no doubt that the menu of varieties available to U.S. corn farmers prior to the advent of modern hybrid corn was anything but static.

This dynamism reflects the nature of maize, a highly variable and mutable plant. Its rate of genetic mutation was exceptionally high.[41] The extent of cross-pollination among corn plants was also very high, over 95 percent, compared with 5 to 40 percent for cotton plants (depending on environmental conditions), and with less than 5 percent for wheat plants. Open-pollinated corn, like cotton, tended to be heterozygous (containing diverse genetic material) and, without careful selection, varieties could change quickly and possibly lose desirable characteristics. Indeed farmers long recognized that they faced a major challenge in preserving the desirable combination of characteristics in their preferred varieties. As John Klippart noted as early as 1860: "There is no plant, whether cereal or other, which so readily hybridizes or intermixes as corn. Everyone who has grown corn, is well aware of the difficulty of keeping the varieties pure."[42] In 1878, William Emerson added, "When a good variety has been established, great pains should be taken to keep it pure and unmixed. Two cornfields planted with different varieties should be widely separated, unless the farmer desires an intermixture.... The old rule is nowhere more applicable than in the matter of seed corn, 'prove all things, and hold fast to that which is good.'"[43]

39 Montgomery, *Corn Crops*, 78. Wallace and Bressman, *Corn and Corn Growing* (1923), 19, state that there are "at least 1,000 varieties."

40 Wallace and Brown, *Corn and Its Early Fathers*, 80.

41 Weatherwax, *Indian Corn in Old America*, 183.

42 Klippart, *Wheat Plant*, 654–5.

43 Emerson, *History and Incidents*, 142.

Maize has long been considered easier to manipulate genetically than either wheat or cotton. Corn differed from cotton by having separate female and male flowers, which made the role of parentage easy to understand and facilitated breeding. It was also relatively easy to exert conscious human control over pollination by detasseling the male flowers from one of the group of plants. The ease of working with corn is reflected in the large and early literature on crossing maize which dates back to the eighteenth century in North America. As examples, in 1716 Cotton Mather of Boston published an article on the effects of cross-pollination, and in 1739 James Logan of Pennsylvania, reported on his pioneering experiments to cross maize. Several other features of the plant and its cultivation made breeding relatively easy. Unlike the small grains which were grown en masse in fields, maize was frequently cultivated as individual plants in hills, making it easier to observe the characteristics of a specific plant. The seeds on a given plant were also clustered together on a separately harvested ear.[44]

The kernel is the part of the maize plant of greatest interest to humans. One thinks of the kernel as a seed, but it is really a fruit surrounding the seed. The grain is comprised of three parts: the pericarp or hull, the embryo cell containing a nucleus formed from the ovule of the mother plant fertilized by a sperm from pollen that traveled down the silk, and an endosperm cell that makes the bulk of the kernel and provides the embryo with nutrients for germination.

Maize, like plants such as peppers and the sweet peas which Mendel investigated, is subject to the xenia effect. That is, the male plant has an immediate effect on the characteristics of the fruit borne by the female plant, so the influence of the male's genetic material can be seen without waiting for the seed to germinate. The endosperm of maize contains a separate nucleus formed from the mother plant fertilized by a second sperm (which is almost always from the same male pollen as the first sperm). This additional sperm affects the development of the endosperm and, thus, the nature of the corn kernel. The xenia effect resulting from the double pollination of each individual corn silk explains why a single ear of corn can contain kernels of different colors.[45]

Xenia effects in corn caught the attention of early colonists. Cotton Mather observed that, when a row of red and blue varieties was grown in the middle of a field of yellow corn, the ears of yellow corn tended to

[44] Weatherwax, *Indian Corn in Old America*, 184.

[45] Mangelsdorf, *Corn*, 7; Weatherwax, *Indian Corn*, 199.

contain some red and blue kernels and, moreover, the prevalence of such kernels was more common downwind from the red and blue row than upwind. Thus, the keen-eyed Puritan published one of the first treatises recognizing sexual activity in maize.[46]

Several distinguishable characteristics of maize grain depend on the development of the endosperm. These include the nature of the starches, the texture, the shape when dried, and the color. The endosperm may be either hard, corneous, and translucent or soft, floury, and opaque (or in some cases waxy). When dried, the corneous and floury tissues contract at a different rate. The nature of the tissues is also related to the nutritional content (e.g., the amount of protein). Pop corns were, almost entirely, and Flint corns predominately, formed from corneous tissue, which creates a hard shell around the relatively small amount of floury tissue present. Flour corn, by way of contrast, contained mostly soft, floury tissue with the corneous tissue evenly distributed on the top and sides of the kernel. Dent corns combined layers of corneous tissue on the sides of the endosperm and floury tissue in the center, widening out at the top. When dried, the floury tissue contracted more than the corneous tissue, resulting in the "dent" at the top of the kernel that gives this type its name. In the other types discussed previously, the tissues were more uniformly distributed and the shape of the kernel remained largely unchanged. For sweet corns, due to the presence of one recessive gene, the carbohydrates chiefly remained sugar in solution rather than starches and, when dried, the kernel wrinkled. The embryo contained the oil of the grain; hence, whether a given plant possessed a high or low oil content depended on the size of the embryo relative to the grain.[47]

The changing relative prevalence of white and yellow corn varieties offers a crude indication of the extent of the mixing of Dents and Flints. Table 3.2 summarizes, by major Census region, state-level data of the production of white, yellow, and mixed varieties of field corns for the 1915–17 crop years.[48] White and yellow corns each made up about 40 percent, with mixed varieties accounting for the remainder. But the

[46] Wallace and Brown, *Corn and Its Early Fathers*, 39, 41.

[47] Weatherwax, *Indian Corn in Old America*, 199–201.

[48] U.S. Bureau of Crop Estimates, "White, Yellow, and Mixed Corn," (December 1915), 83, (February 1917), 17, and (December 1917), 134. Some popular accounts misdate the change. Dale Millis, Morrie Bryant, and Clive Holland, "White Corn Production and Uses," 1, state that "White corn equaled yellow corn production in North America until the 1940s, when the discovery of carotene in yellow corn made yellow corn more popular for animal feed."

TABLE 3.2. *Regional Distribution of Corn Varieties by Color, Percentage Shares of 1915–17 Production*

	Yellow	White	Mixed
United States	42.6	41.9	15.5
New England	74.3	18.3	7.4
Middle Atlantic	60.6	21.1	18.3
East North Central	56.1	31.5	12.4
West North Central	49.7	33.6	16.7
South Atlantic	20.6	68.9	10.5
East South Central	14.7	72.4	12.9
West South Central	27.9	46.6	25.5
Mountain	35.8	34.0	30.2
Pacific	49.5	42.1	8.5

Sources: U.S. Bureau of Crop Estimates, "White, Yellow, and Mixed Corn," (December 1915), 83, and (February 1917), 17.

aggregate numbers hide substantial regional variation in production across latitudes and longitudes. Yellow varieties dominated in the northern states, particularly in the Northeast. (The underlying state data show that virtually 100 percent of the Maine corn crop was yellow.) White varieties reigned in the South. (More than 80 percent of the crop in Georgia and Florida was of this type.) Moving toward the nation's middle latitudes and to the west, there was a greater balance between white and yellow varieties. Mixed varieties also became more prevalent.

A preference for yellow versus white corn highlights an instance when folk practices were subsequently proved to be scientifically sound. Yellow coloration genetically dominates over white. Following the creation of the improved nineteenth- and early twentieth-century hybrids, stockmen tended to prefer yellow corn, believing it to be more nutritious for their animals. Scientists initially found no difference in the carbohydrate values of yellow and white corns and thus argued that the preference was ungrounded. How such a preference for yellow corn arose is unclear, but the nutritional value of yellow versus white corn was being put to the test daily on millions of farms and feedlots. Aesthetics and conformity to convention appear to have played little role in the choice. The grower was not himself consuming the corn and presumably was interested chiefly in obtaining a given nutritional bundle as cheaply as possible.

In the late 1910s, dry-lot swine feeding trials at the Indiana and Wisconsin experiment stations showed that yellow corn was indeed superior to white corn. Scientists traced the differences to the greater quantities of carotene (vitamin A) in yellow corn.[49] So the farmers had been right after all. The release of these findings had a pronounced effect on the type of corn grown. The price differentials, which circa 1915 generally had put a premium on white corns, now shifted to favor yellow corns. Output began to shift toward yellow varieties. The white corn share had fallen to 15 percent in 1944 and to less than 1 percent by the 1970s and 1980s. This transition dovetailed with other developments. Hybrid seed firms directed their investments toward yellow corn and its relative yield rose. Our analysis of county-level data for Nebraska over the period 1941–49 reveals that adoption of hybrid corn was positively associated with the production of yellow corn, both with and without controls for crop year and location.[50] Yellow corn was also considered to be more resistant to drought and other stresses. A further disadvantage of white corns was that their production required isolation to prevent cross-pollination with other corns. If a lot contained just 2 percent of yellow kernels, it might be rejected as commercial white corn. With rising consumption of tortillas and corn snacks in the 1980s, there has been a small shift back to white and even blue corns. The discovery in 2000 that a quantity of genetically modified StarLink yellow corn, which contained Cry9C, an insecticidal protein approved only for nonfood purposes, had mixed into the human food supply also stimulated demand for white corn, because biogenetic researchers have largely left white corn untouched.[51]

Geographic Shifts and the Emergence of the Corn Belt

As with wheat, the geographic center of corn production shifted dramatically over the late nineteenth and early twentieth centuries. Table 3.3 provides the mean and median latitudes and longitudes (in decimal format) of the U.S. corn crop in the Census years from 1840 to 1910. The data are based on a consistent methodology using county-level

[49] Troyer, "Background of U.S. Hybrid Corn," 604; Steenbock, "White Corn vs Yellow Corn," 352–3; Rice, Mitchell, and Laible, *Comparison of White and Yellow Corn*. White corn was also deficient in vitamin D.

[50] Poneleit, "Breeding White Endosperm Corn," 236; Troyer, "Background of U.S. Hybrid Corn," 604; Nebraska Dept. of Agriculture, *Nebraska Agricultural Statistics*, various years.

[51] Millis, Bryant, and Holland, "White Corn Production and Uses"; Lauer, "Management Needs for Specialty Corn Hybrids."

TABLE 3.3. *Geographic Center of Corn Production, 1840–1910*

	West Longitude		North Latitude	
	Mean	Median	Mean	Median
1840	83.8852	84.5379	37.3583	37.7797
1850	83.9456	84.5230	37.8280	38.4249
1860	86.1346	86.5232	38.0847	38.9752
1870	86.9657	87.7985	38.7696	39.5060
1880	89.0205	89.5457	39.4596	40.0122
1890	90.7628	91.6141	39.3210	39.9032
1900	90.3992	90.7477	39.3807	40.1034
1910	90.2376	90.3805	39.4223	40.0727

Sources: 1840–1910 calculated using data from Haines and Inter-university Consortium for Political and Social Research, *Historical Demographic, Economic, and Social Data*, linked to county centroids from U.S. Dept. of Health and Human Services, *Bureau of Health Professions Area Resource File.*

production data linked to each county's centroid.[52] Recall Table 2.2 of Chapter 2, which reported the centers of wheat production.

In 1840, the geographic center of corn production (as measured by the median) was near Richmond, Kentucky. This was far to the west of the center of wheat production, reflecting corn's status as a frontier crop at this time. By 1880, the center of corn production had moved to the vicinity of Springfield, Illinois. In 1890, the median center of production crossed the Mississippi River to a point in Missouri. After 1900, it retreated back to the eastern side of the great river, where it remained at the time of the 1910 census. Over the entire period of 1840 to 1910, the median center of production had moved 352 miles to the west-northwest. By contrast, the center of wheat production had moved over 800 miles west.

This was the era when the very concept of the "corn belt" came into being.[53] According to William Warntz, the first reported use of this term occurred in a July 1882 *Nation* article, and T. N. Carver, an economics professor at Harvard, deserved credit for further "crystallizing the concept of the corn belt" in a series of 1903 articles in *World's Work*. The 1906 USDA *Yearbook of Agriculture* made the first official use of the term. The boundaries of the region gained fuller definition in Oliver E. Baker's maps in the USDA's 1918 *Atlas of American Agriculture* and

[52] Note that our estimates differ from the "classic" centers of production reported in the *Censuses of Agriculture* for the years 1850–1900 and 1920.

[53] An informative, recent treatment of this subject is Hudson, *Making the Corn Belt.*

TABLE 3.4. *Top Ten States Ranked by Corn Output*

	1839	1879	1909
1	Tennessee	Illinois	Illinois
2	Kentucky	Iowa	Iowa
3	Virginia	Missouri	Indiana
4	Ohio	Indiana	Missouri
5	Indiana	Ohio	Nebraska
6	North Carolina	Kansas	Ohio
7	Illinois	Kentucky	Kansas
8	Alabama	Nebraska	Oklahoma
9	Georgia	Tennessee	Kentucky
10	Missouri	Pennsylvania	Texas

Note: The 1879 list is not affected by the split of Virginia and West Virginia. Their combined output would not rank on the top ten.
Source: Derived using data from Haines and Inter-university Consortium for Political and Social Research, *Historical, Demographic, Economic, and Social Data.*

his 1926 *Economic Geography* articles on the "Agricultural Regions of North America."[54] The corn belt later gained an official definition under the USDA's Production and Marketing Administration as the counties where "average corn production (excluding corn used as silage) during the [previous] ten calendar years . . . is 450 bushels or more per farm and four bushels or more for each acre of farm land."[55] By this definition, pockets of Maryland and southern Pennsylvania are included as well as the traditional core of western Ohio, Indiana, Illinois, Iowa, the southern tier of counties in Michigan and Wisconsin, northern Kentucky, southern Minnesota, eastern South Dakota and Nebraska, and northeastern Kansas and Missouri.

Table 3.4 enhances our understanding of the rise of the corn belt and of the regional shift in corn production by examining the changing list of the top ten corn-producing states. The table highlights three years – 1839, 1879, and 1909. In 1839, the leading states were Tennessee, Kentucky, and Virginia, all in the upper South. The remainder of the list is split between states in the South and the Midwest. The modern corn belt did not yet exist. By 1879, the Midwestern states had come to the fore. Illinois and Iowa held the top spots and most of the others belonged to what we now think of as the corn belt. The same leaders were in place in 1909. The

[54] Warntz, "Historical Consideration of the Terms Corn and Corn Belt," 40–5.
[55] Wallace and Bressman, *Corn and Corn Growing* (1949), 13–14.

main change is that states farther west – such as Oklahoma and Texas – pushed the eastern states off the list. The rising stature of these two states, as well as that of Nebraska, is also notable because all three are characterized by harsher and drier climates.

Two related developments contributed to the intensification in American agriculture during the second half of the nineteenth century. The first was the shift from wheat (and other small grains) to corn in the prairie states, with Illinois, Iowa, and surrounding lands coalesced into the corn belt. The second was the shift of the postbellum South from corn to cotton. The slogan "Cotton is King" was a boast in the antebellum South, but corn was actually the region's leading crop in terms of weight, value, and acreage cultivated.[56] As a result, southern plantations and small family farms were largely self-sufficient in food and feed. In the aftermath of the Civil War, the region changed its crop mix to specialize more fully in cotton. Large supplies of corn flowed in from the Midwest. Students of southern history have debated the roles of comparative advantage and of the institutions of sharecropping and debt peonage in leading to these changes. The goal here is to better document the shift and explore some of its implications.

In both the North and the South the shift in crop mix increased labor requirements. In the late nineteenth century, wheat required less labor per acre than corn, which in turn required far less labor than cotton. The differences were significant. According to the USDA, cotton in 1880 used 119 hours of labor per acre, corn 46, and wheat 20.[57] Thus, the shift of acreage from wheat to corn in the prairie states and, more dramatically, from corn to cotton in the South increased the labor-to-land ratio in both regions, which worked contrary to the trend created by mechanization.[58] Undoubtedly, the changes in factor prices, transportation costs, and biological learning contributed to the evolution of crop mixes. If we narrow our attention to exploring the northward shift of the corn–wheat frontier, the significance of biological learning comes into sharper focus.

56 Kemmerer, "Pre–Civil War South's Leading Crop," 236–9.

57 Carter et al., *Historical Statistics*, Tables Da1143–1144, Da1147–1148. The numbers for crop labor requirements have been questioned because the authors provided no documentation and no source notes.

58 Lee Craig and Thomas Weiss argue that there was an increase in labor inputs of women and children in the Civil War decade. This finding is consistent with the changing regional mixes in the composition of output that was a longer run phenomenon. Craig and Weiss, "Agricultural Productivity," 527–48.

The movement of corn production to new environments required biological innovation. M. L. Bowman and B. W. Crossley observed in 1908 that "the cultivation of corn has been gradually extended northward in the United States. Today this cereal is grown successfully, where twenty-five years ago its cultivation was impossible."[59] It is possible to identify specific breakthroughs that facilitated the shift of the corn belt several hundred miles to the north. Of special significance was the work of Andrew Boss, C. P. Bull, and Willet Hays at the University of Minnesota, who developed Yellow Dent Minnesota No. 13 and Yellow Dent Minnesota No. 23. "These varieties had remarkable early ripening properties that reduced the ripening time from 120 to 125 days to about 90 days (for No. 23). These and other early ripening varieties also allowed farmers in the Canadian plains to grow corn for ensilage."[60]

According to Andrew Boss and George Pond, "the development of early-maturing varieties of corn combined with adapted hybrid varieties, and improved cultural practices are steadily drawing the corn belt northward and westward into the Spring Wheat area. Accompanying this movement has been a steady increase in cattle and hog production in the area which furnished the chief outlet for the corn crop."[61] Minnesota 13 was a potent factor in pushing corn grown for grain fifty miles northward in a single decade.[62] We can build on our analysis in Chapter 2 of the shifting spring–winter wheat frontier to quantify these claims regarding the movement of the corn–wheat frontier. Using county-level Census data for 1869 and 1929, we can examine, for longitude groups, at what northern latitude the value of corn production equaled that of wheat production.[63] Given the focus on the northern limit, we restrict the analysis to latitudes above 40 degrees. To compensate for the effects on the sample

59 Bowman and Crossley, *Corn*, 90.

60 Buller, *Essays on Wheat*, 187–90.

61 Boss and Pond, *Modern Farm Management*, 65; Troyer and Hendrickson, "Background and Importance of Minnesota 13 Corn," 905–14.

62 See Troyer, "Persistent and Popular Germplasm," 176; Hays, *Breeding Animals and Plants*, 19, 82. Minnesota 13 was selected over several years from local seed purchased in 1893. It was first released in 1897. A number of even earlier Dents were subsequently developed at experiment stations in Minnesota, North and South Dakota, and Montana. Will, *Corn for the Northwest*, 65, 85–8, 147. Wallace and Bressman, *Corn and Corn Growing* (1928), 204. Will noted that locally bred crosses in Montana outperformed the Dents bred in North Dakota and Minnesota by a substantial margin.

63 These value calculations use prevailing crop national prices: specifically, 105 cents per bushel of wheat and 77 cents for corn in 1929 and 93 cents for wheat and 73 cents for corn in 1869. This is actually biased against the expansion of the corn area because the price of corn relative to wheat in 1929 was lower than that in 1869.

TABLE 3.5. *Corn–Wheat Frontier*

Longitude Range	Northern Latitude (in decimals) of Equal Values of Corn and Wheat Output	
	1869	1929
87–89	42.28	44.12
88–90	42.78	45.28
89–91	42.96	46.52
90–92	42.89	48.86
91–93	42.45	48.23
92–94	42.70	46.86
93–95	43.10	47.67
94–96	42.70	47.48
95–97	42.30	45.84
96–98	41.51	44.60
97–99		43.55
98–100		43.06
99–101		43.28
100–102		42.90
101–103		41.74
102–104		41.09
103–105		40.88

Sources: Derived using data from Haines and Inter-university Consortium for Political and Social Research, *Historical, Demographic, Economic, and Social Data*, and U.S. Census Bureau. 15th Census 1930, *Agriculture*. Vol. 2, pt. 1, linked to county centroids from U.S. Dept. of Health and Human Services, *Bureau of Health Professions Resource File.*

size, we estimate the frontier using regressions on overlapping two-degree longitude bins. The results, reported in Table 3.5, are striking. In 1869, the frontier basically followed a line just south of the 43rd parallel over the longitudes from 87 to 95 degrees and then turned south. By 1929, the frontier was pushed near the Canadian border for these longitudes and remained north of the 43rd parallel up to 100 degrees longitude (the line where the "Great American Desert" began). Thus, an enormous area including most of Minnesota, South Dakota, and Nebraska shifted from the wheat belt to the corn belt over this sixty-year period.[64] If researchers

[64] The northern shift of the corn frontier was celebrated by construction of the famous Corn Palace at Mitchell, South Dakota (43.7034 latitude, 98.0626 longitude) in 1892. The Midwest was home to many such fanciful structures decorated with tapestries of corn. The original was built in Sioux City, Iowa, in 1887. Plumb, *Indian Corn Culture*, 230–2.

had not developed earlier maturing varieties, this shift would not have been possible.

The northward push of Dent cultivation increased agricultural productivity in these new regions. This effect should not be judged by analyzing changes in corn yields per acre in the affected states. A better comparison incorporates the change in yields resulting from the shift from wheat and other small grains to maize. A more complete calculation would include the value of adding corn, a relatively safe crop, to the portfolio of choices available to farmers and of increasing the number of different crops that could be inserted into a farm's rotation. We further note that the movement of Dents to the more northern and arid lands had the effect of reducing measured national corn yields. Average corn yields, as graphed in Figure 3.1, have no statistically significant growth trend over the period before hybrid corn. But these data confound the effects of regional shifts and changing yields in a given region.

Diseases, Insects, and Weeds

Numerous insects, including chinch bugs as well as cutworms, white grubs, wireworms, army worms, cornstalk borers, and seed corn maggots, attacked maize.[65] But one late arrival was particularly serious. The invader most analogous to the Hessian fly and the boll weevil was the European corn borer, *Ostrinia nubilalis* (aka *Pyrausta nubilalis* Hubn). This was a formidable enemy that sapped corn yields primarily through weakening the plant by feeding on its leaves and tunneling into the stalk as larva. The pest at times did direct damage to the ears by causing them to drop prematurely or by feeding on the kernels. The European corn borer remains in the news today because it is the main pest targeted by the controversial, genetically modified, Bt maize. This corn, commercially introduced in 1996, includes a gene from the soil bacteria, *Bacillus thuringiensis* (Bt) subsp. *tolworthi*, which makes a protein that kills European corn borers, as well as southwestern corn borers, black cutworms, and, critics argue, Monarch butterflies.[66]

[65] Myrick, *Book of Corn*, 240–78. As with wheat, the damage assessments to corn from diseases and insects need to be qualified because it is likely that some losses attributed to bad weather or even bad luck were in fact the result of diseases or insects. Shelford and Flint, "Populations of the Chinch Bug," 435–55.

[66] Bt maize is nearly 100 percent effective in combating the European corn borer. Besides attacking field and sweet corn, the borer also damages cotton, sorghum, and a variety of vegetables. See Iowa State University Dept. of Entomology, *European Corn Borer*.

The European corn borer apparently entered the United States in the early 1900s on broom corn from Hungary or Italy, but it was not positively identified until 1917 when it was discovered near Boston, Massachusetts. By 1919–20, the borer was found in two additional clusters: near Schenectady and Buffalo, New York. From these epicenters, the borer swept into Maine, New Hampshire, Rhode Island, Pennsylvania, Ohio, Michigan, and Ontario, Canada, by 1921. By the mid-1930s it had footholds throughout New England, the mid-Atlantic states, the upper South, and the Midwest. In the early years, the borer produced only one generation per year, but by the late 1920s a borer capable of producing two generations per year appeared in the eastern states. This more prolific insect soon spread west to ravage the corn belt. The two-generation borer reached Wisconsin in 1938 and by 1946 had cut a swath all the way to South Dakota. Over time, three- and four-generation borers infested southern corn fields.[67] Table 3.6 offers a better sense of the borer's diffusion.

On its own, an adult moth could fly twenty to thirty miles in a year. But the borer could also travel as a larva within a cornstalk flowing down a river or across a lake; or – more frightening in the modern age – the borer could hitchhike long distances in fresh corn carried by unwitting motorists. In response, the USDA established a quarantine program to stop automobiles from transporting ears of green corn outside the infected areas from July to October. In 1925, the quarantine lines extended through western New York, Pennsylvania, Ohio, and Michigan.[68] By the mid-1920s, alarm began to spread across the corn belt well in advance of the bug. Henry A. Wallace and E. N. Bressman reflected this concern in 1923 when they warned that "the European corn borer has possibilities of causing as much damage to the corn crop as the boll weevil has caused to the cotton crop. There is grave danger that it will reach the heart of the Corn Belt by 1930."[69]

In February 1927, President Coolidge signed a bill allocating $10 million to "eradicate or control" the pest. This was called "undoubtedly . . . the largest insect-control project ever undertaken in the history of the world."[70] Authorities in the United States and Canada actively

[67] Dethier, *Man's Plague*, 202; Davis, "European Corn Borer," 324; Iowa State University Dept. of Entomology, *European Corn Borer*.

[68] Worthley and Caffrey, "Timely Information About the European Corn Borer."

[69] Wallace and Bressman, *Corn and Corn Growing* (1923), 133.

[70] USDA, *Yearbook 1927*, 202–7.

TABLE 3.6. *Square Miles of Territory in the United States and Canada Infected by the European Corn Borer, 1917–46*

	Infected Territory (square miles)	
	United States	Canada
1917	100	
1920	4,500	3,000
1922	7,696	7,690
1924	24,773	18,000
1925	30,237	
1926	93,786	96,650
1927	135,000	114,550
1931	222,000	
1934	265,000	
1940	280,000	
1946	515,000	

Sources: Compiled from USDA, *Yearbook 1920*, 85–104; Caffrey and Worthley, "European Corn Borer and its Control"; USDA, "Progress Report on the Investigations of the European Corn Borer," 9, 13; *Journal of Economic Entomology* 19, no. 2 (1926), 400, and 21, no. 1 (1928), 86, 230, and 25, no. 1 (1932), 12; Baker and Bradley, "European Corn Borer," (1935), 2, and (rev. 1941), 3, and (rev. 1948), 2.

cooperated to stem the threat. With the generous 1927 allocation, Midwestern agricultural officials assembled a large fleet of specialized machinery to destroy stubble and set mandatory stubble cleanup dates in early May (see Figure 3.2). By this date, farmers in the infected areas were obliged to plow under their corn stubble and burn any remnants. The bill established a $4.2 million fund to reimburse growers up to $2 per day to cover the extra costs of the cleanup. More than 300,000 farmers participated and up to 95 percent of the borers were destroyed.[71] During the 1920s and 1930s, USDA entomologists also imported several dozen parasites from Europe and East Asia and experimented with indigenous insects and diseases, but without much effect.[72]

[71] USDA, *Report of European Corn–Borer Control*, 20–4, 42; and Records Concerning the European Corn Borer Control Projects, General Summary Reports 1927, U.S. Federal Extension Service, RG33, Entry 49, Box No. 1, National Archives.

[72] Three insects, the tachinid fly (*Lydella thompsoni*) and two wasps (*Eriboris terebrans* and *Macrocentrus grandii*), were of some value. The parasite research program was a significant undertaking. The UDSA imported about 3 million parasite larvae from across

THE CORN BORER

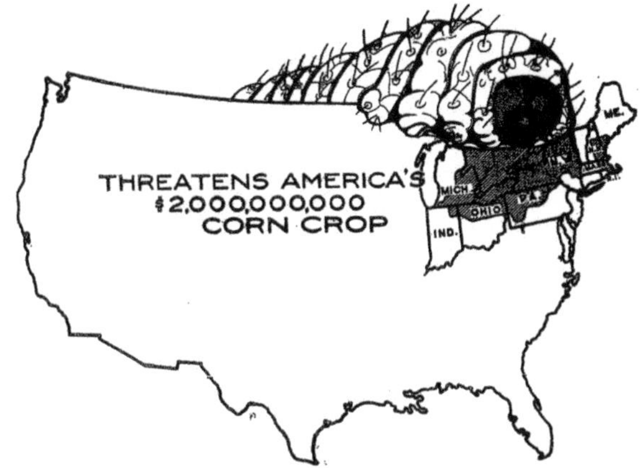

STOP HIM! CLEAN UP BEFORE MAY 1

Infested area in North America equal to total area of Ohio and Michigan.

Infested area increased about 50 per cent in United States in 1926.

Spread in Ohio from 21 townships in 1921 to 525 townships in 1926.

5,000,000 FARMERS EXPECT YOU TO HOLD THE LINE

FIGURE 3.2. USDA–state cooperative corn borer program. *Source:* Records Concerning the European Corn Borer Control Projects, General Summary Reports 1927, U.S. Federal Extension Service, RG33, Entry 49, Box No. 1, National Archives.

The quarantine intensified. In the 1930 season, 226 road and ferry stations set up throughout the Midwest stopped 11.6 million vehicles

Europe and Asia. To feed and colonize these parasites entomologists also imported 27 million European corn borer larvae. Baker, Bradley, and Clark, "Biological Control of the European Corn Borer."

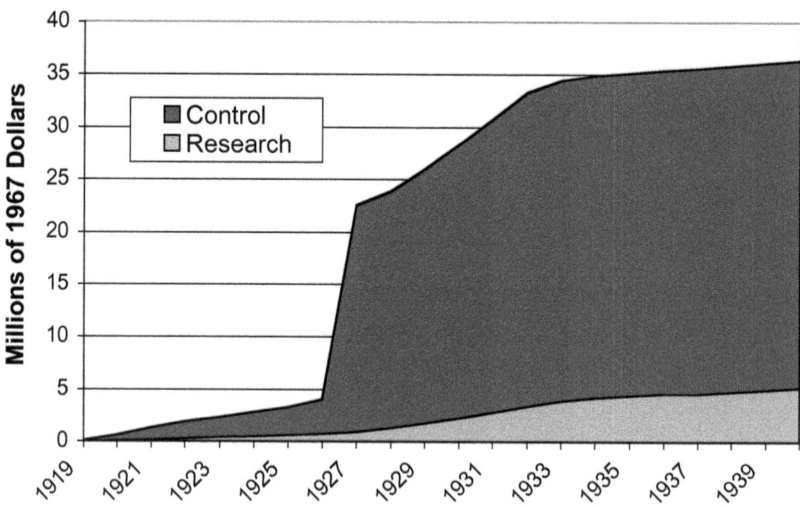

FIGURE 3.3. Cumulative inflation-adjusted expenditures on the federal program to control the european corn borer. *Notes:* Converted to 1967 dollars using CPI, HS series, E 135. Control in 1927 includes $4,662,200 paid to farmers as reimbursements and $4,966,400 in direct expenditures. *Source:* U.S. House Committee on Appropriations, *Agricultural Department Appropriation Bill for 1940*, 717.

and seized 342,800 ears of corn, netting 2,430 borers. The authorities remarked with pleasure that fewer than one thousand motorists refused to stop at their corn borer roadblocks. (Recall this was during prohibition.) Such massive intrusions into private activities have been largely forgotten.

By the late 1930s, the corn borer was clearly evading all attempts to limit its march. Its spread slowed over the late 1920s and early 1930s, but this was more likely due to weather conditions than to controls. According to USDA authorities, the federal government's control program cost about $16 million over its 1919–42 life. This aggregate figure provides little sense of the resources involved (because it is a sum of nominal expenditures in a period of changing price levels), of the timing of the efforts, or of how the expenditures were divided between research and control. The statistics on inflation-adjusted cumulative expenditures presented in Figure 3.3 shed light on these issues. They show that most of the money went to imposing quarantine and that the 1927 appropriation accounted for most of the spending. Overall, the program cost approximately $35 million measured at 1967 purchasing power. These data on government expenditures of course only represent a part of the story because individual farmers also adjusted cultural patterns to limit the borer's damage. The European corn borer could be controlled by plowing under the cornstalks, and considerable effort went into this task.

The government expenditures appear small relative to the losses. For example, J. A. Hyslop calculated that the borer created losses averaging $865,000 per year over the period 1932–36. After World War II, USDA authorities estimated that the insect caused an average loss of $144 million per year over the period 1949–53, or about $185 million at the 1967 price level captured in Figure 3.3. While the corn borer represented a potential threat on par with the boll weevil, its entry after the advent of insecticides and machinery to shred the corn stalks limited damage.[73] As previously noted, the arrival of the corn borer represented the first serious threat to corn producers; thus, until the 1920s, breeding new varieties to resist insects was less imperative for maize than for wheat or cotton.

Maize was generally considered a more healthy and reliable crop than wheat or cotton. Although it was sensitive to drought and summer heat, maize was "more easily protected from its animal and insect enemies than most of the important crops" and "remarkably free from injurious diseases."[74] Corn smut, a fungal disease that was common throughout the country, was the most serious. In rare cases it could wipe out most of a farmer's crop, but in most years it probably only reduced national output by 1 or 2 percent.[75] Maize was also subject to a number of bacterial and fungal rots. Diplodia was the most serious, but by the early 1920s the USDA discovered that treating seed with organic mercury dust controlled this rot.

Although corn suffered less than wheat or cotton from diseases and insects, weeds were a serious problem. As Wallace and Bressman noted, "The wheat crop rusts, the oat crop lodges, the cotton crop has its boll weevil, and the corn crop has its weeds. Good corn growing is a constant battle against weeds." Today an array of herbicides helps control weeds, but in the past this task was largely the duty of the man with the

[73] Hyslop, *Losses Occasioned by Insects*, 17–18; Wallace and Bressman, *Corn and Corn Growing* (1949), 160.

[74] Montgomery, *Corn Crops*, 214, 220; Brewer, "Report on the Cereal Production," 107. Modern hybridization undermined corn's disease resistance. This problem came to light in 1970–71 when southern corn leaf blight, a fungal disease, zeroed in on the 85 percent of corn hybrids bred to possess Texas cytoplasm male sterility. The blight caused losses equal to 17 percent of the total crop in 1970. Ullstrup, "Impacts of the Southern Corn Leaf Blight," 37–50.

[75] Myrick, *Book of Corn*, 278–88.

hoe. Bindweeds, Canada thistle, foxtails, quack grass, and, in the western corn belt, the Russian thistle all increased in severity.[76] Some of the bindweeds (morning glories) and both Canada and Russian thistle were foreign introductions; despite its name, the Canada thistle came from Eurasia. In the words of one Canadian authority, weeds of this "troublesome and aggressive" foreign species found "the new conditions . . . so favorable that they flourish to a greater extent than even in the lands whence they came."[77] This creeping perennial entered North America during the colonial period. Some accounts have it invading eastern New York in 1777 in a manner curiously reminiscent of the Hessian fly – carried in the hay of British General John Burgoyne's horses. The plant soon earned a local reputation as "the worst of all weeds."[78] The thistle chiefly colonized fields by spreading its extensive root system, which deprived other plants of moisture and nutrients. Between locations, the thistle spread through wind, water, and careless human activity. In 1795 Vermont passed legislation requiring that local farmers control the thistle on their land. New York followed suit in 1813, authorizing towns to pay a bounty for destruction of the thistle. The weed had spread across the Alleghenies by the mid-1830s, leading Ohio to mandate in 1844 that its landowners mow any thistle-infested fields. Iowa listed the plant in its first weed law of 1868. By 1900 the plant had spread west across the territory north of the 37th parallel. In its wake, some twenty-four states declared the Canada thistle a noxious weed, many proscribing fines for landowners who allowed the thistle to seed. Mowing down the top growth of established plants to starve the root system and using a rotation of smother crops such as alfalfa or small grains were among the recommended control methods.[79] Diligent farmers immediately hoed new infestations.

In his 1880 survey, Brewer clearly noted that weed problems and the extent of hoeing were a function of a region's date of settlement. In New England practically all corn was hoed and "most of it hoed twice." In New York, "most of the crop is hoed once, some of it twice." In Ohio, only "a little is hoed twice" but most was hoed once. "Passing still farther

[76] Wallace and Bressman, *Corn and Corn Growing* (1923), 59–63.

[77] In parts of the United States, Australia, and New Zealand, the plant, *Carduus arvensis* or *Cirsium arvensis*, is called California thistle. Dewey, "Canada Thistle," 2; Shaw, *Weeds and How to Eradicate Them*, 3–4.

[78] Dewey, "Canada Thistle," 1, 5; Lang and White, "Canada Thistle."

[79] Hayden, "Distribution and Reproduction," 358; Dewey, "Canada Thistle," 1, 5; Wallace and Bressman, *Corn and Corn Growing* (1923), 61–2.

West, ... there is no hand-hoeing at all." The West may have boasted better soils on average than the East, but biological forces also contributed to the comparative advantage of relatively newly settled regions. "Over the most of the prairie regions ... the weeds are not particularly troublesome [due] ... to the newness of the country. The most troublesome weeds are usually imported ones, and their introduction in any new region may be long deferred.... Once in, and once troublesome, they are never again entirely exterminated, but they remain in spite of all war waged against them, and the extent of the damage they cause is inversely as is the care taken for their suppression. Hence, in many districts where one or more species are specially troublesome, hand-hoeing, at first unnecessary, ultimately becomes an absolute necessity."[80] Thus, the growing threat from weeds tended to raise labor requirements over time, making crop yields crucially dependent on the hours of hoeing labor. This observation bears on a larger issue. The simple induced innovation framework identifies output per person as the product of acres per labor hour and output per acre ($Y/L = A/L * Y/A$). This decomposition underscores the assumption that one can reasonably treat yields as being independent of labor inputs (yields depend on biological inputs), and the land-to-labor ratio is independent of yields (it depends on mechanization). Clearly this is an assumption meant to capture the general state of the world and no reasonable person would believe that there are no exceptions. Our example of the relationship between corn yields and hoeing is the tip of the iceberg for a number of crops, because yields were crucially dependent on labor inputs. For this and other reasons discussed in later chapters, identifying mechanization with saving labor and biological innovation with saving land becomes very questionable.

By-Products

As in the case of cotton, over the late nineteenth and early twentieth centuries, the discovery of many new uses for maize increased its value.

80 Brewer, "Report on the Cereal Production," 99. Brewer was a knowledgeable observer, but his emphasis on hoeing may be overstated. By the end of the 1860s, primitive check-row planters were available that used a knotted wire to trigger the seeder. These allowed farmers to cross-cultivate their crop using horse-drawn equipment. By the date of his statement in 1880 such equipment was still crude and probably not widely diffused. The next twenty years saw significant advances which would have accelerated the mechanization of weed control at least outside the South. Johnson, *Farm Inventions*, 73–83; and KINZE Manufacturing, *History of Planting*.

The stalks could be chopped up (together with the ear) and fermented to make silage, which was highly valued as a succulent winter animal feed.[81] The first silos were introduced from Europe into Illinois by Fred L. Hatch in 1873, into Michigan by Manly Miles in 1875, and into Maryland by Francis Morris in 1876. By the 1920s, about one-half million of such structures dotted the nation's countryside. Farmers, in particular those in the Great Plains, also learned less capital-intensive techniques such as making trench silos to produce the improved feed. Over the period 1928–30, roughly 4 percent of total corn acreage was devoted to silage and another 11 percent was used directly for forage. In the eastern dairy belt, corn acreage harvested for silage exceeded that harvested for grain.[82]

It is popular in discussions of mechanization to talk about the backward linkages tying farms to factories, but important nonmechanical linkages to industry have gone relatively unnoticed. Corn had long been employed to make whiskey, but scientific breakthroughs created a growing number of other off-farm uses. The industrial refinement of corn dates to the pre–Civil War era, with the manufacture of corn starch in Jersey City, New Jersey, in 1844. The next major step occurred in 1866, with the production of dextrose from corn starch. In 1882 the first manufacture of refined corn sugar, or anhydrous sugar, broadened the industrial market. In the last decades of the nineteenth century, the corn refining industry developed processes to make use of the fiber, germ, and protein to manufacture commercial animal feeds. The discovery of how to extract oil from the germ led to the commercial production of corn oil in 1889. Further advances in corn sugar chemistry led to the introduction of crystallized corn sugar (crystalline dextrose hydrate) in 1921. A bushel of corn would yield roughly 22 to 33 pounds of sugar. The demand for corn sugar grew rapidly, and by 1928 total consumption had already reached 150 million pounds. During World War I, biochemists discovered how to manufacture acetone and various other solvents from corn mash. In 1928 solvent factories consumed about 11 million bushels, or nearly 5 percent of all the corn sold for off-farm use. By that date corn was also consumed in the manufacture of 112 million pounds of corn oil, 900 million pounds of starch, and 95 million pounds of dextrin.[83] The

[81] Silage typically used the entire plant, as did fodder. Feed using the stalks without the ears yielded stover.

[82] Wallace and Bressman, *Corn and Corn Growing* (1949), 172–3, 178–80, 400.

[83] McMillen, *Too Many Farmers*, 152–4, 177; Corn Refiners Association, *Brief History*.

1930s witnessed an extensive campaign to use maize to produce alcohol fuel for motor vehicles. Much of the rhetoric behind this effort has a very modern ring, although the stated goal was to utilize surplus corn rather than to replace scarce fossil fuels. There were few lasting results.[84] The development of new industrial uses for corn continued into the post–World War II era, most significantly with the invention of high-fructose corn syrup and the production of ethanol, but these steps were a continuation of an inventive process that has been under way since the nineteenth century.

Conclusion

As with wheat, corn production underwent enormous changes before the modern era. There were across-the-board changes in varieties as farmers and breeders succeeded in crossbreeding Dents with Flints. According to Anderson and Brown, "The common dent corns of the United States Corn Belt were created de novo by American farmers and plant breeders during the nineteenth century."[85] The new crosses were essential for the westward and northern expansion of corn production. In addition, there were important changes in the pattern of corn consumption based on new storage and feeding technologies that required considerable additions to the farm capital stock. A key difference that separated corn from wheat and cotton was that insects and pathogens were less of a problem, although this changed with the arrival of the European corn borer. In addition, weeds were more of a problem for corn than for wheat.

The evidence speaks to the importance of biological innovation. As Steckel and others have shown, farmers were careful to plant seed adapted for a specific area. Using data from Illinois field trials, Steckel showed that moving seed north or south 250 miles resulted in yield losses of about 30 percent; other state experiment stations found similar results. More telling was that the Illinois experiments also captured the impact of the most important nineteenth-century development – the previous crossing of Dents and Flints. In these trials, Flints and soft flour corns significantly underperformed the Corn Belt Dents. The work of the Minnesota Experiment Station in the 1890s cut the ripening time of the famous Minnesota

[84] Giebelhaus, "Farming for Fuel," 173–84; Wright, "Alcohol Wrecks a Marriage," 36–66.

[85] Anderson and Brown, "Origin," 137–8. We emphasize that these hybrids were not F1 or F2 crosses from pure lines.

Dents by about 30 days, allowing the northern expansion of the corn–wheat frontier. The Minnesota breeders' achievements were one step in a much longer innovative process. If the European colonizers of North America had simply taken the corn given to them by Native Americans and made no further investment in improving the plant or in matching specific varieties to specific geoclimatic zones, corn would not have become "America's Crop."

This chapter offers insights into the broader development of American agriculture. The spread of corn cultivation typically entailed a significant intensification of production that required more labor per acre of cultivated land. This observation points to the fallacy of focusing on the adoption of laborsaving machinery in any one crop without also looking at shifts in the composition of output. If the induced innovation story were operating at full force, farmers not only should have been mechanizing to save labor, they also should have been shifting production toward the more highly mechanized crops to save even more labor. This did not happen because the shifts from small grains to corn in the North and from corn to cotton in the South show that farmers were opting en masse for significantly more labor-intensive systems. This was a general phenomenon because there were also shifts to highly labor-intensive dairy operations and more intensive methods of animal husbandry in the North. In the far West there was a movement to labor-intensive fruit, nut, and vegetable production.

The chapter also introduced another theme that will come up again in later chapters. The problem with weed control suggests a relationship between labor inputs and yields. This same relationship applied to most crops, and it changed over time. For cotton, the introduction of Johnsongrass and several other weeds in the nineteenth century dramatically increased the labor requirements needed to maintain yields. This example casts further doubt on the common practice of identifying mechanization with saving labor and biological innovation with saving land.

Finally, the chapter bears on the important issues of antebellum regional specialization and the postbellum South's move to specialize more in cotton production. The antebellum South's self-sufficiency in corn depended heavily on the high corn-to-cotton ratios in the border states. The traditional cotton states such as Georgia, Alabama, and Mississippi were relatively specialized in cotton from an early date. What accounted for the postwar increase in specialization? Biological innovation and diffusion played a role. The rise of the corn belt, based on the de novo creation of new hybrids, surely increased the northern states' comparative advantage in corn, encouraging further specialization in the South. This serves as a segue to the next chapter because biological advances in cotton production increased the South's comparative advantage in that crop.

4

Cotton

Varietal Innovation and the Making of King Cotton

Stanley Lebergott summed up the process of innovative activity in antebellum America: "Science was at work, sometimes. Coarse empiricism – try and try again – was at work, many times."[1] Lebergott clearly recognized that biological innovations increased land and labor productivity, and he singled out cotton as a prime example. Hayami and Ruttan were also aware of biological innovations in antebellum cotton production but saw them as an aberration. On the same page that they forcefully reiterate the primacy of mechanical innovations, they observe in a footnote: "This is not to imply that significant advances in biological technology were not achieved." They then quote John Hebron Moore: "All of the farmers of the lower South accepted as their standard strain a new upland cotton developed by Southwestern plant breeders during the early decades of the 1800s. The new variety, the famous Mexican hybrid, improved the yield and quality of American cotton to such an extent that it deserves to rank alongside Eli Whitney's cotton gin in the Old South's hall of fame."[2]

There is considerable support for Moore's claims. According to J. O. Ware, a leading U.S. Department of Agriculture (USDA) cotton expert, the varieties that became the basis for the South's development were a distinctly "Dixie product." "Although the stocks of the species were brought from elsewhere, new types, through [a] series of adaptational

[1] Lebergott, *Americans*, 176.

[2] Hayami and Ruttan, *Agricultural Development*, 209, are quoting Moore, "Cotton Breeding in the Old South," 95–104.

changes, formed this distinctive group the final characteristics of which are a product of the cotton belt of the United States."[3]

This process of molding cotton was repeated over and over again as new varieties were introduced and as production moved into new areas. In Lebergott's words, it was a matter of "try and try again," but with both a calculated purpose and at least a rudimentary understanding of plant breeding. According to Ware, "The vast differences in climate and soil that obtain over the Cotton Belt undoubtedly brought about a kind of natural selection which eliminated many of the kinds that were tried, while others became adapted to the several conditions under which they were grown and selected over a period of years."[4] American upland cotton was a latecomer to the world market, but it proved "much more suitable, than any other kind, for general factory use."[5]

The growth in world cotton production depended on a series of important scientific advances. As history unfolded, the American growers proved more innovative than their counterparts in India, Egypt, and other producing regions. Gregory Clark has shown that, even when textile manufacturers transferred British technology and management lock, stock, and barrel to India, the mills were unable to produce efficiently.[6] Similar problems hampered the development and spread of biological technologies to produce improved cottons. British manufacturers had long pressured authorities to help improve the quality of Indian cotton, but concerted efforts, including the transfer of seeds and the engagement of American managers, led to few successes.[7]

This chapter analyzes the important varietal and cultural changes in cotton production from the eighteenth century to the 1960s. In addition to illustrating the importance of biological innovations, the story advances our understanding of southern economic development. For a period when Gavin Wright and others have decried a lack of indigenous technological advances, the South was the undisputed world leader in

[3] Ware, "Origin," 1.

[4] Ware, "Plant Breeding," 659. Also see Handy, "History and General Statistics of Cotton," 17–66.

[5] This advantage changed over time as varieties developed in the United States were adopted in other regions. Ware, "Origin," 1–2.

[6] Clark, "Why Isn't the Whole World Developed," 141–73.

[7] Indian cottons commanded a much lower price than the American product and could only be used to produce low-quality yarns. See the report of Viscount Morley in Heylin, *Buyers and Sellers*, 6, 66–7. Also Harnetty, *Imperialism and Free Trade*, 82–8.

the creation and diffusion of superior cotton varieties. The achievements of southern cotton breeders rivaled anything accomplished by northern wheat breeders in the nineteenth century. These biological innovations increased the efficiency of slave labor and help explain the pattern of regional development within the South. Our findings inform the debate on the efficiency of southern agriculture and the entrepreneurial role of planters. The economies of scale associated with slave plantations played a major role in the development and protection of new cotton varieties. Our argument also sheds light on the structural inefficiencies that took root following the Civil War.

Our analysis has even broader implications that bear on the very foundations of the Industrial Revolution. The usual story focuses on the inventions of Richard Arkwright, Samuel Crompton, James Hargreaves, and John Kay that mechanized spinning and weaving. Homage is also paid to Eli Whitney, whose saw gin helped ensure a supply of raw cotton to meet the growing industrial demands.[8] However, this account neglects another set of innovators – the farmers and plant breeders who developed cotton varieties suited to the new spinning technologies and, as importantly, to the diverse North American environment. These innovations were essential for sustaining the Industrial Revolution.

The Cotton Plant and Early Introductions

Cottons cultivated in the United States belong to one of two species. Sea Island (*Gossypium barbadense*) was grown primarily along the coasts and on the offshore islands of Georgia, South Carolina, and Florida. Sea Island produced high-quality, long staple fibers (over one and a quarter inches), but was low yielding and difficult to pick. Cottons of the second and more important species (*G. hirsutum*) were commonly referred to as upland cottons because they were grown in the more variable climates away from the coast. As of the turn of the twentieth century, cotton experts grouped the upland varieties into eight general types. Most of these types could be developed to fit specific environmental and economic situations and would be ill suited for other conditions. None of these cottons was native to British North America.[9]

[8] Landes, *Unbound Prometheus*, 202.

[9] Modern genetic research recognizes fifty species of cotton (genus *Gossypium*), of which four major species were domesticated independently in different parts of the world. *G. hirsutum* (upland cotton), which accounts for about 90 percent of the world's production, originated in southern Mexico and Guatemala. It was probably first domesticated in

The introduction of foreign varieties took several forms, but it was rarely haphazard or accidental. The importer – perhaps a migrant, a traveler, a friend or relative, or a scientist deliberately searching for better varieties – noticed something appealing about a specific plant and would harvest its seeds. The recipient of the seeds would plant them in a separate row or even an isolated patch and then select the best of the progeny. The grower would then remove the seeds by hand for replanting the next year. The new variety typically required several generations of selection for the desirable characteristics to stabilize and breed true. Astute planters and scientists often followed the same process when they discovered an unusually desirable plant in their existing fields. In addition, as early as 1857, one scientist, John Griffin, began the laborious process of hybridizing new varieties. This work represented a conscious effort to manipulate the plant's characteristics.

Fresh imports, selections, and hybrids were important in all areas of the country, but new varieties were particularly significant for the expansion of the cotton belt and were essential in the fight against diseases and pests. During a number of infestations, cotton producers avoided what would otherwise have been catastrophic losses by changing varieties.

Adaptation was essential for the successful cultivation of upland cotton. In its native environment in Central America, *G. hirsutum* was a frost-intolerant, perennial shrub with short-day photoperiod response. As a short-day plant, its flowering was triggered when the nights began to grow longer and cooler in the late summer or autumn. This strategy was adapted to a semitropic, semiarid environment where the rains came in the autumn. The greater variation in day length across the seasons at the higher latitudes of the American South meant that the date with the right conditions to trigger flowering occurred later in the year. This meant that many of the introduced cotton varieties either did not mature before the first frost set in or did not flower at all. Initial attempts to

the Yucatan and then adapted for the Mexican Highlands. The origin of *G. barbadense* (Sea Island cotton) is debatable, with Brazil, the Caribbean, and Peru being the leading contenders. The two Old World species, *G. arboretum* and *G. heraceum*, originated in Africa and Asia. Genetic studies of fifty upland cultivars grown in the United States in the 1990s show the overwhelming influence of Mexican and Guatemalan *G. hirsutum*. Wendel, Brubaker, and Percival, "Genetic Diversity in Gossypium Hirsutum," 1291–1310; Brubaker and Wendel, "Reevaluating the Origin of Domesticated Cotton," 1309–26; Brubaker, Bourland, and Wendel, "Origin and Domestication of Cotton," 3–31; Fryxell, *Natural History of the Cotton Tribe*, 3–16, 130–207; Wendel and Cronn, "Polyploidy and the Evolutionary History of Cotton," 139–86.

grow upland cotton in the areas that now constitute the United States faced severe challenges. Success depended on finding a mutation, cross, or variety with the appropriate photosensitivity characteristics. "Following generations of repeated selection, these initial stocks were molded into early maturing, photoperiod-insensitive cultivars adapted for production in the southern United States Cotton Belt."[10] Adaptation was made easier because, as John Poehlman and David Sleper note, the cotton stocks first introduced to the region "were largely mixed populations with varying amounts of cross-pollination and heterozygosity that gave them plasticity and potential for genetic change."[11]

Cotton was purportedly introduced to Jamestown as early as 1607, and over the pre-Revolutionary period southern colonists grew limited quantities, mostly for home consumption. Table 4.1 lists some of the foremost introductions.[12] Other sources offer competing dates, and there was typically a long lag between a specific introduction and its commercial success. Moreover, the introduction of a new type of cotton did not necessarily equate with eventual success, as with the introduction into Jamestown in 1607. The search to find better varieties was truly a global undertaking. Planters experimented with seeds imported from the West Indies, Mexico, Central America, Brazil, Peru, the Middle East, Southeast Asia, and China.

Green Seed

The most common upland variety in the colonial period was Georgia Upland (also known as Georgia Green Seed or green seed) which was first grown in North America in 1733/34 at the Trustee Garden in Savannah. The colonists evidently acquired this variety from Philip Miller's

[10] Poehlman and Sleper, *Breeding Field Crops*, 376; Stephens, "Some Observations of Photoperiodism," 409–18.

[11] Poehlman and Sleper, *Breeding Field Crops*, 376. They further note that "the adjustments were hastened by the contributions of large numbers of early cotton breeders who worked without the genetic guidelines available to cotton-breeders today."

[12] History has failed to remember many innovative farmers, especially blacks and poor whites. Stephens, "Origin of Sea Island Cotton," 391–9. Smith et al., "History of Cultivar Development," 101. Native peoples in the Southwest had grown *G. hirsutum* varieties for over one thousand years by the time of European contact. The Houma Indians planted the first cotton in Louisiana in 1699 after obtaining the seeds from Pierre Le Moyne d'Iberville. Handy, "History and General Statistics of Cotton," 30–3; Lakwete, *Inventing the Cotton Gin*, 22–3. Post, "Domestic Animals and Plants of French Louisiana," 554–86.

TABLE 4.1. *Major Introduction of Foreign Cotton Varieties*

Name(s)	Date of Introduction	Comments
Sea Island	By 1720s	In Louisiana
	By 1786	In the Southeast
Siamese black seed	By 1733	In Louisiana
Green seed	By 1733/34	Georgia and elsewhere
Mexican Burr	1806	A Mexican Highland stock with a cluster phenotype – probably first imported to Mississippi by Walter Burling
Hollingshead	1818	A Mexican introduction
Alvarado	By 1825	
Various varieties	1846–48	Presumed that the new varieties were introduced by the soldiers returning from Mexico
Wyche	1857	From Algeria but presumed to have been developed in Mexico. A parent of Eastern Big Boll type
Mit Afifi and other Egyptian cottons	By 1903	Led to development of Yuma, Pima, and other extra-long-staple cottons. The Egyptian cottons were descendants of American Sea Island. Recent Pima varieties are a result of introgressions from Sea Island, Peruvian, and upland stocks
Kekchi	1905	Discovered by O. F. Cook in Guatemala; a parent of Paymaster type
Durango	1905	Exhibited by the Mexican government at the St. Louis exposition in 1904. F. L. Lewton bought seeds and introduced them in southern Texas. It became popular in Texas and the Imperial Valley of California. It became the first upland cotton to be grown in a one-variety community
Alcala	1907	Discovered by O. F. Cook in Mexico in 1906 and introduced by G. N. Collins and C. B. Doyle in 1907. Became the dominant variety in California

Sources: Compiled from Smith et al., "History of Cultivar Development," 99–171; Ware, "Origin," and "Plant Breeding," 657–744; and numerous plantation diaries and account books.

botanical gardens in Chelsea, England.[13] Miller likely obtained his original specimens from Guadeloupe. If so, the first popular upland variety journeyed from the Caribbean to England to Georgia before diffusing more generally in North America. En route it was tested on at least two experimental farms.

By later standards, early upland cottons such as Georgia Green Seed were low-quality, low-value, short-staple varieties that were difficult to pick. Their seeds were "fuzzy," meaning that the lint clung tenaciously to the seed, making it hard and very labor-intensive to clean. Eli Whitney's saw gin economically removed the seeds of fuzzy seed cotton varieties.[14] Upland cottons evolved as farmers acclimatized the standard Georgia Green Seed to different conditions. Some accounts maintain that by 1800 a new and improved variety, Tennessee Green Seed, had been developed in the Cumberland Valley.[15] Although the evidence is scanty, the claim is that this variety rapidly gained favor elsewhere in large part because it could be picked 20 to 25 percent faster than the older upland varieties.

Sea Island and Creole Black Seed

In the late eighteenth century (1786 is a commonly cited date), planters introduced high-quality *G. barbadense* varieties into Georgia, possibly from the Bahamas or Jamaica.[16] These plants were acclimatized to

[13] Moore, *Agriculture in Ante–Bellum Mississippi*, 13–36. Smith and Cothren, *Cotton*, 102. Other sources have the cotton coming from Philip Nutter of Chelsea. See Handy, "History and General Statistics of Cotton," 31.

[14] Lakwete, *Inventing the Cotton Gin*, 1–96, downplays Whitney's contribution, arguing that the roller gins were capable of handling upland varieties. Sea Island cotton had smooth, black seeds which could be separated from the lint using roller gins. Variants of upland cotton had naked, black seeds as well. But these strains were typically less productive and more disease-prone than the green or brown fuzzy-seeded variants. A small number of alleles control these characteristics and the black trait is recessive. Thus, although it would be technically possible for breeders to create a smooth-seeded upland cotton, such varieties would be associated with lower yields and subject to "invasion" by green seed germplasm.

[15] Watkins, *King Cotton*, 100, 254; Ware, "Plant Breeding," 658–9; Kilpatrick, "Historical," 632–3, excerpt from the *Cotton Book* of G. W. Lovelace, 1817.

[16] A widely accepted account is based on a letter from Patrick Walsh of Jamaica, who asserted that he sent three sacks of seed to Frank Leavet (sometimes spelled Levett). Donnell, *Chronological and Statistical History of Cotton*, 48. Smith's account of the introduction of Sea Island suggests that the variety was native to Brazil. Smith, *Crop Production*, 296. There are at least four other contemporary claims of first introduction. See Watt, *Wild and Cultivated Cotton Plants*, 16–17; Stephens, "Origin of Sea Island Cotton," 391–9; Rosengraten, *Tombee*, 50; Lakwete, *Inventing the Cotton Gin*, 24. Porcher and Fick, *Story of Sea Island Cotton*, 93–9, concluded that the "plants of

become the Sea Island cottons that prospered in the coastal regions of South Carolina and Georgia. Not only did Sea Island cotton produce a longer, finer fiber than upland varieties, its seeds were smooth and could be removed easily with a roller gin – a machine invented in India and widely available in the colonies. Persistent attempts to extend the range of Sea Island by acclimatizing it to grow away from the coast met with little success.

The early Sea Island cottons were a diverse and mutable lot. Ware speculates that "more than one species and no doubt several varieties were brought in during the early period. . . . From these introductions the sea-island growers doubtless developed their own distinct varieties and strains."[17] Contemporary accounts concur. South Carolina planter Whitemarsh Seabrook asserted in 1846 that decades of breeding had resulted in up to fifteen subvarieties of Sea Island cotton.[18] Just as early inventors of textile machinery knew little of the laws of mechanics,

> these early breeders knew little of genetics or the science of plant breeding, but they were artists in knowing their plants. They sought practical ends and concentrated on the development of the long silky types of cotton that the English spinners of the time demanded. Their ideas were based on philosophical biology rather than on scientific biology, taxonomy, or genetics. They felt that environment had considerable effect in producing changes in plants but that heredity transcended all external influences, and that like did really beget like. With this philosophy as a guide, and expertness in observing, sorting, and selecting, they were equipped to build up a great enterprise through plant breeding. Among these growers a particular variety was considered the personal property of the originator and seed was not exchanged or sold unless something better was at hand. The result was the development of many special strains of the finest cotton the world has ever known.[19]

Although it is conventional to think of crops as spreading from east to west, European settlers in the lower Mississippi Valley actually began continuous commercial production of cotton before their counterparts in the Carolinas and Georgia. French colonists experimented with Sea Island cotton in the early 1720s. These efforts created a narrow band of Sea Island production along the Gulf. The repeated failure of Sea Island elsewhere led Louisiana planters to try other options. By 1733 a variety

Gossypium barbadense that came to be called sea island cotton originated in 1786 with its conversion to day-neutral flowering."

[17] Ware, "Plant Breeding," 658; Rosengraten, *Tombee*, 76.

[18] Chaplin, *Anxious Pursuit*, 221.

[19] Ware, "Plant Breeding," 658.

imported from Siam called Creole, or black seed, became popular. Like Sea Island, it had smooth seeds that could be separated from the lint with a roller gin. Creole outyielded Sea Island in this area but produced lower-quality fiber. Creole was difficult to pick due to its small bolls and because the lint clung to the pods. But Creole was superior to the green seed upland cottons in many respects, including price, quality, and ease of handling.[20]

The black seed varieties first developed in the Gulf states moved eastward and diffused widely in the Carolinas early in the nineteenth century. At this juncture the story becomes clouded. The published literature and primary sources dealing with the southeastern states are replete with references to black seed varieties, but it is very difficult to know what was actually being grown.[21] John Hebron Moore and Margaret Des Champs Moore add to this complex picture. In the 1790s, when South Carolina planters were first experimenting with commercial crops of upland cotton, they planted both black seed and green seed varieties. As examples, Evan Pugh planted a small crop of black seed in 1794, but he did not grow cotton on a significant scale until he planted green seed in 1799. Wade Hampton I planted six hundred acres of green seed on alluvial lands near Columbia in 1799, and his success is credited with popularizing green seed throughout the Southeast. However, it took time for the word to spread. In 1803, John Baxter Fraser grew forty acres of black seed on his pineland soils in the Sumter region of South Carolina. The next year Fraser experimented with both black seed and Sea Island, but after the Sea Island succumbed to cold, he replanted ten acres in green seed. In 1805, six years after Hampton's widely cited triumph, Fraser shifted to planting green seed exclusively.[22]

The records of Edmund Bacon of Augusta, Georgia, indicate that seeds of Gulf varieties (presumably Creole) were making their way to the

[20] Moore, *Agriculture in Ante-Bellum Mississippi*, 13–36. Some disagree whether green seed outyielded black seed cotton. Smith and Cothren, *Cotton*, 102–3.

[21] Chaplin, *Anxious Pursuit*, 221. Mendenhall, "History of Agriculture in South Carolina," 62, 100–3, has green seed cotton becoming important among the cultivators of upper South Carolina after 1799 and dominating other types by 1808.

[22] Moore and Moore, *Cotton Culture on the South Carolina Frontier*, 4–36, and telephone interview with John Hebron Moore, 20 April 2005. It is likely that some of the black seed grown in the Southeast evolved independently of the Creole variety of the Mississippi Valley. Little information has survived about some forms of upland black seed; other details vary. For example, Moore notes that Creole was grown as an annual whereas others argue it was farmed as a perennial. Smith et al., "History of Cultivar Development," 102–3.

Southeast. In late 1807 Bacon wrote to his brother-in-law, Colonel Joseph Pannile of Natchez, that he was "very anxious to make an experiment the ensuing year with the best of your Orleans or Mississippi Cotton-seed," and requested that "two hogsheads of the most approved seed in your country" be shipped in time for planting.[23]

After 1793, cotton production soared in response to the booming demand in Great Britain and technical improvements associated with Whitney's gin, but disaster threatened in the second decade of the nineteenth century. According to Moore and a number of older accounts, "cotton rot," first mentioned around 1810, spread through Mississippi and neighboring areas over the next fifteen years, causing devastating losses.[24] In response, growers in the lower South imported seeds from Tennessee and Georgia and increased their plantings of green seed, which was initially less prone to rot.[25] In this region, the transition from Creole black seed to green seed varieties likely represented a case of technological regress that was dictated by the new environmental realities. The situation deteriorated further when the green seed varieties became susceptible to rot. The pace of biological innovation was about to accelerate.

Mexican Introductions and Mexican Hybrids

The first of a long chain of events that would revolutionize southern cotton production occurred in 1806 when a Natchez area planter, Walter Burling, noticed a particularly appealing cotton plant while on a diplomatic mission to Mexico City. Burling packed some of the seeds in dolls and smuggled them out of the country. He passed the seeds on to planter and amateur scientist, William Dunbar, who began the tedious process of experimentation.[26] According to Ware, "All Mexican stocks required

[23] The flow of seed went in both directions. Edmund Bacon, in a letter to his sister, Agnes Pannile of Natchez, 24 December 1815, noted that he would ship her cotton seed of the "choicest-kind." Edmund Bacon Letters, Mss. 2178, Louisiana and Lower Mississippi Valley Collections, LSU Libraries.

[24] Moore, *Agriculture in Ante-Bellum Mississippi*, 31, speculates that the rot was due to a bacterial disease, *Bacillus gossypium Stedman*. But it could have been one of the fungal diseases – *Diplodia*, *Fusarium*, and in wet years *Colletotrichum* – that still plague cotton cultivation in the lower Mississippi River Delta and other damp areas. Bell, "Diseases of Cotton," 572.

[25] Moore, *Emergence of the Cotton Kingdom*, 12–13; Moore, *Agriculture in Ante-Bellum Mississippi*, 13–36; Clairborne, *Mississippi*, 140; Watkins, *King Cotton*, 98–9.

[26] Letter from Dunbar to Green & Wainewright, 2 October 1807, Dunbar, *Life, Letters, and Papers*, 356. Moore, "Cotton Breeding In the Old South," 95–104, and *Agriculture in Ante-Bellum Mississippi*, 32–3; Gray, *History of Agriculture*, 673–7; Collings, *Production of Cotton*, 201; and Watkins, *King Cotton*, 165.

some reselection for adaptation before satisfactory responses under Mississippi Valley conditions were obtained. Fresh seed from Mexico during the first year or two of planting was said to produce no more than one-half crop."[27] Dunbar wrote to his Liverpool agents, Green & Wainewright, that "I have made two small experiments of new Cotton, one is the nankin [sic], which I find very good, the other is a species from Mexico, which I think excellent." He further noted that he had sent a "sample of Each . . . requesting you to get it examined by your manufacturers and other good judges." Consistent with Ware's comment about low yields in the first season or two after introduction, Dunbar noted that he did not believe the Mexican "will yield quite as much to the acre as the common cotton."[28]

By 1820, southern breeders had outcrossed Mexican highland cotton with the green seed, Creole, and possibly even with Sea Island varieties, creating many new genotypes.[29] As of 1820 word of the new cottons had reached Baltimore because an article in the *American Farmer* noted that the new "Mexican cotton" variety was "found to ripen, in a greater degree at one time: has a shorter season than any other cotton known among us: and is collected in much greater quantities, by a labourer, in one day. The cotton wool is represented to hang out of the pods of this species of the plants, and even to drop at times from them."[30] The early success of Mexican cotton stimulated others to seek fresh introductions of Mexican seed, which further increased the genetic material available to breeders. In 1828 Timothy Flint noted that, in Louisiana, Mexican cotton was "getting into common adoption, and the importation of seed from Tampico and Vera Cruz is becoming a considerable business."[31]

The better hybrids that emerged were a vast improvement:

> Its staple was longer and the grade of the lint higher than Creole or Green Seed. It ripened earlier in the fall than any other type then in cultivation in the United States, and it displayed a noticeable tendency to mature many of its bolls simultaneously. Even more importantly, it possessed exceptional picking properties. Its large four or five-sectioned bolls opened so widely upon ripening that their lint could be plucked from the pod more easily than any other known

[27] Ware, "Origin," 50.

[28] Dunbar, *Life, Letters, and Papers*, 356–7. Dunbar's subsequent cultivation of the new crop increased to such an extent that, after his death in 1810, his heirs had ten to twelve bales of cotton to sell under the "Mexican" brand (388, 390).

[29] Smith, *Crop Production*, 296–7.

[30] "Cotton," *American Farmer* (7 July 1820), 116.

[31] Flint, *Condensed Geography*, 24, 523–4.

variety of the staple. Because of this unusual quality, pickers could gather three to four times as much Mexican in a day as they could the common Georgia Green Seed cotton. Most important of all, the Mexican strain was totally immune to the rot, the dreaded plant disease that was then destroying both the Creole and Georgia Green Seed crops in the Mississippi Valley.[32]

James L. Watkins observed that "the average day's picking for a hand was 30 to 40 pounds of black seed and 75 to 100 pounds of the Tennessee Green Seed variety. At first a hand could pick 150 pounds of the Mexican, and this gradually increased to several hundred pounds."[33] Our examination of hundreds of plantation accounts and cotton books confirms there was a dramatic increase in daily picking rates between 1800 and 1850.

In addition to superior yield, quality, picking, and disease resistance, the Mexican cottons had higher lint-to-seed ratios than the varieties they replaced. According to L. C. Gray, "the older varieties yielded lint weighing only 25 per cent of the total weight of seed cotton."[34] For the new varieties, the ratio would have been closer to 33 percent, implying more than a 30 percent premium after ginning. Watkins was so impressed with the improved attributes of the Mexican cotton, he proclaimed that "from an economic point of view the introduction of this seed was second in importance to the invention of the saw gin."[35]

Mexican cotton spread through the lower Mississippi Valley in the 1820s. Seed from Natchez probably first arrived in the Gulf Hills region near Rodney, Mississippi, in 1824. Ware reports that the Mexican cotton first appeared in South Carolina in 1816, but it was not until the 1830s that commercial seed from the Mexican hybrids was being marketed extensively in the Southeast. Mexican cotton was still a novelty in Alabama in 1826 because David Moore, whose plantation was near Huntsville, caused "unusual excitement" when on August 28 and 29 his twenty-seven hands picked "the enormous quantity of 8921 pounds of seed cotton, being an average of 330 to the hand for two days.... It was Mexican cotton, which bears very large bolls, and was beautifully

32 Moore, "Cotton Breeding in the Old South," 95–104, and *Agriculture in Ante-Bellum Mississippi*, 13–36; Gray, *History of Agriculture*, 689–90; Clairborne, *Mississippi*, 140–3.

33 Watkins, *King Cotton*, 13.

34 Gray, *History of Agriculture*, 703.

35 Watkins, *King Cotton*, 13. See also Affleck, "Early Days of Cotton Growing," 668–9; Gray, *History of Agriculture*, 689; and Moore, *Emergence of the Cotton Kingdom*, 28.

opened."[36] As the Mexican cotton spread through this area, picking rates of over 150 pounds a day became common.

By the late 1820s and 1830s, Mexican Highland varieties had diffused into the Southeast. As an example, James Henry Hammond of South Carolina noted in 1833: "planted some Petit Gulf cotton and well pleased with it." In 1834 he observed, "planted half my crop in Petit Gulf.... Shale [*sic*] plant all Petit gulf hereafter."[37] Our investigation of plantation records leads to the same conclusion. By roughly 1820 a movement to superior Mexican hybrids was under way that would revolutionize the American cotton industry. In addition, from the mid-1820s to at least the mid-1830s there was a distinct difference in the modal types of cotton grown in the lower Mississippi Valley relative to those grown in the south Atlantic states. Over this period the western cottons offered higher yields, were of superior quality, fetched higher prices, and were significantly easier to pick.[38] The spread of the new varieties widened the yield and picking efficiency differences that separated the nitrogen-rich alluvial and black soil lands of the New South from the poorer lands in the hill country and the East. Although the early Mexican hybrids were more productive than green seed on almost all types of soils, the productivity differential was far greater in good soils.

Nineteenth-century photographs of pickers working in low cotton and high cotton help demonstrate how the physical formation and fruiting qualities of the cotton plant might dramatically affect picking rates (Figure 4.1). The size of the bolls suggests that both of these photographs show "modern" cottons as of the Civil War era. Neither the green nor the black seed cottons would have had such large plump bolls. The descriptions of the early black seed varieties (which grew tall) depict small bolls the size of a pigeon's egg.[39]

Over the antebellum years, purposeful breeding, mixed with a good dose of serendipity, led to numerous successes as astute farmers developed selections of Mexican Highland stock that had cross-pollinated with green and black seed cottons.[40] By the early 1830s improved hybrids were marketed throughout the South. Among the most popular were Petit Gulf, introduced in the late 1820s by Dr. Rush Nutt of Rodney, Mississippi,

[36] *American Mercury* (10 October 1826), 2.

[37] *Stampp, Records of Ante-Bellum Southern Plantations*, Reel 14, 67.

[38] Watkins, *King Cotton*, 13; Ware, "Plant Breeding," 659.

[39] Moore, *Agriculture in Ante-Bellum Mississippi*, 31.

[40] Smith, *Crop Production*, 297; Ware, "Origin," 12–14.

(a)

(b)

FIGURE 4.1. Picking low and high cotton. The structure of the cotton plant affected picking rates and yields. The top photograph (a) is from 1887 and the bottom photograph (b) is likely from the Civil War decade. Both photos show modern varieties given the relatively large boll sizes. The bolls of earlier varieties were reportedly about the size of pigeon eggs. (a) Courtesy of George Eastman House, photographer S. Fisher Corlies; (b) courtesy of AmericanCivilWar.com.

and One Hundred Seed, bred and distributed by Col. Henry W. Vick of Vicksburg, Mississippi, in the mid-1840s.[41] In 1839 Vick began the annual process of having his most able slaves make special pickings in which they harvested only the finest bolls from the largest and most prolific plants. This cotton was ginned separately and then grown in isolated fields. Vick often ventured into the fields himself in search of valuable mutations and crosses. Vick personally selected the progenitor of the One Hundred Seed cultivar in 1843 from the particularly appealing bolls of a single plant he discovered while visiting another plantation in the delta.[42] He then increased this seed for a few years before marketing it. Petit Gulf and One Hundred Seed spread across a wide area and were noted for high quality, good yields, ease of picking, and rot resistance. Most planters typically bought small quantities of commercial seed, which they then tested and increased themselves. In doing so they had to be very careful to maintain seed quality given the potential for cross-pollination in the fields and seed mixing at the gin. Specialized seed producers such as Vick were particularly diligent to separate their test fields and to clean their gins between runs.[43]

In addition to Petit Gulf and One Hundred Seed, other important varieties produced by antebellum Mississippi Valley breeders were Sugar Loaf (1843), Banana (before 1848), Mastodon (before 1846), and Boyd's Prolific (before 1847). Sugar Loaf was first in the line of cluster types (that is, the plant tended to have multiple bolls at each node on its short fruiting limbs, making the bolls cluster together). Boyd's Prolific was the pioneer variety in the semicluster line (which possessed the clustering habit in a less pronounced form). Varieties such as Mastodon were highly popular in certain regions but were considered "humbugs" in others.[44]

As Joseph Lyman observed in his 1868 book *Cotton Culture* regarding the impact of the Mexican imports and their descendants: "Beginning with the year 1820, and from that time forward, various planters in

41 Moore, "Cotton Breeding in the Old South," 95–104, and *Emergence of the Cotton Kingdom*, 12–16. From reading Moore, one might surmise that the rot disease wiped out the green seed varieties. To the contrary, in 1880 Tennessee farmers reported that green seed varieties were widely grown in the state. Hilgard, *Report on Cotton Production*, 99.

42 Moore, "Cotton Breeding in the Old South," 99–101; *Vicksburg Sentinel* (7 July 1847), 1.

43 Moore, *Emergence of the Cotton Kingdom*, 13; Ware, "Origin," 12–14; Smith, *Crop Production*, 297.

44 Watkins, *King Cotton*, 173, 195, 217, and 243.

different parts of the cotton growing States have devoted themselves to the development and sale of improved varieties of cotton seed, and certain styles of cotton have for two, three, or four years, enjoyed a great, though ephemeral popularity, and, then, as suddenly, been pushed aside for a new reigning favorite. The improvement of a cotton seed as a business, and sale of the improved varieties, has enabled quite a number of prominent and enterprising planters throughout the South to realize handsome fortunes."[45]

Antebellum observers were impressed and at times astounded at the high prices the new seeds brought in the market. Ordinary cotton seed might sell for 25 cents per bushel (when used for manure) and was often simply discarded. By contrast, Petit Gulf seed regularly fetched twice as much. Seeds of choice varieties such as Sugar Loaf, Brown, Hundred Seed, Banana, Multibolus, and Prolific, "under ordinary circumstances, command from one to three dollars per bushel." And when first introduced, Mastodon seed netted five dollars a bushel, Banana "at first sold for a hundred dollars a bushel [and] some paid ten cents apiece [again roughly $100 per bushel] for 'Hogan' seeds."[46]

New varieties gained considerable attention in the southern and even the northern press. As an example, *Harper's Magazine* reported in 1854 that "whenever, by good fortune, a higher-yielding cotton plant appears 'instantly . . . the local newspapers teem with advertisements and commission houses are filled with the magic seed [for marketing].'"[47] Improved cottons of both the Sea Island (from Seabrook) and upland (Petit Gulf and Extra Prolific) varieties were included in the Great Exhibition when it opened in New York in 1853. The writer for the *Times* observed that "although the Crystal Palace contains many articles which make more show, and will attract more attention, it has none which have exercised a greater influence upon the civilization of world" than these cottons.[48]

In the late antebellum period, some of the seed producers who had started as planters, such as Martin W. Philips, conducted various privately funded agricultural experiments and disseminated their findings and latest creations for free. In time, Philips was "drawn into a profitable seed business selling name brands from his Mississippi plantation."[49] Other

45 Lyman, *Cotton Culture*, 121.

46 Ibid., 121, 125. Also see Moore, "Cotton Breeding in the Old South," 100–2.

47 Thorpe, "Cotton and Its Cultivation," 449.

48 *New York Times* (26 July 1853), 1.

49 Norse, "Southern Cultivator," 245–6.

seedsmen, such as David Dickson of Georgia, were commercial operators from the beginning. Much like those of northern manufacturers of agricultural machinery, Dickson's advertisements in the 1850s featured testimonials from satisfied local customers (see Figure 4.2 as an example).[50] The new seeds were not protected by patents, yet the innovations continued to flow. The speed of innovation appears to have been more rapid in the upland cotton sector, where an ethos of greater openness prevailed, than in the Sea Island sector, where planters closely guarded their improved seeds and cultural practices as trade secrets.

The Civil War and the Decline in Cotton Quality

The Civil War is rightly seen as a watershed in southern economic development, leading to the demise of slavery and the rise of sharecropping along with a host of other institutional changes. But the war also interrupted the normal process of experimentation and selection that was vital to maintaining or improving cotton quality. The 1866 *Report of the Commissioner of Agriculture* observed that "the most serious difficulty encountered by cotton-growers, and particularly those who are engaging in such enterprises for the first time since the war, had been found to be poor seed." The report further noted that "for seven years little or no pains have been taken by any cotton-growers to perfect their seed."[51] This concern continued during Reconstruction. In an 1872 article on the deterioration of cotton, the Commissioner of Agriculture reported that many farmers had become careless. "The seed is promiscuously taken from the gin, carelessly thrown upon a heap, where it remains until planting-time, and without regard to and selection of good or indifferent."[52]

In 1873 the Commissioner again reported complaints his office had received from "intelligent planters" that "the quality of cotton offered in our markets had greatly deteriorated within the last twelve years." According to the report, "it seems to be very clear that the present production is, in a large measure, in the hands and under the direction of a less intelligent class of planters, who do not appreciate the importance of a judicious selection of seed, proper cultivation, and especially

[50] David Dickson Advertisement for Boyd's Extra Prolific, 1854, Hargrett Library Broadside Collection, 1850–1859, MS 2622, Hargrett Rare Book & Manuscript Library, University of Georgia Libraries.

[51] USDA, *Report of the Commissioner 1866*, 209.

[52] USDA, *Monthly Report* (May and June 1872), 212.

HIGHLY IMPROVED COTTON!

Boyd's Extra Prolific!

This New and Valuable variety of COTTON, FAR EXCELLS ANY COTTON WE EVER HAD IN THIS SECTION FOR YIELD, as is proven by a Written Report of GEN. WILLIAMSON, CAPT. BASS, MATTHEW WHITFIELD, and other Good Planters who have tested the merit of this Cotton the present year. Every COTTON PLANTER, living in a Short Climate, should Plant of this Cotton, as it produces a crop of SQUARES and BOLLS in near HALF THE TIME of the ordinary Cotton, checking the grouth of the weed and going on to maturity. It should be topped about the 20th of July, and if planted on strong land the side branches should also be topped. Much is lost by Planters for the want of careful selection of Highly Improved Seeds, not only of Cotton, but Corn, Wheat, &c. Planters wanting seed had best apply early. ☞ PRICE OF THE SEED PER BUSHEL, $5.

D. DICKSON.

Covington, Newton Co., Ga., 1854.

CERTIFICATES!

COVINGTON, GA., SEPTEMBER 26, 1854.

We planted this year BOYD'S EXTRA PROLIFIC COTTON, introduced by D. DICKSON, of Covington, which has proved far the most Productive Cotton we ever planted. It is no humbug, but a very Highly Improved variety of Cotton, which we can recommend to our friends with great confidence.

JOHN FRANKLIN, M. C. FULTON, DAWSON B. LANE, of Morgan County,
ALFRED LEVINGSTON, JOSEPH B. SLACK, THOMAS HAMMONDS, of Morgan County,
DAVID COOK, WM. COX, of Morgan County, MATTHEW WHITFIELD, of Jasper County,
HENRY GAITHER, THOMAS W. SIMS, JOHN B. CRIM, of Chattooga County.
THOMAS WYATT, of Jasper County.

COVINGTON, GA., SEPT. 27, 1854

We planted this year "Boyd's Extra Prolific Cotton" by the side of our other Cotton, no difference in the land or cultivation; Boyd's will make at least one third more than the other.

JOHN N. WILLIAMSON,
JOHN W. HINTON.

COVINGTON, GA., SEPT. 29, 1854.

We planted this year "Boyd's Extra Prolific Cotton" by the side of our other Cotton, no difference in the land or cultivation; Boyd's will double the other in production.

JOHN BASS, REUBEN WOODRUFF, JOHN S. WEAVER, JOHN RAY, of Coweta County.

HOUSTON CO., GA., OCTOBER 16, 1854.

I planted, on the same half-arcre the present year, for an experiment, the Seed of some half-dozen of the Best Improved Varieties of Cotton known in this Country, all having the same chance in every respect. I feel confident in saying that "Boyd's Extra Prolific" excelled in Production either of the other varieties at least One-Third, and some much more than that. It is the most Prolific and Closely Bearing Cotton that I have ever cultivated. J. A. MILLER.

COVINGTON, GA., OCT. 7, 1854.

At the request of MR. DICKSON, we examined a patch of "Boyd's Extra Prolific Cotton," planted on the 17th of May, on first and second quality of land, without manure. It is decidedly the heaviest bolled Cotton we ever saw to the age of it, notwithstanding the severe drowth we had. We counted Sixty-Eight grown Bolls on a stalk only Twenty-One Inches High, and 102 grown Bolls on a stalk only 3 Feet High, besides a number of small ones. We know of no other variety of Cotton that would have produced such a crop of Bolls in so short a time.

McKINDREW TUCKER, JOHN B. HENDRICK, PERMEDUS REYNOLDS, COLUMBUS D. PACE,
LEWES ZACHRY, FREDERICK COX, of Whitfield County, BENNETT H. CONYERS, of Cass County,
JAMES THOMAS, of Sparta.

AGENTS.

A few Seed may be had from

J. J. PEARCE, Augusta, Ga., T. STOVALL & CO., Augusta, Ga., CHARLES H. ALLEN, Abbeville, S. C.,
J. R. HIGH, Madison, Ga., J. M. STANFORD & CO., Sparta, Ga., C. J. PEARSON & CO., Eatonton, Ga.,
C. C. NORTON, Greensboro, Ga., T. & J. HIGHTOWER, White Plains, Ga., MATTHEW WHITFIELD & CO., Shady Dale, Ga.,
MORROW & KENNEDY, Monroe, Ga., W. T. & J. C. TURNER, Athens, Ga., W. J. ANDERSON, Fort Valley, Ga.

D. DICKSON.

MESSENGER PRINT, COVINGTON, GA.

FIGURE 4.2. David Dickson's 1854 seed advertisement. *Source:* David Dickson Advertisement for Boyd's Extra Prolific, 1854. Hargrett Library Broadside Collection, 1850–1859, MS 2622, Hargrett Rare Book & Manuscript Library, University of Georgia Libraries.

a careful preparation for the market." Quality suffered for "want of proper discrimination in the market prices between the qualities offered for sale." Structural shifts related to the war and emancipation, coupled with exogenous changes in ginning technology and the arrival of new pests, contributed to the deterioration of the American cotton crop. We will analyze these issues in Chapter 6, but the deterioration did not mean that breeding and other improvement activities had come to a halt. To the contrary, there were impressive achievements, but they were often overshadowed by the new problems created by insects and diseases, and by the structural changes that created adverse incentives to growers.

New American Cotton Types, 1850 to 1920

Cotton breeders today, as in the past, confront a number of trade-offs because improving one plant characteristic often requires sacrificing another desirable quality. Breeders strive for high yields, long staple lengths, soft and strong fibers, good spinning characteristics, ease of picking, high lint-to-seed ratios, whiteness, and more. In addition, breeders work to develop cotton varieties to match local soil and climatic conditions (especially the length of the growing season), to resist specific diseases and pests, to survive high winds, and to appeal to special market niches. The importance of wind resistance became more significant as cotton cultivation moved onto the Texas plains, and the incentive to develop cotton that could be picked more rapidly increased as wages rose. In addition, and more notably, the importance of disease and pest resistance changed radically as new threats, such as the boll weevil, emerged.

The period between the 1850s and the coming of the boll weevil witnessed the development of cotton "types," that is, relatively distinct cotton stocks with specific characteristics suited to particular environmental and market conditions. Experts differed in their exact systems of classifications and most acknowledged that some varieties had characteristics intermediate between those of their ideal types. In a set of influential reports, J. F. Duggar (1907) and Frederick J. Tyler (1910) formalized a schema with eight cotton types. Table 4.2 summarizes the Duggar–Tyler classification.[53]

The South entered the postbellum period with three upland types: the Petit Gulf or long-limbed cottons, which were late-maturing, spreading

[53] Duggar, "Descriptions and Classification," 25–9.

TABLE 4.2. *Characteristics of Major Cotton Groups According to Duggar–Tyler Classifications*

	Long Limbed/ Petit Gulf (a)	**Cluster/ Dixon (b)**	**Semicluster/ Peerless (c)**	**Long Staple/ Allen (d)**	**Early/ King (e)**	**Rio Grande/ Peterkin (f)**	**Eastern Big Boll/ Truitt (g)**	**Western Big Boll/ Stormproof (h)**
Staple Length	Medium	Short	Short	Long	Very short to medium	Medium	Short to medium	Short
Boll Size	Medium to large	Small to medium	Medium	Small to medium	Small to medium	Very small to medium	Large	Large
Seed Size	Medium	Small to medium	Small to medium	Medium to large	Small	Large		
Seed Fuzz	Brown	Gray–brown–greenish	Gray–brown–greenish	White–others, not green	Brown–green	Originally black	Green–brown–white–gray	
Lint ratio	Low	Variable	Variable	Low	Medium	High	Medium	Medium
Maturity	Late	Early	Most early	Late	Earliest	Intermediate	Early	Early
Appearance	Tall and spreading or straggling	Compact with clusters of 2–3 bolls	Erect with semiclustered bolls	Large and spreading or straggling	Small, slender, compact	Slender but no clustering	Spreading to semiclustering	Small plant with bolls having strong lock retention

(*continued*)

TABLE 4.2 *(continued)*

	Long Limbed/ Petit Gulf (a)	Cluster/ Dixon (b)	Semicluster/ Peerless (c)	Long Staple/ Allen (d)	Early/ King (e)	Rio Grande/ Peterkin (f)	Eastern Big Boll/ Truitt (g)	Western Big Boll/ Stormproof (h)
Limbs	Long	Few	Few	Long	Short	Few	Variable	Short
Joints	Long	Short	Short	Variable	Short	Long	Variable	
Origin	Petit Gulf	Sugar Loaf (1843)	Boyd Prolific (1847)	Mexican Highlands	King (1890)	Peterkin (1870)	Wyche (1857)	Bohemian (1860)
Notes	Thrives in lush conditions but unproductive on average soil	Determinant – does not tend to excessive vegetation	Determinant – does not tend to excessive vegetation		Good on poor soils and high yields under best conditions	Tends to excessive vegetation under lush conditions	Tends to excessive vegetation under lush conditions	
Picking		Difficult, trashy	Difficult, trashy	Difficult	Tedious	Easier	Easier	

Notes: The cultivated cotton plant generally has a central stalk and two or three heavy vegetative limbs growing from near its base. Off the limbs and the stalk grow smaller fruiting branches, which bear a flower (and later a boll) at a succession of joints. The length of the limbs and fruiting branches differ across varieties. In the nineteenth century, long-limbed varieties were associated with "want of prolificacy, large bolls, late maturity, long staple" whereas short-limbed cottons had the opposite characteristics. Tracy, "Cultivated Varieties of Cotton," 217. The cluster and semicluster cottons have short branches, so the bolls are close together. The cluster cottons have more than one boll at each node of the fruiting branch. A further difference depends on the distance between joints on the stalk and fruiting branches. Short joints were associated with early maturity.

Sources: Based on Smith et al., "History of Cultivar Development," 99–171; Ware, "Origin"; Tyler, "Varieties of American Upland Cotton"; Duggar, "Descriptions and Classification."

plants that produced long staple fibers and were best suited for fertile lands; the cluster cottons, based on Sugar Loaf (1843) and Boyd's Prolific (1847), which were earlier, more compact plants that produced shorter staple lint; and semicluster cottons, another variant of Boyd's Prolific, with a more moderate tendency for the bolls to cluster. The 1870s saw the development of two additional types – Peterkin and Eastern Big Boll. Three more types gained prominence over the late nineteenth century – early or King, long staple or Allen, and Western Big Boll. Duggar and Tyler included a final category of intermediate or nondescript types.

Peterkin Cottons

Peterkin (which was also known as Rio Grande) was a general-purpose cotton with a high lint-to-seed ratio. In 1870, J. A. Peterkin developed the founding seed of this variety on his farm in Fort Motte, South Carolina. Peterkin relied heavily on seed imported from the "back part of Texas." This represents one of many cases where the flow of germplasm moved from west to east and demonstrates that Richard Steckel's portrayal of farmers carrying their seed and cultural patterns from east to west along a latitudinal belt is undoubtedly incomplete. Peterkin was a medium-staple cotton that was "adapted for poor land and hard treatment" and "very large yields under the best conditions." Yet its bolls were "small and tedious to pick."[54]

Eastern Big Boll Cottons

Eastern big boll types had roots in the antebellum era. In 1857, Wyche, a recent German immigrant to Georgia, acquired a packet of cotton seed through his brother in Algeria, which he planted at his Oakland, Georgia plantation. This Algerian import was an American cotton that most likely had been taken to Africa from Mexico. Because of the Civil War, "little attention was given to the culture of the new cotton and it was practically lost."[55] Following the conflict, seedsmen J. F. Jones and Warren Beggarly noticed a single patch of the unusual Algerian cotton and began selecting and multiplying the seed. Beggarly introduced his Beggarly's Big Boll in the early 1870s. Jones undertook further breeding and sold the resulting seed as Jones Improved. In the mid-1870s J. S. Wyche (son of the aforementioned pioneer Wyche) began to market his

54 Tyler, "Varieties of American Upland Cotton," 23, 25.
55 Ware, "Origin," 14. Wyche's first name does not appear in the record.

Mortgage Lifter seed that also was derived from the Algerian cotton. A number of other seedsmen developed further variants in the 1880s. Later varieties of eastern big boll cottons, including David R. Coker's Cleveland and Wannamaker lines, gained large market niches in the post–boll weevil era. Another notable member of this type, Half-and-Half, was developed in 1904 by H. H. Summerour of Duluth, Georgia, and gained popularity after 1911. Its name referred to its high lint ratio, which was over one-half, whereas the typical ratio was about one-third. This high quantity came at the cost of quality because the staple length of "Half-and-Half" was only between five- and seven-eighths of an inch.[56]

The rise of big boll cottons after the 1880s was frequently attributed to their greater ease of picking in an era of growing labor scarcity. Tyler observed in 1907:

> At present the big-boll is the most widely grown and popular, and its supremacy will probably be permanent unless a successful picking machine is invented. Cottons of this group are more easily and quickly picked than the smaller boll varieties and when conditions permit will be grown in preference to them for that reason alone. In many parts of the cotton belt labor conditions are such that picking is done by the smaller farmer and his family with very little hired help. On the larger plantation, even when pickers are plentiful, it is often necessary to pay a little more for small-boll picking, and in Texas the difference often amounts to 25 cents per hundred pounds – a strong argument in favor of large-boll cottons.[57]

Big boll cotton was especially popular in Texas and Oklahoma where wages were relatively high. This popularity was due in part to the stormproof qualities of the western types, but the eastern big bolls were also prevalent in these states.

The use of big boll cottons apparently took off after World War I. In 1927 William Price Wood, Jr. (scion of a Richmond, Virginia seed firm owner) noted:

> The biggest element in . . . raising cotton is the expense of picking. . . . The ease or difficulty of picking a variety has become the controlling factor in determining what varieties will pay to grow. The thing which decides how easy or hard a variety is to pick is the size of the boll. Very small boll varieties, which were popular before the war, are now becoming obsolete. Large boll varieties, which require much less labor to pick, are becoming very popular. . . . King cotton is an

[56] Ware, "Origin," 26.
[57] Tyler, "Varieties of American Upland Cotton," 21.

example of the small boll type. It is one of the earliest of the cotton varieties, and for some twenty–five years has been one of the favorites of the northern border of the cotton belt. However, it is now fast becoming unpopular.... Cleveland Big Boll is a new variety that is steadily becoming more popular. It has a very large boll and takes about half as much labor to pick as King.[58]

The rise of big boll cotton, as with the pre–Civil War adoption of Mexican types, represents one of many instances where biological technologies had a large impact on labor productivity.

Upland Long Staple or Allen Types

A major undertaking in the nineteenth century was the development of long-staple varieties suitable for upland conditions. Apart from a variety created in the 1860s by John Griffin, who succeeded in hybridizing Sea Island with upland cotton, all of the long- and extra-long-staple varieties grown in the United States were selected from Mexican varieties such as Alvarado, which was first imported sometime before 1825. A succession of high-quality long-staple cottons were selected and developed in the more fertile southern regions during the slave era. Among the more celebrated were Belle Creole and Jethro, developed by Vick near Vicksburg; Jones Wonderful, bred by J. H. Jones near Herndon, Georgia; and Peeler, which originated in Warren County, Mississippi. Great diligence was required to maintain these high-quality varieties, and many deteriorated rapidly when grown more generally. In 1879, J. B. Allen of Port Gibson, Mississippi, discovered an exceptional stalk from which he selected and increased the seed. Allen cotton became the standard variety of American upland long-staple cotton until the boll weevil era. In all, Ware alludes to more than fifty other long-staple varieties of upland cotton developed in the post–Civil War era. In addition to having staple lengths in excess of one and one-eighths inches, these cottons were late to mature. This made them vulnerable to the boll weevil, which accounts for their mass extinction early in the twentieth century.[59]

Most long-staple upland cotton varieties originated in the Mississippi Delta or arose from stocks that had come recently from that area. When these varieties were planted elsewhere, even in rich soils, they tended to

[58] Wood, "Economic Study of the Southern Seed Prices," 37–9. Helms, "Just Lookin' for a Home," 205–6, provides evidence that laborers received 42 percent per pound more to pick King than to pick big boll cotton.

[59] Ware, "Origin," 50–64.

perform poorly. This represents but one example of the need to match varieties with environmental conditions. As another illustration, both the eastern and western big boll types did poorly in delta conditions. "These types, under highly fertile and excessively humid or wet conditions, would produce much branching and dense foliage, but little or practically no fruit."[60] With one known exception, all long-staple upland varieties descended from older upland cottons which in turn traced their lineage to Mexican imports. That exception was the cotton developed by John Griffin of Greenville, Mississippi.[61]

Griffin's pioneering work deserves special mention because it represents an early application of methods that came to symbolize modern plant breeding. Griffin began his quest to hybridize Sea Island with an upland big boll green seed variety in 1857. Ware describes Griffin's procedure:

> Selection was practiced on the parents for 5 years before hybridization was started, and then was also continued on the recurrent Green Seed parent and of course, among the hybrid generations while the steps in backcrossing were taking place. The first generation hybrids, due to interspecies hybrid vigor were 12 to 16 feet high and very unfruitful. These F1 plants were backcrossed to the Green Seed parental selections by use of the pollen from the latter. The offspring of each succeeding backcross was pollinated by the constantly improved Green Seed plants for five years. Because of the gradual disappearance of hybrid vigor and the continuous introduction of more of the Upland characteristics by backcrossing, the plants were reduced in size and the fruitfulness brought up to the level, practically, of the Green Seed parent. Each successive cross from the Green Seed parent made on the hybrid material was to stalks least resembling the Sea Island form, but coming from stalks which the year before had most nearly approached the Sea Island lint. The variety was established about 1868 after some 10 years of work, but in order to keep it pure selection was practiced without intermission until after the end of the century.[62]

John Griffin's son, M. L. Griffin, continued this work and the variety prospered on good lands along the Mississippi River. This variety was

[60] Ibid., 72.

[61] Ibid., 50–74. There is another, but later, exception. The USDA, through a long process of selection, developed the long-staple Pima in the first decades of the twentieth century from seeds imported from Egypt. These fine Egyptian cottons were in large part the descendents of American Sea Island. Niles and Feaster, "Breeding," 201–31.

[62] Ware, "Origin," 62. John Griffin was one of many farmer-scientists who developed valuable hybrids in the mid-nineteenth century.

one of the South's finest upland long-staple cottons, but it did not mature early enough to survive the boll weevil.[63]

Early or King Cottons

King Cotton was developed by T. J. King of Louisburg, North Carolina, from "a stalk of very prolific cotton" found in his field of Sugar Loaf in 1890.[64] The type matured extremely early in the season, in as little as ninety days after planting. This property initially made King popular along the northern rim of the cotton belt, including Tennessee, North Carolina, and Virginia. Offsetting this advantage, the type possessed extremely small hard-to-pick bolls and yields that varied greatly with soil quality. As the boll weevil devastated late-maturing varieties, farmers began adopting the early type. By the 1920s, early or King cottons had become very common throughout the cotton belt.

Western Big Boll, Stormproof, and Texas Big Boll Types

The types known as Western Big Boll, Stormproof, and Texas Big Boll cotton were noted for two characteristics. They were resistant to shedding or breaking in high winds, and they were relatively easy to pick because of their large bolls. Whereas the eastern big boll cottons likely evolved from the Mexican variety imported by Wyche via Algeria, Texas Big Boll cotton likely evolved out of varieties imported directly from the dry plains of northern Mexico. The process of selection was similar in both regions. "Under the conditions of the great climatic change, pronounced environmental shock was effective in breaking up or isolating favorable responding genotypes. These better balanced and, therefore, more fruitful forms were readily recognized by growers who would save the seed from them. In this way desirable plant habit having the necessary production characteristics for the new adaptation or ecological area in question was established."[65]

Mexican stocks imported into different parts of the South thus took on different characteristics, presumably due to different origins but also due to breeding to fit local environmental conditions. The first Texas Stormproof variety of note was called Supak or Bohemian, in honor of the German immigrant who developed the variety around 1860. Probable

[63] Ware, "Plant Breeding," 667.

[64] Tyler, "Varieties of American Upland Cotton," 63. Ware, "Origin," 43, states the parent "sugar loaf" variety was a very old North Carolina variety, "not the old Mississippi cluster form of the same name."

[65] Ware, "Origin," 83.

derivatives of this variety were Meyer and Texas Stormproof. These three varieties gained wide acceptance in Texas, and Texas Stormproof was distributed extensively across the South. In addition, these varieties provided the germplasm for breeders such as W. L. Boykin and A. D. Mebane, who developed improved Western Big Boll lines. In 1869, Boykin commenced a decade-long program of carefully selecting Meyer seed from the best plants on his farm near Terrill, Texas. Around 1880 he began planting his improved Meyer among Moon, a long-staple variety, in a quest for a favorable hybrid. To breed storm-resistant cotton, Boykin attached a string with a one-pound weight to the tip of the locks and then held up the boll by the slender stalk holding the fruit. He selected seeds only from bolls with stalks that didn't break under the pressure. Boykin's cotton was similar in appearance to Meyer, easy to pick, and exceptionally storm resistant. It had a high seed-to-lint ratio with a lint length greater than one inch.

Mebane began studying cotton near Lockhart, Texas, in the mid-1870s. Over the next quarter-century he bred cotton in pursuit of a number of characteristics, including storm and drought resistance, higher lint ratios and yields, and larger easy-to-pick bolls. He succeeded in most of these areas and in the process changed his cotton's appearance, creating a stocky and compact plant that would not whip around in the wind. The high cotton so prized in the Mississippi Delta was a detriment in the windswept plains. When the boll weevil entered Texas, Mebane's variety became especially important because it was early to mature. Its success in weevil-infested areas led Seaman A. Knapp to name it "Triumph." Breeders created many other Western Big Boll varieties in the pre–World War II era. Much of this effort focused on satisfying the critical need for early varieties. Western Big Boll represented the last of the types in the Duggar–Tyler categories still cultivated around 1900. Emerging challenges, specifically the boll weevil, destroyed many of the favored varieties and encouraged the introduction of replacements.

New Mexican and Guatemalan Introductions

By the early twentieth century, scientists seeking to combat an insect or disease often explored the region of the pest's origin to find resistant plants or predators. Following this model to combat the boll weevil, the USDA dispatched researchers to scour Mexico and Guatemala. In 1902, O. F. Cook discovered a promising, early fruiting variety in the Kekchi Indian region of eastern Guatemala. Moreover, Cook observed that the population of boll weevils in the area appeared to be held in check by predatory ants. Cook brought back both the Kekchi cotton (along with numerous other samples from Guatemala) and colonies of ants for testing. Cook's ants received great fanfare, and Teddy Roosevelt even mentioned them in his 6 December 1904 State of the Union Address. Alas, the insects failed to survive. As was typical of most introductions, the Kekchi also at first appeared to be a failure. In early trials the cotton did not acclimate well in Texas, growing eight to ten feet high, and it was often sterile. But after years of selection, breeders eliminated many of the undesirable traits and eventually succeeded in developing commercial varieties, the most important of which was Paymaster-type cultivars, which resulted from a complex series of crosses of Kekchi with several other varieties. Paymaster would come to dominate western production and account for about 21 percent of U.S. acreage in the late 1950s. In 1995 it still comprised 24 percent of U.S. cotton acreage.[66]

Cook and his colleagues made several trips, crisscrossing Central America and collecting native upland cottons. In June 1906 Cook stumbled upon a single plant growing by the roadside in eastern Chiapas that had a longer and denser fiber than any big boll cottons then grown in the United States. After studying local cottons and interviewing farmers, Cook deduced that the prized cotton was not cultivated in the immediate area. Torrential rains blocked his attempt to explore remote areas, and he failed to discover more samples. Adding to his problems, the seeds that he had picked from his one plant rotted. Cook returned empty handed. In December 1906, G. N. Collins and C. B. Doyle took up the quest to find the home of Cook's chance discovery. They tracked the cotton to the village of Acala, where they acquired seeds for trials in the United States.[67] Breeders soon discovered that Acala did not offer any special advantage in combating the boll weevil. However, the variety's fiber qualities warranted further work. Beginning in 1907, the laborious process of planting and repeatedly selecting the best plants ensued, first at several locations in Texas and later at test sites in Oklahoma (1914), California (1917), Arizona (ca. 1924), and New Mexico (ca. 1926). Throughout this process breeders worked to adapt strains for specific areas. Although Acala varieties were adopted across the West, the type had its greatest impact

[66] Smith et al., "History of Cultivar Development," 134–5; Roosevelt, *State of the Union Address*, 6 December 1904.

[67] The USDA imported and extensively tested a number of other promising varieties from Mexico at about this time. Many failed to perform and were discarded. Among the more important was Durango from northern Mexico, which was adapted to southwest conditions and grown extensively in the Imperial Valley of California.

in California, where it became the only variety planted on any scale for over forty years.[68]

Distribution of Varieties

A sense of the magnitude of the change in cotton seed stock across types in the pre–boll weevil era can be garnered from Table 4.3. It summarizes both cross-sectional differences and temporal changes in cotton types. These data should be interpreted with care because they are compiled from two imperfectly comparable sources. The 1880 figures, from Eugene Hilgard's classic U.S. Census Office study on cotton production, assemble responses from correspondents in cotton-growing counties to the question "What variety do you prefer?" The 1907 figures are collected from data in Frederick Tyler's extensive county-level survey on "Varieties of American Upland Cotton." Neither Hilgard nor Tyler present the acreage planted in each variety (and it is likely that the vast majority of cotton was grown in both years from gin-run seed some distance from breeder's stock). Furthermore, the sizes of the samples are quite different and it was not always possible to match the varieties mentioned in 1880 to Tyler's categories a quarter-century later.[69] There were distinct regional concentrations – the early group in the northern belt, the long-staple cottons in the Delta states, and the Stormproof cottons in Texas and Oklahoma. Peterkin, by contrast, was grown throughout the cotton belt. A second pattern is the shift over time from the Petit Gulf, cluster, and semicluster groups to the newer types. The ratio figure in the far right column shows the total percentage of the new varieties (upland long staples, Peterkin, early, and the big bolls) to the total percentage of the old varieties (Petit Gulf, cluster, and semicluster). As a rule, the old varieties comprised 75 to 80 percent of the total observations in the 1880 census and the new varieties 20 to 35 percent. By 1907 the relationship was reversed. Indeed, according to Tyler the long-limb Petit Gulf category had effectively become extinct, due in large part to mixing with other varieties.

The underlying data of Hilgard and Tyler reveal another crucial change. Over the late nineteenth and early twentieth centuries, importers

[68] Collings, *Production of Cotton*, 206–13; Turner, *White Gold Comes to California*, 40–1; Ware, "Origin," 116–27.

[69] Watkins, *King Cotton*, 81, 129, 146, 151, and 243, listed Petit Gulf and variants as the most common upland cotton in South Carolina, Florida, Alabama, Mississippi, and Arkansas in the late 1840s.

TABLE 4.3. *Relative Prevalence of Cotton Types in 1880*

		Percent of Total											
		Long Limbed/ Petit Gulf (a)	Cluster/ Dixon (b)	Semi-Cluster/ Peerless (c)	Long Staple/ Allen (d)	Early/ King (e)	Rio Grande/ Peterkin (f)	Eastern Big Boll/ Truitt (g)	Western Big Boll/ Stormproof (h)	Other/ Intermediate (i)	Identifiable Observations (j)	Total Observations (k)	New-to-Old Ratio (l)
Virginia	1880	33.3	0.0	33.3	0.0	0.0	0.0	33.3	0.0	0.0	3	6	0.5
	1907	0.0	0.0	37.5	0.0	50.0	0.0	0.0	0.0	12.5	8	8	1.3
North	1880	13.9	72.2	8.3	0.0	5.6	0.0	0.0	0.0	0.0	36	43	0.1
Carolina	1907	0.0	9.2	13.8	8.4	14.9	14.6	29.8	5.1	4.1	751	751	3.2
South	1880	11.1	55.6	22.2	0.0	0.0	0.0	11.1	0.0	0.0	9	10	0.1
Carolina	1907	0.0	8.7	10.5	12.9	6.6	20.9	31.8	3.7	4.8	836	836	3.9
Georgia	1880	6.8	83.0	10.2	0.0	0.0	0.0	0.0	0.0	0.0	88	92	0.0
	1907	0.0	8.7	9.8	7.3	8.9	11.3	44.8	4.9	4.4	1,802	1,802	4.2
Florida	1880	0.0	25.0	50.0	0.0	25.0	0.0	0.0	0.0	0.0	4	5	0.3
	1907	0.0	9.0	9.0	17.0	6.0	14.0	33.0	9.0	3.0	100	100	4.4
Alabama	1880	33.3	33.3	16.7	0.0	16.7	0.0	0.0	0.0	0.0	6	7	0.2
	1907	0.0	7.4	9.7	8.9	10.7	8.6	42.4	8.0	4.3	1,070	1,070	4.6
Mississippi	1880	34.0	38.0	18.0	0.0	2.0	4.0	4.0	0.0	0.0	50	59	0.1
	1907	0.0	6.6	10.6	19.9	8.7	9.8	33.0	6.6	4.8	958	958	4.5
Tennessee	1880	24.3	18.6	30.0	11.4	14.3	1.4	0.0	0.0	0.0	70	76	0.4
	1907	0.0	5.1	8.7	17.7	20.2	5.4	28.9	9.4	4.7	277	277	5.9
Missouri	1880	50.0	0.0	0.0	0.0	50.0	0.0	0.0	0.0	0.0	4	5	1.0
	1907	0.0	2.4	14.6	17.1	29.3	4.9	22.0	7.3	2.4	41	41	4.7

(continued)

TABLE 4.3 *(continued)*

		Percent of Total											
		Long Limbed/ Petit Gulf (a)	Cluster/ Dixon (b)	Semi-Cluster/ Peerless (c)	Long Staple/ Allen (d)	Early/ King (e)	Rio Grande/ Peterkin (f)	Eastern Big Boll/ Truitt (g)	Western Big Boll/ Stormproof (h)	Other/ Intermediate (i)	Identifiable Observations (j)	Total Observations (k)	New-to-Old Ratio (l)
Arkansas	1880	25.0	21.4	17.9	10.7	10.7	0.0	10.7	0.0	0.0	28	31	0.5
	1907	0.0	6.9	10.2	14.7	13.3	10.8	28.9	10.6	4.6	972	972	4.6
Louisiana	1880	20.6	32.4	35.3	2.9	0.0	5.9	2.9	0.0	0.0	34	37	0.1
	1907	0.0	7.3	10.7	16.1	8.9	10.8	28.7	9.7	7.8	628	628	4.1
Texas	1880	17.4	22.0	9.2	2.8	4.6	0.0	5.5	38.5	0.0	109	126	1.1
	1907	0.0	5.7	3.9	3.6	13.3	5.0	29.1	34.8	4.5	2,122	2,122	9.9
Oklahoma	1880	0.0	0.0	0.0	100.0	0.0	0.0	0.0	0.0	0.0	1	2	–
	1907	0.0	0.8	5.9	4.6	16.3	7.9	30.5	31.8	2.1	239	239	13.6
Kansas	1880	0.0	0.0	0.0	0.0	0.0	0.0	0.0	0.0	0.0	0	1	–
	1907	0.0	0.0	0.0	0.0	0.0	0.0	100.0	0.0	0.0	1	1	–
United	1880	19.0	40.7	17.0	3.6	5.7	1.1	3.2	9.5	0.0	442	500	0.3
States	1907	0.0	7.1	9.0	10.1	11.3	10.2	34.1	13.5	4.6	9,805	9,805	4.9

Notes: New-to-Old Ratio= $(d+e+f+g+h)/(a+b+c)$. The data for 1880 are generally the number of counties where the correspondent names a variety of a given type in answer to the question: "What variety do you prefer?" Exceptions include South Carolina and Alabama, where correspondents are not identified with specific counties. In these states the total number of responses is used. Note that national figures are based on the total number of responses without weighting for the relative importance of the states in U.S. cotton production.

Sources: U.S. Census Bureau. 10th Census 1880, *Report*; Tyler, "Varieties of American Upland Cotton."

and breeders contributed to a tremendous increase in the number of cultivars (varieties) grown in the United States. Careless cultural and ginning practices added to the proliferation of new varieties. The numbers should dispel any notion of a biologically static past. In 1800 only a handful of varieties were grown in the United States; by 1880 there were 58 varieties enumerated, by 1895 there were 118, and by 1907 more than 600.[70]

In the postbellum South, new cotton varieties proliferated but soon lost their distinctive advantages under mass cultivation. As Ware put it, a "very high percentage" of the varieties "come and go within a rather brief period."[71] Almost every contemporary authority highlighted the rapid turnover in varieties under cultivation. Of the 58 varieties reported in the Tenth Census (1880), "only 6 were commonly in cultivation in 1895," and none were grown by the mid-1930s. Of the 118 varieties Samuel Tracy listed in 1895, only 2 were still present in 1925, and of the 600 varieties Frederick Tyler enumerated in 1907, fewer than 25 were in existence in 1925 and "only 9 were cultivated extensively." The problem was that "much of the benefit gained by bringing in new varieties and by the excellent breeding work that was done by the Department of Agriculture, private breeders, and the State experiment stations, has been lost by the failure to perpetuate the best strains and varieties and to keep them free from admixture with inferior kinds."[72] The evolution of cotton ginning technology contributed to this failure to maintain varieties.

Cotton Seed and the Evolution of the Cotton Gin

A significant advance came with the development of commercial uses for cottonseed. Following the introduction of the Mexican Highland varieties, cotton plants produced roughly two pounds of seed for every pound of lint. Today, that seed is an input for numerous valuable products, including oil (such as Crisco and Wesson), linters (the small fibers adhering to the seed that are useful in explosives and batting), and hulls (which are crushed into meal for animal feed). Remarkably, in the eighty years after Eli Whitney's invention of the saw gin, cotton growers found few commercial uses for the seed. Almost all farmers saved a small fraction (5 to 10 percent) for replanting purposes and many farmers utilized the

[70] Smith, *Crop Production*, 297; Collings, *Production of Cotton*, 236–47; Brown and Ware, *Cotton*, 48.

[71] Ware, "Plant Breeding," 712.

[72] Ibid., 696.

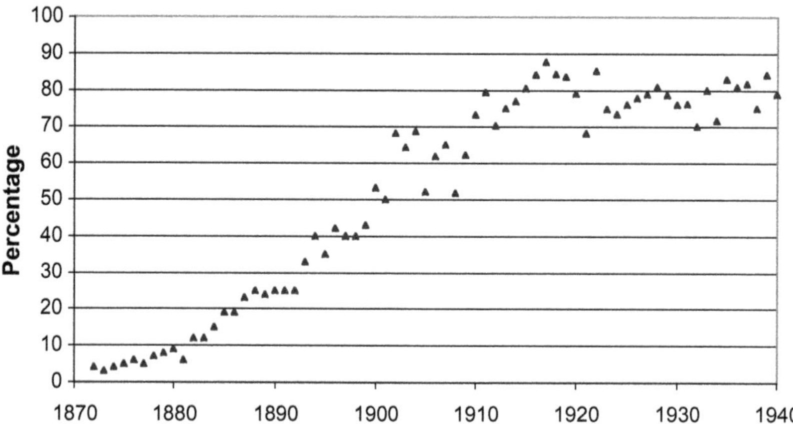

FIGURE 4.3. Fraction of U.S. cottonseed crushed for commercial purposes, 1872–1940. *Sources:* U.S. Census Bureau. *Cotton Production in the United States 1931–32*, 72–73, and *1965*, 52.

seed as a fertilizer, but most seed was considered waste and imposed a serious environmental problem. Despite decades of experimental efforts to crush seed commercially, only seven cottonseed mills were thought to have existed in the United States on the eve of the Civil War. Accompanying the transformation of southern agriculture after 1866, the cotton economy began to turn this waste product into a valuable by-product. Figure 4.3 charts the rising fraction of U.S. cottonseed crushed for commercial purposes from 1872 to 1939. By the 1920s, roughly 80 percent of U.S. cottonseed was being processed. At this time, farm revenues from cottonseed comprised about 13 percent of the combined earnings from seed and lint.[73] The increase in milling operations was linked to the concentration of ginning.

Although all accounts of the Cotton South pay homage to the invention of the cotton gin, few accounts note how Whitney's primitive device evolved or understand the impact of changing gin technology on the quality of cottonseed. Prior to 1850 the typical plantation gin was animal powered and processed only three to four bales per day of cotton grown in the gin's immediate neighborhood. With the spread of steam power and other important innovations, gin capacity increased. A breakthrough occurred in the mid-1880s with Robert Munger's invention of "system ginning," which employed pneumatic and mechanical conveyance

[73] Moloney, "Cottonseed," 431–63; Wrenn, *Cinderella of the New South.*

technologies and multiple stands of gin equipment. This represented one of the major technological advances in the New South. By 1900 the prototypical modern ginnery, containing four gins of seventy saws each, could process forty to sixty bales of cotton per day and some were capable of handling one hundred fifty bales per day. The new system gins were much more efficient than the older methods, but their complicated machinery and larger clientele led to the unintended consequence of substantially increasing the problem of seed mixing.[74]

The literature of the early twentieth century often associated the deterioration of seed quality with the breakup of the slave plantation system. This structural change undoubtedly accelerated the shift of ginning off the plantation to larger public (or custom) gins. Formerly slaves would pick cotton on sunny days and gin it on the farm on rainy days. After emancipation, the freeman would typically haul his wagon, designed to carry about a single bale, to one of the neighborhood public gins that appeared in large numbers after 1865. Technological forces also drove the process of increasing scale as steam engines replaced draft animals as the power source. This shift was well under way before the Civil War. Steam power was in use in public gins by 1840 and became prevalent in plantation gins after 1854.

But dating the deterioration of seed quality to 1865 seems too early because most public gins remained small. The typical public gin of the immediate post–emancipation period was capable of handling six bales of cotton a day, compared with three to four bales for the plantation gins. Consequently, opportunities for mixing seed remained limited. The most dramatic changes came with the advent of the system era of cotton ginning over the period 1878–84. As with many industries, cotton ginning experienced a series of important innovations over the mid-nineteenth century. Besides the adoption of steam power, the technical changes included the use of steam packing in place of screw presses in compressing operations (1878), the shift from single-stand gin batteries to multiple-stand equipment (late 1870s and early 1880s), the development of pneumatic seed cotton and lint handling equipment to replace hand labor (early 1880s), and the introduction of a stream of improved mechanical seed cotton cleaning, feeding, and other auxiliary equipment (from the early 1880s on). In 1883, H. M. Munger integrated these

[74] Bennett, *Saw and Toothed Cotton Ginning*, 7, 32; Aiken, "Evolution of Cotton Ginning," 199–206; Roper, "Cotton Ginning," 338–9; Ballard and Doyle, "Cotton–Seed Mixing," 1–3.

technologies to produce "the world's first complete ginning system," a multistand, steam-driven gin housed in a large, two-story structure that was connected by pneumatic tubes to separate seed cotton and seed storage houses and that was capable of handling the crops of twenty farmers at time. Many other producers of gin and compress equipment followed Munger's lead and multibattery system gins began to dot the New South landscape. According to Charles Bennett, a leading USDA authority on the evolution of post-harvest cotton machinery, it was after the 1880s that "the large number of single stand ginneries at plantations rapidly gave way to custom gins at the crossroads of the communities."[75] Chapter 6 will further develop our argument that changes in ginning technologies adversely affected seed quality well into the twentieth century.

Conclusion

The history of the American cotton industry over the nineteenth century is simply at odds with the claim that "the advances in mechanical technology were not accompanied by parallel advances in biological technology. Nor were the advances in labor productivity accompanied by comparable advances in land productivity."[76] To the contrary, a stream of revolutionary biological innovations vastly increased both land and labor productivity in an era of few mechanical innovations (except in processing and transportation). Between 1800 and 1860 the actual amount of cotton an average slave picked in a day increased about fourfold, while cotton yields and quality also increased. Contemporaries were unanimous in attributing the bulk of these improvements to the diffusion of Mexican hybrids. The impact of biological innovations, however, was even greater because in the absence of new varieties the observed yields and picking rates would have plummeted due to the spread of cotton rot and other diseases and pests that first entered the Cotton South in the antebellum years.

By the mid-twentieth century, professional scientists had become essential to research and development programs, but the achievements of modern science should not blind us to the enormous contributions of earlier generations of innovators. Lebergott noted that a hazy line often separated science and coarse empiricism. This insight certainly applies to the agriculturalists who imported, acclimated, and developed the numerous

[75] Bennett, *Saw and Toothed Cotton Ginning*, 32.
[76] Hayami and Ruttan, *Agricultural Development*, 209.

distinct types of cotton grown in the United States and then matched these general types and specific varieties to the cotton belt's diverse geoclimatic conditions. In total this story parallels the history of wheat production and reinforces our thesis that biological innovations were an important source of productivity growth in the pre–World War II era. In addition, the cotton story questions the general treatment of biological innovations as land-saving and mechanical innovations as laborsaving.

The development of superior Mexican cottons in the 1810s and 1820s and the subsequent adoptions of Petit Gulf, big boll, Stormproof, and disease-resistant varieties repeatedly changed the map of U.S. cotton. Revealed preference speaks forcefully to the economic significance of the progression of varieties. In the pre–Civil War era, the new cotton varieties had an enormous and highly visible impact on yields and picking efficiency. All future discussions of the slave economy will have to account for this productivity revolution. In the postwar era, a higher percentage of biological innovation went to maintain production in the face of growing threats from insects, diseases, and weeds. This was especially true in the decades following the arrival of the boll weevil, when there was a wholesale conversion to early fruiting varieties. The next two chapters carry the story of cotton through the travails of the late nineteenth century and explain the industry's revival in the post–World War I era. This too is largely a story of biological as opposed to mechanical innovation.

5

Cotton and Its Enemies

Like wheat farmers, cotton growers had to run hard just to stay in place. Sometimes, despite intense efforts, they lost ground. The boll weevil is the best known of all cotton pests. This insect has been the inspiration for books, songs, and even a statue. Historians of the South often focus on the boll weevil to the virtual exclusion of cotton's many other enemies. Due to its lasting imprint on the region and to the large and often contradictory literature on the insect, we too will focus on the weevil. Nevertheless, our emphasis should not imply that other insects, diseases, and weeds were of little significance. To the contrary, as discussed in Chapter 4, Mexican varieties were introduced to combat rot. This was but a single skirmish in the ongoing battle against diseases, insects, and weeds. This battle impacted both land and labor productivity, because the pest problem increased significantly over the nineteenth century. Thus, a mere tracing of the trends in yields, labor requirements, and cotton quality fails to capture the impact of biological innovations, just as plotting wheat yields seriously undercounted the importance of technological changes. Recent evidence helps illustrate the magnitude of the disease and pest problems.

For the period 1991–97, the Cotton Disease Council estimated that diseases caused a mean annual loss to the potential U.S. cotton production of over $1 billion – about 12.3 percent of the potential crop of 2.45 million bales. Over roughly the same period (1990–95), insects caused a mean annual loss of about 7.2 percent of the potential crop. Thus, at the end of the twentieth century, cotton farmers were losing nearly 20 percent of their crop to insects and diseases in spite of decades of modern scientific research and enormous expenditures on fungicides, insecticides, and modern disease and insect management schemes. In fact, the Disease Council reported that between the 1950s and the 1990s the percentage of the potential crop lost to disease changed little – the hundreds of millions of dollars spent on research and preventatives merely allowed cotton farmers to stand still in their battle against diseases. Had farmers taken no precautions and continued to plant now obsolete varieties, the losses would have been catastrophic.

Over the period 1952–97, the most serious problems were, in order of importance, seeding diseases, nematodes, boll rots, *Verticillium* wilt, *Phymatotrichum* root rot, *Fusarium* wilt, and bacterial blights. Seeding diseases still ranked high even though, by the 1990s, virtually all planting seed was chemically treated to reduce such problems. Diseases were constantly evolving and mutating, thereby thwarting control attempts. In addition, some of the diseases on the Disease Council's list were relatively new to the United States. As an earlier example, *Verticillium* wilt was first found in California in 1927, but by the mid-1930s it was a growing problem in the Mississippi Delta, where infection rates soared as high as 40 percent.[1] The key point is that, without significant investments to maintain production (there is much more involved than just yields), disease losses would have been dramatically higher than reported and many otherwise highly productive areas would have had to abandon cotton production. The story of *Fusarium* wilt illustrates the process of biological learning near the dawn of the twentieth century.

Fusarium wilt is of special interest because, at the very time the boll weevil was marching eastward out of Texas, the *Fusarium* wilt pathogen was spreading from the Atlantic coast into the southeastern states – a dire turn of events as the varieties of cotton that proved wilt resistant were not effective in combating the weevil and vice versa. Many hard-running farmers began to rapidly slide backward. *Fusarium* wilt is caused by a fungus that is present in some soils and enters the cotton plant through its roots. An early indication of the disease is a yellowing of the plant's lower leaves. Eventually the leaves shrivel, the plant growth is stunted, and yields suffer. In heavily infected areas, crop losses can exceed 70 percent. By the end of the nineteenth century large sections of South Carolina, Georgia, and Alabama were heavily infected, and the fungus was spreading rapidly. It was particularly serious in wet years

1 Smith and Cothren, *Cotton*, 130–2, 145–6, 554–93; Holley and Arnold, "Changes in Technology," 88; Brown, *Cotton* (1938), 319.

and in sandy and loam soils. Growing nonresistant cotton varieties on wilt-infested soils increased the severity of the infestation. Once in the soil, eliminating the fungus was economically infeasible. Normal crop rotations did not help. Agronomists and farmers tried, unsuccessfully, to combat the wilt with massive doses of fungicide.[2] The solution lay in finding and developing varieties that were resistant to the disease.

Farmers in the infected areas gravitated toward varieties (including Hard-Shell, Woods Improved, Jackson Limbless, and Toole) that demonstrated at least some wilt resistance. In 1895 W. A. Orton of the U.S. Department of Agriculture (USDA) began work on wilt resistance in upland cotton. Using the seeds from a single plant thriving in an otherwise badly wilt-infected field of Jackson Limbless cotton, he created Dillon, the first systematically selected wilt-resistant variety. Orton and his associates also developed Dixie, another successful wilt-resistant variety.[3] J. O. Ware describes the process: "As to the starting point of Dixie, Orton in 1901 collected seed of some healthy plants, presumably Peterkin growing in an infested cotton field...near Montgomery, Alabama."[4] After several years of repeated selection, and of growing resistant plants of different varieties next to each other to facilitate cross-pollination, Orton propagated the seeds from one exceptional plant, which he thought was a hybrid, to create Dixie.

Dillon and Dixie were the first of a long list of improved wilt-resistant varieties. Major advances occurred after Coker's Pedigreed Seed Company focused its research on developing wilt-resistant varieties. In 1941 the company released Coker 100 Wilt, which by the early 1950s accounted for more than 95 percent of the cotton grown in Virginia and the Carolinas.[5] Its popularity shows that farmers were indeed running fast to fight disease.

Numerous insects also threaten the cotton plant. Today the bollworm, not the boll weevil, tops the list of injurious insects.[6] In 1879 the USDA

[2] Ware, "Origin," 41; Brown, *Cotton* (1938), 306–19; Atkinson, "Diseases of Cotton," 287–92. The disease was known as "Frenching," presumably because it was strange and unnatural.

[3] Brown, *Cotton* (1938), 311; Ware, "Origin," 31–42. John Milton Poehlman and David Allen Sleper, *Breeding Field Crops*, 10, credit Orton's achievement as one of the pioneering efforts to breed for disease resistance in any crop.

[4] Ware, "Origin," 42.

[5] Ware, "Origin," 42–3; Olmstead and Rhode, "Hog-Round Marketing," 469.

[6] Leonard, Graves, and Ellsworth, "Insect and Mite Pests," 489–551.

published a five-hundred-page report on cotton insects that focused on the cotton worm and the bollworm.[7] J. Henry Comstock analyzed the introduction and spread of these insects in the United States, their life cycles and feeding habits, the effects of weather, estimates of crop losses, possible predators, and remedies. The cotton worm (also called the caterpillar) probably entered the United States immediately before 1800. Local reports of serious damage appeared after that date, with widespread infestations occurring in 1804, 1825, 1838, 1840, 1846, and 1847. Thomas Affleck described the 1846 infestation in Mississippi: "'At this moment every field... south of Vicksburg, is stripped of everything but the stems, the larger branches, and a few of the first bolls, already too hard for the worms' powerful mastication.'"[8] (See Plate 5.1 for the life stages of the cotton worm.)

Information improved after 1865 with the publication of USDA surveys. These recorded widespread infestations of both the cotton worm and the bollworm in most years between 1866 and 1876; few areas escaped damage. Paralleling the experience of wheat, the cotton growing environment deteriorated. Comstock noted that, "Both the caterpillars and the boll-worm have been infinitely more injurious than in the time before the war."[9] Noted entomologist Charles V. Riley agreed, claiming that "the losses in localities of heaviest production, or where the fields are numerous and contiguous, is nearly double what it is where the fields are more isolated." In addition, the "highest average of loss is sustained in the southern portion of the belt, as in Florida and southern Texas."[10] The USDA estimated that in 1866 the caterpillar destroyed half the crop in Louisiana, 42 percent in Alabama, 40 percent in Texas, and 30 percent in Mississippi. Losses were high in 1868, 1871, and 1872, but 1873 proved the worst year ever, with serious damage from Texas to Virginia.

Planters invested heavily in fighting the worm. The most basic recommendation in the antebellum press was to send slaves into the fields to pick the worms by hand – plantation records show that many planters

7 Comstock, *Report upon Cotton Insects*. As an indication of the concern with these insects, this is one of three volumes totaling over 1,100 pages. Howard, "Insects Which Affect the Cotton Plant," 317–18.

8 As quoted in Howard, "Insects Affecting the Cotton Plant," 24. Also see Riley, *Fourth Report*, 23–34. National data show that cotton production dipped in 1838, 1840, and 1846, but not in 1825.

9 Comstock, *Report upon Cotton Insects*, 28.

10 Riley, *Fourth Report*, 2–4.

followed this advice. "A hand might, with exceptional diligence, go over an acre in *fifteen or twenty days!*" Another practice was to turn turkeys loose in the cotton fields. Farmers also destroyed heavily infested patches to kill the caterpillars. Patented moth traps using poisoned sweets were used "quite extensively throughout the South." Numerous schemes to combat the worm, such as lighting bonfires and flying white flags upon which the moths would deposit their eggs, proved humbugs.[11]

The tide began to turn in the mid-1870s. A series of dry years reduced caterpillar numbers, and southern farmers began experimenting with arsenic poisons. In 1871 "Thos. W. Mitchell, of Richmond, Tex., obtained a patent for its use against Cotton Worm." Riley tested Paris Green, a commercial arsenic-based compound that was proving successful against the Colorado Potato Beetle. In May 1873 he strongly recommended "the green" to cotton growers, and the USDA soon offered instructions on using arsenic. The general adoption of arsenical poisons in the southern regions of the cotton belt greatly diminished cotton worm damage after the late 1870s. But as cotton worm attacks receded, new threats arose.[12]

Over the period 1875–1900, the bollworm was the most destructive cotton pest. It belonged to a family of pests that included the corn worm, the tobacco budworm, and the tomato worm. The bollworm was first mentioned in the United States in 1841, but it did not attract much attention until the mid-1850s. By 1871 it was "a by-word in all the Southern cotton-growing States." Control efforts met with little success as poisons and illuminated traps proved ineffective. Farmers tried planting trap crops such as corn near their cotton. Hand picking the worms off cotton was "out of the question," whereas picking them off corn "is eminently practicable"[13] (see Figure 5.1). Machines designed to brush the insects off cotton plants and then destroy them also proved ineffective. The most important laborsaving machines were the patented spraying and dusting equipment, which complemented more effective insecticides.[14] By the time the boll weevil arrived in the 1890s cotton farmers already had considerable experience fighting pests.

[11] Ibid., 4, 34–7, 128–38.

[12] Howard, "Insects Which Affect the Cotton Plant," 318, 328; Comstock, *Report Upon Cotton Insects*, 38–46; Riley, *Fourth Report*, 4, 34–7, 138–53.

[13] Howard, "Insects Which Affect the Cotton Plant," 328–34; Comstock, *Report Upon Cotton Insects*, 287–95; Riley, *Fourth Report*, 255–384; Leonard, Graves, and Ellsworth, "Insect and Mite Pests," 501–2.

[14] Comstock, *Report Upon Cotton Insects*, 255–6.

FIGURE 5.1. Before modern insecticides workers picked insects by hand. Courtesy of the Mississippi Department of Archives and History.

The Boll Weevil

The boll weevil would not head the Red Queen's 1940 scorecard of insects detrimental to agricultural production – the Hessian fly and chinch bug would likely top the list. But in terms of the publicity and lore the boll weevil generated, its place in history as American agriculture's most destructive pest is secure. The boll weevil was native to Mexico and Central America. Most accounts assert that it entered the United States in 1892 near Brownsville, Texas, and thereafter advanced forty to one hundred sixty miles a year.[15] By 1922 it had swept up the Atlantic seaboard and infested more than 85 percent of the cotton belt.

The weevil feeds almost exclusively on cotton. The female deposits her eggs in the squares (the structures encasing the young buds) and young

[15] See Giesen, "South's Greatest Enemy," 24–5, for an account of the insect's activities in Mexico. Scientists had collected specimens near Veracruz as early as 1840. The USDA set 1892 as the date when the insect became officially established in the United States, and this has since been taken as the date of entry. But the first USDA entomologists on the scene concluded in 1895 "that the boll weevil had probably been present in the Brownsville area for as long as 10 years." Stavinoha and Woodward, "Texas Boll Weevil," 453–4.

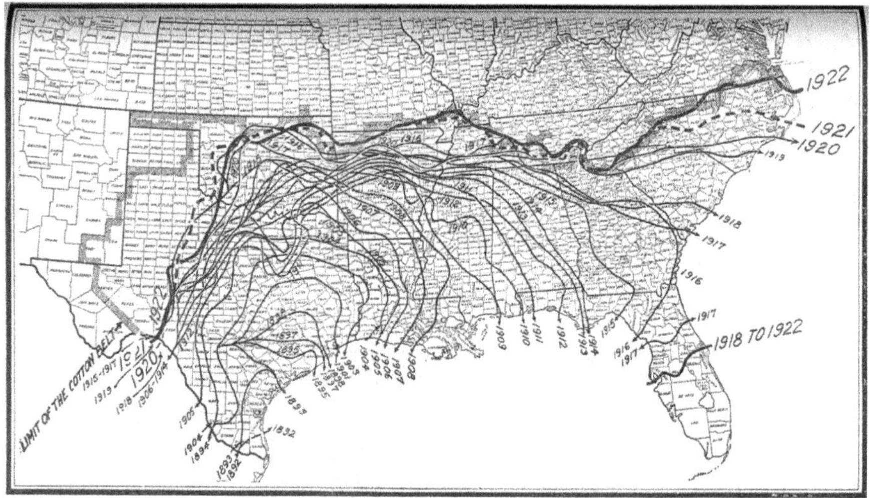

FIGURE 5.2. Land area invaded by boll weevil, 1892–1922. *Source:* Hunter, "Boll-Weevil Problem," 3.

bolls. The larvae feed on the inside of the squares and bolls, causing them to shed. The fiber in the remaining bolls is often stained or otherwise damaged. Female weevils can produce one hundred to three hundred eggs and typically deposit only one egg to a square. Depending on the weather, weevils can produce two to eight generations in a year. Few weevils survive their winter hibernation (diapause stage), but even a small number can cause serious problems. Under ideal conditions, one pair of weevils can generate millions of progeny in a single season. Fortunately such conditions are rare, and recent studies show rates of population increase per generation of one- to tenfold. In the 1950s the USDA reported that only about 5 percent of weevils hibernating in the cotton belt survived and many of those died before the squares formed. In addition, even after eggs are laid, heat, dry weather, birds, and parasites can further limit the weevil population. The insect thrives in years with heavy rain in the spring and summer.[16] Cotton farmers and scientists observed these general patterns and then set out to identify methods of limiting weevil populations. (See Plates 5.2 and 5.3 for photos of a boll weevil and the damage it can cause.)

The spread of the boll weevil classically has been displayed using maps of the land area invaded, such as that in Figure 5.2. Such maps

[16] Brown, *Cotton* (1938), 339–46; Gains, "Boll Weevil," 501–5; Head, *Management and Control.*

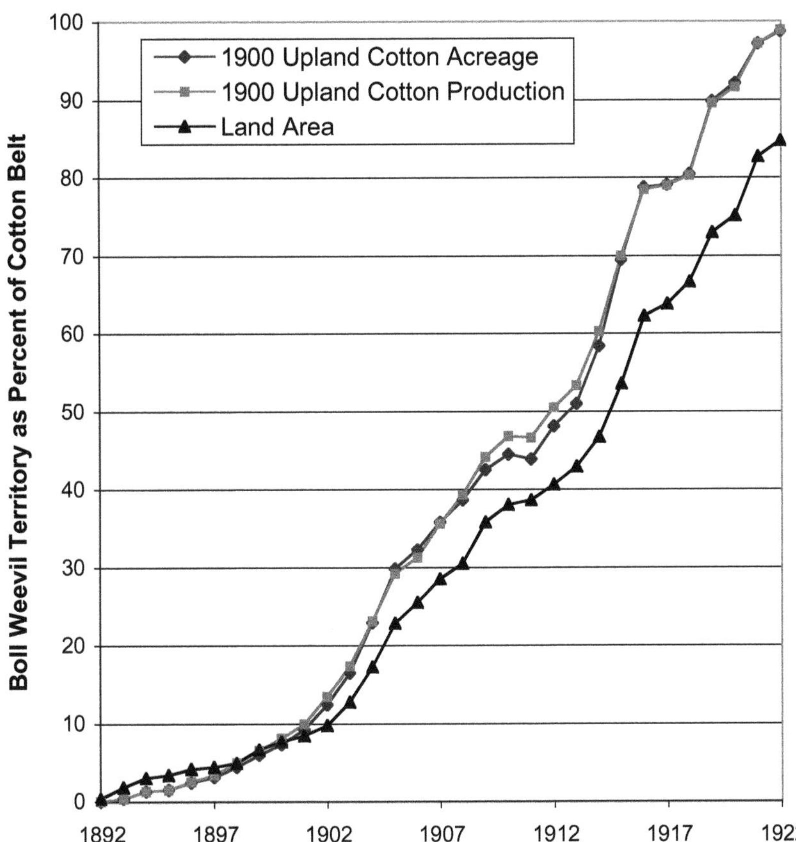

FIGURE 5.3. Spread of the boll weevil, 1892–1922. *Source:* Constructed from information in Figure 5.1 and Haines and Inter-university Consortium for Political and Social Research, *Historical, Demographic, Economic, and Social Data.*

do not reveal the weevil's effect on cotton production, because many areas attacked were not growing much cotton and, in addition, large areas on the fringes of the cotton belt were never invaded. To correct for these problems, we have assembled data showing when the weevil invaded each county weighted by the county's cotton acreage and production as reported to the 1900 Census.[17] The series are graphed in Figure 5.3, which also shows the land area covered – the usual measure of the boll

[17] The use of the 1900 Census may understate the insect's effect in southern Texas, where it was already reducing yields. A further complication is that the weevil arrived during its August migration and did not do much damage in the initial season of contact. Adding two to three years to the date listed provides a better sense of land area under the weevil's thrall.

weevil's progress. As the figure shows, after 1900 the traditional land area measurement significantly understates the weevil's importance in the cotton belt. As an example, those areas still free from the weevil in 1922 accounted for 13 percent of the land mass but produced less than 1.5 percent of the 1900 crop.

Our new production- and acreage-weighted series of the weevil's spread show an acceleration in diffusion in the period 1898–1905. This was the period when the insect's path of destruction made its eastward turn. By 1907, the weevil had crossed the Mississippi River. Weather conditions in 1915 and 1916 were exceptionally favorable and the pest engulfed most of Georgia, leaped into the Florida cotton fields to the south, and threatened the Carolinas to the north. The advance slowed over the next two years, and the weevil did not finish its geographical conquest until the 1921–22 season. There was a brief period between 1908 and 1913 when the production and acreage series diverge. In these years, the weevil attacked the Mississippi Delta where yields were relatively high, pushing the production series above the acreage series.

Control Efforts

In 1894 and 1895 the UDSA sent a number of entomologists, including C. H. T. Townsend, E. A. Schwartz, L. O. Howard, and Charles W. Dabney, to southern Texas to study the weevil. From this beginning, the USDA, various state agencies (including the Texas Boll Weevil Commission established in 1903), private companies, amateur scientists, farmers, and numerous quacks sought ways to limit the insect's damage. Insecticides proved ineffective. Efforts to erect quarantine buffers also came to naught. An early proposal in Texas to establish a fifty-mile-wide cotton-free zone ran into legislative resistance.[18] By the time Georgia had adopted quarantine measures in 1904, the weevil was too well established in the South to pause for long.

Many worthy ideas on how to coexist with the weevil diffused rapidly. Entomologists recommended the adoption of early maturing varieties, the destruction of stalks and brush, the use of fertilizers to hasten ripening, early planting, and thorough cultivation.[19] It did not take farmers long to switch to earlier maturing varieties. The boll weevil entered Robertson

[18] Helms, "Just Looking for a Home," 56–7.

[19] Agricultural scientists would still be proposing recommendations advanced in the 1890s more than forty years later. Brown, *Cotton* (1938), 347–53; Howard, "Insects Affecting the Cotton Plant," 1–31; Hunter, "Methods of Controlling," 1–15.

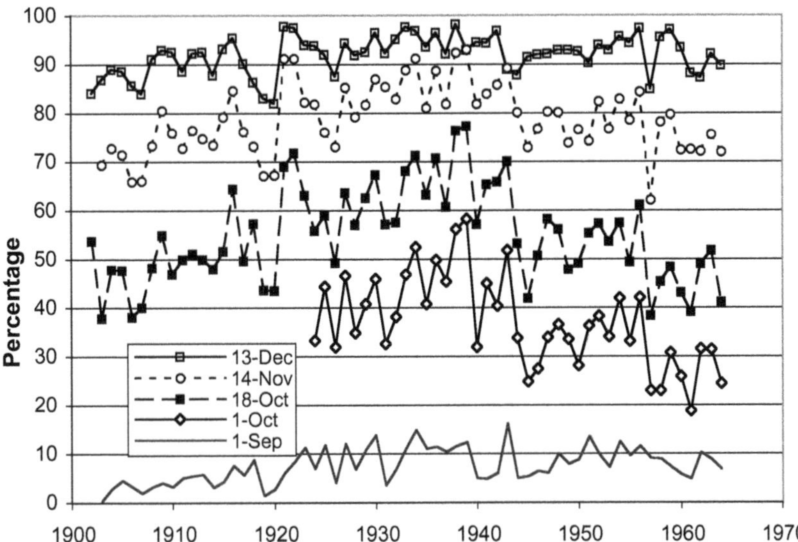

FIGURE 5.4. Cumulative percentage of U.S. cotton ginned by selected dates, 1902–61. *Source:* U.S. Census Bureau, *Quantity of Cotton Ginned*, and, *Cotton Production in the United States*, 1906–1962.

County, Texas, between 1898 and 1901. By 1901 farmers in that county were importing seed from northern Texas, and by early 1904 the Dallas Jobbers' Cotton Association had imported nineteen carloads of seed from North Carolina. According to Douglas Helms, "One estimate held that Texas farmers 'imported thousands of car loads of short staple cotton seed' in the rush to adjust to weevil destruction."[20]

We can offer a crude quantitative indication of the movement to earlier ripening varieties by charting the dates of cotton ginning as displayed in Figure 5.4. In the early years (1902–07) of the twentieth century, less than 45 percent of U.S. cotton was ginned before the eighteenth of October. By the period 1934–39, almost 70 percent was ginned by that time of the year. With the application of chlorinated hydrocarbon insecticides such as DDT in the post–World War II period, there was a shift back to later maturing varieties. During the period 1959–64, the share ginned before the mid-October date had returned to around 45 percent. It is likely that the spread of the mechanical cotton picker, which reduced the number of pickings, also contributed to earlier ginning.

[20] Helms, "Revision and Revolution," 109–11; Giesen, "South's Greatest Enemy," chapters 4 and 5, provide a detailed account of the adjustment process in the Delta.

Over the period of the weevil's spread, cotton production was moving onto the high plains of Texas and Oklahoma as well as shifting to the irrigated fields of Arizona, California, and New Mexico. The spread of the boll weevil accelerated this trend, because the far West was weevil free.[21] But in Arizona, California, Oklahoma, and New Mexico, ginning occurred much later than the national average, and consequently the regional shift of production meant that the trend toward early ginning in the Cotton South was even more rapid than implied in Figure 5.4. The move to early varieties was also apparent within individual states. For example, our regression analysis of state-level data using fixed effects (not reported here) reveals that over the period 1902–40 the arrival of the boll weevil led to a 17 percent increase in the share of cotton ginned before the eighteenth of October.

Long-staple cotton typically was late maturing and, thus, was especially hard hit. The destruction of Sea Island cotton production offers a graphic example of the impact of the boll weevil (Figure 5.5). The weevil reached the Atlantic coast by 1917. Between 1914 and 1917 annual production of Sea Island cotton averaged more than 92,000 running bales; in 1919 it fell to 7,000 bales; and over the period 1920–24 the crop averaged about 2,000 bales – it never recovered.[22] The destruction of Sea Island cotton mirrored a more general phenomenon.

A swath of fertile cotton lands from Texas to the Atlantic seaboard was largely denuded of its prime long-staple cottons. For more than a hundred years breeders had selected and acclimated cottons for specific areas, and in just a few years this work was lost. A number of qualities, including fiber characteristics, picking ease, and storm resistance, lost importance in the face of one overriding concern – early maturation. Picking efficiency and quality suffered as vast areas abandoned 1 1/8-inch big boll cotton to grow varieties with small bolls and very short, 5/8-inch staples. About fifty long-staple varieties ceased to be commercially viable and in most cases died out altogether.[23] This represents only the tip of the ecological

[21] Kent Osband downplays the impact of the weevil on the westward expansion of cotton production. In fact the weevil greatly accelerated this movement. The USDA's research program in the western states was dedicated to breeding and promoting varieties that would help offset the loss of the longer cotton belt varieties. Osband, "Boll Weevil versus King Cotton," 627–43.

[22] Collings, *Production of Cotton*, 222.

[23] Brown, *Cotton* (1938), 339–55; Helms, "Revision and Revolution," 110–11; Ware, "Origin," 50–81, 95–7. The extinction was nearly complete. A long-staple cotton named Sunflower was the only variety of "the old Mississippi Valley series" to survive the devastation. It became a parent for most of the important long-staple varieties later developed.

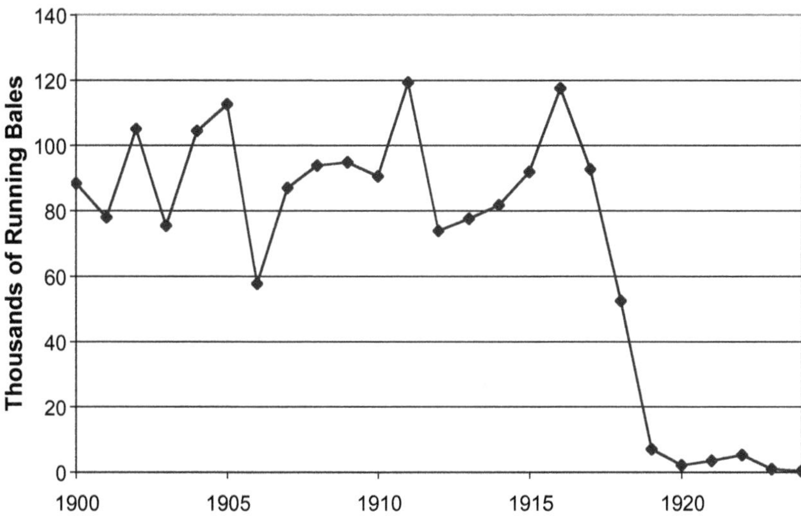

FIGURE 5.5. Production of Sea Island cotton in the United States, 1900–24. *Source:* Collings, *Production of Cotton*, 222.

disaster because many high-quality, mid-staple varieties followed the long staples into extinction. As discussed in the next chapter, the conversion of large areas away from long-staple cotton production had a significant impact on grading and marketing practices, which led to a further deterioration in quality. In the 1930s, scientists reversed these trends by developing mid-staple varieties, thereby changing the map of U.S. cotton yet again.

In addition to adopting earlier ripening varieties, farmers sought other means to diminish weevil-induced losses. One of these was to plant earlier in hopes of earlier crop development. After the weevil struck Robertson County, Texas, in 1903, area farmers moved their planting date up by about twenty days. Farmers in some areas purportedly planted a month earlier than had been the custom. Cotton growers were thus following in the footsteps of wheat farmers who, more than one hundred years earlier, had shifted varieties and cultural methods to fight the Hessian fly.[24]

Among the most important recommendations in the integrated pest management (IPM) system was to burn or plow under the cotton stalks immediately after harvest to reduce the number of weevils before they hibernated. Repeated experiments showed that destroying the stalks

[24] Brown, *Cotton* (1938), 351–3; Helms, "Revision and Revolution," 112–14. In other areas, farmers attempted to plant later in hopes of depriving weevils emerging from hibernation of food.

significantly reduced the next year's damage and generated higher yields. In one USDA experiment an isolated group of Swedish cotton farmers followed the prescribed stalk destruction policy and, relative to control test plots, harvested an extra 600 pounds of seed cotton. But according to Helm's careful investigation of this issue, the practice was not widely adopted. One drawback was that it required a great deal of labor to cut the stalks while green rather than after they had died and dried out. The advent of the tractor made the job much more manageable (once again a machine aided biological innovation). Another drawback was that weevils could migrate to nearby fields, meaning that a farmer who destroyed his green stalks would not capture the full benefit of his investment, nor be fully protected, unless his neighbors followed suit. Success required a community effort. USDA scientists understood this externality problem and in 1896 urged Texas state officials to enact legislation to establish mandatory stalk destruction dates. The individualistic Texans turned a deaf ear. Other control recommendations – destroy volunteer cotton, clean up trash, locate fields away from the woods, and use more fertilizer to hasten ripening – addressed the same goal: invest resources to lessen the weevil's chances of surviving the winter.[25]

Many historians indict the old villains of the Cotton South – the labor relations system and absentee landlords – for lax eradication policies, claiming that croppers and tenants had little incentive to do the extra work needed to control the pest. In 1910 less than 50 percent of share renters stayed in one place for more than two years.[26] Helms builds on this theme: "The close of the picking season ended the obligation between landlord and tenant if all debts were paid. Tenants planning to move had no reason to turn under the stalks. In cases where there seemed to be success in getting owners to destroy stalks, the tenants often did not participate. Reports from Sumter County, South Carolina, indicate that most owners had plowed under their stalks in 1922, but that tenants in the county had 'done nothing towards destroying cotton stalks.'"[27] Helms further notes that southern agricultural leaders proposed longer-term contracts to give tenants greater incentives to improve cultural methods, but that little came of this idea. Some landlords did succeed in adjusting contracts to require tenants to "carry out cultural control measures as directed."[28]

[25] Brown, *Cotton* (1938), 351–4; Helms, "Revision and Revolution," 118–20.

[26] Wright, *Old South, New South*, 93.

[27] Helms, "Revision and Revolution," 119.

[28] Ibid., 122.

Much commentary suggested that tenants working in the presence of the weevil needed more supervision.

It is difficult to interpret this testimony. Were landlord–tenant relations so hidebound that simple mutually beneficial contractual adjustments were infeasible? Were tenants lazy or otherwise resistant to extra work *for extra pay?* Were many landlords indifferent to the new possibilities? A remaining possibility is that some of the new cultural systems simply were not cost-effective for all situations. If so, the failure to adopt them could represent efficient behavior. Even so, landlords, who wanted their tenants to work harder without extra pay, might point an accusing finger.

The measure that would later become the main line of defense – poisons – was not available during the first wave of destruction. This was not for want of trying, as both farmers and entomologists experimented unsuccessfully with the bromides used against other insects. Poisons such as Paris Green gained popularity even though USDA tests showed that they were not cost-effective. The grubs feeding inside the squares were well protected from poisons and the foliage gave the adults considerable shelter from contact poisons. In 1908 the *Journal of Economic Entomology* devoted much of its inaugural issue to boll weevil control. In the same year William Newell experimented with dusting (as opposed to spraying) the plants with a powdered lead arsenate formulation.

The first really effective poison arrived in 1918 when the USDA's B. R. Coad developed a calcium arsenate mixture for dusting. The calcium helped the poison adhere to the plant, making the poison more accessible to the weevils. In 1919, 3 million pounds of calcium arsenate dust were sold in the South, and sales increased to 10 million pounds by 1920. The discovery of an effective poison was only part of the story because application methods also had to be perfected. After numerous experiments, the USDA recommended that farmers raise a large dust cloud at night or in the early morning and let it settle while dew was still on the plants. There were also trials involving dusting machinery that ranged from hand dusters, to mule- and tractor-towed devices, to airplanes. Calcium arsenate was costly and beyond the reach of many farmers. Moreover, the expense was a function of the acres dusted as opposed to the amount of cotton produced – the average cost varied inversely with yields. This meant that dusting was often only cost-effective on lands with high yields. In addition, arsenic had the predictable side effects of fouling water, poisoning farm animals (and farm workers), contaminating other crops, and leaving a residue on finished cotton goods. In spite of these externalities, southern states encouraged the use of the arsenic-based compounds. As one example, the Georgia State Board of Entomology bought the poison in bulk to be sold at cost to the state's farmers.[29]

Although dusting was widely practiced, the rates of weevil damage remained high. Something new was needed. Research on new chlorinated hydrocarbon (CHC) pesticides led to a fundamental breakthrough in weevil control, or so it seemed. In Georgia, work was undertaken on benzene hexachloride and chlorinated camphene around 1946, on toxaphene in 1947, and on other compounds, including DDT, in 1948. Across the South, test results showed good weevil control and higher yield. The new pesticides were rapidly adopted, especially for larger operations. But in 1954 weevils resistant to CHCs appeared in Louisiana; the Red Queen's pesticide treadmill had just accelerated. CHCs accounted for 98 percent of the pesticides sold in 1954 but only 34 percent by 1958. In the mid-1950s, new organic pesticides entered the trade, capturing a market share of 20 percent in 1956 and 55 percent in 1957.[30]

The boll weevil literature indicates that, besides adjusting cotton growing practices, farmers also switched to other crops and abandoned large tracts of land. These issues need more study, but our preliminary tests of these hypotheses using county-level data suggest that cotton acreage plummeted and stayed down when the weevil arrived in a given locale, but that total land in agriculture changed little. The overall increase in cotton production in the South resulted largely from the introduction of new lands. The weevil certainly stimulated researchers such as George Washington Carver to experiment with alternative crops such as peanuts, sweet potatoes, soybeans, and pecans.[31]

Weevil Damage Assessments

There is considerable controversy about the financial magnitude of boll weevil losses. On one hand, traditional USDA studies generate large estimates of the aggregate losses, often on the order of $300 million (in current dollars) annually. And the conventional view was that the insect destroyed "between one-third and one-half of the crop in newly infested areas."[32] On the other hand, Kent Osband presents considerably smaller

[29] Brown, *Cotton* (1938), 348–52; Helms, "Technological Methods," 291; Haney, Lewis, and Lambert, "Cotton Production," 8–11.

[30] Haney, Lewis, and Lambert, "Cotton Production," 16–18; Douglas Helms, "Technological Methods," 295, offers slightly different numbers on pesticide sales.

[31] Lange, Olmstead, and Rhode, "Impact of the Boll Weevil," 25–9; Holt, *George Washington Carver*, 222–5, 318.

[32] Manners, "Persistent Problem of the Boll Weevil," 25.

TABLE 5.1. *Estimated Weevil Damage*

Locality	Period	Percent Reduction
Talluah, LA	1920–34	32.2
Florence, SC	1928–35	23.6
Oklahoma (Eastern)	1928–35	32.8
Mississippi (Hill Section)	1934–36	10.8

Source: Hyslop, *Losses Occasioned by Insects*, 4–5.

numbers. And as James Giesen observed, "southerners were growing more cotton in 1921 than in 1892."[33] There is no doubt that, in the two or three seasons after the weevil arrived in an area, local farm and business communities were hit hard. In addition, extensive changes in farming practices and varieties grown imposed costs not captured by production and yield data. The Bureau of Agricultural Economics (BAE) began estimating annual boll weevil losses in 1909. Over the period 1921–35, the estimated reduction in yield for the United States (excluding the weevil-free far West) ranged from a high of 31.2 percent in 1921 to a low of 4.1 percent in 1925 and averaged about 13 percent overall.[34] As the BAE economists understood, the price effects of the output decline offset some of the losses to farmers' income.

There are several problems with production loss estimates; they tend to measure the declines in production without regard to the efforts that farmers took to diminish their losses. A Bureau of Entomology and Plant Quarantine (BEPQ) study conducted across the South takes us a step closer to a measurement of the weevil's impact on output. The BEPQ compared "the yield in plots where the boll weevil was controlled with that in untreated plots."[35] The results, summarized in Table 5.1, suggest that average physical losses would have been much higher than the BAE estimates.

But even these data understate the effect of biological learning, because the cotton varieties that the BEPQ researchers planted in the "untreated"

[33] Brown, *Cotton* (1938), 345; Osband, "Boll Weevil versus King Cotton," 627–43; Giesen, "South's Greatest Enemy," 2. Comparisons between production in 1892 and 1921 are problematic because both years had short crops. But taking a longer view also indicates rising cotton acreage and output.

[34] U.S. Bureau of Agricultural Economics, "Statistics on Cotton," 67–80.

[35] J. A. Hyslop, *Losses Occasioned by Insects*, 4–5, also raises the question of increased cost of production but only gives a rough estimate of dusting for the period 1926–30. On average in this period farmers dusted more than 3.2 million areas. At an estimated cost of two dollars per acre and assuming one-half of the dusting was directed at the boll weevil, about $3.2 million was spent per year to dust the weevil.

plots had survived the weevil's first shock. In addition, the researchers moved up the time of planting on these plots and varied other cultural practices. The gap between the test and control plots would have been far greater if the scientists had planted past varieties that had already proven vulnerable to the weevil and if they had employed pre-weevil cultural practices.

Osband noted that, for all the damage done after the weevil first arrived in a particular locale, the Cotton South as a whole was resilient. "Cotton farmers learned to cut their losses to the weevil: They changed their cultivation methods, harvested sooner and applied poisons. After the initial shock, every state witnessed a decline in weevil losses and resurgence of cotton production ... the weevil seems a symbol less of King Cotton's collapse than of its perseverance."[36] Based on his assumptions about the elasticities of demand and supply (including foreign supply), Osband estimated that the aggregate revenue loss to southern cotton producers was a modest 2 percent. His estimates assumed that land taken out of cotton earned a smaller return in other uses. From this macro perspective, the higher cotton prices greatly benefited foreign producers and hurt consumers everywhere. Even within the South, some producers initially benefited while others suffered.

None of these estimates adequately captures the true value of biological innovation in combating the weevil. To form a correct measure, one should ask what would have happened to cotton production and southern income (after allowing for cropping adjustments) if biological technologies had remained static. But just as farmers and scientists growing wheat in Minnesota did not plant winter wheat varieties or varieties known to be susceptible to rusts, cotton growers and scientists in the South did not continue to use pre-weevil cotton varieties and methods. In fact the weevil effectively drove older and higher-quality varieties into extinction. The fact that the estimates of weevil damage are so low highlights the significance of biological learning and innovation in a region of the country normally considered technologically and institutionally backward.

Epilogue

The boll weevil – the scourge of American cotton – may soon be relegated to the pages of history. As of 2003, the boll weevil had been eradicated in Virginia, the Carolinas, Georgia, Alabama, Kansas, Florida,

[36] Osband, "Boll Weevil vs King Cotton," 628.

California, New Mexico, and Arizona. In addition, Baja California and Sonora, Mexico, are now weevil free. The pest has also been expelled from large pockets of Texas, Arkansas, Louisiana, Oklahoma, Mississippi, and Tennessee.[37] This success has been made possible by the intensive and focused use of chemical pesticides and pheromone traps that attract the weevils. But pesticides and traps are only part of the story.

The other ingredient necessary for success has been collective action to overcome the free rider problem. In many areas cotton farmers sprayed fifteen to twenty times a year to control the weevil. In the 1970s, roughly 50 percent of all pesticides used by American farmers were aimed at the boll weevil.[38] But as long as one farmer was lax and weevils remained in the area, the insect could reestablish itself in subsequent years. Since the time of Townsend's pioneering work in the 1890s, entomologists had preached the need for collective action to combat spillover effects, but the required cooperation was difficult to obtain in the highly individualistic rural South. Recall that one of the early, albeit temporary, successes with collective action was in a colony of Swedish farmers in Texas. It seems at least plausible that the USDA chose to conduct its community-wide stalk destruction program with the Swedish settlers because they had more social cohesion – what economists call social capital – than existed in most communities. In the early 1970s, a chance discovery underscored the importance of collective action. Scientists working on a pilot area-wide eradication program in southern Mississippi found that "a single untreated acre of cotton provided a source of infestation for at least 1,800 surrounding acres."[39] To rid a region of the insect required that all land be included, and this required government fiat to force free riders to participate.

Research published between 1959 and 1971 showed that, with new techniques, boll weevils could be exterminated. Based on a better understanding of the winter diapause behavior of the boll weevil, entomologists recommended intensive spraying in the late fall to reduce the following year's population.[40] Further research fine-tuned the procedure. As an example, tests showed that spraying seven times in the fall reduced the following spring's weevil population from an average of 8,800 to 25 per acre. Large-scale tests conducted in Texas (1964 and 1967) and in

37 U.S. Animal and Plant Health Inspection Service, *Boll Weevil Eradication*.

38 Haney, Lewis, and Lambert, "Cotton Production," 20–2.

39 Ibid., 19.

40 Ibid., 18; Perkins, "Boll Weevil Eradication," 1044–50.

Mississippi, Alabama, and Louisiana (1971–73) convinced scientists and cotton industry leaders that it was possible to eradicate the boll weevil from the United States if large geographical areas were treated as a unit to help prevent reinfestation from nearby lands.[41]

A blue ribbon committee sent an eradication program recommendation to the Secretary of Agriculture in December 1973. In 1978, after a prolonged debate in Congress, the Ford Administration commenced a three-year program designed to expel the boll weevil from Virginia and North Carolina. The program's success in those states led to its adoption elsewhere. A closer look at the efforts in Georgia provides a sense of how the program worked.[42]

Georgia farmers passed an eradication referendum in 1986 and the program began in September 1987. The referendum required a two-thirds vote of cotton farmers and, once passed, became binding on all cotton growers. The program involved intensive fall spraying with an average of 8.4 pesticide (guthion) applications per acre by the end of December. The trapping phase commenced in April 1988 with the placement of nearly 350,000 pheromone traps (one per acre). In the following two years the fall spraying increased, with more than 12 treatments in 1989, and the trapping continued as before. By the fall of 1991, the weevil population had declined significantly and the number of fall sprayings fell to an average of 2.7. In addition, fewer traps were needed and they were concentrated where the weevil population persisted. The use of insecticides continued to decline as the program went into its maintenance phase in 1992. The average number of insecticide applications fell from 14.4 in the pre-eradication era to 3.4 in the years after 1992. Average cotton yields per acre increased from 482 pounds in the pre-eradication period to 656 pounds in the post-eradication years. At the same time the average acreage in cotton increased more than fivefold.[43]

The modern boll weevil eradication program is much more than an interesting postscript to our story of the early history of biological innovation in cotton production. It takes us full circle, back to the first years of the boll weevil invasion. Townsend, the USDA entomologist sent to southern Texas in 1894, had immediately recommended a quarantine with a fifty-mile cotton-free zone along the southern border of Texas to

[41] Haney, Lewis, and Lambert, "Cotton Production," 18–20.

[42] U.S. Animal and Plant Protection Service, *Boll Weevil Eradication*.

[43] Haney, Herzog, and Roberts, "Boll Weevil Eradication," 260–1; Haney, Lewis, and Lambert, "Cotton Production," 22–3.

stop the spread of the weevil. He also recommended early harvesting, the mandatory destruction of stalks to kill overwintering adults, crop rotation, and the flooding of fields where possible. Although his recommendation for a quarantine zone was endorsed by the USDA leadership, Texas Governor James Hogg failed to respond to their admonitions. In 1895 Assistant Secretary of Agriculture Charles W. Dabney met with the new Texas governor Charles Culberson and representatives of the state legislature. Dabney proposed banning cotton growing from the Nueces River to the Colorado River, a zone encompassing about one-third of the cotton area of Texas. He also advised the state to enact legislation mandating stalk destruction and to appoint a state entomologist with a staff to enforce quarantines. Once again, state officials failed to act.[44] Other states did impose quarantines after the weevil began its march. For example, Georgia enacted its first quarantine in August 1904, but it was too late. The weevil already had too large a foothold to prevent entry. In 1909, entomologist W. E. Hinds lamented the lack of action in Texas. He estimated that $20,000 spent when the weevil first appeared could have eradicated the pest and "prevented its reaching the principal cotton producing portions of the country."[45]

As we have noted, history is littered with "what might have beens." Would a protective zone have worked? It is hard to know. But in light of what was soon learned about the weevil's impact, creating a cotton-free corridor was surely worth a try. The refusal to do so demonstrates the difficulty of forging a collective response. State and federal attempts to overcome the free rider problem via collective action were not limited to the boll weevil or to cotton. Since the early nineteenth century, states had imposed quarantines to limit the movement of sick animals, and in 1865 the federal government devised policies to prevent the introduction of rinderpest.[46] The USDA leadership recognized the weevil's threat very early and showed remarkable foresight in its recommendations; Texas officials, however, chose to do nothing. The early history of the boll weevil in the United States represents a clear case of a national problem handled first at the state level, where spillover effects received insufficient weight. In 1895, if Louisiana or Georgia farmers could have voted, they

[44] Helms, "Technological Methods," 287; Stavinoha and Woodward, "Texas Boll Weevil," 454–6. Texas did appoint an entomologist in 1899.

[45] Stavinoha and Woodward, "Texas Boll Weevil," 457; Hinds, "Facing the Boll Weevil," 87–99.

[46] Houck, *Bureau of Animal Industry*, 29.

would have supported a quarantine in Texas – but whether they would have been willing to tax themselves to pay for it is far more problematic.

By the time the boll weevil eradication program commenced in the 1970s the USDA had, through its campaigns against tick fever, bovine tuberculosis, hog cholera, and a number of plant diseases and pests, amassed considerable experience with organizing broadly based collective action programs. The USDA had also successfully intervened in the cotton industry to overcome the free rider problem and negative externalities in the production, grading, and marketing of cotton to counteract the decline in cotton quality. The next chapter examines these programs.

6

The Other Revolution in the Cotton Economy

Cotton's Revival in the Twentieth Century

During the twentieth century American cotton farming evolved from a backward sector to a highly productive industry. Most histories of the modernization of the cotton industry have concentrated on mechanization and the end of sharecropping, ignoring any changes in biological technologies.[1] But the lack of attention to new cotton varieties should not suggest that no changes occurred. To the contrary, from 1928–32 to 1958–62 the growth in cotton yields actually outpaced that of corn (see Figure 6.1). This chapter analyzes the role of government policy and, in particular, the one-variety improvement movement and the Smith–Doxey Act in promoting the diffusion of new, high-performing cottons.

The timing of the development and diffusion of new cotton varieties roughly parallels the well-known diffusion of hybrid corn. But the cotton story is far more intriguing because of the role of government policy in modernizing one of the most backward sectors of American agriculture and because of the interaction effects of biological and mechanical technologies. Once new varieties of corn became available, the story of diffusion was largely the result of market interactions between individual farmers and private seed companies (apart from extension service educational campaigns). By comparison, the cotton industry was long plagued

[1] Wright, *Old South, New South*, 226–38; Street, *New Revolution in the Cotton Economy*; Day, "Economics of Technological Change," 427–9; Musoke and Olmstead, "Rise of the Cotton Industry," 385–412; Whatley, "Labor for the Picking," 905–29; Holley, *Second Great Emancipation*, 124–9.

This chapter is a refined version of our article, "Hog-Round Marketing, Seed Quality, and Government Policy: Institutional Change in U.S. Cotton Production, 1920–1960." *Journal of Economic History* 63, no. 2 (2003): 447–88.

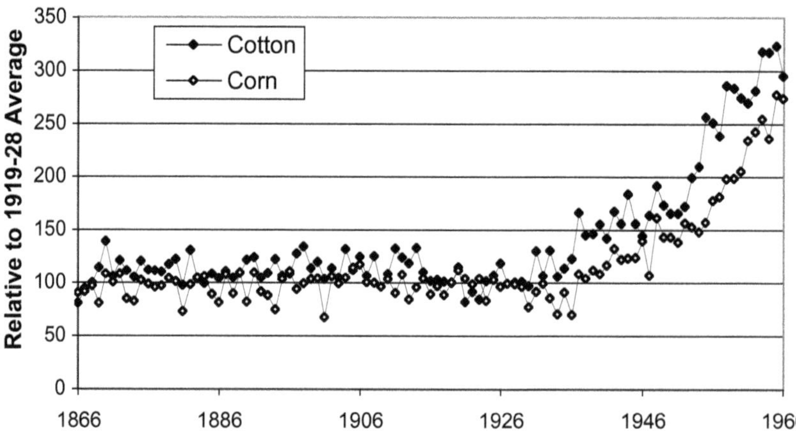

FIGURE 6.1. Yield per acre of U.S. cotton and corn, 1866–1995. *Source:* Carter et al., *Historical Statistics*, Tables Da693–694 and Da755–756.

by chronic problems of market failure that dulled the incentives for both seed breeders and individual farmers. To overcome negative externalities in production and a "lemons problem" in marketing, the U.S. Department of Agriculture (USDA) and state officials orchestrated a campaign to create one-variety communities and provide impartial grading services. These initiatives played a key role in the adoption of new biological technologies.

Yet Another Burden of Southern History

The 1921 USDA *Yearbook* is representative of an extensive literature bemoaning yet another burden of southern history: "According to the testimony of the cotton trade in Europe as well as in the United States, the quality of the American cotton crop has deteriorated in recent decades."[2] The available quantitative evidence supports this claim. Table 6.1 collects data on the staple length of U.S. cotton by state, circa 1880, 1913, and

[2] Doyle, Meloy, and Stine, "Cotton Situation," 400. Also see Johnson, *Cotton and Its Production*, 53–4. The 1866 USDA *Report of the Commissioner of Agriculture*, 209, dates the problem to the war itself. "The most serious difficulty encountered by cotton-growers, and particularly those who are engaging in such enterprises for the first time since the war, had been found to be *poor seed*." The report further noted that "for seven years little or no pains have been taken by any cotton-growers to perfect their seed."

TABLE 6.1. *Average Staple Lengths (in thirty-seconds of an inch)*

	1880	1913	1928–30
Alabama	32.9	29.8	28.4
Arizona			32.8
Arkansas	33.2	30.4	31.2
California			33.8
Florida			28.8
Georgia	34.1	30.6	29.0
Louisiana	34.2	31.4	31.0
Mississippi	33.5	31.4	32.8
Missouri			31.0
New Mexico			33.5
North Carolina	33.9	29.3	29.7
Oklahoma		32.6	29.7
South Carolina	39.5	29.8	30.4
Tennessee			29.9
Texas	34.4	31.4	30.0
Virginia			29.2
Other			33.7
United States	34.3	30.8	30.3

Note: United States is the average of available states weighted by output.

Sources: Compiled from Hilgard, *Report on Cotton Production*; Taylor, "Relation Between Primary Market Prices"; U.S. Bureau of Agricultural Economics, "Grade, Staple Length," 20–1.

1928–30.[3] In every state for which data are available, the staple length in 1913 and 1928–30 was shorter than it was in 1880. Based on the national weighted average, staple length fell by over 12 percent between 1880 and 1930.[4]

The consequences of the deterioration in quality were very serious given rising competition from foreign cotton, rayon, and other synthetic

[3] Cotton length was one of the most important factors in pricing cotton. Cotton classers divided samples into six different color classes. Within each color there was a range of grades. For white cotton there were nine specific grades that captured factors such as the existence of foreign matter, the cotton's color quality, and "ginning preparation," which included the roughness, nappiness, and stringiness of the fibers. With high-volume-instrument testing, it is now possible to cheaply determine important characteristics. Cox, "Cotton Classing and Standardization," 320–3.

[4] Determining exactly how the market valued this decline is slightly more complicated. Over the period 1928–30, the average price of cotton in ten central markets varied as

fibers. The causes of quality decline were twofold. First, the invasion of the boll weevil, beginning in 1892, led farmers throughout the cotton belt to discard late-maturing varieties, which were most susceptible to the pest. "In this way many excellent varieties of long-staple upland cotton and practically all of the better types of medium-staple were lost within a comparatively short time, to be replaced by the early, rapid-fruiting types brought in from the northern parts of the belt."[5] But the boll weevil was only part of the problem. As indicated in Table 6.1, staple length also declined between 1880 and 1913 in areas not yet hit by the weevil, such as North Carolina. Contemporaries noted that cotton culture was burdened by an interlocking set of production and marketing problems that both hampered the ability of and reduced the incentives for individual farmers to maintain and improve cotton quality. As noted in Chapter 4, cotton production was plagued by negative spillovers that made it difficult to maintain the "genetic purity" of the seed supply.[6] These technical difficulties were exacerbated by post–Civil War institutional changes, in particular the breakup of plantations into small production units and the increased importance of public gins. In addition, the prevalence of price pooling through what was known as the hog-round system muted incentives to produce high-quality cotton. A "vicious circle" thwarted efforts to improve the crop and reduced demand for quality seed. This, in turn, reduced incentives for seed breeders to invest in research and development (R&D), reinforcing the low-level equilibrium trap.[7]

On the production side, problems of maintaining purity arose because cotton is subject to cross-pollination. The incidence of cross-pollination varied greatly depending on the variety, the weather conditions, the distance between fields, and the population of insects (especially bumblebees). When cotton was cultivated in small fields located near the

follows (staple length is in thirty-seconds of an inch and the price premium is taken to be 100 for 26/32):

Length	28	30	32	34	36	38	40
Price premium	108	111	115	120	125	133	158

USDA, *Agricultural Statistics 1936*, 84. Note that the marginal value of increasing staple length was greater at the ends, especially at the high end, than at the middle of the distribution. Due to these nonlinearities, the decreased market value due to the decline in staple length over the period 1880–1930 was larger than a comparison of the prices of mean qualities would indicate.

5 Ware, "Plant Breeding," 661.

6 "Purity" is obviously a loaded term. In this context, it means that the seed was relatively homogeneous and would be expected to yield descendents with similar characteristics.

7 Burges, "Break This Vicious Circle," 5, 6, 29.

woodland habitat of feral bees – conditions common across much of the South in 1900 – cross-pollination rates could exceed 40 percent. But when it was grown in large one-variety fields that were frequently sprayed with insecticides, as was common in the Mississippi Delta by the 1950s, the cross-pollination rates were only a few percent.[8] The median rate of natural crossing between alternate rows was typically about 8 percent.[9]

Maintaining pure seed lines became an increasingly serious problem after the Civil War with the emergence of public gins and changes in ginning technology. According to the USDA roughly 90 to 95 percent of the seed used to plant the U.S. cotton crop in the 1920s and early 1930s was of mixed gin-run quality – seed returned to farmers by the gin.[10] Even when farmers purchased seed rather than using their home-grown product, the "outside" seed was often simply gin-run seed from other areas. Post–Civil War structural changes and the advent of Robert Munger's integrated gins (see Chapter 4) increased the market areas of gins and made it harder for farmers to protect their seed.

Seed mixing in successive gin runs was a serious problem that separated cotton from corn and other crops that were prone to cross-pollinate in the field.[11] The USDA estimated that "seed from a farmer's first bale at the gin contains 26 percent of the seed from the preceding bale."[12] Investigators reported cases of farmers receiving seed from the previous four farmers who used the gin. Given the practices of the day, gin operators were apt to indiscriminately return seed to farmers even if the growers requested return of their own seed. Thus, "the farmer as an individual finds himself practically powerless when he attempts to establish and maintain a pure stock of cotton."[13] This contributed to the rapid turnover in varieties and the resulting dissipation of many of the benefits to research.

[8] Brown, *Cotton* (1927), 165–6. Early efforts to grow Durango and Egyptian cottons in California failed because of cross-pollination with shorter-staple cottons. Cook, "One-Variety Cotton," 10–11; McGregor, "Insect Pollination of Cultivated Crops," 171–90D; Simpson, "Natural Cross-Pollination in Cotton."

[9] Brown and Ware, *Cotton*, 158.

[10] Doyle, "Multiplicity of Varieties," 108, and "Cotton Growing in One-Variety Communities," 264; U.S. Bureau of Markets, "Marketing Cotton Seed For Planting," 65.

[11] U.S. Bureau of Plant Industry, *Better Cottons*, 955–8; Doyle, Meloy, and Stine, "Cotton Situation," 400. Most farmers brought only one or two bales to the gin at a time, which magnified the problem of intermingling of seed. Cook, "One-Variety Cotton," 13.

[12] U.S. House Committee on Agriculture, *Study*, 957.

[13] Doyle, "Cotton Growing in One-Variety Communities," 264. Virts, "Change in the Plantation System," 171, notes that larger plantations "were more likely to own their own gins, preventing both cotton fiber and seed from mixing with lower quality cotton from other farms."

There are numerous accounts of promising varieties being destroyed by cross-pollination and mixing at the gin. As a prominent example, in 1912 Roland M. Meade first selected a prized variety in fields around Clarkville, Texas. This variety (Meade) yielded lint longer than 1 1/2 inches and black seeds practically devoid of fuzz. The seed was taken to the Sea Island areas of South Carolina, increased, and sold in that region. There it produced a staple length averaging 1 5/8 inches and showed exceptional uniformity. "Meade was on the way to becoming a striking success. More than 10,000 acres were grown between 1920 and 1922, but mixing of seed and planting in close proximity to fuzzy-seeded upland varieties resulted in a rapid contamination in the stocks, the mixed fiber was rejected by the trade, and the variety was largely abandoned after 1925."[14] Cross-pollination and seed mixing during ginning reduced the ability to cultivate high-yielding, high-quality varieties.[15] Purchasing commercial seed was an expensive proposition. Data from the early 1920s indicate that the commercial product cost 2.5 to 4 times as much as gin-run seed and that improved seed sold by breeders cost 6 to 8 times as much.[16]

Coupled with these production externalities were pervasive marketing imperfections. These too were, for the most part, post–Civil War phenomena. According to contemporary reports, local markets in the South failed to provide sufficient rewards for producing higher quality cotton to warrant the added expense.[17] Complaints about middlemen are common to all agricultural commodities, but in this case the criticisms went beyond habitual grousing. The grading and marketing system in place around 1900 was one of the most complicated and controversial aspects of the whole cotton production process.[18] Accurately grading individual

[14] Ware, "Plant Breeding," 690–1.

[15] Several forces affected seed quality. For example, if farmers chose their seed stock from a random sample from the gin, it would have the negative effect of selecting strains with a high seed-to-lint ratio.

[16] The ordinary ratios are based on data for the period 1920–22 from USDA, "Cotton Seed for Planting Purposes," 49, 59, and U.S. Crop Reporting Service, *Prices Paid by Farmers*, 143. The improved seed ratios are based on prices found in Coker's catalogs. For examples see the price lists in *Coker's Pedigreed Seed Company Catalog* (Spring 1918), insert, and (Spring 1927), 14–29. In 1927 new releases cost farmers about $3.00 per 30-pound bushel plus shipping charges. In addition to the aforementioned reasons, southern poverty, tenancy, low investments in education and extension activities, and high interest rates may have discouraged investments in better seed.

[17] Cook, "One-Variety Cotton," 12, 36; Cook and Martin, "Community Cotton Production," 4; Crawford, "Point Buying of Cotton," 376–86.

[18] Virts, "Efficiency of Southern Tenant Plantations," 390–1.

bales of cotton in local markets was prohibitively expensive, given the technology of the day. The use of mixed or gin-run seed added complications because there would be "considerable variation in quality and length of lint" within a single bale.[19] As a result, the use of pooling contracts was widespread. Cotton was generally sold in small local markets on the hog-round or "on point" system, meaning buyers graded a small sample of bales and then paid one average price for all the cotton in that market.[20] The cotton would then be shipped to regional markets, where highly trained specialists would grade a sample from each bale in special rooms with proper lighting, temperature, and humidity. After grading, the cotton would be assembled in larger running lots of roughly similar-quality bales for sale to the cotton mills. There was a regional division of labor among mills; some demanded better grades of cotton and produced higher quality output than others.[21] Once a given mill had adjusted its machinery it required a uniform staple length to run efficiently – a difference of 1/32 of an inch could be significant.

The hog-round system encouraged farmers to market lower quality cotton. This helps explain the rising importance of relatively high yielding but short staple varieties such as "Half-and-Half." One might expect pooling contracts to break down as individuals who produced higher quality goods demanded a higher price. As predicted, some plantation owners did sell directly in central markets where their crop could be graded separately.[22] But most tenants and small-farm operators lacked the economies of scale, information, and perhaps savvy to mitigate the

[19] Darst, "Cotton-Seed Production," 190–1.

[20] Cook and Doyle, "One-Variety Community Plan," 132. Elsewhere Cook noted that "'the practice [pooling cotton in local markets] is old and longstanding, so that nobody now alive can be blamed for starting it.'" As quoted in Coruthers, "One-Variety Cotton Communities," 13. The nomenclature used to describe the cotton market varies among authors. Garside, *Cotton Goes to Market*, 179.

[21] Wright, *Old South, New South*, 133–5.

[22] Virts, "Efficiency of Southern Tenant Plantations," 390–3. Nancy Virts argued that these general marketing problems help explain the persistence of the plantation system. Ibid., 387–8. In addition to marketing efficiencies, plantations also provided a form of vertical integration and scale economies that protected seed quality by reducing cross-pollination and seed mixing problems. For example, Delta and Pine Land Company (D&PL) generally grew only one commercial variety at a time. When the company grew other varieties as part of its breeding program, it was meticulous in separating them in the field and at the gin. For this reason, plantations represented a market solution to the externality problems, which begs the question of why land in plantations was declining. Edmonds, "Around the Clock," 40–3, 80. Burges observed that only shipments of one hundred bales or more could command special treatment with respect to quality. Burges, "Break This Vicious Circle," 5, 6, 29.

problem. Such farmers suffered from problems of unequal bargaining power and asymmetric information because they likely faced only one or two buyers in local markets. As George Akerlof's 1970 "lemons model" would suggest, high-quality cotton varieties were being driven from the market.[23]

Testing for Market Failure

Although the "lemons problem" is much discussed, important real-world examples are rare in the literature. Thus, the claim that grading and marketing problems seriously distorted cotton production incentives should be of considerable interest. Numerous researchers investigated the relationship between price and quality in local and regional markets across the South. Studies of local markets in Arkansas (1913–16), North Carolina (1914–16), Texas (1926), Alabama (1926 and 1927), and South Carolina (1925–27) all found that local prices varied little with quality.[24] The two most definitive cotton pricing reports were published by the USDA in 1936 and 1939. L. D. Howell and John S. Burgess monitored individual transactions in more than one hundred local markets between 1928 and 1933, and they independently classed 300,000 bales. They compared the prices received in local markets for cotton of a given quality with those prevailing in central markets for the same quality on the same day. (Neither buyers nor sellers in the local markets knew the results of Howell and Burgess's classifications at the time of sale.) The survey found that the local market price differentials for various staple lengths were far smaller than the differentials in the central markets (which better reflected the value cotton spinners placed on quality). Panel 1 of Table 6.2 offers a summary measure of the price differentials by staple length prevailing in the local and central markets over the period 1928–33. "For the 5-year period, on an average, premiums for staples longer than 7/8 inch in local markets amounted to only 17 percent of those in central markets and varied from only 12 percent for 15/16-inch cotton to 34 percent for 1 1/8-inch cotton. At the other end of the spectrum, discounts for cotton shorter than 7/8 inch in local markets amounted to only 6 percent of those quoted in central markets for cotton with a staple length of 13/16 inch."[25] The price signals given to farmers in local markets systematically failed to reflect the incentive structure being generated in central markets.

[23] Akerlof, "Market for Lemons," 488–500.

[24] As reported in Howell and Burgess, "Farm Prices of Cotton," 2–3.

[25] Ibid., 21.

TABLE 6.2. *Average Price Differential Between 34 and 28 Staple Cottons, 1928–36 (in cents per pound)*

	Season	(1) Local Markets	(2) Central Markets	(3) Ratio (1)/(2)
Panel 1	1928	0.45	1.65	0.27
	1929	0.37	1.98	0.19
	1930	0.18	1.55	0.12
	1931	0.23	1.03	0.22
	1932	0.12	0.80	0.15
Panel 2				
Market without PCS	1933	0.16	1.00	0.16
	1934	0.21	1.17	0.18
	1935	0.21	0.97	0.22
	1936	0.58	1.53	0.38
Market with PCS	1933	–	–	–
	1934	0.38	1.33	0.29
	1935	0.54	0.95	0.57
	1936	0.84	1.52	0.55

Notes: PCS = public classification system; 34 staple cotton refers to cotton classed as 34 thirty-seconds of an inch in length; similarly 28 staple cotton refers to cotton classed as 28 thirty-seconds of an inch.

Sources: Howell and Burgess, "Farm Prices of Cotton"; Howell and Watson, "Cotton Prices in Relation to Cotton Classification."

Farmers who sold short-staple cotton were overpaid, whereas those who marketed longer staples were shortchanged.[26]

Howell and Leonard J. Watson's study covering the period 1933–36 took the research a step further.[27] In addition to comparing local and central markets, they compared local markets that offered impartial public classification services (PCS) with those lacking such services. Panel 2 of Table 6.2 shows the differentials by staple length for bales sold in the two types of local markets compared with equal-quality cotton sold on the same day in the central markets. (Note that the timing of sales in the local markets with PCS and those without PCS differed, making it necessary to report two series for central market differentials.) Howell and Watson found that the quality differentials in the central markets were more closely reflected in local markets with PCS than in those without

[26] For the assertion that local cotton graders systematically cheated small farmers, see the testimony of Representative Hampton P. Fulmer in U.S. House Committee on Agriculture, *Hearings* (1928), 1–5.

[27] Howell and Watson, "Cotton Prices in Relation to Cotton Classification," 1–54.

such services.[28] As an example, the local markets with public classers captured 56 percent of the central market premiums for cotton of 1 1/16 inch (relative to 7/8 inch), whereas markets without public classers captured only 30 percent. Compared with the period 1928–32, farmers who sold in both types of markets were increasingly receiving greater quality differentials, especially at the higher end.[29]

Other evidence testifies to widespread market failures in cotton pricing. There were several cases in which farmers trucked inferior cotton long distances to markets that were known for selling high-quality cotton to capture the local premiums. Such dumping eroded the premiums for all growers. This behavior, coupled with the exceptionally detailed and careful studies on local and central market pricing and grading practices, strongly supports the assertions of contemporary cotton specialists. The hog-round system was indeed widespread, and it resulted in a "lemons problem" that contributed to a decline in cotton quality.[30]

The One-Variety Community Movement

Early in the twentieth century, USDA scientists intensified their breeding and extension projects aimed at improving the yields and quality of U.S. cotton. Initially, these efforts paralleled similar campaigns for other crops.[31] Researchers soon realized that it would not be sufficient to develop and distribute small quantities of better seeds; rather their campaign would have to change the complex institutional structure to reduce negative externalities and better align local prices with those in regional markets. According to the father of the one-variety community movement, O. F. Cook, "the method of [seed] distribution that was first projected did not result in establishing commercial supplies of pure seed. Several of the varieties that were developed and distributed in the early years of the cotton-breeding work were lost completely before the system of distribution was changed."[32] To counter these problems, Cook developed

[28] It was also found that price variability, conditional on quality, was lower in markets with PCS.

[29] Garside, *Cotton Goes to Market*, 176–84.

[30] For a discussion of these natural experiments see Olmstead and Rhode, "Hog–Round Marketing."

[31] For an example of these efforts in the wheat industry see Chapter 2.

[32] Cook, "One-Variety Cotton," 8. Among the early advocates of the one-variety concept was the *Wall Street Journal*; see, for example, "Southern Farm Needs Analyzed," (21 December 1926), 6.

an ambitious program to develop better cotton varieties, improve cultural methods, standardize cotton classification systems, advance new seed treatment processes, and train qualified cotton graders.[33] Biological innovation required institutional innovation.

At the heart of the program was a utopian scheme to fundamentally change the way cotton was grown, ginned, graded, and marketed in the United States. To succeed would require a "new association of ideas" to alter how farmers thought about their community.[34] Instead of each individual farmer choosing his own variety, the new system would be built on a cooperative structure in which cotton farmers would organize one-variety communities. The USDA, in conjunction with state authorities, would provide education, guidance, and standardized contracts. Cook and his fellow reformers envisioned communities ranging in size from a group of farmers using one gin to those encompassing an entire state. In addition to producing, ginning, and marketing only one variety of cotton, the communities would be responsible for increasing (and in some cases breeding) pure seed for their members.

Cook first suggested the idea of one-variety communities in 1909 and subsequently developed the concept in a 1911 article.[35] At first the USDA concentrated its one-variety campaign in the newly irrigated cotton regions in the far West, promoting the idea in conjunction with the distribution of a number of recently developed long- and medium-staple varieties. At times local USDA scientists withheld the distribution of the new seed until a one-variety structure was in place. The USDA launched the first one-variety community in 1912 with the distribution of Yuma cotton in the Salt River Valley of Arizona.[36] At about the same time, a single-variety community growing Durango was organized in the Imperial Valley of California. After 1920 Acala, which the USDA had

[33] By the mid-1930s, the USDA had initiated genetic and breeding research programs in every important cotton-producing state. Ware, "Plant Breeding," 665.

[34] Cook, "One-Variety Cotton," 33.

[35] Cook, "Cotton Improvement," 397–410, and "Local Adjustment," 41. The one-variety movement literature emphasized the benefits but paid little attention to the costs associated with individual farmers who lost their freedom to tailor cultural practices to fit their particular growing situations. Even the few more balanced accounts typically argued that under the prevailing adverse conditions each farmer stood to gain from adopting the community's improved seed. Willis, "One-Variety Cotton Community Organization," 3–4.

[36] The North Carolina Agricultural Experiment Station began work on community production around 1915, but it is not clear if any communities were actually formed at that time. Coruthers, "One-Variety Cotton," 75–9.

introduced from Mexico in 1907, became a popular one-variety cotton in many western areas.[37] The initial efforts were often loosely structured. As an example, to gain access to Acala seed, growers in Riverside County organized the Acala Cotton Growers' Association of the Coachella Valley in 1920. By 1923 the region's farmers had voluntarily planted Acala on over 96 percent of their cotton acreage. It was only after the fact, in order to prevent the mixing of seeds at gins and cross-pollination, that Riverside County gave legal protection to the district by passing an ordinance in 1924 that declared the county a pure seed district.[38] These western initiatives, as well as scattered efforts in southern states, generally met with mixed results as farmers and USDA officials experimented with varieties and structures. Most one-variety districts reported higher yields and increased premiums, but for a number of reasons (including problems of free riding from nearby farmers and inadequate supplies of the one-variety seed to serve a given area) many of the early districts were short-lived.[39]

A giant step in the one-variety campaign occurred in May 1925, when California enacted legislation that organized eight San Joaquin Valley counties and Riverside County into a one-variety community. The new law represented a triumph for W. B. Camp, the USDA's California cotton specialist who had been sent west in 1917 to promote one-variety production of high-quality varieties. The law, along with the institutional structure that evolved in 1926, would define the initial development of the state's cotton industry. Thereafter, only Acala could be planted, harvested, or ginned in a district of well over four million acres. Even the possession of non-Acala seeds was illegal (except at a few research stations). The USDA's Cotton Research Center at Shafter became the de facto sole Acala breeder in the state as the USDA successfully strove to keep private seed breeders out of the Central Valley. Under this system, Shafter's "head breeder" held enormous power, overseeing a research program that for the next sixty years would be the only source of cottonseed for most of California. To increase and market the seed bred at

[37] Ware, "Plant Breeding," 697, 689. After Acala was introduced it took researchers several years to select and develop outstanding strains suitable for commercial use. Durango was another recent Mexican introduction and Yuma and Pima were the product of USDA breeding programs in the Southwest and depended largely on crosses of Egyptian cultivars with Sea Island cotton.

[38] McKeever, "Community Production of Acala Cotton," 29.

[39] For a discussion of inadequate supplies of planting seed, see Coruthers, "One-Variety Cotton," 112.

Shafter, growers organized the California Planting Cotton Seed Distributors in 1926. Most specialized accounts credit the one-variety system with contributing significantly to California's high cotton yields which, over much of the twentieth century, were roughly double the national average. (Actually many factors, such as climate and irrigation, contributed to the state's yield advantage.) These accounts also credit the one-variety community with helping California growers earn quality premiums for their relatively uniform, medium-staple product.[40]

John Constantine, Julian Alston, and Vincent Smith argue that in the 1970s and 1980s the legislation artificially limited California's production, resulting in higher prices for the state's cotton.[41] Landowners in regions most suitable for Acala production (the community variety) benefited, while other Central Valley farmers experienced yield losses or abandoned cotton. Constantine, Alston, and Smith concluded that the legislation became increasingly inefficient and by the late 1970s was costing growers collectively more than 10 percent of the annual value of the state's cotton output. The law remained in force because a faction of California farmers who benefited from the legislation had captured the system's administrative apparatus.[42] Constantine, Alston, and Smith provide a valuable perspective on the recent history of the California one-variety law, but they say little about the early history of the state's experience and ignore a far larger, but more short-lived, southern one-variety movement. Understanding the movement outside of California not only provides a fresh perspective on the sources of southern development but also helps in reevaluating the California experience.[43]

[40] For example, see Turner, *White Gold Comes to California*, 55–94.

[41] Constantine, Alston, and Smith, "Economic Impacts," 951–74.

[42] Ibid. Constantine, Alston, and Smith did not address the question of whether or not the law was efficient in its early decades. Oklahoma farmers were concerned about yield losses due to the inability to fine-tune varieties to local conditions. Campbell, "One-Variety Cotton in Oklahoma," 5–19, and "Comparisons of One-Variety," 7–33. Other critics of one-variety communities feared there could be catastrophic losses should a new disease appear for which the limited number of varieties lacked resistance. Karl S. Quisenberry, "The Role of Public and Private Agencies in Cotton Improvement," 1–8, Dallas TX, 2 February 1954, D&PL Company Records, Box 15, Miscellaneous, Joint Cotton Breeders Policy Committee file (2/8), 1953–54. Manuscript Collections, Mississippi State University Libraries (MSUL).

[43] An Acala one-variety movement started in the Rio Grande and Pecos rivers areas of New Mexico in 1922. There was no statewide one-variety legislation, but by the early 1930s Acala constituted more than 95 percent of the cotton grown in the state. Between 1922 and 1932, the state's yields increased from 201 to 412 pounds of lint per acre (roughly on par with what occurred in California), with local observers giving most of

There were fits and starts in the one-variety movement in the traditional Cotton South before 1930, but few lasting accomplishments. However, the USDA intensified its efforts, initiating one-variety campaigns throughout the South in the period 1931–32. Almost all contemporary studies of early one-variety programs reported immediate increases in yields and quality, along with greater financial returns to farmers.[44] The movement started in Georgia in 1931. By the end of 1934, there were forty-five communities in various stages of development, with twenty-five others in the planning stage. The one-variety producers were immediately rewarded with higher yields, along with quality and length premiums valued at about $7.13 per acre. In Oklahoma, the one-variety movement began in earnest in 1932. By early 1933 there were six communities with over 25,000 acres and 11,000 participating farmers.[45] In Mississippi, fourteen communities were organized in 1931, with the number growing to thirty-three in 1932; six of these were countywide organizations. By 1937 there were 197 communities in the state, with members receiving an estimated average increase in revenue (stemming from increased yields and premiums) of $8.71 per acre.[46] A similar transformation of cotton production was taking place across the South during the 1930s.

From its humble beginnings in the 1930s, the one-variety movement took off. Table 6.3 pieces together key indices of the extent of the movement for the years 1934–49. By 1946 there were about 2,275 one-variety communities, which produced roughly one-half of the entire cotton output of the United States. Table 6.4 shows that California was only a small part of a much larger one-variety movement. In fact, in 1946 California accounted for less than 2 percent of the community members, less than

the credit to the community production system's pure seed program. Coruthers, "One-Variety Cotton," 106–8; Leding, "Community Production of Acala Cotton"; and U.S. Bureau of Agricultural Economics, "Statistics on Cotton," 82.

44 U.S. House Committee on Agriculture, *Hearings* (1928), 135. The two exceptions appear to have occurred in Florida and Oklahoma. Florida farmers evidently made a poor choice of varieties. Studies in Oklahoma reported significant quality increases with an added revenue of about $2.31 per bale for the 1932–33 crop. But in the period 1933–34, the estimated added benefit to one-variety production fell to a mere $0.33 per bale. Ballinger and McWhorter, "Results Achieved by One-Variety Cotton," 68–71. Also see Porter, "Toward Standardized Cotton Production," 21–2.

45 Coruthers, "One–Variety Cotton," 67–72, 80–8. Coruthers gives the membership and acreage for only five of the six communities. See also Westbrook, "One-Variety Community Cotton Production," 3–8; and Bledsoe and Westbrook, "History and Progress," 16–19, who provide the $7.13 estimate.

46 Willis, "One-Variety Cotton Community Organization," 1–2; Coruthers, "One-Variety Cotton," 60–3.

TABLE 6.3. *Cotton in One-Variety Communities, 1934–49*

Year	Counties Participating		Communities Participating	Grower Members (thousand)	Production of Adopted Varieties			
					Acres		Bales	
	Number	Percent			Number (thousand)	Percent	Number (thousand)	Percent
1934					589	2	nd	nd
1935	161	19	331	nd	788	3	571	5
1936	234	28	511	nd	1,470	5	1,112	9
1937	312	38	730	nd	2,453	7	1,883	10
1938	425	33	1,056	nd	2,284	9	1,445	12
1939	495	62	1,516	132	2,987	12	1,656	14
1940	548	70	1,922	185	4,518	18	2,742	22
1941	550	71	2,116	229	6,239	27	3,367	32
1942	577	75	2,564	292	7,614	33	4,570	37
1943	549	77	2,544	306	8,869	40	4,771	43
1944	581	80	2,194	299	7,226	36	4,762	39
1945	500	72	1,800	319	7,071	40	4,172	45
1946	485	70	1,601	310	6,808	39	4,350	50
1947	nd	nd	1,963	331	8,537	40	5,659	48
1948	531	77	2,275	353	11,549	50	9,511	64
1949	546	79	2,422	426	13,500	49	9,500	59

Sources: Compiled from U.S. House Committee on Agriculture, *Study*, and *Research*, 753; USDA, *Report of the Administrator of the Production and Marketing Administration 1948*, 319; Brown and Ware, *Cotton*, 58; U.S. Federal Extension Service, *Report of Cooperative Extension Work*, 21; and U.S. Agricultural Marketing Service, *Report of the Chief 1941*.

TABLE 6.4. *Cotton Growing in Standardized One-Variety Communities: State Totals for 1946*

	Counties Participating				Production of Adopted Varieties			
					Acres		Bales	
Year	Number	Percent	Communities Participating	Grower Members (thousand)	Number (thousand)	Percent	Number (thousand)	Percent
Alabama	53	83	263	39,225	610	40	386	48
Arizona	6	100	9	954	151	100	143	100
Arkansas	30	57	229	10,788	580	36	546	44
California	7	100	7	5,509	339	100	435	100
Georgia	84	81	241	38,417	574	47	315	57
Kentucky	2	67	2	496	10	97	8	97
Louisiana	26	70	29	22,238	404	51	133	53
Mississipi	61	81	185	49,605	867	38	410	37
Missouri	8	100	126	7,654	272	89	272	89
New Mexico	5	100	5	4,117	130	98	145	98
North Carolina	33	69	38	27,800	332	58	259	62
Oklahoma	29	59	51	13,871	271	27	78	30
South Carolina	26	62	30	26,554	421	46	329	47
Tennessee	22	79	29	26,898	345	57	291	57
Texas	91	57	354	35,955	1,499	25	508	36
Virginia	2	33	3	265	3	13	1	9

Source: U.S. House Committee on Agriculture, *Study*, 962.

5 percent of the acreage in one-variety communities, and about 10 percent of community output in the United States. The Golden State was different because of the size of the participating farms and the legal rigidity of the system, not because its farmers were banding together to overcome negative externalities and to capture economies of scale in grading, information, and marketing.[47]

A clearer image of the microstructure and daily operations of the southern one-variety communities may be distilled from numerous descriptions in experiment station and cotton trade publications. A small group of local farmers working with a county extension agent typically initiated the organizing effort. The agent would provide a set of standardized bylaws for a "Cotton Improvement Association" to be established as a nonprofit, unincorporated cooperative association. Membership was voluntary and involved no fees or dues. Under some bylaws, membership was "open to any cotton grower" who agreed to the one-variety regulations; in others, new members were admitted with the approval of existing members. Conditions for exit also varied. In some agreements members could withdraw at any time, and those who failed to comply with community rules were automatically dropped without penalty. Other bylaws specified a five-year membership term. In almost all bylaws the association's membership periodically selected by majority rule (on a one-member, one-vote basis) the variety to be grown and elected a small board responsible for the daily operations. The association also formed a relationship with a local gin. If only a fraction of the local growers chose to enter the association, the one-variety community contracted with the ginner to set aside specific days or specific machines to handle members' crops with special care. Thus, the southern one-variety communities were neither as compulsory nor as formal as in the California model.[48]

The communities acquired foundation seed from private breeders such as Delta and Pine Land Company (D&PL), Stoneville, or Coker or a state experiment station. A common arrangement was to purchase annually one bushel of foundation seed (enough to plant one acre) for every hundred acres of cotton in the community. A small number of selected growers planted this seed in isolated fields, harvested and ginned the

[47] According to E. C. Westbrook, "One-Variety Cotton Communities," 17, in 1956 all one-variety communities were voluntary except those in California.

[48] Willis, "One-Variety Cotton Community Organization"; Westbrook, "One-Variety Community Cotton Production"; Leding, "Community Production of Alcala Cotton"; Wasson, "One Variety Cotton Improvement Association"; Lowery, "Cotton Improvement"; Rains, "Cotton Improvement."

resulting seed cotton in a manner to ensure purity, and then distributed the so-called first-year seed to other members at set prices that were well below the market price of the foundation seed. The other members agreed to plant at least one-tenth of their acreage with this first-year seed, producing sufficient second-year seed for their remaining acreage in the next season.[49] To help maintain purity, all of the cotton grown from the foundation seed was to be ginned under close supervision before any of that grown from the first-year seed, which in turn was ginned before that grown from the second-year seed. The resulting third-year seed was to be sold to the oil mill. This plan "provides for a continuous flow of new, pure breeder or foundation seed into the community each year and a continuous outflow of old seed to the oil mill."[50] These policies essentially reestablished the types of careful practices enlightened plantation owners had followed before the Civil War. Figure 6.2 shows a certified seed depot, and Figure 6.3 shows African American farmers queued to obtain the seed.

Smith–Doxey Cotton Classing

During the mid-1930s, the one-variety advocates turned their attention to reforming cotton marketing and grading. In 1938 the Smith–Doxey Cotton Classing Act, which was meant to complement the traditional one-variety communities, went into effect.[51] The law made free-market news service and cotton classing available to members of all organized cotton improvement groups. Figure 6.4 shows cotton classers at work in Memphis, Tennessee. The act was to be largely self-supporting (through the sale of the sample material) and would benefit almost everyone up and down the marketing chain except perhaps local buyers and graders. Smith–Doxey classification cards became accepted within the trade, cutting marketing costs by reducing the need to repeatedly resample and regrade bales every time they changed hands. The primary aim was to

[49] This method of expanding the seed supply represented a large-scale collective implementation of the 1–10–100 technique recommended by extension agents and seed companies to individual farmers for maintaining pure seed. *Coker's Catalog* (Spring 1917), 16. For a detailed guide on how to organize a one-variety community see Bode Hughes, "Organizing Communities," D&PL Company Records, Box 15, Miscellaneous (2/8), Manuscript Collections, MSUL.

[50] U.S. House Committee on Agriculture, *Study*, 960.

[51] U.S. Senate Committee on Agriculture and Forestry, *Authorizing*, 1–3; U.S. House Committee on Agriculture, *Letter*, 3164.

FIGURE 6.2. Weighing certified cottonseed in North Carolina in 1938. A group of African American farmers had collectively purchased this seed. *Source:* Library of Congress, Prints & Photographs Division, FSA/OWI Collection [LC–USF34–008314–C]. John Vachon, photographer.

narrow the discrepancies between grading and pricing in local and central markets. Under the Smith–Doxey program, farmers could mail cotton samples to one of thirty-one central locations established throughout the cotton belt and within a few days receive by return mail a government certified "green card" specifying the cotton's grade, length, and so forth. However, there was a catch – to qualify for the free services a farmer had to be a member of an organized cotton improvement group with at least ten members. These Smith–Doxey districts were typically much less formal than one-variety communities, and in some cases they simply represented an agreement between a group of farmers and a ginner to provide special care in handling the group's single variety. The Smith–Doxey districts played no role in breeding, increasing, or marketing seed. A bureaucratic difference was that Smith–Doxey groups were organized out of the USDA Agricultural Marketing Service, whereas the one-variety community project was under the aegis of the Bureau of Plant Industry, Soils, and Agricultural Engineering.

FIGURE 6.3. African American farmers waiting to receive their certified seed. There were 139 wagons in the queue. *Source:* Library of Congress, Prints & Photographs Division, FSA/OWI Collection [LC–USF34–008315–C]. John Vachon, photographer.

On the ground, the cooperative extension service administered both programs.[52]

A pair of USDA beltwide surveys of classification practices and the use of information, taken in the 1935–36 and 1947–48 crop years, provide a sense of the changes wrought by the Smith–Doxey Act. The first survey, conducted by John W. Wright, revealed just how uninformed most growers were when they sold their cotton.[53] This survey of 101 local markets found that 36 percent of growers sold their cotton with no information about general market prices except their price offer, and 60 percent of growers (accounting for 60 percent of the crop) sold their cotton without knowing its grade or staple length. Even when growers reported knowledge of their cotton's quality at the time of sale, the most common source

52 Betts, "Green Card Pays Off," 13–16; U.S. Agricultural Marketing Service, *Report of the Chief 1941*, 36–9.

53 Wright, *Marketing Practices*, 20–3, 60–2.

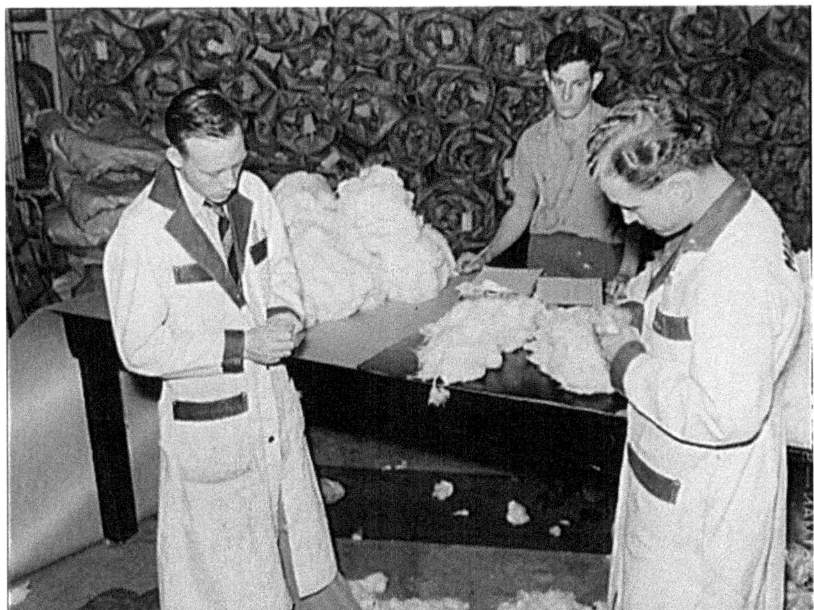

FIGURE 6.4. Classing cotton at the Mid-South Cotton Growers Association, Memphis, Tennessee, in 1939. *Source:* Library of Congress, Prints & Photographs Division, FSA/OWI Collection [LC–USF34–052552–D]. Marion Post Wolcott, photographer.

of this information was the buyer. Less than 10 percent of cotton was classified by impartial parties such as the USDA, licensed classers, warehousemen, factors, or ginners.[54] There was widespread dissatisfaction with the poor state of market information. Nearly 60 percent of cotton growers reported a willingness to maintain a self-supporting sampling service to provide official classification. Most ginners surveyed (84 percent) also favored such a service. By way of contrast, most first-buyers disliked the idea and reported a disinclination to base their purchases on official classes.[55]

[54] Ibid., 20–3. In addition to surveying growers, the study also queried first-buyers regarding their practices. These buyers reported that relatively little cotton – only 11 percent – was purchased without any effort to classify the cotton in individual bales. The author expressed skepticism, however, about the thoroughness of the first-buyers' own classification efforts. Only one-third of first-buyers owned or had access to a copy of the official cotton standards (29–30).

[55] Wright, *Marketing Practices*, 60–2.

The 1947 follow-up study revealed substantial improvements in grower awareness.[56] In the 1947–48 crop year only about 25 percent of growers, who accounted for about 15 percent of U.S. cotton output, sold without having independent information about general cotton prices. Only 45 percent of growers (with 30 percent of the cotton) sold their crop without knowing its quality (down from the 60 percent of producers and output in the 1935–36 season). In the period 1947–48, 52 percent of the cotton crop was sold by growers who had obtained quality information from impartial sources compared to just 9 percent of the crop in 1935–36. The spread of the Smith–Doxey system accounted for much of the change. In the period 1947–48, 40 percent of the crop received green cards (Form 1 classifications) by the time of sale.[57] The study concluded that growers in 1947 "generally occupied a stronger bargaining position than in 1935" when "most growers reported knowing neither the market price nor the quality of their cotton at the time of sale."[58]

The Smith–Doxey program apparently went a long way toward correcting the problems highlighted in the cotton pricing studies of the 1920s and 1930s. For example, based on surveys conducted in twenty-four local markets across the cotton belt during the 1951–52 to 1953–54 seasons, William Faught concluded that "prices to growers in markets where cotton is sold on the basis of Smith–Doxey cards . . . reflected central market differentials for grade and staple rather fully and accurately. In markets where growers did not have or did not use reliable quality information in the sale of their cotton, local prices reflected little, if any, of the central market differentials."[59]

In local markets where most "cotton was sold on the basis of Smith–Doxey cards," the price reflected, on average, 78 percent of central market differentials, whereas in local markets where "cotton quality information was not readily available to growers" the price reflected only 3 percent

[56] Soxman, *Marketing of Cotton*. This study covered 98 of the 101 local markets analyzed in the 1935–36 study.

[57] Wright, *Marketing Practices*, 12, 69. The use of the Smith–Doxey system was unevenly distributed across the cotton belt and over farms of different sizes. Virtually all cotton growers in Arizona, California, and New Mexico received Form 1 classing, as did about one-half of the growers in Arkansas, Oklahoma, and Texas and one-quarter of the growers in Alabama, Georgia, and Mississippi (16). Larger growers were far more likely than smaller growers to sell on the basis of Form 1 classification (12, 16, 62, 69). Many buyers expressed an unwillingness to use the government classification system. Faught, "Cotton Price Relationships," 28.

[58] Soxman, *Marketing*, 2, 69.

[59] Faught, "Cotton Price Relationships," 3.

of differentials.[60] Officials at the Texas Agricultural Extension Service concluded that "Smith–Doxey has probably done more than anything else to breach the traditional system of hog round buying."[61]

As a result of the advantages of the Smith–Doxey program, participation rapidly increased. Table 6.5 provides summary data of the growth of the program between 1938 and 1952. By 1951 about 65 percent of American cotton was being graded under this system. The diffusion of this new classing system proceeded at a pace rivaling that of the era's better known mechanical and biological technologies. In the 1954 season just under three-quarters of the U.S. crop was classified under the program. And from the mid-1960s to the present, Smith–Doxey cotton has represented roughly 95 percent of the crop. There appears to be little doubt that the Smith–Doxey program had a significant impact on narrowing the gap between local and central markets in cotton classing.[62]

The one-variety community movement and the Smith–Doxey program ran counter to, and at the same time reinforced, the larger Agricultural Adjustment Administration (AAA) acreage reduction and price support programs which became fixtures of the cotton economy in 1933. The main push of the early AAA programs in 1933 was to plow up cotton land and later to restrict acreage to deal with the problem of "overproduction." The one-variety community movement helped increase yields and total output. On the other hand, the movement also encouraged higher quality production, increasing the competitiveness of American cotton in world markets and raising farm incomes. Paradoxically, the AAA's early

[60] Ibid., 14–15, 26–7. As is often the case in the American federal system, experiments initiated by individual states subsequently provided a model for national programs. This appears to have been the case with the Smith–Doxey classification system. In the early 1920s, at the urging of local Farm Bureaus, the California State Department of Agriculture began a PCS. In addition, farmers in some areas organized selling agencies to handle bulk sales. The combination of these programs resulted in a price premium of "1 1/2 to 3 cents a pound over that [cotton] of similar grade sold independently in the local yard on the same day." The fact that California Farm Bureau members, who in the main would have been relatively educated and informed producers, requested and benefited from such a program suggests that the benefits for most southern producers could have been substantial. Blair, "Grade and Staple," 628–31.

[61] Texas Agricultural Extension Service, *New Agriculture*, 37.

[62] Fortenberry, "Story of Cotton." From Brown and Ware's account it is likely that many one-variety communities were also Smith–Doxey groups but that Smith–Doxey groups generally were not enumerated as one-variety communities. In addition, Brown and Ware noted that after 1948 Smith–Doxey groups "have taken the place of one-variety communities in many areas, especially in large portions of the main Cotton South." It is likely that by 1946 well over 50 percent of U.S. cotton production came from one or the other of these forms of community organizations. Brown and Ware, *Cotton*, 83.

TABLE 6.5. *Farmer Participation in the Smith–Doxey Cotton Grading Program*

Crop Year	Cotton Improvement Groups: Number	Cotton Improvement Groups: Members	Samples Classed (1,000 bales)	Share of U.S. Cotton Production
1938	312	18,589	84	0.7
1939	918	64,399	265	2.3
1940	1,573	128,216	1,531	12.4
1941	2,511	278,782	2,520	24.0
1942	2,465	281,100	3,567	28.7
1943	2,459	281,493	3,337	30.1
1944	2,410	321,284	4,037	34.4
1945	2,444	343,000	2,888	33.0
1946	2,515	343,700	2,574	30.3
1947	2,453	346,500	4,300	37.3
1948	–	371,061	8,067	55.3
1949	–	497,064	10,456	65.1
1950	–	507,873	5,215	53.2
1951	–	495,391	9,844	65.3
1952	–	515,711	9,382	62.0
1953	–	–	12,700	77.0
1954	–	–	–	–
1955	–	–	–	81.0
1956	–	551,077	11,200	85.0
1957	–	–	–	–
1958	–	–	–	93.0
1959	–	–	–	95.0
1960	–	–	–	96.0
1961	–	699,632	13,703	96.0
1962	–	691,670	13,510	91.0
1963	–	678,749	14,016	92.0
1964	–	–	–	–
1965	–	–	14,311	96.0

Sources: Betts, "Green Card Pays Off," 13–16; U.S. Office of Marketing Services, *Report 1942/43*, 111, *1943/44*, 44–7, *1944/45*, 20–4; USDA, *Report of the Administrator 1946*, 36, *1947*, 33–4, *1948*, 39, *1950*, 11–14, *1951*, 13, *1952*, 19–20, *1953*, 11; USDA, *Report of the Secretary 1956*, 35, *1957*, 39, *1959*, 41, *1960*, 41, *1961*, 41, *1962*, 41, *1966*, 90; U.S. House Committee on Appropriations, *Agricultural Department Appropriation Bill 1954*, 1734, *1955*, 1031, *1963*, 1249, *1964*, 1344, *1965*, 84, *1967*, 548.

price support programs gave many cotton farmers an incentive to reduce quality at the same time that AAA officials were touting the benefits of higher quality production. This is because "up to 1938 cotton loans were made at a flat rate regardless of grade and staple length," thereby generating a form of Gresham's law with bad cotton driving out good.[63] Starting in 1938 loan differentials became based on the price differences of each staple length and grade in ten spot markets, thus giving farmers a stronger incentive to produce higher quality cotton. By making grading services widely available, the Smith–Doxey Classing Act helped facilitate this change in loan policy.

The Revolution in U.S. Cotton Production

There were many quantitative indices of the revolution in U.S. cotton production, including changes in cotton quality and in varietal concentration. Just as the one-variety advocates had planned, there was an almost immediate increase in the staple length in one-variety districts compared to that of nearby areas. The aggregate data on the staple length of U.S. cotton reflected these developments. Figure 6.5 shows that between the periods 1928–33 and 1945–49 the average length of U.S. upland cotton increased by about one-eighth of an inch, or four staple lengths. The percentage of upland cotton 29/32 of an inch and shorter fell from over 50 percent in 1928–32 to less than 14 percent in 1944–49. Between the periods 1928–30 and 1946–47 the percent equal to or greater than one inch increased from about 22 to 73 percent. Mississippi has traditionally been known for producing high-quality cotton. Writing in 1950, J. F. O'Kelly noted that "twenty years ago only 31 per cent of the cotton produced in Mississippi was 1 to 1–1/32 inches. Currently 92 per cent of the State's cotton is in this staple range."[64] At the other end of the scale, cotton with staple less than one inch fell from 45 percent in the period 1928–30 to about 2 percent of Mississippi's production in 1946–47. Similar progress occurred across the cotton belt. As an example, "the South Carolina crop went from an estimated 20 percent cotton stapling 15/16

[63] Shepherd, *Agricultural Price Control*, 64–72. Although Shepherd is correct for many pre-1938 years, it appears that the AAA loan programs in 1933 and 1934 offered two rates – one rate for cotton classed as low, middling, or better of 7/8-inch staple and another for cotton less than 7/8-inch staple. Still, even the two-rate system was a far cry from the myriad loan rates in effect in 1938 and after. Richards, *Cotton and the AAA*, 213–30.

[64] O'Kelly, "Cotton Varieties and Breeding," 36–7.

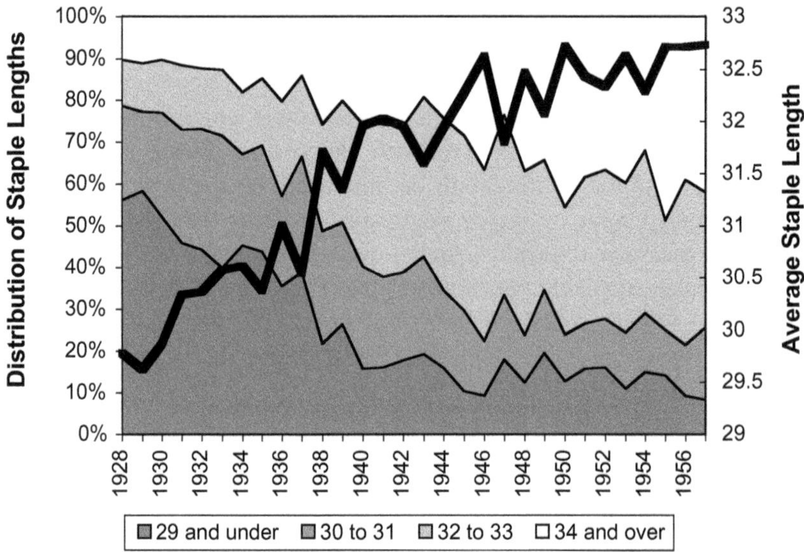

FIGURE 6.5. U.S. cotton staple-length distribution, 1928–57. *Source:* U.S. Economic Research Service, "Statistics on Cotton," 126.

inch and longer in length in 1926 to over 97 percent of such lengths in 1943. With this increase in length there has also been an increase in per acre yield."[65] A number of factors such as changing cultural practices surely contributed to the change in cotton quality and yields. But the rapidity of the change (with local yield increases and longer staples reported within a couple of years following the introduction of a community) and widespread contemporary testimony point to the one-variety movement as an important catalyst for the changes.[66]

In addition to promoting the production of longer staple cottons, the one-variety program contributed to a dramatic decline in the number

[65] Testimony of Dr. George J. Wilds, President, Coker's Pedigreed Seed Co., Hartsville, SC, in U.S. House Committee on Agriculture, *Cotton: Hearings*, 399.

[66] Ware, "Plant Breeding," 662; Brown and Ware, *Cotton*, 84. Our assessment of the premiums and discounts suggests that the economic significance to farmers of increased fiber length was probably much less than the impact of higher yields. According to O'Kelly, "Cotton Varieties and Breeding," 51, "varieties producing a medium staple length (1 to 1 3/32 inch) have been the favorites for at least two decades. In many of the Cotton Belt states 75 to 95 per cent of the cotton produced is in this length range. This fact has considerably reduced the discounts in the markets for lengths just shorter than one inch and has greatly increased the premiums paid for lengths 1 1/8 inches and longer."

of varieties of cotton grown in the United States. One-variety advocates saw this decline in biodiversity as a positive step. One of the USDA's initial goals was to significantly reduce the number of varieties in order to eliminate inferior cottons, to reduce the problems of cross-pollination and gin mixing, and to promote a standardized product for spinners. No one really knows how many varieties and strains of cotton were being grown in the American South in 1930. In 1907 Fredrick J. Tyler listed more than 600 varieties; given the tendency for the number to increase due to mutations and cross-pollination, it is likely that substantially more existed at the dawn of the one-variety movement. E. C. Westbrook claimed that there were about 300 varieties being grown in Georgia alone in 1930.[67] Many of these so-called varieties were undoubtedly just different local names for the same variety, but the exact number is not as important as the general magnitude in relation to what existed after the one-variety movement picked up steam. With rare exceptions, such as the Acala communities in the West, no single variety dominated a given region or state in 1930. This situation changed rapidly.

In 1954 only ten varietal types (a variety such as Acala had several strains) accounted for more than 77 percent of the cotton grown in the United States, and five pure varieties accounted for almost 52 percent of the nation's crop. A single variety (Deltapine 15) made up 25.5 percent of all U.S. cotton acreage. Contemporary reports credited the one-variety campaign with playing a major role in creating the concentration of varieties. As an example, according to the 1947 congressional hearings on cotton quality, "the one-variety program has reduced the number of varieties grown and standardized the entire crop of the organized areas on a few improved high-yielding varieties."[68] William Brown and J. O. Ware were equally emphatic that "the cotton-varietal-standardization movement... has practically made over the situation in cotton varieties

[67] Westbrook, "One-Variety Cotton Communities," 16. Westbrook's assertion that 1,200 varieties had been grown in the United States in 1930 is probably a misreading of Ware. A 1947 congressional report on cotton asserted that, prior to the one-variety cotton movement, more than 500 were grown. U.S. House Committee on Agriculture, *Study*, 955.

[68] U.S. House Committee on Agriculture, *Study*, 960; Brown and Ware, *Cotton*, 97. With the introduction of one-variety communities, the decline in the number of varieties happened fairly rapidly in local areas. As an example, the county agent for Carroll County, Georgia, reported that between 1933 and 1937 the land in one variety increased from a few acres to 25,000 acres and the number of varieties grown in the county had dropped by half. Wiley, "Cotton Improvement in Carroll County, Georgia," 46–7.

in America and has thereby contributed greatly to quality improvement of cotton."[69]

State-level data offer a clearer sense of the movement toward varietal concentration, because, as shown in Table 6.6, by the early 1950s several states had effectively become de facto one- or two-variety enclaves. Some of these enclaves transcended state boundaries. As an example, Coker 100 Wilt made up more than 95 percent of the cotton grown in South Carolina, North Carolina, and Virginia. In many states the extent of de facto one-variety production far exceeded the production of official one-variety communities. Even in states with a greater number of varieties, such as Texas, a given region likely had a high concentration of a specific commercial variety whether or not there was a formal association of farmers. In 1952, 35 percent of U.S. cotton was ginned in counties where one variety comprised 90 percent or more of acreage, and 46 percent came from counties where one variety accounted for 75 percent of the acreage.[70]

Across the South, local studies reported net benefits to one-variety community members similar to those we reported earlier. In addition, USDA scientists generated a number of more global estimates. In 1943 the Bureau of Plant Industry, Soils, and Agricultural Engineering estimated that one-variety producers were receiving an additional return of $7.50 per acre. In 1945, C. B. Doyle reported a benefit of about $7.00 per acre to participating growers. He further reported that the USDA had invested $800,000 from 1911 through 1944 in creating the one-variety community system, and this investment had generated an annual return in excess of $56 million in 1944 alone. In 1950 the USDA estimated that one-variety communities had generated an increased value to growers in the "old belt" over the period 1938–45 of $260 million. The USDA reported cumulative expenditures on one-variety community development up to and including 1945 of less than $1 million. In its presentations to Congress during the late 1930s and early 1940s, the USDA showcased the one-variety movement as one of its high-profile programs, consistently reporting annual net benefits to participants in the range of $5 to $7 per

69 Brown and Ware, *Cotton*, 98.

70 Compiled from U.S. Production and Marketing Administration, *Cotton Varieties Planted 1950–1952*. With the concentration in varieties came a parallel concentration in the number of seed breeders and seed distributors. By 1961, "four large companies produce the cotton seed that is used on 90 percent of the planted acreage in the Southern and Southeastern States." Waddle and Colwick, "Producing Seeds of Cotton," 188.

TABLE 6.6. *Cotton Variety Concentration by State in 1954*

State	Percent of Acreage Planted to: One Variety (percent)	Two Varieties (percent)
Alabama	37	68
Arizona	81	86
Arkansas	62	77
Georgia	60	80
Illinois	72	90
Kentucky	83	98
Louisiana	84	89
Mississippi	76	82
Missouri	40	70
Nevada	100	100
New Mexico	57	66
North Carolina	98	99
Oklahoma	26	46
South Carolina	95	97
Tennessee	57	78
Texas	11	21
Virginia	95	96

Source: Compiled from Brown and Ware, *Cotton*, 53–6.

acre. This was a substantial increase given that the average value of cotton lint per acre was $33.90 over the period 1935–44.[71] Another important indicator that the one-variety movement contributed to the revolution in cotton production is the widespread support it received beyond the USDA, the Extension Service, and grower communities. Representatives of the cotton textile industry, prominent breeders, leading shippers, and southern bankers all lauded the movement's contributions.[72]

In light of the subsequent developments in government policy regarding intellectual property rights of genetic materials, it is important to

[71] Porter, "Toward Standardized Cotton," 21–2; Doyle, "One-Variety-Cotton Communities"; U.S. House Committee on Agriculture, *Research*, 754–5, *Hearings* (1938), 317–19, 938–40, and *Hearings* (1943), 328–29; Carter et al., *Historical Statistics*, Tables Da755–757.

[72] Merrill, Macormac, and Mauersberger, *American Cotton Handbook*, 116; U.S. House Committee on Agriculture, *Cotton: Hearings*, 399; "One-Variety Cotton Program," 103–4.

understand the incentive structure and position of private seed companies regarding one-variety communities. An innovative seed company faced a number of interrelated problems. Such a firm's primary contribution was its investment in R&D to produce new plant varieties; its value added in cleaning or providing seed treatments was generally secondary. As principally a seller of intellectual property, a seed company had to be able to exert market power and price above marginal cost to recoup its sunk R&D expenses. But even if a company could exercise market power, it faced the problems of a durable-goods monopolist – namely, it created its own competition – to an especially severe degree. In the textbook view, a durable-goods monopolist suffers from the following time inconsistency problem: it has an incentive to sell initially to high-demand buyers at a high price and in subsequent periods to sell at lower prices to attract the lower-demand buyers. But this threat causes the high-demand buyers to lower the initial price they are willing to offer. The preferred solution for the durable-goods monopolist is to lease a good – that is, sell the services rather than the good outright. If this proves infeasible, the firm may reduce the intertemporal competition by reducing the durability of the good (planned obsolescence). Obviously this reduces the value of the good to the buyer and the firm must weigh this negative price effect against the positive effect of sustaining its market power to determine the optimal level of durability.[73]

For seed producers, the problem of intertemporal competition is especially severe because the seed possesses the natural ability to produce multiple offspring. (For cotton, the multiple was on the order of 20 to 1.) Hence, a farmer could purchase commercial seed to meet a fraction of his requirements and, within a few seasons, raise enough for his whole operation and have a surplus to sell to his neighbors. The seed companies could partially offset this latter form of competition through quality control guarantees and branding. But in an environment with weak intellectual property protection, a company could not easily or effectively prevent its gene stock from serving as the basis of a competitor's "improved" variety. These forces help explain why the desideratum for commercial seed breeders was a product without the ability to reproduce naturally – a seed such as an F2 hybrid or one with a terminator gene.

[73] These are only a few of the alternatives available. Another preferred solution for the durable-goods monopolist is to creditably commit to a price schedule or to provide buyback offers. Bulow, "Durable–Goods Monopolists," 314–32, and "Economic Theory of Planned Obsolescence," 729–49.

In recent decades cottonseed companies were among the most vocal opponents of the one-variety law in California.[74] Breeders also opposed more ambitious New Deal plans for the South. Shortly after the AAA was established, leaders of the Cotton Section developed a program for the South, patterned on the California model. This was a relatively centralized scheme that would have empowered the directors of the state experiment stations to choose one variety of cotton for communities in their states. The directors were supposed to consider the diverse growing conditions in defining community boundaries, but the hope was that large areas, possibly entire states, might convert to a single variety. The plan also called for an expansion of government breeding and seed distribution activities. According to Cully Cobb, who headed the AAA's Cotton Section, USDA Secretary Henry A. Wallace had tentatively signed off on the program, but at the last minute David R. Coker convinced Wallace to scuttle it.[75]

But it would be wrong to conclude that such opposition was a constant. During the heyday of the one-variety community movement in the South, the major seed companies embraced the effort and lent their support in spite of the communities' seed multiplication and distribution policies. In his 1938 article, "Break This Vicious Circle Which Shuts You Out from Cotton Seed Sales," which appeared in the inaugural issue of *Southern Seedsman*, Austin Burges argued that the "opening wedge must be one-variety communities" and that assisting their development "means heavy extra profits for the seed dealer."[76] George J. Wilds, president of Coker's Pedigreed Seed Company, spread the same message in his 1944 congressional testimony: "The one-variety community is the best solution for all of us interested in cotton."[77] Seed company marketing policies conveyed the same theme. The USDA noted that "the 1947 catalog of the largest commercial cottonseed breeding firm in the Southeast states" (presumably Coker) contained a strong endorsement of one-variety communities,

[74] As an example, in 1964 D&PL purportedly encouraged dissident growers in the San Joaquin Valley to campaign for the repeal of California's one-variety law. Camp, *Cotton, Irrigation, and the AAA*, 172. D&PL began to push its breeder's rights outside of the California context. In the case of *Delta and Pine Land Co. v. Peoples Gin Company* (1982), the U.S. District Court in Mississippi ruled that the Plant Variety Protection Act of 1970 prevented the cooperative gin from arranging the sale of collectively ginned cottonseed between its member farmers. Kloppenburg, *First the Seed*, 146.

[75] Cobb, *Cotton Section*, 96–9. Camp, *Cotton, Irrigation, and the AAA*, 135–8.

[76] Burges, "Break This Vicious Circle," 5, 6, 29.

[77] U.S. House Committee on Agriculture, *Cotton: Hearings*, 400.

asserting that they had been of great value to breeders, growers, and manufacturers. Coker also adjusted its breeding and marketing program to support cotton standardization. Other breeders jumped on the bandwagon: "To further promote standardized production two other large commercial breeders in the Mississippi Valley [most likely D&PL and Stoneville], who furnish the foundation planting seed for the great bulk of the one-variety communities in the Central and Eastern States, have adopted the policy of retaining the same varietal name for their new stocks from year to year, thus simplifying the continued operation of the one-variety developments."[78] Other evidence suggests that up to the early 1950s the major seed companies saw the growth of one-variety communities as a bonanza to increase sales.[79] Dr. C. W. Manning, an early breeder with the Stoneville Pedigreed Seed Company, recalled that his firm gladly sold to one-variety communities, knowing full well that they planned to increase the seed and supply it to local farmers. This meant "the company had to put more salesmen on the road."[80] Evidently in this evolutionary stage in the development of the cottonseed business some market was better than no market.

Before World War II, the leading private breeders were typically small and near the margins of commercial viability as stand-alone operations. Coker's Pedigreed Seed Company of Hartsville, South Carolina, the "South's Foremost Seed Breeders," appeared chiefly to have been the farm-improvement hobbyhorse of its wealthy, public-spirited owner,

[78] U.S. Agricultural Research Service, *Report of the Administrator 1947*, 302.

[79] Well before the onset of the one-variety community movement, Coker had encouraged farmers to buy seed for a seed patch and then use the resulting seed to plant their entire crop the following year. Coker catalogs also contained farmer testimonials describing how they made money increasing and selling the improved seed to neighbors. While initially beneficial to seed companies, such policies created competition for the firms' future sales. Seed certification programs and later the Plant Variety Protection Act helped reduce this form of competition. Webb, "Private Cotton Breeding in the Southeast," 522–34.

[80] Telephone interview by authors with Dr. C. W. Manning of Leland, Mississippi, 1 February 2002. Manning's statement referred to the period around 1950. Early C. Ewing, Sr., the head breeder at D&PL, also linked the increased popularity of improved seeds with "the phenomenal growth of one-variety communities, one-variety gins, and one-variety farms." Ewing, Early C. "History of Cotton Varieties," D&PL Company Records, Box 9, History: Published Material, Manuscript Collections, MSUL. The support that the seed breeders in the 1940s and 1950s gave the one-variety communities is somewhat analogous to the support book writers and publishers might provide public libraries in areas dominated by illiteracy. From the book trade's commercial standpoint, it would be better if each reader bought the book and could not resell it or share it with others (although advertising it by word of mouth would be welcome).

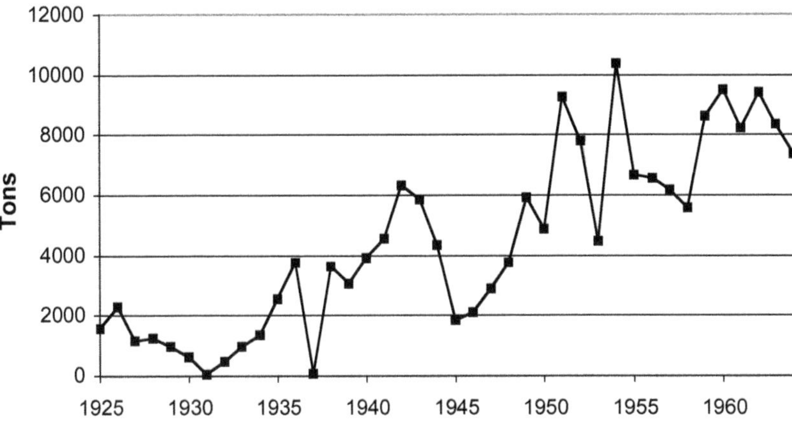

FIGURE 6.6. Sales of planting cottonseed of the Delta and Pine Land Company, 1925–64. *Source:* Compiled from Annual Statements and President's Reports, D&PL Company Records, Box 2, Manuscript Collections, MSUL.

D. R. Coker. The firm's weak financial record over the 1920s and 1930s led Coker to consider handing over the operation to a philanthropic organization, such as the Rockefeller Foundation, to support as a southern improvement project.[81] The leading commercial breeding operation in the mid-South was a division of the D&PL, a 38,000-acre plantation in the Mississippi Delta. Building on its success in creating early maturing, high-yielding, high-quality seed for its own lands, D&PL became a major seed supplier. According to *Fortune*, the company sold "more cottonseed to planters than any other single world agency" in the mid-1930s.[82] Yet for all of D&PL's prominence, its sales over the decade 1925–34 averaged only about 1,060 tons per year, which represented less than 1 percent of the seed planted for the entire U.S. cotton crop. Figure 6.6 shows that this situation changed significantly after the one-variety movement took hold.[83] Despite a 62 percent reduction in U.S. cotton acreage and a more than 50 percent decline in seeding rates (discussed later in this chapter), D&PL sales over the period 1955–64 were greater than seven times those prevailing 30 years earlier.

[81] Rogers and Nelson, *Mr. D. R.*, 152–73, 197; Coclanis, "David R. Coker," 105–14.

[82] "Delta & Pine Land Co.," 158. It is likely that quasi-public agencies in California and Texas distributed more seed than the D&PL.

[83] For example, bad weather wiped out the seed crop in 1937. Later the company subcontracted seed production to reduce climatic risks. Compiled from Annual Statements and President's Reports, D&PL Company Records, Box 2, Manuscript Collections, MSUL.

Evidence on the source of seed supply indicates the vast majority of seed used before 1930 was gin-run. In the 1920s and early 1930s only 5 to 10 percent of cotton planting seed came from breeders and dealers. By 1955, purchased seed made up 74 percent of the cottonseed used for planting.[84] Notably, 70 percent of the purchased seed and 52 percent of all planting seed in 1955 was comprised of certified seed.[85] In 1971, only about 19 percent of cotton farms nationally (with 19 percent of acreage) planted seeds they grew themselves. The percentages were even lower when Oklahoma and Texas are excluded. These states followed a low-input, low-quality, and low-yield system of production. Outside of these states, 15 percent of farms (with 13 percent of acreage) reported using homegrown seed. At this time, certified seed was planted on 64 percent of cotton farms nationally (and 74 percent of those outside Oklahoma and Texas).[86]

The one-variety community movement was part of a more general campaign to improve seed varieties, which centered on the activities of the International Crop Improvement Association (ICIA). The association was chartered in 1920 with members from across the United States and Canada. The aims were to limit fraudulent practices in the seed business, prevent the loss of valuable varieties as a result of contamination by other varieties, and develop international standards for seed identification and distribution. The ICIA developed field and laboratory standards, regulations to ensure proper isolation and handling of seed breeding and increase programs, a system of uniform seals and tags to identify pure seed, and uniform definitions for classification and certification. The ICIA also successfully lobbied to obtain legal backing for its standards and established uniform cottonseed certification standards in 1926.[87]

[84] U.S. Census Bureau. Census of Agriculture 1954, v. 3, 19. The survey shows that 583,000 farmers purchased 194,100 tons of cottonseed for planting. This implies that purchased seed was used on more than two-thirds of cotton farms. The total amount of cottonseed used for planting comes from U.S. Economic Research Service, "Statistics on Cotton," Table 192.

[85] The data on certified seed come from *Report of Seed Certified 1954*, 47–50 and 102. The certified seed data omit seed produced by government agencies in California (and likely in other states) and, thus, understate the total production of high-quality seed. The shares reported in this text include our estimates for the pure seed used in California.

[86] U.S. Census Bureau. Census of Agriculture 1969, v. 5, pt. 3, 29–34.

[87] A specific terminology evolved where "foundation seed" was that developed by the breeder; "registered seed" represented the first year's multiplication of foundation seed (under tightly controlled conditions); and "certified seed" represented the multiplication of registered seed (again, under controlled conditions). Hackleman and Scott, *History of Seed Certification*, 1–67. See U.S. Bureau of Markets, "Marketing Cotton Seed for

The timing of the takeoff of the adoption of improved cottonseed varieties clearly predated adoption of the mechanical picker (see Table 6.7). In 1940, when the mechanical picker was in its infancy, there were 98,000 acres approved to produce certified cottonseed. At prevailing seed yields (472 pounds per acre over the period 1946–48) and seeding rates (32 pounds per acre), the output of this acreage would have been sufficient to plant less than 6 percent of U.S. cotton land (outside California). By the period 1952–54, an average of 577,000 acres had been approved, producing sufficient certified seed for more than half of U.S. cotton land (outside California). Moreover, the average quality of noncertified seed also increased, because it was likely to be only a generation or two removed from certified seed. After the mid-1950s the number of approved acres fluctuated, but the percentage of the crop planted with certified seed continued to grow because, with improved seed varieties, delinting (see the following section), and improved mechanical seeders, the amount of planting seed required per acre of cotton declined substantially.[88]

The Demise of the One-Variety Movement

By the mid-1950s the USDA had deemphasized its one-variety community campaign in the South. After 1952, the USDA Bureau of Plant Industry abandoned the program and began closing many of its regional offices. By the end of 1954, the agricultural extension services assumed responsibility for the program.[89] In roughly the same period, the Agricultural Marketing Service also deemphasized the requirement that farmers be members of a cotton improvement group, opening up Smith–Doxey classification to farmers who contacted their county agents.[90] From the USDA's perspective, the program had served its purpose – educating farmers about the importance of growing high-quality cotton – and a number

Planting," 73–4, for a discussion of the severe problems concerning quality standards in the market for planting cottonseed circa 1920.

[88] Hackleman and Scott, *History of Seed Certification*, 53. Acres planted come from U.S. Economic Research Service, "Statistics on Cotton," 63. By 1980, nearly all planting seed outside the High Plains was certified.

[89] Westbrook, "One-Variety Cotton Communities," 17. Beginning in 1945 in most areas of the South, the state extension services folded the one-variety program into the new "Seven-Step Cotton Program." This program also addressed emerging issues such as cotton mechanization and chemical application. U.S. Federal Extension Service, *Report of Cooperative Extension Work 1946*, 29.

[90] U.S. Agricultural Marketing Service, "Get Your Green Card," 4–5. The Smith–Doxey services continue to this day, although the program is now called Form 1 classification.

TABLE 6.7. *Acreage and Production of Certified Cotton Seed, 1940–57*

Year	Acres Approved for Certification (thousands of acres)				Certified Cottonseed Production (thousands of tons)			
	Foundation	Registered	Certified	Total	Foundation	Registered	Certified	Total
1940				98.0				
1945				296.0				
1946				297.7				70.4
1947				436.5				94.5
1948				506.3				129.4
1949				558.9				49.1
1950	10.4	169.1	271.0	450.5	2.9	34.1	70.1	107.1
1951	6.5	225.9	406.9	639.3	1.4	44.5	95.7	141.6
1952	9.5	277.4	414.6	701.5	2.8	76.2	104.2	183.2
1953	7.7	216.0	343.4	567.1	2.5	80.7	89.2	172.5
1954	4.9	183.7	272.9	461.4	1.4	58.5	85.5	145.4
1955	8.4	62.5	294.3	365.2	3.1	21.7	98.9	123.6
1956	6.9	120.9	246.3	374.1	2.4	34.0	61.6	98.0
1957	7.5	94.4	201.8	303.7	3.3	23.7	83.1	110.1

Note: Acreage and output exclude California activity.

Sources: Hackleman and Scott, *History of Seed Certification*, 53; *Report of Seed Certified, 1946–1958*.

of technological and institutional changes made the one-variety concept less appealing. The maturation of a commercial seed industry able to supply abundant quantities of high-quality foundation seed was an important contributor to the demise of one-variety communities in the traditional Cotton South. With the increased presence of quality private breeders and strict new seed certification systems, the seed increase activities of one-variety communities became unnecessary. Thus, the South took a different path than California, where a legally entrenched bureaucracy, with its own internal seed breeding program, prevented competition from private breeders. Lacking in-house research and breeding operations, southern one-variety communities had always been dependent on private breeders or experiment stations for their foundation seed.

The development of high-quality varieties which gained favor over wide areas was just one of a series of economic and technological changes that made one-variety communities obsolete in the South and may help account for the finding that California's system was inefficient by the late 1970s. Among the most important of these changes was the development and diffusion of acid delinting (and later other chemical treatments) of planting seed. This technology would eventually strengthen the position of commercial seed companies, increase on-farm productivity, and facilitate the mechanization of the last major bastion of hand labor in the production of cotton.

When upland cotton is ginned, the seeds remain "fuzzy" because the gin fails to remove all of the lint. Throughout the ages farmers planted fuzzy seed and then chopped (thinned) the cotton plants to obtain an even stand. Fuzzy seeds worked poorly with mechanical seeders because they would clump together and clog the machines. Clumping also made it difficult to obtain a well-spaced, uniform stand, whether the seed was planted by hand or by machine. More precise planting to a row greatly reduced the need for chopping, increased yields, and allowed for more efficient machine cultivation.[91] The solution was to use one of several technologies to delint the cotton. In California, the use of machines to delint planting seed dates to the beginning of the industry and was widespread by the 1940s. By the 1950s the technology had gained popularity in the Cotton South, and it remained the most common form of delinting to the early 1970s.[92] Essentially the cotton was re-ginned using special machinery

[91] Alexander, *Arkansas Plantation*, 81; Hopper and McDaniel, "Cotton Seed," 299.

[92] Camp, "Cotton Culture," 8. Machine delinting in the South was common earlier because the linters had value, especially during World War I when they were used to make munitions. Agelasto et al., "Cotton Situation," 381–3. Evidence on the use of delinted

designed to remove most of the lint. This helped with machine planting, but the remaining lint still made it difficult to obtain an even stand; thus, the need for chopping continued. The next stage was to expose the mechanically delinted seed to an intense flame to burn off the remaining lint. This improved the seeds' handling characteristics but not sufficiently for precision metering during planting. The ultimate solution was to use one of several acid processes to chemically delint the seed. Besides allowing farmers to mechanize seeding operations and dispense with chopping, acid delinted seed offered several other advantages. Delinting (and indirectly the more even spacing of plants) allowed cotton to come to a stand earlier – a real plus given the threat of the boll weevil. In addition, acid delinting reduced plant diseases and greatly increased germination rates. For these reasons, farmers needed much less planting seed.[93]

Delinting cottonseed with acid was an unpleasant and hazardous task. Experiment station reports provided farmers with detailed instructions on how to prepare the acid and soak the seeds, noting especially the caution: "Never add water to the acid, as this causes a violent reaction."[94] For all the benefits, the cost of the acid and the unpleasantness of the task sharply limited the number of cotton farmers who adopted the delinting technology. But H. P. Smith saw the handwriting on the wall when he noted in 1950 that "most cotton growers plant regularly ginned seed which are covered with fuzzy lint. Mechanization may be influencing the trend toward delinted seed."[95] With the development of improved acid technologies and the advent of the mechanical cotton harvester in

seed for planting is sparse, but D&PL's standard practice as early as 1922 was to sell mechanically delinted seed. Delta and Pine Land Co. of Mississippi, Salisbury Cotton, D&PL Company Records, Box 1, Oral History, Manuscript Collections, MSUL.

93 Cherry and Leffler, "Seed," 531–3; Hancock, "New Method of Delinting," 1–2. Accounts differ on the decrease in planting seed required per acre. N. I. Hancock notes savings of 20 percent, but Donald Chrichton Alexander notes that planting delinted cotton "utilizes less than one third the amount of seed needed for ordinary cotton production." Between the 1930s and the 1990s, seed planting rates outside the High Plains declined from about 34 pounds per acre to as low as 8 pounds per acre. Alexander, *Arkansas Plantation*, 81.

94 Sherbakoff, "Improved Method of Delinting Cotton," 2.

95 Brown's comment that "some authorities recommended delinting cotton seed that are to be used for planting purposes" suggests the lack of adoption in 1938. Brown, *Cotton* (1938), 212. In 1943, Alexander still asserted that delinting was relatively expensive. Alexander, *Arkansas Plantation*, 81. In the mid-1950s Basil G. Christidis and George J. Harrison, *Cotton Growing Problems*, 310–11, continued to recommend machine rather than acid delinting. They noted that the latter processes were only "occasionally" used. Smith, "Cultural Practices," 144.

the 1960s, the acid processes began to compete more effectively with machine delinting.[96] In 1970 P. R. Smith estimated that 95 percent of U.S. planting seed was delinted, with acid delinting accounting for 23 percent of the planting seed in California, about 90 percent in Texas and Arizona, 15 percent in the mid-South, and 40 to 50 percent in the Southeast.[97]

The adoption of delinting (and especially acid processes) reflected the interaction effects of mechanical and biological technologies because the diffusion of one reinforced the demand for the other. It was in the interest of farmers to have ample labor during the peak season. Thus, as an example, adopting tractors that would save labor in plowing would only exacerbate the imbalance between the peak and nonpeak needs, potentially leading to labor shortages during the peak period. Farmers who adopted a mechanical picker had added incentive to reduce chopping-labor requirements, and planting acid-delinted seed made that possible. Moreover, the new planting technologies made the mechanical picker much more efficient. If plants are too widely spaced, they develop woody branches that hinder machine performance. With delinted cotton and a mechanical seeder, farmers could achieve a thick, uniform stand suitable for efficient machine operation and could eliminate most of the labor required for chopping. The result was that whereas acid-delinted seed was rare in 1950, it had started to gain acceptance in the 1960s and was nearly universal by the late 1970s.[98]

Seed treatments were not the only way biological technologies interacted with machines. With the coming of harvest mechanization, cotton

[96] Elliot, Hoover, and Porter, *Advances in Production*, 125–6.

[97] Smith, "Introductory Remarks," 90, claimed that 70 percent of the planting seed in Georgia was acid delinted. Leaders in the development of the California cotton industry maintain that acid delinting came much earlier than Smith claimed. In Arizona acid delinting appears to have been gaining wide favor as early as 1938, and at least one commercial delinting plant was in operation by that date. "Much Delinted Seed," *Arizona Producer* (15 March 1938), 4. Beginning in 1938 the firm of Feffer–Wharton regularly advertised its acid delinting services. By 1962, 90 percent of D&PL seed sales in Arizona were acid delinted. Feffer–Wharton, 1 March 1938, 4 and 15 March 1938. Sales Department Review, April 1962, D&PL Company Records, Box 19, Manuscript Collections, MSUL.

[98] Waddle and Colwick, "Producing Seeds of Cotton," 190. The higher yields that came with improved seed and delinting directly stimulated the mechanization of the harvest by allowing the fixed cost of the machine to be spread over a larger volume of output. Musoke and Olmstead, "Rise of the Cotton Industry," 402. Advances in chemical delinting technology included a dry system in which hydrochloric acid was mixed with sulfuric acid to form a gas that reacted with and crystallized the fuzz on the seed. The seeds were then treated to remove the crystallized lint. The various steps of the wet acid process were also integrated and mechanized in large delinting facilities.

breeders once again began redesigning the physical appearance of the plant to make it more amenable to machine picking. There were many challenges. Breeders initially sought to select varieties with a growth habit yielding a medium-stature plant with a strong central stem and spreading branches above ground level, more uniform fruiting, medium to early maturation, smooth leaves and modified bracts to produce less trash, and fluffy bolls that were detached from the burr easily by the machine but not by wind or rain. Later efforts added the requirement of successful cultivation under more dense spacing – this produced plants without the woody branches that interfered with the machine's efficient operation. To reduce trash further, it would become standard practice to chemically defoliate cotton plants. Cotton thus represents another clear case where mechanization and biological innovation were not separable in ways that the conventional wisdom asserts.[99]

The new cottonseed delinting technologies offered significant economies of scale because the production of delinted seed typically was concentrated in a few plants in any one state.[100] The adoption of delinted seed had implications for the growth of the commercial cottonseed industry that were similar to the effects of hybridization to seed-corn producers. Corn farmers had to purchase new seed every year because pure-line F2 hybrids lost their vigor in a single generation. But the improved cotton varieties were not F2 hybrids and, thus, farmers could recycle seeds, significantly reducing the demand for new commercial seeds. But delinting dramatically increased the economic benefits of purchased seed relative to gin-run seed and stimulated farmers in most regions to buy seed annually. This single technological change greatly reinforced the benefits that cottonseed companies received in 1970 with the passage of the Plant Variety Protection Act.[101]

[99] Neely, "Challenges of Cotton Mechanization," 21; Colwick and Williamson, "Harvesting to Maintain Efficiency," 442–4; Harrison, *History of Cotton in California*, 83.

[100] Around 1970 there were only four or five acid delinting plants in California, and only three in Mississippi. In recent decades the production and delinting of planting seed for the entire South has largely moved to the arid West, where weather conditions generally ensure that the seed will not get wet. Many delinting operations have moved to Indian reservations, outside the reach of the Environmental Protection Agency. However, new technological advances in developing a foam delinting process have greatly reduced the negative environmental side effects.

[101] Delinting was not quite as effective from the seed companies' perspective as F2 hybrids because farmers could in principle take recycled seed to delinting companies or these companies could buy and sell delinted seed. But the process did raise transactions costs for farmers and the chemical companies became easy targets for legal actions after 1970.

Along with delinting came a number of other chemical seed treatments to control diseases and insects. The result was that cottonseed in the 1960s not only was of higher quality in terms of its yield potential than seeds of the 1930s, but the modern seed embodied numerous other valuable technological features. These included greater responsiveness to nitrogen-rich fertilizers, which were falling in price. Thus, by the late 1970s (the date for Constantine, Alston, and Smith's estimates for California), a new division of labor had been firmly established. In most regions the majority of farmers bought delinted certified seed nearly every year. The practice of planting one's own seed obtained at the gin had become a rarity. New seeds were produced in tightly controlled isolated areas to help guarantee purity. These changes in seed technology, certification, and marketing, along with the development of superior varieties that effectively captured the market in whole regions or states, ended the need for formal one-variety communities. It mattered little if seed was mixed at the gin if it was not intended for planting. In addition, the problem of cross-pollination in the field was minimized by the annual purchase of new seed – the de facto one-variety production – and the decades of insecticide use that reduced the density of insects. The old production externalities that had haunted the industry were no longer an issue. In addition, the nearly universal use of Smith–Doxey classification services, and later the adoption of extremely accurate High Volume Instrument testing devices, largely solved the problems of classing cotton. Thus, with the major exception of California, one-variety communities simply faded away, having served their purpose of promoting the transition to better cottons and improved cultural and marketing practices.

Conclusion

In 1957 James Street published his classic, *The New Revolution in the Cotton Economy*. For Street, the cotton revolution involved a wholesale transformation of cotton production – a transformation that was critical for the modernization of the southern economy. Street's story of technological change focused largely on mechanization, with only an occasional mention of cotton breeding and improvement activities, usually in the context of how breeders were assisting in the drive to develop plant qualities conducive to mechanizing the harvest.[102] The message of

[102] For example, see Street, *New Revolution in the Cotton Economy*, 112–13, 147–8; and McClelland, *Sowing Modernity*, 64–93.

this chapter is to emphasize the importance of the other revolution in the cotton economy. Biological and structural rather than mechanical, this revolution led to a fundamental change in the source of the seed supply, in the varieties of cotton grown, and in the way cotton was classed and marketed. One of America's leading cotton breeders, J. Winston Neely, offered an assessment of recent accomplishments as of the late 1950s:

> The progress made by cotton breeders is truly phenomenal. Yields have been markedly increased. Varietal resistance to diseases, resulting from breeding programs, has made profitable the growing of crops where non-resistant varieties would fail completely. The quality of fiber produced by improved varieties has been greatly increased. Characteristics of plant type, growth habit and fiber quality of many varieties have been altered by breeding, to the extent that they are much better adapted to cheaper and better methods of planting, culture, harvesting and processing. We would be growing far less cotton today if we had to depend upon varieties grown only a few years ago.[103]

These changes to seeds and their complementary fertilizer inputs, rather than the arrival of the mechanical picker, accounted for the rough tripling of American cotton yields and the significant increases in average staple length between 1930 and 1960.[104] As Neely noted, new biological technologies interacted with mechanical technologies, thereby reinforcing the drive to increase southern agricultural productivity. Because of these interaction effects with mechanical technologies, the new biological systems had a far greater effect on reducing labor demand than analogous biological innovations in the grain sectors. Thus, biological innovations represent a hitherto unrecognized factor contributing to the changes in farm tenure and scale that transformed the southern landscape in the mid-twentieth century.[105]

What fundamentally separated the biological revolution in cotton from what transpired in corn and most other crops was the greater role that the USDA played in orchestrating institutional innovations. In the nineteenth and early twentieth centuries, USDA research programs played a key role in improving the quality of the seed supplies of most major crops. USDA scientists searched the globe for useful varieties, and government breeding, testing, and outreach programs made an enormous contribution to

103 Neely, "Cotton Breeders Pave the Way," 74.

104 As with corn, other factors such as the application of improved fertilizers, herbicides, and insecticides, clearly contributed to cotton yield increases.

105 The causality ran both ways because the decline in tenancy and consolidation of plantations undoubtedly also hastened the adoption of the new biological technologies.

the transformation of the seed supplies available to American farmers.[106] In addition, the USDA focused on improving the efficiency of seed markets. Seed embodies a complex array of technical characteristics that are difficult for an individual farmer to assess. Evaluations about the relative performance of different varieties and the quality of a given batch of seed (for example, was it cleaned and stored properly, etc.) usually must wait until the harvest and, given annual variations in growing conditions, several years may be required to make a reliable assessment. As a result, the development of a seed market requires particularly good information and mechanisms for providing guarantees to farmers and building trust between seed buyers and sellers.[107] For most crops, the USDA (often following the lead of individual states) helped create markets by developing national procedures, laws, and agencies for testing and certifying the genetic and physical characteristics of seed. But in the case of cotton, the task of market development was complicated by the exceptional problems of maintaining pure seed supplies, and the failure of the system of local cotton classing to reflect the relative premiums and discounts prevailing in central markets. These problems, combined with the relative lack of education and the more depressed conditions in the Cotton South, created formidable barriers to technological diffusion which in turn dampened the incentive for private breeders to invest in creating improved cotton varieties.

The one–variety community movement and Smith–Doxey classification system were the vehicles for a comprehensive reform program that included educational campaigns and seed certification systems. The goal was to fundamentally redesign both the production and marketing of cotton in order to shock the cotton economy out of the prevailing low–productivity equilibrium.[108] Within a few years of the beginning of these efforts, local reports were touting improved yields, longer staple lengths, and greater price premiums. These improvements all took place before the advent of liquid ammonia fertilizers. By the 1940s, private seed breeders had embraced the one-variety movement. Incredibly, to gain a toehold in the fledgling seed market, breeders actually encouraged individual farmers and organized community groups to purchase small quantities of

106 Olmstead and Rhode, "Red Queen."

107 Tripp, "Institutional Conditions for Seed Enterprise Development," 24.

108 The problems of maintaining pure seed supplies and of grading cotton were not unique to American growers. Other countries, including Egypt, Brazil, Argentina, and India, experimented with one-variety communities in the 1930s and 1940s.

commercial seed and then increase it and sell the resulting seed (embodying the breeder's intellectual property) themselves. In 1949, more than 400,000 cotton farmers, producing roughly 60 percent of American cotton output, belonged to one-variety communities. This represented one of the largest and most successful cooperative movements in American agricultural history. Institutional and technological changes, along with the development of private-sector seed companies, gradually eroded the advantages of one-variety systems. In the South they simply faded away. In California, the one-variety law and institutions were harder to dispose of, and the system persisted into the 1990s, long after it evidently had become obsolete.

7

That "Stincking Weede of America"

The Evolution of Tobacco Production[1]

Tobacco has held a special place in American agriculture since the time of Jamestown. Few commodities better demonstrate the importance of biological innovation because successful tobacco culture was particularly dependent on the discovery of types and cultural methods suited to specific soils and climatic conditions. In keeping with recent economic models of product differentiation, the growth of the tobacco trade was associated with the emergence of an expanding number of localized clusters of producers focusing on highly specialized products.[2] For example, the extraordinarily rapid growth in demand for cigarettes in the twentieth century led to wholesale shifts in the types of tobacco grown.

The story of the birth of commercial tobacco production in this country hardly needs retelling. When the Virginia Company established the Jamestown colony in 1607, its adventurers lacked the know-how and organization even to feed themselves in the new environment. They were also without a suitable export. By 1610 the settlers began growing a harsh tobacco, *Nicotiana rustica*, which they had obtained from the indigenous population. In 1612, John Rolfe devoted a small plot to experimenting with "Spanish" seed of the *N. tabacum* species smuggled in from Trinidad or Caracas, Venezuela. Rolfe "took the pains to make trial partly for the love he hath a long time borne (for tobacco) and partly to raise [a] commodity" for the settlers. Cured like hay as heaps in the

[1] Thomas Cornwallis, "that Stincking Weede of America," cited in Menard, "Tobacco Industry," 109.

[2] Krugman, "Increasing Returns," 460–79.

sun, the Jamestown product was "pleasant, sweet and strong" and commanded extraordinarily high prices in England. A great boom ensued. In 1616, the colony exported 2,500 pounds of leaf. Upon returning to the colony in 1617, Samuel Angall "found but five or six houses, the church downe, the palizados (palisades) broken, the bridge in pieces, the well of fresh water spoiled, . . . the marketplace, the streets and all other spare places planted with tobacco . . . the Colonie dispersed all about planting Tobacco."[3] By 1618 Thomas Lambert had discovered that curing the tobacco on lines rather than in heaps enhanced its quality, further fueling the industry's growth. Exports soared from 20,000 pounds in 1618–19 to 500,000 pounds by 1627.

The early Virginia colonists grew two variants of *N. tabacum* – Orinoco, a plant with sharp-pointed leaves which yielded a strong product, and Sweet Scented, a plant with rounder finely veined leaves which yielded a milder product. Exactly when and from where these varieties were introduced is unknown. According to Charles Gage, the Orinoco "probably originated in the proximity of the river in South America from which its name is derived."[4] These two varieties provided the founding germplasm for most of the commercial tobacco later cultivated in the United States. Early settlers, possibly transplanted Virginians, carried tobacco culture to southern Maryland around 1635. Initially this region grew principally Orinoco.[5] Later in the eighteenth century, at a date now difficult to determine, the Maryland planters adopted a light, air-cured variety – apparently derived from Sweet Scented – that came to be known as the Maryland Broadleaf.[6] In 1729, Maryland Governor Benedict Leonard Calvert observed: "In Virginia and Maryland, . . . Tobacco,

[3] Smith, *Generall Historie of Virginia*, 123.

[4] Gage, "Historical Factors," 44.

[5] Carr, "Diversification in the Colonial Chesapeake," 344; Killebrew, *Report*, 89. Geographical specialization began early. Russell Menard notes that in the seventeenth century "the valuable sweet-scented tobaccos (were) grown between the James and Rappahannock rivers; the high quality oronocco (was) grown north of the Rappahannock and on the upper Eastern shore: the low-grade tobaccos (of the oronocco type) were grown on the lower Eastern Shore, south of James River, and the upper reaches of the Bay." The Sweet–Scented was grown in the York basin. Menard, "Tobacco Industry," 112.

[6] Garner, Allard, and Clayton, "Superior Germ Plasm," 813–14; Middleton, *Tobacco Coast*, 108–9. The date of arrival of Broadleaf, also called Kite's Foot, in Maryland is uncertain. But we know that in 1794 Jedidah Morse (father of Samuel Morse) mentions as "peculiar to Maryland . . . the bright kite's foot tobacco, which is produced at Elkridge, on the Patuxent, on the western shore." This is just north of the Potomac River. Morse, *American Geography*.

as our Staple, is our all, and Indeed leaves no room for anything Else."[7] At this time, Virginia and Maryland exported 40 million pounds of the weed.[8] The Chesapeake colonies were built on smoke.

Distinctive Features of Tobacco

Tobacco, for lack of a better phrase, is a medicinal herb. It provides consumers with neither nutrients nor fibers to clothe themselves. Instead, it provides pleasure-inducing and habit-forming nicotine. Tobacco has been consumed in a variety of ways, initially through direct ingestion into the mouth or nose and now principally through combustion and inhalation. The leaf's chemistry is highly complex, leading today's cigarette makers to employ chemists and other high-technology workers. The consumer experience with tobacco has always been shaped by the manner in which the plant was grown, cured, processed, and manufactured, because hard-to-measure qualities such as flavor and aroma are much more important for tobacco than for most other cultivated crops.[9] As a result, tobacco has never been as simple a commodity as corn, wheat, or cotton. The importance of quality is manifest in the greater cross-sectional price variation apparent in markets for tobacco than in corn or wheat markets. Higher quality tobaccos regularly sold for ten times the price of ordinary grades. The potential for substantial variation in quality led to the development of a consignment system where the buyer had a relational contract with the producer, to the use of auction sales for small open lots, and to the elaboration of a complex system of "types" characterized by the leaf's major region of origin, variety, method of cultivation, and principal use. These market responses were reinforced by the early introduction of state inspection and grading, which date to 1730 in Virginia and 1747 in Maryland.

Tobacco differed from most other staple crops in that the commercial product was the plant's leaves. For corn and wheat, the most valuable product was the seed grain itself; for cotton, it was the fiber surrounding the seed. Successful production of these staples required fostering of their reproductive processes. For tobacco, production and reproduction were separate. American farmers learned early that they could stimulate

[7] Calvert, "Letter," 602.

[8] Tobacco Institute, *Virginia & Tobacco*, 1–2, 10–12. This series includes a wealth of information and misinformation. Tilley, *Bright-Tobacco*, 5–7.

[9] Garner, Allard, and Clayton, "Superior Germ Plasm," 785–830.

and shape the growth of the tobacco leaf by removing or "topping" the upper part of the plant to eliminate the flowers. Topping increased the size, weight, and nicotine concentration of the lower leaves. Along with topping, farmers often pruned the bottom leaves and suckers. These labor-intensive cultural practices required considerable skill and greatly affected the tobacco's quality. The decision as to when and how to top and prune varied substantially over time, place, tobacco variety, and the intended product. As a consequence of topping, tobacco farmers had to select their seed deliberately, which involved either maintaining a separate seed patch or buying commercial seed from specialized producers. But a little went a long way; a single plant produced one million or more seeds, enough for planting at least a dozen acres. Tobacco is mainly a self-pollinating plant, with cross-pollination rates of less than 5 percent.[10] Whereas a field of tobacco of a single broad "variety" might contain a high level of genetic diversity, a single plant was relatively homozygous.

Tobacco, in its Central or South American region of origin, grows as a perennial, and the plant is intolerant of cold weather and wet or waterlogged soils.[11] In the tobacco belts of North America, tobacco is grown as an annual crop on well-drained soils and is harvested before the early frost. The term "annual crop" is somewhat of a misnomer because producing tobacco typically required more than twelve months. In the traditional crop cycle, farmers planted the seed in small, fertile, protected seedbeds around the beginning of January. After the threat of frost had passed and spring rains had sufficiently moistened the soil, the cultivators carefully transplanted the seedlings into hills in the main fields. This operation generally occurred in May, but could last into June. Next came months of nearly constant weeding (and constant rebuilding of the hills), followed by topping, pruning, and suckering. Beginning in September (and before the first frost), the leaves were cut as they ripened. In some times and places the entire stalk was harvested at once; in others the leaves were taken individually or in pairs. As Timothy Breen noted, colonial farmers were reluctant to call the cutting period the "harvest" because much work remained to be done before the tobacco could go to

[10] Garner, *Production of Tobacco*, 87; Garner, Allard, and Clayton, "Superior Germ Plasm," 787; Hutchens, "Tobacco Seed," 67; Tso, "Seed to Smoke," 11.

[11] T. C. Tso, "Seed to Smoke," 10, notes that a crop year of 120 frost-free days is one of tobacco's "basic requirements." W. W. Garner, "Tobacco Culture," 1, notes that the "tobacco plant readily adapts itself to a wide range of conditions, except that it does not thrive in an excessively wet or waterlogged soils."

market.[12] Through the winter and into the spring, the leaves were hung in special barns to cure, so that they would lose moisture without becoming too dry and brittle. After curing, the tobacco was stripped, stemmed, and prized; that is, the leaves were detached from the stalk (if necessary), the larger stems were removed, and the leaves were carefully layered and pressed into a round hogshead. In the spring, more than a year after the seeds were planted, the hogsheads could be rolled to market.

The critical details of tobacco farming, such as when to transplant, when to harvest, and how to cure, were special skills gained from tacit learning and experience rather than from books. Tobacco culture required more care and expertise than was common in most staples production. Inasmuch as tobacco was an "American" crop, it was the indigenous population who initially taught the Europeans about proper plant spacing, the advantages of topping, and the use of smoke in curing. European planters, in turn, passed these and other lessons to enslaved Africans. Many of these techniques changed radically as tobacco varieties evolved and as cultivation moved into new areas.

Tobacco was an incredibly labor-intensive crop. For the period 1910–14, the U.S. Department of Agriculture (USDA) estimated the labor requirements for tobacco at 356 hours per acre, compared with 116 for cotton, 35.2 for corn, and 15.2 for wheat. And in contrast to other staples, labor use per acre of tobacco increased over the first half of the twentieth century. USDA statistics indicate that, by the period 1950–53, the labor requirements per acre for wheat had fallen to 4.4 hours (a decline of 71 percent), for corn to 13.1 hours (63 percent), and for cotton to 70 hours (40 percent). For tobacco, hours per acre rose to 467, an increase of 31 percent.[13] Mechanization did not play much of a role in tobacco culture until after World War II.

Given the differential rates of technological change before 1910, the crop-specific labor requirements would have been different in earlier periods. Referring to tobacco's labor-intensive nature, Lord Calvert observed in 1729 that the crop "requires the Attendance of all our hands, and Exacts their utmost labour, the whole year round."[14] The careful work of Lorena Walsh documents the dynamics of labor productivity in the first two centuries of commercial tobacco production. Until roughly 1690,

[12] Breen, *Tobacco Culture*, 49–50. For a discussion of practices in tobacco culture around 1800, see William Tatham's account reproduced in Herndon, *William Tatham*, 7–333.

[13] Hecht and Vice, "Labor Used for Field Crops," 4–5.

[14] Calvert, "Letter," 602.

output per worker grew substantially as "planters learned how to handle more tobacco plants per worker and developed improved strains." As an example, in her sample of plantations and farms, the average reported output of tobacco per laborer soared from 345 pounds in the period 1623–29 (N = 8) to 1,564 pounds in 1680–99 (N = 28). After 1690, output per laborer fell as "a result of diminishing returns to available natural resources" and changes in the management and composition of the labor force. The output per worker averaged about 829 pounds in her 1770–74 sample (N = 150). Walsh concludes, "As settlements matured, limits to the maximum amount of tobacco a worker could produce were reached and did not change much until the twentieth century."[15] Mechanization had a hand because some of the early increases in tobacco output per worker were due to the shift from hoe to plow cultivation.[16] Despite this role for mechanization, tobacco creates a puzzle for the induced innovation hypothesis. This extremely labor-intensive crop drove growth in the Chesapeake in the 1600s when Virginia and Maryland were among the most labor-scarce outposts of European settlement on the globe.

Westward Movement of Tobacco

In the colonial era, commercial tobacco production was concentrated in the Chesapeake. When the Anglo Americans moved into Kentucky, Tennessee, and southern Ohio, they carried the seed, cultural techniques, and curing practices of the Chesapeake with them. In particular, the Maryland Broadleaf evolved into the red Burley tobaccos grown in the West in the antebellum period. Tobacco was also among the first export crops of the early French settlements in the lower Mississippi Valley. By the 1790s, production of the weed had extended as far north as Natchez. "Cultivated upon alluvial soils, badly harvested, and cured in poorly-appointed houses, it presented a rough, bony appearance, though full of gum and highly charged with nicotine." Production declined after 1787 with the appearance of higher quality tobaccos from the Ohio River Valley in the New Orleans market following James Wilkinson's expedition.[17]

[15] Walsh, "Plantation Management in the Chesapeake," 394. Terry L. Anderson and Robert Paul Thomas, "Economic Growth," 382–3, concluded that tobacco output per worker at least tripled between 1623 and 1700 without shifting from hand to machine production methods.

[16] Carr and Menard, "Land, Labor, and Economies," 407–18.

[17] Killebrew, *Report*, 181.

TABLE 7.1. *Geographic Center of Tobacco Production, 1620–1910*

	Median		Mean	
	Latitude	Longitude	Latitude	Longitude
1620	37.3167	78.2833	37.3167	78.2833
1840	37.2716	79.8297	37.4332	82.3220
1850	37.2924	81.5433	37.5849	83.0208
1860	37.3457	81.0615	37.7159	82.6824
1870	37.5450	85.6409	37.9891	83.7174
1880	37.7561	84.2061	38.2081	82.9223
1890	37.8797	84.5697	38.2052	83.7374
1900	37.5105	84.1732	37.9768	82.7561
1910	37.6336	84.2061	37.9214	82.8732

Louisiana was home to Perique, a distinctive tobacco grown in Saint James Parish. Almost every aspect of the production of this strong, dark tobacco, including cultural techniques and choice of nitrogen-rich soil, differed from Anglo American practices.[18] The elaborate curing methods developed after 1755 were most unusual. Beginning about ten days after hanging the whole plant, a handful of leaves were plucked, wrapped into loose twists, and put under 7,000 pounds pressure for twenty-four hours. The press was then opened and turned, allowing the leaves to reabsorb the extracted juices. This process was repeated every few days for several months. "At the expiration of three months the tobacco is cured and emits a rich, spirituous flavor, which has been imparted to it by the re-absorption of the aerated juices." The individual leaves were then separated, aired, and stacked between two cloths, which were then repeatedly rolled and tightly bound with a coil of rope, after which the tobacco was ready to sell. Perique represented one of many specialties in the tobacco market.[19]

Table 7.1 charts the geographic movement of tobacco production from 1840, when Census production data first become available, to 1910. As a point of reference, Jamestown is included as the center of commercial production for 1620. The data for 1840–70 indicate a change in longitude, which undoubtedly was a continuation of the pre-1840 westward shift. The largest change occurred during the Civil War. The turmoil

[18] Hahn, "Making Tobacco Bright," 30; Gray, *History of Agriculture*, 69–73; Killebrew, *Report*, 80–2.

[19] Killebrew, *Report*, 81–5.

disrupted production in the traditional eastern belt of Virginia, Maryland, and North Carolina, leaving Kentucky to emerge as the leading tobacco-producing state. The wartime stimulus to cigar leaf production in the Connecticut Valley worked to partially offset the shift in longitude. It is notable that, after 1840, tobacco witnessed far less western movement than did corn, wheat, and cotton production. There was also less change in the environmental conditions (such as annual precipitation) in which tobacco was typically grown than there was for the other three major staples. But this does not mean that biological adaptation was any less important.

As tobacco moved into new areas, farmers encountered "marked differences in soil type and in climate" which "greatly changed the properties of the tobacco" and its commercial value.[20] As early as 1693, John Clayton observed that "the same sort of Seed in different Earths, will produce Tobacco much different, as to goodness."[21] More than most crops, tobacco's growth habit, size, and commercial quality proved sensitive to the natural and cultural environment. These "influences greatly affected its adaptability for use in different forms, the product of one section, for example, being especially suitable for making smoking or chewing tobacco but perhaps not producing so acceptable a cigar as that of another section."[22] Over time, farmers learned to adapt their methods of cultivation and curing to the local environment. As W. W. Garner and his colleagues at the USDA aptly noted, "Thus, through a process of gradual evaluation tobacco culture has become highly specialized, each producing district furnishing a distinctive type of leaf especially adapted for certain uses, based ultimately on the tastes and preferences of the consumer. It is the accumulated experience of three centuries of tobacco culture that each of these types can be produced only under certain conditions of soil and climate, by using certain varieties of seed, and by employing special methods in growing and handling the crop."[23]

Once the tobacco belt became established, the crops' yield per acre was relatively flat. Figure 7.1 graphs USDA data on yields from 1866 to 1970. There was a modest upward trend into the early 1900s, followed by a

[20] Garner, Allard, and Clayton, "Superior Germ Plasm," 810, 786, 791; Garner, "Tobacco Culture," 1.

[21] Clayton, "Continuation of Mr. John Clayton's Account," 943.

[22] Garner et al., "History and Status," 405–6.

[23] Ibid.

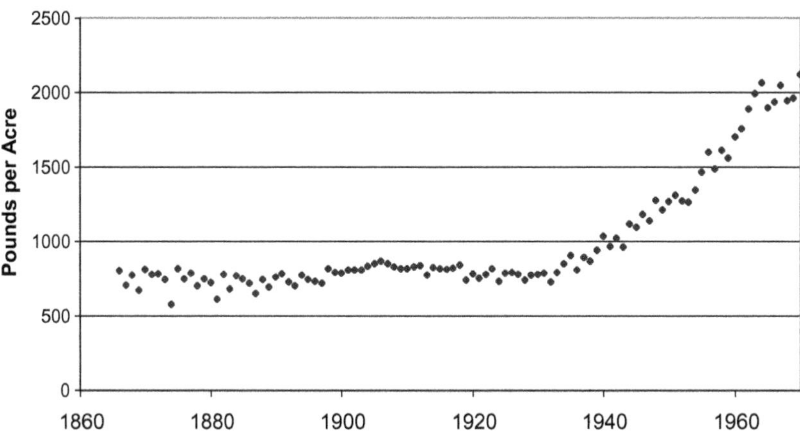

FIGURE 7.1. Tobacco yield in pounds per acre, 1866–1970. *Source:* Data downloaded from U.S. National Agricultural Statistics Service. *Quick Stats.*

decline before yields took off in the 1930s and beyond. These patterns are roughly similar to those holding for wheat, cotton, and maize. Due to the relative importance of quality over quantity in tobacco production, trends in yields per acre are less meaningful than for other staples. It was possible to increase yields per acre by planting on rich soils or by adding fertilizer, but this could affect the leaf's flavor, aroma, and processing characteristics in undesirable ways. The reduction in quality and market value often more than offset the increase in yield. The fixity of location and near constancy of yields should not lead us to downplay biological innovation, because the second half of the nineteenth century witnessed the development of the distinct tobacco types that would dominate the market to the present day.

Development of the Major Types

The development of American tobacco, at least after 1840, was largely intensive rather than extensive. Instead of spreading out across the country, tobacco cultivation took the direction of increasing quality and product differentiation. In 1922 the USDA noted that "the United States leads the world not only in the total production of tobacco but also in the number and diversity of distinctive types produced."[24] By the latter half

[24] Ibid., 396.

of the nineteenth century, American farmers, aided more than a little by serendipity, had developed five major tobacco types. The original American commercial type, dating from Jamestown, was dark fire-cured tobacco. This method of curing used smoke from slowly burning hickory or other woods to dry the leaf and impart strong flavors. Orinoco and its derivation, Pryor, which diverged in the early 1800s, were the principal fire-cured varieties. Large broad leaves, heavy in body, deep in color, oily, and with high nicotine content became the desirable characteristics for fire-cured tobacco plants.[25] A second type was dark air-cured. This tobacco was cured in barns without artificial heat. Lighter leaves came to be desirable for this purpose. Both dark types were exported to be used for snuff and plug chewing. The other three major types were "Bright Leaf," grown in North Carolina and southwestern Virginia, Cigar Leaf from Connecticut, and White Burley from the Ohio River Valley. A brief account of these latter three types highlights the importance of biological learning.

Today, Bright Leaf tobacco is synonymous with "flue-cured" tobacco and is a major ingredient in cigarettes. The exploding popularity of cigarettes after 1890 was due in large part to the perfection of lighter tobaccos, which produced smoke that could be inhaled into the lungs. Before these lighter tobaccos, "most smokers were content to puff their cigarettes without inhaling, like cigar and pipe smokers." Inhaling led to more rapid addiction, stimulating the demand for the mild tobacco.[26]

The development of Bright Leaf tobacco did not involve simply substituting one variety of seed for another. Rather it required a new way of doing business that, in Nannie Tilley's words, "ushered in a new era for the weed in North Carolina and Virginia." She further notes that "with the development of Bright Tobacco came far-reaching and radical changes in the industry. In order to produce the coveted leaf, farmers willingly discarded cultural techniques which had long sufficed for dark, fire-cured tobacco and experimented with various new schemes until methods for handling the seed bed, for harvesting and curing the leaf, and for rotating crops bore little resemblance to ante-bellum practices." The new tobacco also inspired revolutionary changes in tobacco marketing and manufacturing. The crucial ingredients in this process were the soil and curing methods, not the seed. "Pryor or Orinoco, if grown on rich lowlands, will produce a heavy dark leaf with a strong nicotine content, while the

[25] Garner, Allard, and Clayton, "Superior Germ Plasm," 811.
[26] Prince, *Long Green*, 49.

same varieties on light siliceous soils soon reverse their characteristics by yielding a thin, fine-flavored, yellow leaf."[27]

Starting with Rolfe's experiments, planters sought to produce lighter, milder, and sweeter-smelling tobaccos. Shortly after 1650 Edward Digges of the York River area of Virginia took one of the first steps toward the development of a lighter variety. Digges grew his crops on lowly regarded sandy loam soil. His "E. Dees" brand of yellow tobacco became renowned for its pleasant smell and fine taste and commanded a superior price. Gradually other farmers learned that "gray" as opposed to rich dark soils produced the finest, highest priced weed. Some succeeded in growing a yellow tobacco that commanded an "extraordinary" price. By the mid-1840s yellow or golden tobacco became so common that "Maryland" became the moniker for this unusual tobacco in Europe. Some Ohio farmers had also succeeded in producing a highly regarded yellow-leafed tobacco by 1825, and Kentucky planters had started experimenting with yellow tobacco by the 1840s. The importance of soil types became more apparent when tobacco culture moved into the new western uplands of Virginia and North Carolina, where the region's sandy, infertile soils produced leaf with an especially mild flavor. By the 1860s this area was fast becoming the most preferred tobacco-growing region in the South.[28] To accentuate the differences, farmers in the region altered their cultural practices, including topping plants higher and later, to produce the desired thinner leaves.

The ascendancy of Bright Leaf also depended on improved curing methods. The most important changes involved the switch from wood to charcoal and ultimately the adoption of flues, which allowed curers to more accurately regulate the heat and protect the leaves from developing strong flavors imparted by smoke and dust. Flue-curing also resulted in a higher sugar concentration in the leaf. Curing tobacco was the most delicate step in the production process, and "a trifling inattention, may at a critical moment, reduce a barn of the finest yellow tobacco to the lower grades."[29] A large dose of serendipity most likely gave rise to charcoal as the fuel of choice for fire-curing yellow tobacco. On a stormy night in 1839, Stephen Slade (a slave of Caswell County, North Carolina farmer Abisha Slade) fell asleep while tending the curing fires in a barn filled with tobacco. Upon awakening he discovered that his fires were

[27] Tilley, *Bright-Tobacco*, 3–11, quote on 11.

[28] Ibid., 6–11.

[29] Killebrew, *Report*, 117–18.

nearly out and that there was no dry wood to revive them. The quick-thinking slave substituted some nearby charcoal and to his amazement the tobacco kept turning yellower as the curing progressed. This tobacco sold for four dollars a pound – four times the market's average price. Charcoal curing quickly gained favor in the immediate neighborhood of the Slade property, but it did not take off elsewhere until Abisha Slade began publishing articles in 1856 detailing the cultivation, harvesting, and curing practices (including precise temperatures for different stages of the process). Changes in relative prices and in curing methods stimulated a significant shift in the quality of the tobacco marketed in North Carolina. In 1869, 36 percent of the state's crop was judged low-quality "shipping leaf." A decade later only 15 percent fell into this low-end category.[30]

In Virginia, farmers had begun experimenting with flue systems as early as 1809.[31] The most prominent inventor was Dr. Davis G. Tuck of Halifax, Virginia, whose detailed designs, patented around 1830, remained popular for over a century. The use of brick, stone, or iron flues dated to the 1860s, and "a general adoption" began about 1872. Flue-cured tobacco was thus a relatively new product when it became the basis for the expansion of the "tobacco road" in northern North Carolina and southern Virginia. In 1879 flue-curing had superseded other methods of curing in this region. As a result, the production of light, mild tobacco was transformed from a "hit-or-miss" affair to one governed by "a formula that would insure a 'yellow' cure."[32] Manuals instructed farmers on several flue techniques, explaining the advantages of each depending on the stage of the drying process, the characteristics of the leaves, the outside temperature, and so forth.[33] A leading example was Robert L. Ragland's *On the Cultivation and Curing of Tobacco*, which was published in multiple editions under various names from the 1860s through the 1890s. Ragland of Hyco, Virginia, was a leading breeder and seedsman in the Bright Leaf trade as well as an early promoter of tobacco fertilizers.[34]

[30] Tilley, *Bright-Tobacco*, 22–32; Peedin, "Flue-Cured Tobacco," 104; Killebrew, *Report*, 115. There are many versions of the "sleeping slave" story. Many accounts date the event to the 1850s on the farm of Eli and/or Elisha Slade and attribute the breakthrough to a slave named Samuel or Peter in some accounts. We rely on Nannie Tilley's impressively documented rendition.

[31] Tilley's account builds in part on the pioneering work of Joseph Clarke Robert, *Tobacco Kingdom*, 44–6.

[32] Herndon, *William Tatham*, 411.

[33] Killebrew, *Report*, 118; Tilley, *Bright-Tobacco*, 18–21.

[34] Ragland, *On the Cultivation and Curing of Tobacco*, 1–10, and *Tobacco*, 1–8.

The advent of flue-curing technologies and the growing market for Bright Leaf led to the spread of production into the Pee Dee region of South Carolina in the late 1880s. Seventy years earlier this region had all but abandoned tobacco as a cash crop. The tobacco revival was not an accident – rather, leading agriculturalists, concerned with the region's dependency on cotton, promoted a return to tobacco production. Most prominent in this movement was Frank Mandeville Rogers of Darlington County. In 1884, guided by advice from correspondents in North Carolina and Virginia, Rogers began experimenting with Orinoco on a small plot. In the following years Rogers increased the size of his plots and hired two experienced North Carolina tobacco growers to teach him the intricacies of growing and curing Bright Leaf. These were the first of many "tobacco instructors" who taught Pee Dee farmers the trade. Rogers gave seminars and demonstrations to neighbors and touted the new crop's virtues in the press. Drawn by Roger's accounts of high profits, other farmers experimented with a number of tobacco varieties, including Cuban cigar leaf. In 1886 the state commissioner of agriculture organized experiments with selected farmers in every county, recording "the types of seed and fertilizer used, as well as methods of planting, cultivating, harvesting, and curing." The commission offered a prize of one hundred dollars for the best tobacco. Rogers' lemon-colored tobacco won easily against weak competition – two-thirds of the entries were deemed to be of very poor quality. Many farmers still had to discover the best varieties and methods for their particular conditions and to master the difficult art of curing. But the commissioner's statewide trials helped energize the tobacco movement. Over the 1890s, "tobacco fever" swept many areas of South Carolina. In 1879, the Palmetto State had produced less than 46,000 pounds of tobacco; but by 1899, Bright Leaf production had grown to almost 20 million pounds (over 92 percent in the Pee Dee region), and by 1919 to 79 million pounds.[35]

Impressive changes also occurred in the fourth major type – the cigar leaf tobaccos which came to be identified with northern areas. Indigenous populations had long grown the weed in the Northeast. Early attempts by English and Dutch colonists to cultivate *N. tabacum* in the Connecticut Valley had produced a pungently flavored leaf that was not well received in Europe. The colonists later introduced a Virginia variety that evolved into "Shoestring," which had long, narrow leaves. Tobacco remained a

[35] Prince, *Long Green*, 51–62, 71, 85.

minor farm product in the Northeast until the market for cigars emerged after 1820.[36] In 1833 (or 1830 in alternative accounts), an East Windsor farmer, B. P. Barber, introduced Maryland Broadleaf. After he successfully adapted it to local conditions, the tobacco industry in the Northeast took off. In the Connecticut Valley's long, humid summer days and sandy soils, Broadleaf produced large, thin, elastic leaves with good burning qualities and a pleasant taste. Local strains of this type, known as Connecticut Broadleaf or Seedleaf, rapidly replaced the inferior Shoestring variety. Broadleaf strains spread to Pennsylvania and Ohio by the 1850s and to Wisconsin by the 1870s, founding their cigar leaf sectors.[37] Around 1870, a new introduction, known as Havana or Spanish Seed, became an important competitor of Seedleaf across the cigar leaf producing districts. This new variety, grown under different names in different districts, was selected from an imported Cuban tobacco and apparently possessed a high degree of genetic diversity. By 1879, J. B. Killebrew distinguished seven varieties of Seedleaf, of which he judged Connecticut the finest.[38]

The next important change in the cigar leaf sector began in 1896 when a USDA scientist in Florida started experimenting with seed from Sumatra. Sumatra produced such excellent wrapper leaves that American growers successfully sought tariff protection. The experiment revealed that the imported seed failed if grown in the open sun but, if shaded by trees, produced thin high-quality wrappers. By 1898, a system of growing the Sumatran seed under cotton-cloth shades proved successful. In 1900, the shade-growing technology was transferred from Florida to the Connecticut Valley. A "shade craze" erupted in the early 1900s, as northern farmers paid extravagant prices for Florida seeds, which often yielded unsatisfactory results. A longer period of experimentation, under the guidance of the U.S. Bureau of Plant Industry, helped define the boundaries of this system, which yielded a more sustained prosperity.[39] These innovations allowed American wrappers to compete with the Sumatran product.

[36] Ramsey, *History of Tobacco Production*, 113–14, 125–30; Anderson, "Growing Tobacco," 1–5; Russell, *Long, Deep Furrow*, 378–9.

[37] This is based on the USDA account. For alternative accounts of the introduction of cigar leaf tobaccos to Pennsylvania and Ohio, see Fletcher, *Pennsylvania Agriculture, 1640–1840*, 165–6; and Killebrew, *Report*, 132.

[38] Garner, Allard, and Clayton, "Superior Germ Plasm," 814–16; Killebrew, *Report*, 16.

[39] Shamel, "Improvement of Tobacco," 436–7; Stewart, "Production of Cigar-Wrapper," 1–12; Russell, *Long, Deep Furrow*, 456, 502–3.

White Burley was the fifth major type of tobacco. By 1820, planters in southern Ohio and Kentucky grew a red Burley tobacco descended from the old Maryland Broadleaf. Red Burley cured to a cinnamon colored leaf used for chewing plugs. A succession of better Burley strains replaced older types and by the 1830s leaf quality had improved markedly. In the "Burley" counties (Adams, Brown, and Clermont) of Ohio, production jumped from 131,000 pounds in 1840 to 1.8 million pounds in 1860.[40] By one standard account, two tenant farmers, George Webb and Joseph Fore, of Brown County, Ohio, acquired some Little Burley seed from across the Ohio River in Kentucky in the spring of 1864. Most of the seed grew well, but in one part of the seedbed, the plants appeared sickly and had a "dirty yellow" color. Assuming that those plants were diseased, Webb and Fore initially left them in the seedbed. Eventually, as seedlings were in short supply, they transplanted the suspiciously pallid plants. After taking root, the sports thrived. What's more, the "freak tobacco" – White Burley – brought excellent prices in the Cincinnati market.[41]

Within fifteen years of the discovery of White Burley, "four-fifths of the plug tobacco used in the east, north, and west is made from this variety, and its introduction and culture had worked one of the most remarkable revolutions known to the agriculture of this country." In Kentucky, growers in the "White Burley district... abandoned every other variety."[42] As well as commanding a higher price, White Burley required less labor and was cheaper to produce. It could be harvested quickly by cutting the stalk whole and could be processed quickly by air curing. Eventually, the new variety took the name Burley strictly for itself. This absorbent, chlorophyll-deficient tobacco served as an ingredient in chewing tobacco, snuff, and, later, cigarettes. White Burley was "an outstanding exception to the general rule that light open soils tend to produce a thin leaf, light in weight and color, weak in aroma, and mild, while heavy soils

[40] Axton, *Tobacco and Kentucky*, 50–1.

[41] Other accounts offer a slightly different timeline, which dates White Burley to as early as 1860 in Kentucky. See Axton, *Tobacco and Kentucky*, 69; Killebrew, *Report*, 126–7; "Our Tobacco Jubilee," *Louisville Courier-Journal* (18 September 1885), 2; Van Willigen and Eastwood, *Tobacco Culture*, 11–12.

[42] Killebrew, *Report*, 74. The manuscript census documents Webb's rising fortunes. In the 1860 census, a George Webb is listed as a 27-year-old farmer in Pleasant Township of Brown County, Ohio, with $300 in personal wealth and no real estate, which is consistent with tenancy status. In 1870, he was living with his family in Foster Township in Bracken County, Kentucky, and had acquired $2,000 in real estate and $675 in personal wealth. U.S. Census Bureau. 8th Census 1860. *Population Schedules. Ohio. Brown County*, and *Kentucky. Bracken County.*

produce thick, heavy, strong leaf of dark colors."[43] As demand shifted from chewing tobacco to cigarettes, the process of adaptation continued. Burley farmers shifted from drooping-leaf strains to so-called "stand-up strains," which had thinner leaves better suited for cigarettes.[44]

The end result of this evolutionary process was the elaboration of a series of types officially defined by the USDA in the interwar years. Few changes occurred thereafter and these definitions were adopted internationally. Types were categorized chiefly by use and method of curing, but not strictly by seed variety. Orinoco, descended from the original variety grown in the Chesapeake, was used as seed for a range of different types. Table 7.2 lists the types, their uses, the varieties of seeds used in the mid-1930s, and their relative importance as measured by acreage in 1929. Only 16 percent of the acreage was planted in types (that is, varieties grown and processed using specific cultural and curing techniques) that were in standard use in 1840. Adding to the picture, Figure 7.2 maps the areas where each type was chiefly grown. Note that, given yield and price differentials, acreage does not fully capture the value of the crop. By the interwar years, clusters of production of highly specific varieties had clearly emerged and solidified.

Fighting Decline

Tobacco cultivation is renowned for exhausting the soil. The early Chesapeake planters practiced a form of slash-and-burn agriculture – clearing new fields, planting tobacco for a few years, turning perhaps to other crops, and then allowing the land an extended period of fallow.[45] Tobacco and new lands were virtually synonymous in the colonial period and beyond. Various reasons have been offered for this close association. Avery Craven argues that colonial legal restrictions, which set a maximum number of plants or leaves per worker, created an impetus to cultivate on the best soils, where the plants and leaves would be larger. These practices led to the rapid depletion of existing soils and a continual movement to new soils.[46] As Thomas Jefferson wrote to George Washington in 1793, "We do not manure because we can buy an acre of new

43 Garner, *Production of Tobacco*, 353.

44 Garner, Allard, and Clayton, "Superior Germ Plasm," 820–1; Gage, "Historical Factors," 44.

45 Walsh, "Plantation Management in the Chesapeake," 393.

46 Craven, *Soil Exhaustion*, 31–3, 40–1, 56–8.

TABLE 7.2. *Principal Commercial Types of Tobacco, 1930*

U.S. Types	Name	Chief Use	Variety	Acreage (Thousands)
21–24	Fire-Cured	Export, snuff, and plug wrapper	Orinoco	213
35, 36, 37	Dark, Air-Cured	Chewing plug, export	Orinoco and One Sucker	74
32	Maryland	Cigarettes and export	Maryland Broadleaf	32
61–62	Cigar–Wrapper (shade–grown)	Cigar wrapper	Cuban, Florida 301, Round Tip	13
51, 52, 54, 55	Cigar Binder	Cigar binder	Conn. Broadleaf, Havana Seed	58
41–44	Cigar Filler	Cigar filler	Penn. Seedleaf, Ohio Seedleaf, Zimmer Spanish, Little Dutch	69
11–14	Flue-Cured	Cigarettes, pipe and chewing tobacco, export	Orinoco	1,134
31	Burley	Cigarettes, pipe and chewing tobacco	White Burley	417

Note: Total acreage was 2,016,000 and includes a number of minor types.
Sources: Garner, Allard, and Clayton, "Superior Germ Plasm," 811; USDA, *Yearbook 1930*, 706–7.

land cheaper than we can manure an old acre."[47] John Taylor of Caroline noted that "agriculture in the South does not consist so much in cultivating lands as in killing it."[48] Instead of abandoning their Virginia homes and moving west, Taylor suggested that tobacco planters apply manure

[47] Betts, *Thomas Jefferson's Farm Book*, 194.
[48] Lebergott, "Demand for Land," 190.

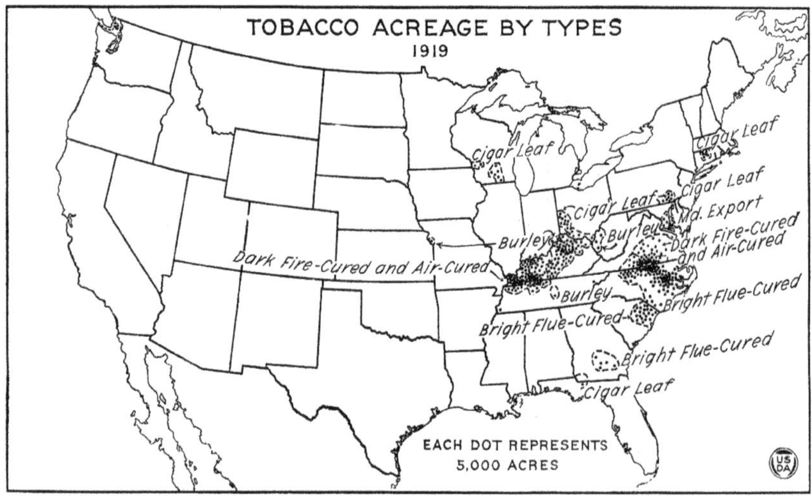

FIGURE 7.2. Tobacco acreage by types, 1919. *Source:* USDA, *Yearbook 1922*, 410.

more generously. Finding this remedy ineffective, Edmond Ruffin later advocated the use of marl, a source of lime that offsets the soil's natural acidity.

The effect of fertilizer on tobacco production is highly contingent on the specific conditions. In the North, farmers applied fertilizer liberally, which had a positive effect on quality and yields. Fertilizer caused earlier ripening, which was a key to success in Connecticut and other northern centers of production.[49] As long as planters in the Chesapeake were growing stronger, darker tobacco, they utilized alluvial soils and rotated to fresh land every few years. This helped ensure an adequate supply of nitrogen and other nutrients. But the movement to the lighter, milder tobaccos changed the calculus. Success here depended on "starving" the tobacco of specific nutrients. Farmers, first in Maryland and Virginia and then in the Carolinas, discovered that nitrogen-poor soils were the ticket to success. The pioneering Bright Leaf farmers even abandoned the rotation schemes involving wheat and clover that were used to regenerate the richer soils. The application of fertilizer was contraindicated.[50]

In the postbellum period, there was a distinct change in cultural practices, and even Bright Leaf farmers turned to fertilizer. Tilley notes that

[49] Killebrew, *Report*, 243, 263; Herndon, *William Tatham*, 432.

[50] Tilley, *Bright-Tobacco*, 181.

chemical fertilizers were "in general use" in the Bright Leaf areas by the end of the 1870s. Firms such as Blue Anchor and Virginia–Carolina began selling chemical fertilizers in the 1870s. In 1877, the new North Carolina experiment station began analyzing chemical fertilizers and initiated field trials four years later.[51] Taking five Bright Leaf counties, the expenditures on fertilizer increased tenfold in real terms between 1879 and 1919. The growing scarcity of land may help explain this reversal in practice, but there were also significant advances in understanding the effects of fertilizer composition on the quantity and quality of output.

The composition of the fertilizer was of critical importance. Animal manures contained chloride in quantities that adversely affected the commercial qualities of cured tobacco. Bright Leaf farmers experimented with Peruvian guano in the 1870s but found it unsatisfactory (it was too rich in ammonia and lacked potash). Tobacco had an unusually high potash requirement, but supplying it via the commonly used form, mutriate of potash, created similar leaf-quality problems – as a result, sulfate of potash was preferred.[52] Excessive use of lime led to frenching, a disorder which caused a curling of the leaves. Phosphoric acid promoted rapid maturation but too much led to premature ripening. The tobacco plant, like many crops, proved to be "very sensitive to nitrogen nutrition." As noted, high quantities of nitrogen in the soil produced tobacco that, when cured, tended to be dark and have "a strong and pungent smoke." Low quantities caused the leaves to yellow prematurely and yielded paler tobacco when cured. "An increase in nitrogen will increase the yield of tobacco, but quality is often reduced at high levels."[53]

Garner, America's foremost authority on tobacco, summarized the state of knowledge regarding fertilizer in the early twentieth century. The value of tobacco output per acre was generally "sufficiently high to justify considerable expenditure for fertilizer." But their "rational use" was "a complicated problem because of the marked effect" that fertilizers had on product's quality.[54] Strongly flavored tobaccos, such as those used for cigars, did best in terms of both quality and yield when supplied with abundant nutrients, whereas lightly flavored tobaccos, such as those used in cigarettes, "possess best quality only when grown with a

[51] Ibid., 155–72. Some firms "mislabeled" worthless products, which led to state laws requiring manufacturers to align the content of their fertilizers with the specifications on their labels.

[52] Worthen, *Farm Soils*, 325–7.

[53] Flower, "Field Practices," 80–1.

[54] Garner et al., "History and Status," 420–1.

restricted supply of plant food that serves to limit the yield."[55] Given the price structure, pursuing higher yields at the expense of quality was less advantageous for tobacco than for most other crops.[56]

Like the producers of the other crops we have considered, tobacco growers faced an evolving insect and disease environment. Despite the fact that nicotine was widely used as an insecticide, the tobacco plant was subject to damage from an array of insects. Even a small amount of damage to the leaves could significantly reduce revenues. The most destructive insects were hornworms – *Phlegethontius sexta* and *P. quinquemaculata.* Garner noted that "in most areas, especially in the South, culture of tobacco would be impossible" unless their ravages were controlled. As Killebrew observed in his discussion of Ohio: "The hornworms are more numerous some seasons than others, and in 1877 and 1878 about one-third of the crop was very much damaged by them. Tobacco-growers, especially those of the Miami valley, are unusually industrious and energetic enough to prevent serious injury to their crops, but now and then their vigilance is severely taxed." Producing tobacco required a great deal of care and attention, which could be unpleasant at times, but those engaged in the plant's cultivation found few tasks as disagreeable as "worming," that is, removing the worms by hand.[57] Powdered arsenate of lead could also be used to control hornworms. Farmers faced additional challenges from wireworms, cutworms, and other insects.

As with most crops, the introduction of new pathogens and the pervasiveness of diseases after years of cultivation in a given area increased the severity of tobacco plant diseases over time. By the 1930s diseases claimed about 15 percent of the crop annually, but without biological learning, losses would have been much heavier.[58] Many diseases that caused great damage in the field originated in the seedbed. Even before diseases were well understood, farmers burned trash or wood on their seedbeds to cleanse the soil. By the 1920s farmers were experimenting with formaldehyde and acetic acid to sterilize seedbeds and control diseases, weeds, and insects. But the only method that came into general

55 Garner, *Production of Tobacco*, 101. Good results were generally attained when conditions allowed "rapid, uninterrupted growth" of the tobacco plant. USDA, *Yearbook 1922*, 418.

56 Garner et al., "History and Status," 422.

57 Garner, *Production of Tobacco*, 281; Van Willigen and Eastwood, *Tobacco Culture*, 108; Killebrew, *Report*, 146.

58 Garner, *Production of Tobacco*, 234.

FIGURE 7.3. Steaming tobacco seedbeds in Kentucky in the 1940s. As with other aspects of tobacco production, this was a very labor-intensive process. *Source:* Courtesy of the J. Winston Coleman Kentuckiana Collection, Transylvania University Library.

use was an elaborate soil steaming system that employed a steam boiler, pipes, and an inverted sheet-metal pan about six feet wide and ten feet long. Figure 7.3 shows a steaming crew at work. Treating each small parcel for at least thirty minutes raised the soil's surface temperature to 212 degrees Farhenheit, which was sufficient to kill most pathogens, weed seeds, and insects down to 6 inches.[59] "Most plant growers who have used the steaming method find it pays from the weed-killing standpoint alone." But the major benefit was the control of "damping off or bed rot, black root rot, root knot, mosaic, wildfire, and black fire." A thriving market emerged with contractors providing steaming services at one to two dollars per hundred square feet.[60]

After 1880 the tobacco-growing environment took a decided turn for the worse. Granville Wilt, caused by *Bacterium solanacearum*, was among

[59] Ibid., 116–19.

[60] Johnson, "Steam Sterilization," 2, 11, and "Tobacco Diseases," 3–7. The inverted pan method was invented around 1900 to kill nematodes. Gilbert, "Root Rot of Tobacco," 34–42; Lucas, *Diseases of Tobacco*, 22–6.

the most serious of the emerging threats. The disease was first observed on the farm of B. F. Stern in Granville County, North Carolina, in 1881 and was scientifically identified and named in 1903.[61] The disease caused wilting and rapid death. The bacteria took "possession of the soil, prohibiting successful tobacco culture in succeeding years, and in sections where tobacco is the chief, possibly the only, profitable money crop the advent of the disease has caused great depreciation in farm values."[62] Areas with sandy and sandy loam soil proved especially susceptible. The only remedy was extended retirement of the fields from tobacco production. Another new threat was Black Shank, a fungal disease that caused stunting and a black discoloration on the stalk. Black Shank first appeared in Java in 1893 and entered the United States at Quincy, Florida, in 1915.[63] By 1930, it had engulfed most of the tobacco belt. Quincy was also the location where Blue Mold, or Downey Mildew, first struck. This fungal disease appeared in 1921, died down until early 1931, then reappeared in Florida and spread as far as Louisiana and Virginia by the end of that year, to Maryland by 1932, to Pennsylvania and Tennessee by 1933, and to Connecticut by 1937. Farmers learned to associate Blue Mold with cool, wet weather and to be watchful for the telltale signs of infection – the shriveling of seedlings and the development of yellow leaf spots on more mature plants. Yet another new threat was Wildfire, a leaf-spot disease that first appeared in North Carolina around 1917 and quickly spread to most other producing areas.[64] The first line of defense against most of these new diseases was crop rotation, but more control options were needed.

As with wheat and cotton, tobacco farmers frequently complained that the plant itself "ran down." As early USDA plant breeders noted, "without careful selection and breeding there is no crop which so quickly deteriorates in yield and quality." As of 1904, it was "a well-known fact that the proportion of poor-grade tobacco" was "increasing, resulting in a corresponding loss to the growers" in the long-established districts. The decline in yield and quality was in part "due to the lack of systematic and careful seed selection."[65] However, on the positive side, "no

[61] Hairr, *Harnett County*, 131; Smith, "Granville Wilt of Tobacco."

[62] Stevens and Hall, *Diseases of Economic Plants*, 300; Lucas, *Diseases of Tobacco*, 452–3.

[63] Garner, *Production of Tobacco*, 241; George B. Lucas, *Diseases of Tobacco*, 166, has Black Shank first appearing in southern Georgia.

[64] Garner, *Production of Tobacco*, 253; Shoemaker and Shew, "Fungal and Bacterial Diseases," 183–96; Garner et al., "History and Status," 425.

[65] Shamel, "Improvement of Tobacco," 435.

crop . . . responds so readily to breeding as tobacco."[66] Scientific breeders, at least since Darwin, devoted substantial effort to understanding tobacco as a plant model. Garner and Harry A. Allard discovered the photosensitivity of plants during their study of tobacco.

Until 1936 cigar tobacco varieties had "received more attention from breeders than the other commercial types of leaf. Beginning about 1900, experiments in the production of improved varieties through hybridization and selection were undertaken in several of the cigar-tobacco areas." Achieving desired leaf characteristics without upsetting curing qualities proved difficult. But by 1921 in New England, Edward M. East and Donald F. Jones produced Round Tip – a cross of Connecticut Broadleaf and Sumatra. East and Jones trumpeted their new creation as "Made to Order," bred to meet the specifications of cigar smokers, manufacturers, and tobacco farmers. It also was resistant to black root rot, but unfortunately it proved susceptible to Black Shank.[67] To that point, scientific breeders had done little to improve flue-cured varieties.[68] Over the following decades, tobacco breeders searched for disease-resistant strains of tobacco. At the North Carolina experiment station, E. G. Moss and Thomas E. Smith produced a wilt-resistant tobacco. In 1931, W. B. Tisdale of Florida bred a cigar wrapper variety relatively resistant to Black Shank. Later work identified varieties resistant to both Granville Wilt and Black Shank.[69] Today "(a)lmost all flue-cured and burley crops are produced from modern disease-resistant varieties."[70]

Conclusion

Tobacco is the crop most closely associated with America's birth as a nation. It is the staple responsible for the success of the Jamestown settlement; it was the first production activity where bound labor was employed and where the use of black slaves began; it was the cash crop that generated the wealth of many of the founding fathers and, indeed, of four

[66] Shamel and Cobey, "Tobacco Breeding," 8. "Improvement in methods of culture, curing, and fermentation have resulted in the production of tobacco having an increased value, but the most important factor in the development of more valuable tobacco has been the production of improved varieties by seed selection and breeding. The production of these improved varieties adapted to local soil and climatic condition has made possible the rapid development of the industry."

[67] East and Jones, "Round Tip Tobacco," 51–6; Babcock, "Recent Progress," 393–400.

[68] Garner, Allard, and Clayton, "Superior Germ Plasm," 816, 819.

[69] Garner, *Production of Tobacco*, 244; Tilley, *Bright-Tobacco*, 186.

[70] Miller and Fowlkes, "Dark Fire-Cured Tobacco," 165.

of the first five presidents. Tobacco was the first commercial crop grown on a large scale west of the Appalachians. And for most of the colonial period and for key regions thereafter, tobacco was the leading export. Like many of the plants and animals important to this country's agricultural success, tobacco's introduction depended on smuggled genetic material.

Yet tobacco fits few of the stylized accounts concerning American agricultural development. Tobacco production was highly labor-intensive and resistant to mechanization, especially of the harvest. Many of the mechanical devices subsequently adopted were used in pre-harvest (cultivating) or post-harvest (curing) operations. Its development, at least after 1840, took an intensive rather than extensive form. Rather than push the tobacco frontier onto new lands in the West, farmers engaged in growing and processing specialized products suited to more narrow geographic niches. The failure of tobacco to fit the stylized laborsaving account of American agricultural history parallels California's experience.

8

California

Creating a Cornucopia

California is the nation's leading agricultural state, renowned for a multiplicity of specialty crops ranging from grapes to tree crops to vegetables. Although the state's seal depicts Minerva, implying an economy born fully grown and imbued with wisdom and knowledge, nothing could be further from the truth. The first several generations of European colonizers struggled to learn to grow extensive crops such as wheat and to adapt livestock to the state's varied landscapes. The switch to the state's signature intensive crops (fruits, nuts, vegetables, and cotton) was first and foremost an exercise in biological learning as farmers experimented with new crops and gradually matched them to soils and microclimates. In the process they made massive investments in redesigning the landscape by controlling surface water, pumping ground water, and leveling fields. California farmers also faced a fearsome array of pests and diseases that increased in severity with the introduction of new invasive species and with the greater concentration in production of specific crops. In response the state became a world leader in creating scientific and administrative infrastructures to protect its crops. These initiatives generated significant positive spillover for farmers elsewhere. Taken in its totality the state's history provides powerful evidence of the importance of biological innovation.

Extensive Crops in the Nineteenth Century

When disgruntled miners left the gold fields, they found a nearly ideal environment for raising wheat – great expanses of fertile soil and flat terrain combined with a climate of rainy winters and hot, dry summers. By the mid-1850s, the state's wheat output exceeded local consumption. As production continued to expand, California began shipping grain to European markets, setting a pattern of integration into world markets that has characterized the state's agriculture ever since. Over the late nineteenth century, California's grain operations evolved quite differently from that of the family farms of the American North. The image of lore is of vast tracts of grain, nothing but grain, grown on huge bonanza ranches in a countryside virtually uninhabited except at harvest and plowing time. While this picture is clearly overdrawn, it contains many elements of truth. By the standards of the eastern United States, California grain operations were quite large and most employed laborsaving, scale-intensive technologies.[1] But the operations also utilized different biological technologies – growing different wheat varieties with different cultural techniques – than did their eastern brethren. In fact, when eastern farmers migrated to California they had to relearn how to grow the crop. With the mild winters of California, farmers learned that it was advantageous to extend the season of continuous growth by sowing spring-habit wheat in the fall, which was unheard of in the East.

According to contemporary accounts, after learning to cultivate Sonora and Club wheats in the 1850s, 1860s, and 1870s, Californians invested little to improve cultural practices, introduce new varieties, or even maintain the quality of their seed stock. Decades of monocrop grain farming, involving little crop rotation, fallowing, fertilizer, or deep plowing, mined the soil of nutrients and promoted the growth of weeds. The grain also deteriorated in quality, becoming starchy and less glutinous. These unsustainable "soil mining" practices may well have been an economically rational land-use policy under the high interest rates prevailing in the state in the mid-nineteenth century, but with sharply declining yields and quality, wheat ceased to be a paying crop in many areas and was virtually abandoned (as indicated in Figure 8.1).[2] This experience illustrates what happens in the absence of continual biological innovation.

Figure 8.1 documents the changes in California's crop mix by showing how cropland harvested in California was distributed across selected

[1] The search for large-scale, laborsaving technologies culminated with the perfection and diffusion of the combined grain harvester in the late 1880s and the 1890s. See Olmstead and Rhode, "Overview of California Agricultural Mechanization," 86–112.

[2] Shaw, "How to Increase the Yield of Wheat," 255–7; Blanchard, "Improvement of the Wheat Crop," 1–5.

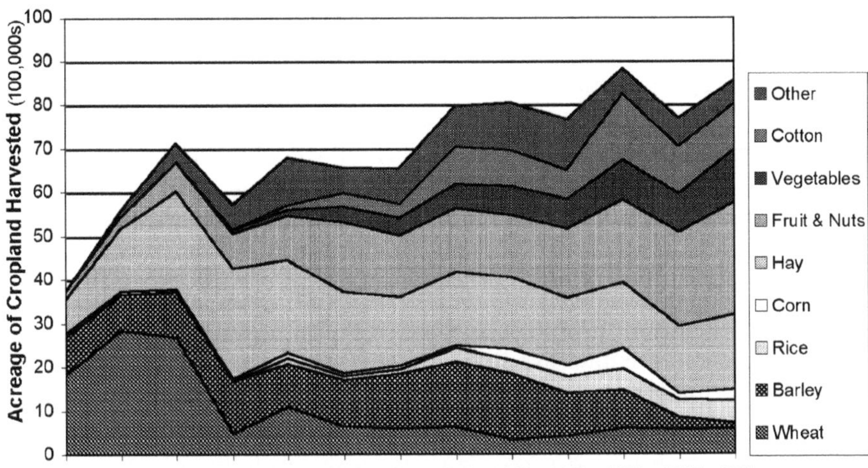

FIGURE 8.1. Distribution of California cropland harvested, 1879–1997. *Sources:* Constructed from U.S. Census Bureau. 9th Census 1870, *Statistics*; Ibid. 10th Census 1880, *Agriculture*, Vols. 3, 6; Ibid. 12th Census 1900, *Census Reports*, Vol. 5; Ibid. 14th Census 1920, *Reports*, Vol. 5; Ibid. 15th Census 1930, *Agriculture*, Vol. 2. Reports by States, pt. 3, Western States; Ibid. 16th Census 1940, *Agriculture*, Vol. 1, *Statistics for Counties*, and State Reports, pt. 6, *Mountain and Pacific States*; Census of Agriculture 1950, Vol. I. *Counties and State Economic Areas*, pt. 33, *California*; Ibid. Census of Agriculture 1959, *Final Report*, Vol. 1, pt. 48, California, Vol. 2, *Statistics by Subject*; U.S. Dept. of Agriculture. Census of Agriculture 1997, Table Ca 1–01; Craig, Haines, and Weiss, *U.S. Censuses of Agriculture*.

major crops over the period 1879–1997. In 1879, wheat and barley were grown on more than 75 percent of the state's cropland, whereas the combined total for intensive crops was around 5 percent. By 1929, the picture had changed dramatically – wheat and barley accounted for about 26 percent of the cropland harvested and the intensive crop share stood at around 35 percent. In absolute terms, the acreage in the intensive crops expanded by more than tenfold during this half-century while that for wheat and barley fell by more than one-third.

The Growth of Specialty Crops

Between the 1880s and 1914, the California farm economy swiftly shifted from large-scale ranching and grain-growing operations to smaller scale, intensive fruit cultivation. By 1910, California had emerged as one of the world's principal exporters of grapes, citrus, and various deciduous fruits. The rapid growth continued even after these crops were well established. By 1930, production of most of these crops had increased several hundredfold since 1880 and three- to fivefold from the levels of the early 1910s. In 1919, California produced 57 percent of the oranges, 70 percent of the prunes and plums, over 80 percent of the grapes and figs, and virtually all of the apricots, almonds, walnuts, olives, and lemons grown in the United States. The share of intensive crops in the value of total output climbed from less than 4 percent in 1879 to over 20 percent in 1889. By 1909, the intensive share reached nearly one-half; by 1929, it was almost four-fifths.

Another vantage point on the state's transformation is offered in Table 8.1, which provides key statistics on the evolution of California agriculture between 1859 and 1997. Almost every aspect of the state's development after 1880 reflected intensification. Between 1859 and 1929, the number of farms increased by about 700 percent. The average size of farms fell from roughly 475 acres per farm in 1869 to about 220 acres in 1929, and improved land per farm dropped from 260 acres to about 84 acres over the same period. Movements in cropland harvested per worker also fell. The land-to-labor ratio fell from about 43 acres harvested per worker in 1899 to 20 acres per worker in 1929. It is especially noteworthy that the shift to more labor-intensive crops overwhelmed the effect of mechanization.[3]

Explanations for the causes and timing of California's structural transformation have long eluded scholars. The traditional literature presents a laundry list of causes, including (1) increases in demand for income-elastic fruit products in eastern urban markets; (2) improvements in transportation, especially the completion of the transcontinental railroad and the development of refrigeration; (3) slumping world grain prices; (4) the spread of irrigation and the accompanying breakup of large land holdings; and (5) the increased availability of "cheap" labor. Yet a careful investigation of the transformation yields a surprising result – much of the credit for the shift to intensive crops must be given to exogenous declines in real interest rates and to "biological" learning

[3] After 1909, cotton and sugar beets became important, contributing to the impressive rise of the intensive share in the 1910s and 1920s. See Rhode, "Learning, Capital Accumulation," 773–800; Olmstead and Rhode, "Induced Innovation (1993)," 100–18.

TABLE 8.1. *California's Agricultural Development*

	No. of Farms (thousands)	Land in Farms (thousands of acres)	Improved Land (thousands of acres)	Cropland Harvested (thousands of acres)	No. of Farms Irrigated (thousands)	Irrigated Land (thousands of acres)	Ag. Labor Force (thousands)
1859	19	8,730	–	–	–	–	53
1869	24	11,427	6,218	–	–	60–100	69
1879	36	16,594	10,669	3,321	–	300–350	109
1889	53	21,427	12,223	5,289	14	1,004	145
1899	73	28,829	11,959	6,434	26	1,446	151
1909	88	27,931	11,390	4,924	39	2,664	212
1919	118	29,366	11,878	5,761	67	4,219	261
1929	136	30,443	11,465	6,549	86	4,747	332
1939	133	30,524	–	6,534	84	5,070	278
1949	137	36,613	–	7,957	91	6,599	304
1959	99	36,888	–	8,022	74	7,396	284
1969	78	35,328	–	7,649	51	7,240	240
1978	73	32,727	–	8,804	56	8,505	311
1987	83	30,598	–	7,676	59	7,596	416
1997	74	27,699	–	8,543	56	8,713	260

Sources: Constructed from Taylor and Vasey, "Historical Background of California Farm Labor"; U.S. Census Bureau. 15th Census 1930, *Agriculture*, Vol. 4, *General Report*; U.S. Census Bureau. *Census of Agriculture 1959*, Vol. 1, *Counties*, pt. 48, *California*; U.S. Census Bureau. 20th Census 1980, *Census of Population*, Vol. 1, pt. 6; USDA, *1997 Census of Agriculture*, Table 1; U.S. Census Bureau. 21st Census 1990, *Census of Population*. Sec. 1; U.S. Census Bureau. 22nd Census 2000, *Census 2000 Summary File 3*, Table QT-P27; Craig, Haines, and Weiss, *U.S. Censuses of Agriculture*; Rhode, "Learning, Capital Accumulation," 797.

about how to grow horticultural crops in the California environment.[4]

Isolated from America's financial markets, California farmers faced high interest rates, which discouraged capital investments. Rates fell from well over 100 percent during the Gold Rush to about 30 percent around 1860 and then down to 8 to 12 percent by 1890. Contemporaries well understood that the implications of falling interest rates for a long-term investment, such as an orchard, were enormous.[5] The break-even interest rate for the wheat-to-orchard transition was about 10 to 13 percent, which conformed closely to the levels prevailing in California when horticulture began its ascent.

The second key supply-side force was the increase in horticultural productivity associated with biological learning. Yields of major tree crops nearly doubled between 1889 and 1919. When the Gold Rush began, the recent immigrants knew little about the region's soils and climate. As settlement continued, would-be farmers learned about soils, climate, and the availability of water. This sometimes required overcoming deep-seated prejudices.[6] Fruit growers engaged in a costly and time-consuming process to identify the best plants and cultural practices for particular locales. Horticulturalists introduced varieties from around the world, and local plant breeders created new varieties. Most notably, the legendary Luther Burbank, who settled in Santa Rosa in 1875, developed hundreds of varieties of plums and other fruits over his long career.[7]

The ability of California farmers to compete with low-cost European growers offers stark testimony to the importance of biological learning.[8] California horticulturists relied on several strategies, including product differentiation, better organization, and a more intensive use of scientific inputs, to offset the impact of higher wages and transportation costs. For most crops, American producers rapidly gained a reputation on both sides of the Atlantic for quality and consistency that was unmatched by their European counterparts. Quality mattered. For many crops the

4 Rhode, "Learning, Capital Accumulation," 773–800.

5 Burns, "Pioneer Fruit Region," 11.

6 Rhode, "Learning, Capital Accumulation," 795.

7 Tufts et al., "Rich Pattern of California Crops," 206; Hodgson, "California Fruit Industry," 337–55. Many plant breeders thought that Burbank's public image outshined his accomplishments. Burbank applied an intuitive approach to mass selection that depended on his powers of observation and memory. Rather than follow the latest advances in plant science, he believed in vitalism and the inherence of acquired characteristics. See Dreyer, *Gardener Touched with Genius*, 216–30.

8 Morilla Critz, Olmstead, and Rhode, "Horn of Plenty," 316–52, and "International Competition," 211.

PLATE 2.1. A cloud of rust spores in a grain field. Courtesy of the UC Statewide IPM Project. Photo by Jack Kelly Clark.

PLATE 3.1. The mixture of Northern Flint (left) and Southern Dent or gourdseed (right) led to the creation of Corn Belt Dents. *Source*: Russell and Sandall, *Corn Breeding*, Fig. 5, Courtesy of University of Nebraska-Lincoln, Dept. of Agronomy and Horticulture, 2004.

PLATE 5.1. The life stages of the cotton worm as depicted in Riley's 1885 treatise. *Source:* Riley, *Fourth Report of the United States Entomological Commission.*

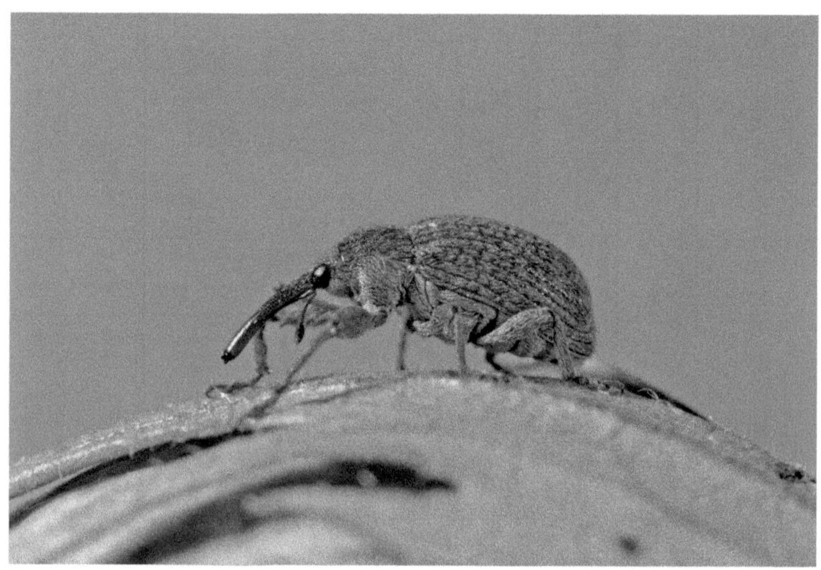

PLATE 5.2. Adult boll weevil. Courtesy of the UC Statewide IPM Project. Photo by Jack Kelly Clark.

PLATE 5.3. Cotton damaged by the boll weevil. Courtesy of the UC Statewide IPM Project. Photo by Jack Kelly Clark.

PLATE 8.1. Vedalia beetle attacking cottony cushion scale. This beetle saved the California citrus industry. Courtesy of UC Statewide IPM Project. Photo by Jack Kelly Clark.

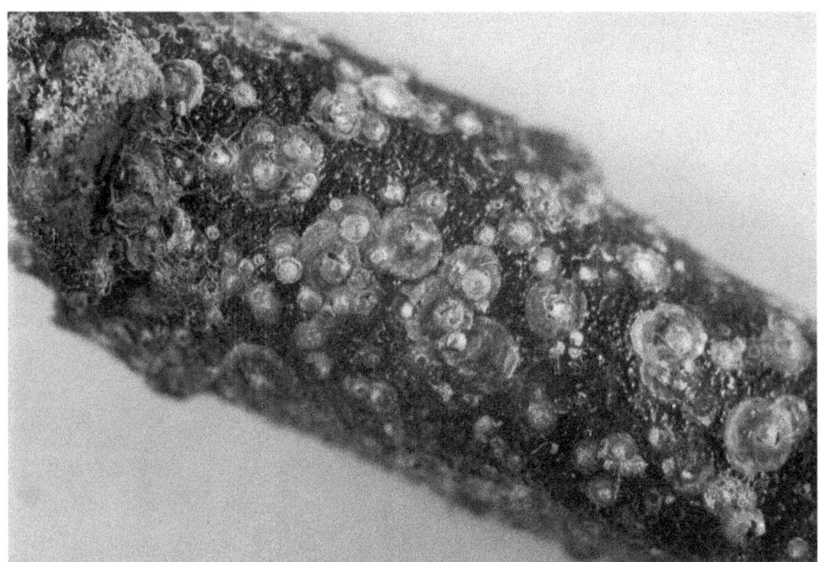

PLATE 8.2. San Jose scale colony on old wood. Courtesy of UC Statewide IPM Project. Photo by R. E. Rice.

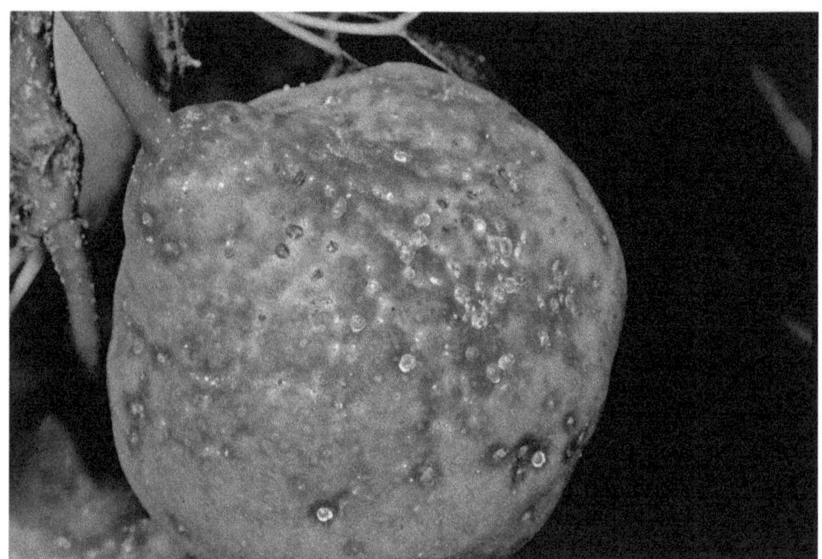

PLATE 8.3. Pear infested with San Jose scale. Courtesy of UC Statewide IPM Project. Photo by Jack Kelly Clark.

premium for California fruit was two to four times that of their European competitors.[9]

A review of the introduction and spread of a few horticultural crops illustrates the extent of the learning process and points to the distinctive and arguably more advanced practices that California producers had developed. Many Europeans studied the "California Model" but had great difficulty building the coordinated infrastructure and transplanting the farm production methods essential for its adoption. The biological technologies that were the basis for California's competitive advantage simply did not travel well.[10]

Grapes and Wine

Over the course of the first century of Anglo settlement, California farmers and vintners gradually learned how to perfect their products. Biological learning took several forms. Farmers went through an arduous process to discover where and how to cultivate grapes for winemaking. Growers then had to perfect their skills and adopt new varieties to improve the quality of their product (and they also had to develop winemaking skills). Finally, they had to adapt quickly to fend off the diseases and insects that threatened their vines.

Many early observers recognized California's potential to become America's best wine-growing region. For example, the 1860 Census of Agriculture observed: "In California alone, it is stated, there are five millions of acres well adapted to grape culture. Here is something to reflect upon, and to give hope for the future."[11] Californians were already reflecting or, more aptly, dreaming. In 1854 the founding editor of the *California Farmer* noted, "California is destined to become a mighty vineyard – her wine presses running over with wine."[12] But even with all its promise, there was a vast divide between dreaming and doing.

A long view of the growth of the state's wine industry is offered in Figure 8.2, which graphs the annual wine and brandy output (in millions of gallons) from 1865 to 1916.[13] California was the leading winemaking state in 1850, with production of a mere 58,000 gallons. This small level of output made up about one-quarter of all U.S. production on farms. By

9 See U.S. House Committee on Ways and Means, *Tariff Hearings 1921*, 2027, 2035; and Morilla Critz, Olmstead, and Rhode, "Horn of Plenty."

10 For a detailed discussion of the California Model in citrus, see Karlinsky, *California Dreaming*.

11 U.S. Census Bureau, 8th Census 1860, *Agriculture*, clxi.

12 Butterfield, *History of Deciduous Fruits*, 30.

13 Olmstead and Rhode, "Quantitative Indices."

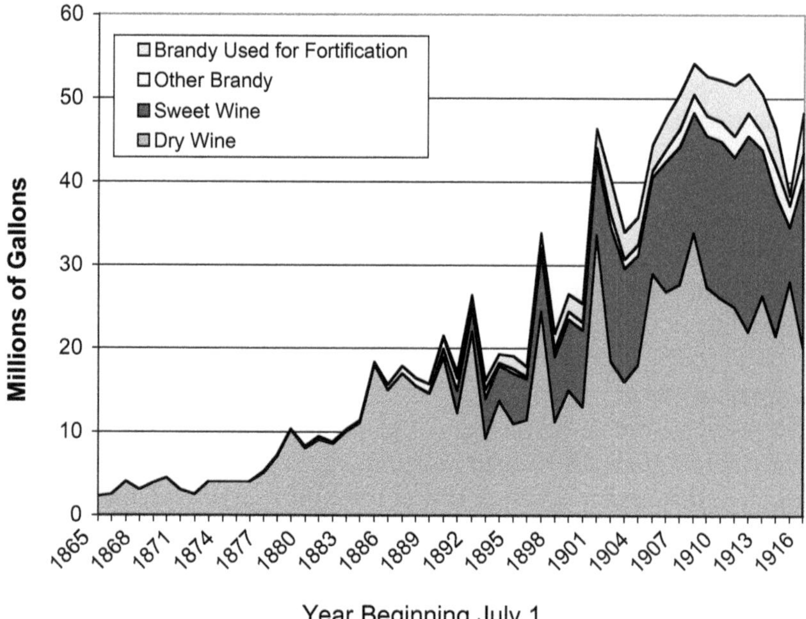

FIGURE 8.2. California wine and brandy production, 1865–1916. *Source:* California State Board of Agriculture, *Statistical Report*, various years.

1870, California produced more than six-tenths of the national output. The history of the efforts to grow wine grapes and to produce quality wine in California represents a classic story of biological learning on a massive scale as thousands of farmers tried numerous varieties and cultural methods over a wide range of growing environments.

Grapes suitable for wine did not exist in California prior to the second half of the eighteenth century, when the Franciscans introduced the Mission grape. This variety thrived and became the primary ingredient for wine production in California until the 1880s. Commercial winemaking first took root in the Los Angeles area, which reportedly had 53,000 vines by 1818 and 100,000 by 1831. Jean Louis Vignes, a French émigré, was the first in a long line of wine-industry entrepreneurs. Vignes began importing European varieties for his estate in the 1830s. In the pre–Civil War era numerous other prominent Los Angeles–area growers imported new cuttings from Europe for varietal trials.[14]

[14] Pinney, *History of Wine*, 237, 245–6, 250, 258, 284, 289, and 346; Carosso, *California Wine*, 7–37.

Beginning in the late 1850s, a number of large ventures emerged. The Los Angeles Vineyard Society was founded in 1857 in what is now Anaheim. By the mid-1870s the colony was producing about 1.25 million gallons of low-quality wine – mostly from Mission grapes. The San Gabriel Wine Company, founded in 1882, was another highly capitalized venture which was briefly the largest winery in the state. The firm attracted investors from England and built a winery with a capacity of one million gallons. A decade later the San Gabriel Wine Company quit the business without ever producing quality wine.[15]

By 1900 Southern California had been producing wine commercially for roughly eighty years. This record was marked by many more failures than successes. The setbacks were not primarily due to a lack of a market, labor problems, or inadequate capitalization; rather the firms failed because they never learned how to produce even moderate-quality table wines. Moving along the biological learning curve was not a simple matter. In addition, the advent of Pierce's disease doomed the industry in large sections of Southern California. But even without this scourge, geoclimatic factors favored the north. Northern California's more varied terrain and cooler and wetter climate offered numerous niches capable of producing complex and well-balanced table wines. In the south, the heat produced grapes with excessive sugar and inadequate acid content for quality wine production. Biological learning pulled the industry northward. The share of the state's wine produced by the San Francisco region rose from 11 percent in 1860 to 57 percent by 1890. By comparison, Los Angeles County's share fell from 66 to 9 percent.[16]

The movement to Northern California required considerable trial and error. In the 1830s and 1840s small vineyards were planted in the Sierra Nevada foothills, the Sacramento Valley, and the Bay Area. As the Gold Rush subsided in the late 1850s, many prospectors with experience in the vineyards of Europe applied their skills in the northern region. French émigrés in the Santa Clara Valley were leaders in introducing new varieties and methods. Pierre Pellier made several voyages to Europe to acquire prized cuttings and is credited with importing the first Grey Riesling, French Colombard, and Folle Blanche. Louis Prevost planted about 60 different varieties, and by 1857 Antoine Delmas boasted 105 varieties among his 350,000 vines. Scores of visionaries, including Sam Brannen, Leland Stanford, and Remi Nadeau, imported thousands of cuttings from

[15] Pinney, *History of Wine*, 304.

[16] Ibid., 313.

across Europe and set their sights on building premium vineyards. There were few successes, and Stanford and Nadeau's enterprises were abject failures. Smaller, less capitalized growers did not need to travel to Europe or import their own cuttings because secondary markets soon developed, with some nurserymen in the 1860s advertising more than one hundred varieties.[17]

California's most famous wine industry pioneer was the flamboyant Hungarian immigrant Agoston Haraszthy, a self-proclaimed count and "Father of California Wine."[18] His story encompasses many of the lessons of the development of viticulture in California and the importance of biological learning in the ultimate success of the state's agriculture. Haraszthy's first stop was in Wisconsin, where he planted vines and built a large wine cellar. Not surprisingly, he failed. He next settled in San Diego in 1850, but again gave up on his vineyards. His third try was in San Francisco, where the cold, foggy weather doomed his vines. This was followed by a venture in the mountains of San Mateo County just south of San Francisco, where he planted thirty acres of vines in 1854 with a similar outcome. After considerable research, he purchased 560 acres in Sonoma County in 1856 and, on this fifth attempt, found land suitable for growing wine grapes. Haraszthy was no fool – far from it. Like Stanford, Nadeau, and so many other entrepreneurs, Haraszty repeatedly failed because he knew little about the specific conditions required for success.[19]

Haraszthy's Buena Vista Winery became one of California's landmark estates. By the end of 1857 he had planted 26,000 vines, and by 1859 he was evaluating roughly 280 varieties. But what sets Haraszthy apart (aside from his flair for self-promotion) was his tour of Europe in the summer and fall of 1861. Armed with a commission from the California state government and a letter of introduction from the U.S. Secretary of State, Haraszthy explored the prime winemaking areas of France, Germany, Italy, and Spain. In all, Haraszthy and his agents, who scoured Morocco, Anatolia, and the Crimea, shipped about 100,000 vines (with roughly three hundred varieties) to his California estate. Although many of the varieties Haraszthy imported undoubtedly already existed in California, his endeavor nonetheless stands out as a major scientific feat

[17] Carosso, *California Wine*, 16–17; Pinney, *History of Wine*, 262–5; Tufts et al., "Rich Pattern of California Crops," 206; Butterfield, *History of Deciduous Fruits*, 29–30. Butterfield maintains that the great majority of the varieties in California by the mid-1930s had arrived in the state by 1861.

[18] A useful biography is McGinty, *Strong Wine*.

[19] Rhode, "Learning, Capital Accumulation," 773–800.

that advanced the growth of the California wine industry. Building California's vineyards was indeed a global enterprise.[20]

In spite of all of the attempts to import and propagate better varieties for winemaking, the Mission grape continued to be the dominant variety until the 1880s. Even under the best of conditions, this grape could only yield an ordinary table wine, condemning most of the state's output to the lower end of the quality spectrum. Its long popularity was most likely a result of inaccurate biological knowledge. Growers mistakenly transferred short pruning methods that worked for the heavy-bearing Mission grape to lighter-yielding premium varieties. With short pruning, the newly imported varieties simply would not produce a profitable crop even if planted in a suitable region. Many frustrated growers, unable to disentangle the sources of their failure, assumed that the premium vines were not matched to their locale and reverted to the Mission grape.[21]

Raisins

In 1851 Agoston Haraszthy reportedly brought the first vines especially suited for producing raisins to California. The next two decades witnessed considerable experimentation. In 1855, Antoine Delmas of San Jose imported the white Muscat of Alexandria grape, and in 1861 the ever active Haraszthy, now residing in Sonoma, imported the Muscatel de Gordo Blanco, the leading variety of the late nineteenth century. However, it was not until the late 1860s that G. G. Briggs and R. B. Blowers of Yolo County (near Sacramento) established the first commercially successful raisin enterprises in California. Over the 1870s and 1880s, raisin cultivation spread to hotter areas in the San Joaquin Valley and to Riverside, Orange, and Los Angeles counties. In 1873 G. Eisen first introduced the Muscat grape to the Fresno region for raisin production, and by the late 1880s Fresno became the center of production. There was a sharp decline of raisin production in the southland associated with the spread of a mysterious epidemic beginning in 1883 (later known as Pierce's disease).[22]

[20] Hutchinson, "Northern California," 30–7; Pinney, *History of Wine*, 275–7; Carosso, *California Wine*, 38–48.

[21] "Gen. Naglee of San Jose was one of the first to point out the value of long pruning for certain varieties. By changing from low to high pruning he was able to increase his crop from 100,000 pounds to 400,000 pounds." Butterfield, *History of Deciduous Fruits*, 31–2; Pinney, *History of Wine*, 347.

[22] California State Board of Agriculture, *Statistical Report 1911*, 144.

As knowledge accumulated, there was a growing realization that the San Joaquin Valley was especially well suited for raisin production. M. T. Kearney, an industry leader, noted that "ten or fifteen years ago raisin vineyards were planted in many parts of California, and in the northern and southern districts to a far greater extent than in the central district, but experience has shown that the country around Fresno is peculiarly favored in the matter of climate, and the raisin-growers there find their crops mature and are ready to pick several weeks earlier than in adjoining counties. This difference in time is of great importance, because it enables the grower to cure his crop in the sun and with much less risk of damage from early rains."[23]

By 1916, Fresno County produced more than 78 percent of the state's crop, and neighboring Tulare and Kings counties accounted for an additional 15 percent. The concentration of production around Fresno was one of many important developments. In the early 1870s, George Pettit of San Francisco invented a machine to remove grape seeds and stems, improving quality and reducing the shipping weight by more than 10 percent. The device was popularized by William Forsythe, a pioneering Fresno grower, in the 1880s and 1890s. By 1904–06, roughly one-half of state's raisin crop was processed using seeding machines.[24]

The introduction and adaptation of seedless raisin grapes from Asia Minor would revolutionize the raisin industry. As one of his many "firsts," Haraszthy introduced Sultanas in 1861. Of greater long-term significance for the industry, in 1878 William B. Thompson of Yuba City obtained cuttings of an improved variety, the Sultanina, from the famed Ellwanger–Barry nursery of Rochester, New York.[25] Honoring the local man's efforts and adding to confusion over appellation, the Sutter County Horticultural Society renamed the newly imported variety "Thompson Seedless." A further irony regarding its naming was that Thompson's neighbor, John P. Onstott, was the first farmer to see the potential of this white grape variety in commercial raisin production. By 1900, Onstott had established a nursery business, with branches in Yuba City, Fresno, and Los Angeles, that sold millions of cuttings throughout

[23] California State Agricultural Society, *Transactions 1899*, 13; California State Board of Agriculture, *Statistical Report 1918*, 183.

[24] California State Board of Agriculture, *Statistical Report 1911*, 155, and *Statistical Report 1919*, 167.

[25] Other sources list the date as 1872, 1873, or 1876 as well as call the variety *Lady de Coverly*. See, for example, *California State Historical Landmark*, no. 929. For an account of the Ellwanger–Barry Nursery, see McKelvey, "Flower City," 121–69.

California.[26] The rapid adoption of this new variety points to its superiority and to the pace of biological innovation. On the eve of World War I, about 76 percent of California's crop was Muscats, 14 percent Thompson Seedless (Sultaninas), and 9 percent Sultanas. Roughly twenty years later, Thompson Seedless grapes made up almost 90 percent of the state's raisin output.[27] Thus, over the span of a few decades there were extensive changes in both the regions of production and the varieties grown.

California raisin production grew from about 120,000 pounds in 1873 to 90 million pounds in 1893 and then to 180 million pounds in 1914. As an indicator of California's rising competitiveness, by the mid-1900s, the United States had shifted from the status of a net importer to that of a net exporter of raisins. California's rapid growth propelled it past traditional producers such as Spain and Turkey. California accounted for more than 38 percent of the volume of world commercial output by 1909–13 and for more than 63 percent by the 1920s.[28] The ability of California farmers to outcompete the Mediterranean growers in spite of significantly higher labor and shipping costs was due primarily to biological and organizational innovations.[29]

Oranges

Orange production dates to the Spanish era. The first "substantial" orange grove in California was planted at the San Gabriel mission in 1804. The first commercial orchard was established thirty years later in Los Angeles by William Wolfskill, using graftings from this mission grove. Wolfskill continued to play a pioneering role, shipping the first train car of oranges east in 1877.[30] The state's modern orange industry dated from the mid-1870s, when at least three parties independently introduced the Valencia orange, which matured in the summer months,

[26] See entries "Thompson Family" and "John Paxton Onstott," in Delay, *History of Yuba and Sutter Counties*, 371–2 and 388–93. "A New Grape," *Los Angeles Times* (1 October 1890), 5.

[27] California State Board of Agriculture, *Statistical Report 1911*, 146, 151; *1918*, 183; *1921*, 217; Tufts et al., "Rich Pattern of California Crops," 209–11; Christenson, "Raisin Grape Varieties," 38–47; Colby, "California Raisin Industry," 49–108.

[28] For an analysis of raisin production see Morilla Critz, Olmstead, and Rhode, "Horn of Plenty," and "International Competition."

[29] As evidence of the quality difference, imported raisins and currants often arrived with stems attached and "mixed with a great deal of dirt and sand." U.S. House Committee on Ways and Means, *Tariff Hearings 1909*, 3937.

[30] Lawton, "History of Citrus in Southern California," 8; and Webber, "History and Development of the Citrus Industry (1943)," 37.

into Southern California. The most publicized case involved A. B. Chapman of San Gabriel, who imported orange trees from Thomas Rivers of London, England, in 1876, but several Riverside nurseries also shipped in Valencia trees from Florida at about this time.[31]

Of greater importance was the birth and growth of the navel orange sector. As we have seen with many important plant varieties, a great thicket of lore has grown up around the introduction of the navel orange to the United States. Indeed, the fruit's name – alternatively the Bahia, Washington navel, Riverside navel, or the Tibbets orange – reflects the choices of the storytellers. A standard history runs as follows. The Portuguese introduced the orange to Brazil from India. Around 1822, a Portuguese gardener in Bahia, Brazil, noted unusual seedless fruit on a single limb of an orange tree of the "laranja Selecta" variety. This outcome was likely a mutation or bud variation. Recognizing the value of this sport, the gardener decided to propagate the new strain, which spread in the Bahia region.

Learning of the Brazilian strain that produced superior seedless fruit, William O. Saunders, the USDA's Chief Horticulturist (not the Canadian wheat breeder), obtained a shipment of the trees in 1868. But all of the specimens were dead upon reaching Washington, DC. A second shipment in 1870 fared better and buds were propagated in a USDA greenhouse. The USDA began distributing the progeny to growers in Florida (where they did poorly) and California, including sending two budded trees in 1873 to Eliza Tibbets. She had been a friend of Saunders in Washington, DC, prior to moving to Riverside, California. Tibbets planted the trees in December of 1873. (An alternative account ascribes a greater role for Eliza's spouse, Luther.)[32] A display of Tibbets' oranges at the first California Citrus Fair in Riverside in early 1879 aroused great interest among the southland's growers. Soon neighboring farmers planted cuttings for their own orchards. The high initial demand for these distinctive winter oranges spurred on the citrus boom. By 1933, the two trees planted

[31] Shamel, Scott, and Pomeroy, "Citrus-Fruit Improvement," 2–3.

[32] Shamel, "Washington Navel Orange," 434–45. For the alternative accounts, see interview with Luther Calvin Tibbets, 20 January 1900, and letter from Minnie T. Mills to W. A. Taylor of the USDA, 29 April 1933, Washington Navel Orange History File, Box 4. Archibald Dixon Shamel Papers, UC Riverside Libraries; and McClain, "New Look at Eliza Tibbets," 452. For the controversy over the fruit's introduction, see Roistacher, "History of the Parent Washington Navel Orange Tree," 58–65; Moore and Moore, "When was the Navel Orange Imported?" 219; and Mills, "Luther Calvin Tibbets," 127–61.

in Riverside in 1873 had become the parents of about 10 million navel orange trees growing in the Golden State.[33]

California's citrus industry profited from Florida's misfortune. A succession of freezes in 1894–95 killed most of the citrus trees in the Sunshine State.[34] In the few years immediately preceding the great freezes, California supplied 37 percent of the two states' combined orange output and 44 percent of the total lemon production. The freezes' impact on the Florida lemon industry was particularly severe and the state never recovered. A decade after the freeze, California produced more than 88 million pounds of lemons whereas Florida only generated 548,000 pounds. Biological learning came at a high price for Florida lemon growers. Florida's orange industry did recover, so the impact of the freeze on California was not as permanent as with lemons. California's share of orange production fell from 93 percent in 1896–98 to 63 percent two decades later. By 1911 the state produced more than one billion pounds of oranges.

Scientific advances helped propel this meteoric rise in production. The collection of systematic meteorological and soil records was particularly important for the citrus industry because of the fruits' vulnerability to freezing weather. California also experienced bad freezes (for example, in 1879 and 1913), but none as destructive as that in Florida. The California freeze of 1879 revealed a quirk in the state's environment – a narrow thermal belt in the foothills ringing the Central Valley that was suitable for growing subtropical fruits. This important discovery opened the way for the eventual growth of the orange industry in the valley. In 1910 the bulk of the state's orange industry was still firmly entrenched in Southern California, led by San Bernardino, Los Angeles, and Riverside counties. But by this time the Central Valley already had more than 1.1 million orange trees, about 16 percent of the state's total. Even at this early date the handwriting was on the wall because by 1913, 60 percent of the state's nonbearing trees were in the Central Valley.

33 A. D. Shamel, "A Brief History of the Origin and Introduction of the Washington Navel Orange, 17 February 1933," mimeo, Box 1, File 26. Archibald Dixon Shamel Papers, UC Riverside Libraries; Rhode, "Learning, Capital Accumulation," 794–5; Tufts et al., "Rich Pattern of California Crops," 217–23.

34 A by-product of the freeze was that the USDA closed its Florida Subtropical Laboratory (started in 1892) and recalled its plant scientists Herbert J. Webber and Walter T. Swingle. Webber was to play a leading role in California after the University of California established its Citrus Experiment Station at Riverside in 1907. See Webber, "History and Development of the Citrus Industry (1967)."

USDA scientists also played an important role in the industry's growth. In 1904, the department sent pomologist G. Harold Powell to help address the decay of citrus in transit due to blue mold.[35] His efforts led to the development and adoption of the so-called "Powell methods of handling." The key scientific discovery was that even small nicks and bruises on the skin introduced the mold. Powell innovated special picking and packing methods that protected the fruit. Prior to 1909, California farmers lost an average of 12 percent of their oranges in transit to mold. With the new handling methods, losses were cut substantially and the overall quality of the salable fruit also increased.[36]

Another stimulus came in 1909 when the USDA transferred Archibald Shamel to California to work on citrus breeding. Shamel soon noted the importance of bud variation, that is, the existence of differences within trees of the same variety, even within plant material cloned from the same ancestry. One tree was not the same as all others. The phenomenon, due to genetic mutation among other causes, represented both a challenge and an opportunity. It was a challenge because reproducing existing stock without careful selection could lead the strain to "run out" or deteriorate; but it was an opportunity because new sports appeared that allowed the selection of improved strains. The Bahia navel orange is a leading example of such a promising sport.

In the 1910s and 1920s, Shamel developed a breeding program to improve citrus through careful bud selection. In this endeavor, he advocated the use of scorecards, similar to those he had used to judge corn while in Illinois, to evaluate desirable strains of fruit. While many, including Secretary of Agriculture Henry A. Wallace and agricultural historian Deborah Fitzgerald, have harshly criticized scorecards as promoting irrelative or even counterproductive traits in corn, Shamel noted that he had "never found better material for such study than citrus fruits."[37] This claim is quite reasonable given that humans were the final consumers of citrus fruits (versus hogs for corn) and consumers likely cared about myriad characteristics, including taste, size, color, and lack of scarring.

After World War II, Valencia production in Southern California declined due to a combination of urban sprawl and a viral disease. Citrus production increasingly moved to the southern and eastern edge of the San Joaquin Valley. There was also a change in varieties, with most of the

[35] See Moses, "G. Harold Powell," 119–55; and Sackman, *Orange Empire*, 135–40.

[36] Powell, *Bureau of Plant Industry*, 35.

[37] Shamel, "Citrus Performance Records."

plantings in navel oranges. By 2000, the three most important producing counties – Tulare, Kern, and Fresno – were all in this area.[38]

Figs

The history of fig cultivation in California offers a special insight into the need for biological learning. As with many specialty crops, fig cultivation began in the Mission period. Indeed, the so-called mission fig remained the most important variety until the late nineteenth century. In the 1880s, Californians began planting two new varieties of fig, the White Adriatic and the Smyrna. Early attempts to produce the high-value Smyrna figs failed because the growers did not understand how to pollinate their trees. In 1890, George C. Roeding of Fresno began experimenting with artificial caprification – pollinating the blossoms with the aid of a goose quill. This method yielded fruit but proved prohibitively expensive. By 1899, with the help of Walter T. Swingle of the USDA's Bureau of Plant Industry, Roeding imported the fig wasp (*Blastophaga grossorun*), which was essential for successful natural pollination. This led to the first commercial crop of Smyrna figs in 1901. Efforts to expand Smyrna fig production were hindered by the spread of a disease (endosepsis) that rotted the fruit.[39] By 1916 California had produced more than 16 million pounds of figs, but about 80 percent were of the Adriatic variety because of the inability to overcome production problems with the superior Smyrna fig.

Common Threads

We have touched on three fruits – raisins, oranges, and figs. All of California's fruit crops have similar stories, with several common threads. In all cases, success required a long period of tinkering and experimentation. The biological exchange was a global process. California attracted immigrants from all over the world who brought a diverse pool of biological knowledge suitable to the state's varied environmental niches. This helps explain why so many different plants were introduced and tried in so short a period.

[38] Rhode, "Learning, Capital Accumulation," 794–95; Wallschlaeger, "World's Production and Commerce in Citrus Fruits," 66–7.

[39] Tufts et al., "Rich Pattern of California Crops," 225; Roeding, "Fig in California," 96–101; USDA, *Yearbook 1897*, 316–18. California State Board of Agriculture, *Statistical Report 1918*, 163–5; U.S. House Committee on Ways and Means, *Tariff Schedules and Hearings 1913*, 2886–7, and *Hearings on Tariff Readjustment 1929*, 337, 346–53, 358.

One biological feature – the ability to propagate asexually – was also common to these horticultural crops and helps account for their rapid spread. Growers could take samples or cuttings of their preferred vines and trees to replant on a large scale. Indeed, it often proved advantageous to mix and match scions with suitable root stocks of a different type via grafting or budding. And for many fruit products such as citrus and grapes, varieties without mature seeds were of greater commercial value.[40] The possibility of asexual propagation meant that once a desirable strain was discovered, clones with identical (or nearly identical) genes could be generated with relative ease. Using cuttings instead of seedlings reduced problems of deterioration and variability.[41] There were fewer steps backward for every step forward, and the potential rate of propagation also accelerated. These characteristics encouraged the expansion of a specialized nursery trade where improved varieties could be developed and multiplied.

The technical ease of asexually replicating garden and fruit plants made it particularly easy for others to appropriate an improved strain, thereby making it more difficult for plant breeders to recoup their financial investments. This problem led the federal government to enact the world's first plant patent law in 1930. It granted a seventeen-year monopoly to breeders of asexually reproduced plants (other than tubers). Thomas Edison, who lobbied for the law, asserted that patent rights "will, I feel sure, give us many [Luther] Burbanks."[42]

Combating Pests

We have made numerous references to plant diseases in our discussion of specific crops. When California gained statehood in 1850, the area

[40] "Seedless" fruits were produced by several biological mechanisms. Stenospermocarpy, which was at work for Thompson Seedless grapes, involved normal pollination with the subsequent abortion of the fertilized embryos. Immature seeds remained in the fruit but were of negligible size. Parthenocarpy, which was at work for navel oranges, involved the production of fruit without fertilization. The fruit was truly seedless. Mutants with this trait faced survival problems in a natural environment but, with human intervention, could reproduce via cloning.

[41] Allard, *Principles of Plant Breeding*, 36. Asexual reproduction presented problems of its own. Even when farmers exercised great care, viral infections, which led to degeneration, were common in cloned material. In addition, mutations could occur in the process of cell division, leading to genetic variation. Hartman and Kester, *Plant Propagation*, 192–3, 206–9.

[42] U.S. Senate Committee on Patents, *Plant Patents*, 3; "Bill Before Hoover Grants Plant Patents," *Washington Post* (25 May 1930), 1, 5.

was relatively free of pests and plant disease problems. The uncontrolled importation of biological materials changed all that, and by the 1870s and 1880s a succession of mysterious invaders were taking a terrible toll, threatening the commercial survival of most horticultural commodities. The concentration of agricultural production, an efficient transportation system, and the absence of natural parasites or predators, which often existed in older agricultural regions, created an ideal setting for the spread of exotic pests and diseases. Nurseries further contributed to the difficulties by incubating diseases and spreading infected plants. Among the worst offenders that caused serious damage before 1900 were phylloxera, Pierce's disease, "San Jose scale, woolly apple aphid, codling moth, cottony cushion scale, red scale, pear slug, citrus mealybug, purple scale, ... pear and apple scab, apricot shot hole, peach blight, and peach and prune rust."[43] Thus, within a few decades California's farmers went from working in an almost pristine environment to facing an appalling list of enemies in an age with few effective methods for cost-efficient, large-scale pest control.

As a frontier state, California was also largely devoid of the political, scientific, legal, and commercial infrastructures needed to combat the new threats. The spread of diseases and pests prompted collective action and research efforts. For crop after crop the creative efforts of farmers, scientists, and government agencies overcame the "free rider" problem to protect large-scale commercial horticulture.[44]

There was a general pattern to the appearance, spread, and eventual control of new pests and diseases. At first the afflictions were not well understood and the losses were often catastrophic. This led to the tearing out and burning of orchards, to quarantines, to the development of chemical controls, to a worldwide search for parasites to attack the new killers, and often to the development of new cultural methods and improved varieties that were resistant to the pests or diseases. The University of California and government scientists spearheaded these efforts, making stunning breakthroughs. Of the more than one thousand diseases and insects that attacked the state's vineyards and orchards, we focus on four of the more serious threats – phylloxera, Pierce's disease, cottony cushion scale, and San Jose scale.[45]

43 Smith et al., "Protecting Plants from Their Enemies," 245.

44 Morton, *Winegrowing in Eastern America*, 27–31.

45 This analysis of pests and diseases draws heavily from Iranzo, Olmstead, and Rhode, "Historical Perspectives on Exotic Pests," 55–67.

Phylloxera

Phylloxera is an aphid endemic in the eastern United States. The insect feeds on a vine's roots, weakening and eventually killing the plant. Phylloxera was first identified in Europe (where it had been accidentally introduced on imported American rootstock) in 1863. Soon giant bonfires fueled by dead vines became commonplace in grape-growing regions. Phylloxera first gained notice in California in the mid-1870s.[46] Between 1873 and 1879, more than 400,000 infested vines were dug up in Sonoma County alone. By 1880, phylloxera outbreaks had occurred in all of the state's wine-growing regions except Los Angeles. By the early 1890s nearly all of the vineyards of Napa Valley were infected.[47] The future looked dire.

Advances in scientific knowledge eventually gave growers the upper hand in the battle against phylloxera, but the costs were staggering. Experiments conducted in both France and the United States during the 1870s and 1880s investigated hundreds of possible chemical, biological, and cultural cures. Most techniques, including applying ice, toad venom, and tobacco juice, proved ineffective.[48] The idea of grafting European vines onto American rootstocks to resist phylloxera was advanced in the 1860s and 1870s as a result of the pioneering works of Charles V. Riley in Illinois and Missouri, Eugene Hilgard in California, and George Husmann in Missouri and California.[49] This remedy proved economically feasible for most growers, although it required huge investments. The majority of the vines of Europe and California were torn out and the lands replanted with European varieties grafted onto American rootstocks. The achievement in controlling phylloxera represents a major biological feat – today's viticulturalists owe their success to scientific advances made in the nineteenth century.

Once the general principle of replanting on American rootstocks was established, scientists on both sides of the Atlantic continued the tedious

[46] According to Pinney, *History of Wine*, 343, "the disease had been discovered as early as 1873 in California," but this was when it was first positively identified by the Viticultural Club of Sonoma. Carosso maintained that the "disease was known to have existed in California before 1870," and vines on the Buena Vista estate probably had shown signs of infestation as early as 1860. See Carosso, *California Wine*, 109–11; Butterfield, *History of Deciduous Fruits*, 32.

[47] Carosso, *California Wine*, 111, 118; and Pinney, *History of Wine*, 343.

[48] Ordish, *Great Wine Blight*, 64–102.

[49] Morton, *Winegrowing in Eastern America*, 30–1; Ordish, *Great Wine Blight*, 21, 103; Carosso, *Califonia Wine*, 113–27; Pinney, *History of Wine*, 392–5.

process of trial and error. The key challenges were to discover which American varieties were in fact more resistant to phylloxera, which would graft well with European varieties, and which would flourish in a given region.[50] In addition, grafting techniques had to be perfected.[51] By 1880, Husmann wrote that "millions upon millions of American cuttings and vines had already been shipped to France." George Ordish estimated that France, Spain, and Italy required about 35 billion cuttings to replant their vineyards. This represented roughly 12 million miles of cane wood – enough to circumnavigate the earth about five hundred times.[52] Contrary to the conventional wisdom that Americans of this era were not even interested in biological innovation, U.S. scientists were among the global leaders in developing the technology that saved the world's viticulture.

In California, the phylloxera threat played a major role in generating the political support necessary to fund the institutions that would contribute immensely to the state's agricultural productivity. Most important was the work of the University of California College of Agriculture. In 1880 the state legislature expanded university funding to combat phylloxera. Under Eugene Hilgard's leadership, the university spearheaded a number of research and outreach programs, including the dissemination of French learning. Also in 1880, the state founded the Board of State Viticultural Commissioners as a direct response to the epidemic. The board quickly followed its own independent course of research and extension.[53] Yet, in the 1880s, the battle against phylloxera was still in its infancy. The general principles were understood, but knowledge of the best procedures and varieties for each microregion of the state had to be laboriously accumulated, and the costly process of ripping out vines and transplanting onto the recommended rootstocks was only beginning.

[50] "Resistance" is not a sure thing. When replanting onto apparently identical resistant rootstock it is expected that about 20 percent of the plantings will be susceptible to phylloxera. In addition, insects evolve over time to overwhelm plants that had previously been resistant. Thus, the initial spread of phylloxera represented a watershed in the history of grape growing and ever since it has been necessary to develop new resistant varieties to stay ahead of the insect.

[51] As an example, the first U.S. varieties shipped to France were labrusca and labrusca–riparia hybrids that had a low resistance to phylloxera. In California the initial recommendation that growers use *Vitis californica* for rootstock proved to be a mistake. Pinney, *History of Wine,* 345, 394; Carosso, *California Wine,* 125; Ordish, *Great Wine Blight,* 116–19.

[52] Pinney, *History of Wine,* 345, 39 2–95; Carosso, *California Wine,* 125–6; Ordish, *Great Wine Blight,* 114–15.

[53] Pinney, *History of Wine,* 344.

It was not until 1904 that the USDA initiated a systematic program of testing throughout the state. By 1915 about 250,000 acres of vines had been destroyed, but relatively little land had been replanted with resistant rootstock.[54] That task remained on the horizon.

Pierce's Disease

The history of phylloxera tells a story of how biological learning allowed farmers to coexist with a pest, albeit at enormous cost. The story of Pierce's disease is altogether different. It represents a frightening case study in which early research efforts offered little help. The disease systematically destroyed vineyards in what was the heart of the state's wine industry, dramatically altering the fortunes of thousands of farmers. Farmers in the infected areas had no recourse but to abandon their vineyards and search for other crops.

The story begins in the Santa Ana Valley at the German colony of Anaheim, now in the shadow of Disneyland's Matterhorn. This agricultural community started with the organization of the Los Angeles Vineyard Society in 1857 with a capital stock of $100,000. After overcoming early organizational problems, the settlement began to flourish. The first vintage in 1860 yielded about 2,000 gallons. Production increased rapidly, reaching over 600,000 gallons in 1868. By 1883 the region was home to fifty wineries with about 10,000 acres of vines and a production of about 1.2 million gallons of wine.[55] Prospects for the Southern California wine industry looked bright.

However, an unknown affliction, originally termed Anaheim disease, changed everything. In 1883,

> the vineyard workers noticed a new disease among the Mission vines. The leaves looked scalded, in a pattern that moved in waves from the outer edge inwards; the fruit withered without ripening, or, sometimes, it colored prematurely, then turned soft before withering. When a year had passed and the next season had begun, the vines were... late in starting their new growth; when the shoots did appear, they grew slowly and irregularly; then the scalding of the leaves reappeared, the shoots began to die back, and the fruit withered. Without the support of healthy leaves, the root system, too, declined, and in no long time the vine was dead. No one knew what the disease might be, and so no one knew what to do. It seemed to have no relation to soils, or to methods of cultivation, and it was not evidently the work of insects.[56]

[54] Ibid., 342–5.
[55] Ibid., 290–4.
[56] Ibid., 292.

Within a few years most of the vines had died. The disease soon spread to neighboring areas and contributed to the eventual demise of commercial grape culture in Southern California. Every manner of spraying, dusting, and pruning method was tried without success.

Identifying the disease was a slow process and, more than one hundred years later, farmers are still waiting for a cure. At first several growers mistakenly thought the vines might be succumbing to phylloxera. In August 1886, F. W. Morse, a University of California chemist who had been working on phylloxera, inspected the Santa Ana Valley, but could not identify a cause.[57] In 1887 the USDA dispatched one of its scientists, F. L. Scribner, and eminent French researcher Dr. Pierre Viala, to the infected area. Scribner concluded that the disease was not fungal but appeared in the roots; Viala argued that a parasite might be at fault.[58] In 1888, the Board of State Viticultural Commissioners hired a "Microscopist and Botanist," Ethelbert Dowlen, who after years of work tentatively, but mistakenly, concluded that a still unidentified fungus caused the disease.[59] Numerous other experts came and went, but the vines kept dying.

The USDA dispatched another scientist, Newton B. Pierce, to Santa Ana in May 1889. After extensive research that included a five-month stint in France investigating known vine diseases, Pierce rejected most popular theories. In 1891 he concluded that the disease was hitherto unknown, that it was probably caused by a microorganism, and that there was no known cure. By this date the wine industry had disappeared from the Santa Ana Valley.[60]

Pierce's study closed the investigations into this vine disease for almost half a century. The hiatus was partly due to the difficulty of the task but also because the malady mysteriously ceased being a serious problem. As a postscript, the identification of the causal agent and vector responsible for the disease has only been achieved in recent years, but the disease itself is still only partially understood. It is caused by the bacterium *Xylella fastidiosa* and is transmitted by a number of leafhoppers, most importantly the newly introduced glassy-winged sharpshooter. The disease, which is

[57] Smith et al., "Protecting Plants from Their Enemies," 266; Gardner and Hewitt, *Pierce's Disease*, 6–12; Carosso, *California Wine*, 128.

[58] Gardner and Hewitt, *Pierce's Disease*, 14–15.

[59] Pinney, *History of Wine*, 307; Gardner and Hewitt, *Pierce's Disease*, 18–96. Dowlen reportedly had studied botany at the South Kensington School in London with Thomas Huxley and billed himself as a French expert on vine disease.

[60] Smith et al., "Protecting Plants from Their Enemies," 267; Pinney, *History of Wine*, 313.

inevitably fatal, kills the plant by inhibiting its ability to utilize water and other nutrients. At present, short of attacking the vector (which most scientists think is at best a delaying action) there still is no effective method to control the disease.

Cottony Cushion Scale

The campaign against the cottony cushion scale (*Icerya purchasi*) represents one of the truly fascinating stories in California's agricultural development. The cottony cushion scale sticks in bunches to the branches and leaves of citrus and has devastating effects if uncontrolled. This scale was first observed in California in 1868 in a San Mateo County nursery (near San Francisco) on lemon trees recently imported from Australia. The scale hit the citrus groves of Southern California in the early 1870s, and by the 1880s the damage was so extensive that the infant industry appeared doomed. Growers burned thousands of trees and helplessly watched their property values fall.[61]

Farmers tried many chemical controls, including alkalis, oil soaps, and arsenic, but the insect continued to multiply. Apparently the waxy covering of the scale protected it from these liquid poisons. In desperation, the USDA and the University of California experimented with fumigation, a costly endeavor involving enclosing the trees in giant tents and then pumping in toxic gases. The Culver Fumigator displayed in Figure 8.3 was one of many contraptions invented to facilitate the process. Carbon disulfide came into use in 1881, and by 1890, hydrocyanic acid had emerged as the most promising treatment. Potassium cyanide, sodium cyanide, liquid hydrocyanic acid, and calcium cyanide all gained favor at one time or another.[62]

American scientists, aware that cottony cushion scale was present in Australia yet did little damage, turned their attention to discovering why. They surmised that the scale was native to Australia and that natural predators limited its spread. Incredibly, bureaucratic and financial obstacles initially prevented the USDA from sending one of its scientists to Australia. Undaunted, Charles V. Riley, then Chief of the USDA Division of Entomology, and Norman Colman, the Commissioner of Agriculture, persuaded the U.S. State Department to allocate two thousand dollars to send USDA entomologist Albert Koebele to Australia as part of a delegation to the 1888 International Exposition in Melbourne. Koebele's

[61] Stoll, "Insects and Institutions," 227–8.
[62] Smith et al., "Protecting Plants From Their Enemies," 244–61.

FIGURE 8.3. Culver Fumigator. The vedalia beetle (see Plate 8.1) ended the need to fumigate for cottony cushion scale, but farmers continued to fumigate to kill other pests. *Source:* Riley, "Injurious and Beneficial Insects," 608.

true mission was to search for predators of the cottony cushion scale. He hit the jackpot in October 1888 with the discovery of a ladybird beetle (vedalia or *Rodolia cardinalis*) feeding on the scale in a North Adelaide garden. Koebele sent a shipment of twenty-eight ladybird beetles to Los Angeles-based USDA entomologist, D. W. Coquillet – many more shipments would follow. Coquillet studied and multiplied the insects and by the summer of 1889 the beetles were being widely distributed to growers. Within a year after general release, the voracious beetle had reduced cottony cushion scale to an insignificant troublemaker, thereby contributing to a threefold increase in orange shipments from Los Angeles County in a single year. According to one historian of this episode, "the costs were measured in thousands and the benefits of the project were undetermined millions of dollars."[63] (See Plate 8.1, which shows a vedalia beetle eating cottony cushion scale.)

This success encouraged Koebele to make another journey to Australia, where he discovered three more valuable parasites helpful in combating the common mealybug and black scale. Other entomologists

[63] Ibid., 249–50; Graebner, "Economic History of the Fillmore Citrus Protection District," 30–4; Doutt, "Vice, Virtue, and the Vedelia," 119–23; Marlatt, "Fifty Years of Entomological Progress," 8–15; Seftel, "Government Regulation," 386–7.

made repeated insect safaris to Australia, New Zealand, China, and Japan, as well as across Africa and Latin America. There were many failures, but by 1940 a number of new introductions were devouring black scale, yellow scale, red scale, the Mediterranean fig scale, the brown apricot scale, the citrophilus mealybug, the long-tailed mealybug, and the alfalfa weevil. In addition, scientific investigations led to improved methods of breeding various parasites so that they could be applied in large numbers during crucial periods.[64] As with Koebele's initial successes, the rate of return on these biological ventures must have been astronomical.

San Jose Scale

San Jose scale (SJS; *Aspidiotus pernicious)* was first discovered in the United States in the early 1870s at the San Jose orchard of James Lick. Lick, a wealthy real estate developer best known for the observatory he funded, was an avid collector of exotic plants. His avocation would prove to have a high cost. Most accounts suggest that SJS hitched a ride on trees Lick imported from Asia. (Other accounts name Chile.) From Lick's property SJS soon spread to nearby farms, becoming prevalent throughout the San Francisco Bay Area and California's Central Valley by 1880. The fact that San Jose was a center for commercial nurseries undoubtedly hastened the scale's spread. At first, farmers were slow to respond to the new scale, in part because the pest took time to multiply and, due to its innocuous appearance, growers tended to attribute their losses to other causes. By 1880, farmers and scientists recognized SJS as a grievous problem.[65] In the late 1880s, the insect hit the deciduous orchards of Southern California, leading the Los Angeles Horticultural Commission to warn in 1890 that "if this pest is not speedily destroyed it will utterly ruin the deciduous fruit interests of the Pacific Coast."[66] Alarm spread nationally after the scale was discovered in Virginia in 1893. The USDA's first extensive survey (1896) found that the pest had established a foothold in at least fourteen states east of the Rockies in addition to seven states in the West.

The pest attacks all deciduous fruit trees, many ornamental and shade trees, and selected small fruits, especially currants. The scale infests all parts of the trees above ground. On mature trees the scale scars and shrivels the fruit, in many cases rendering it worthless. It can also stop

64 Smith et al., "Protecting Plants from Their Enemies," 250–5.

65 Marlatt, "San Jose Scale," 156. In 1880 the scale received its official name of *Pernicious.*

66 Howard and Marlatt, "San Jose Scale," 1–5.

growth and cause a systemic decrease in vigor, thereby reducing the yield of the tree. Eventually the tree dies prematurely, albeit long after it has become economically unprofitable. If left untreated, most varieties of fruit trees infested at the nursery would not survive to a bearing age. In the 1870s little was known about SJS and the technologies for dealing with it were not yet developed. Thus, as was the case when phylloxera began destroying the world's vineyards, the very future of the deciduous fruit industry seemed in doubt. Hundreds of thousands of trees were destroyed, property values in infected areas suffered, the development of new orchards temporarily stalled, and the agricultural press lamented the deterioration in fruit quality.[67] (See Plates 8.2 and 8.3 for photos of SJS.)

From the perspective of hindsight, the response to this and other new pests of the period was truly remarkable. Scientists at the University of California and the USDA were methodical in their search for biological and chemical controls. Coupled with these efforts, a new chemical industry with its own research, manufacturing, and sales forces came into being, which in turn led to the development of the modern agricultural spray equipment industry. The relatively little attention that SJS receives today is a testimonial to the success of these efforts. However, writing in 1896, L. O. Howard and C. L. Marlatt observed that "[t]here is perhaps no insect capable of causing greater damage to fruit interests in the United States, or perhaps the world, than the San Jose, or pernicious, scale."[68] In 1902, Marlatt noted that "the fears aroused by this insect have led to more legislation by the several States and by various foreign countries than has been induced by all other insect pests together."[69] Just when California producers were beginning to access international markets, the SJS threat led Canada and many European countries to restrict American fruit imports. In California, SJS was a driving force behind the creation of the Board of State Horticultural Commissioners in 1883 and the passage of the state's first horticultural pest control and quarantine law.[70] National legislation, however, was long delayed. In 1897 Howard, chief of the USDA's Entomology Division, pressed for a federal quarantine act, but the opposition of nurserymen and their congressional allies stifled the effort. It was not until 1912 that such powers were granted. Canada

[67] Marlatt, "San Jose Scale," 155–74; Quaintance, "San Jose Scale," 1–3.
[68] Howard and Marlatt, "San Jose Scale," 9.
[69] Marlatt, "San Jose Scale," 155.
[70] Morilla Critz, Olmstead, and Rhode, "Horn of Plenty," 316–52, and "International Competition," 199–232; Marlatt, "San Jose Scale," 157; Smith et al., "Protecting Plants From Their Enemies," 245–7.

moved faster, passing the San Jose Scale Act in 1898, which prohibited the importation of nursery stock from infected countries. By 1900 the federal and provincial governments relaxed their importation bans, requiring instead that imported nursery stock be fumigated with hydrocyanic acid. Both federal and regional governments attempted to wipe out the scale that had already entered Canada. As an example, in 1897 the Ontario legislature authorized "inspectors to enter any nurseries or orchards and burn infested trees. In 1898 inspectors had ordered the destruction of an estimated 41,000 trees."[71]

The fight to control the scale took two separate and at times competing tracks – the biological and the chemical. The discovery of biological controls was a high priority for the USDA. "The importance of discovering the origin of this scale arises from the now well-known fact that where an insect is native it is normally kept in check and prevented from assuming any very destructive features, or at least maintaining such conditions over a very long time, by natural enemies, either parasitic or predaceous insects or fungous or other diseases."[72] By careful observation and deduction the USDA's entomologists eliminated, one by one, Australia, New Zealand, the Hawaiian Islands, and Africa. After spending the period 1901–02 exploring the rural areas of Asia, Marlatt concluded that China was likely the source of the scale. Marlatt also found a native predator – the Asiatic ladybird beetle (*Chilocorus similis*) – that feasted on the scale. Marlatt sent boxes of the beetles to his experimental orchard in Washington, DC, but of the thirty that survived the journey, only two made it through the first winter. Reinforced by fresh imports from Asia, the USDA's beetle population was increased and studied. Subsequently, some twenty other insect predators were identified.[73] Although biological controls initially appeared promising, they did not pan out. A. L. Quaintance of the USDA noted that "the combined influence of these several agencies [insects that fed on the scale] is not sufficient to make up for the enormous reproductive capacity of this insect [SJS]."[74]

Without reliable biological controls, farmers turned to lye solutions as the primary defense against SJS. The early practice was to strip the rough bark from the tree and then scrub the tree with a caustic solution.

[71] Castonguay, "Naturalizing Federalism," 7–14, quote from 13; Powell, *Bureau of Plant Industry*, 91.

[72] Marlatt, "San Jose Scale," 158.

[73] Ibid., 155–74; Quaintance, "San Jose Scale," 11–13.

[74] Quaintance, "San Jose Scale," 11.

FIGURE 8.4. Two hand-pump spray rigs at work in a pear orchard circa 1900. *Source:* Floyd Halleck Higgins Collection, 1796–1982. Courtesy of Special Collections, UC Davis Libraries.

An alternative was to drench the trees' foliage using a handheld syringe sprayer (much like contemporary insecticide sprayers sold for household use) or home-rigged force pumps designed for moving water (which the lye solutions rapidly corroded). Eventually farmers learned that applying the chemicals during the dormant season was effective and did less damage to their trees. Lime–sulfur sprays gained favor around 1886. The formulas were soon perfected and standard commercial preparations increasingly replaced homemade concoctions.[75] But by 1908 the insect had developed resistance to the lime–sulfur pesticide – the first instance of the phenomenon of pesticide tolerance observed in the United States – leading to the introduction of new chemical treatments. As Figure 8.4 suggests, spraying could be a nasty task. The developments in the chemical industry and the spray equipment industry in the fight against SJS

[75] Smith et al., "Protecting Plants from Their Enemies," 255–6.

would prove invaluable in fighting other pests. In addition, many cultural methods learned in the fields, such as short pruning and shaping of trees to facilitate pest control, proved valuable in improving quality and reducing harvest cost.[76]

Collective Action

These battles against plant pests and diseases represent classic cases of a geographically dispersed and economically diverse population trying to grapple with the problems of externalities and public goods in a democratic society. Externalities are present when the costs and benefits derived from an individual action are not completely borne or captured by the agent undertaking the action – in this case an agent's actions positively or negatively affect other economic actors. As a result there is a gap between the costs and benefits to an individual agent (the private costs and benefits) and those to society as a whole (the social costs and benefits). The public goods problem arises from the lack of rivalry and excludability in consumption.[77] A successful eradication plan for a pest such as SJS required that all the orchards in an infected area be protected to prevent infestation. Because pest control displays characteristics of a public good and has positive externalities, leaving it to private individual initiative would likely encourage too little pest control, as reflected both in investments in research and in the farm-level application of prevention and eradication methods. In this situation, there is a case for public authorities to intervene by coordinating and leading individual efforts to generate a collective societal response.

Under these conditions, the finance of eradication programs by voluntary contributions allows individuals to benefit even though they do not contribute to the cost of the program, and they may not even cooperate with the pest control measures. This in turn creates a demand for collective action to employ the power of the state (or some form of

[76] Marlatt, "San Jose Scale," 156.

[77] There is "rivalry" in the consumption of a good or service when the consumption by one agent prevents others from enjoying it as well. This is not the case for a pest control program. Two farmers can simultaneously enjoy a plan's benefits without imposing additional costs on each other. "Excludability" exists when one can limit the access to a good. This is true of most goods sold in the marketplace. However, when a pest control program is under way it may be hard to exclude any one farmer from benefiting from eradication efforts on nearby farms.

contractual authority) to coerce compliance in both the financing and the operation of the control programs. Such actions necessarily limit individual freedom. In a democratic and market-oriented society, enacting such infringements on property rights can be a difficult and costly process. The fact that farmers not only acquiesced but also actively campaigned for such controls offers strong testimony as to the severity of the threats to their livelihood.

As discussed earlier, most of the diseases had been introduced from other parts of the world and were therefore unknown in California when the problems arose. To eradicate the disease from their private holdings, individual growers would have had to make enormous investments to develop basic and applied research programs and eradication methods. Given costly information and the small scope for expected private benefits, such investments would probably have been unprofitable for individual growers. Despite the substantial monetary losses, from the growers' individual economic point of view it would have been more efficient to let the disease destroy their crops and perhaps shift to less intensive production processes or to other crops. In fact, this was what happened after the arrival of Pierce's disease, when vine growers of the Anaheim and San Gabriel valleys abandoned vines and planted citrus trees.

On the benefit side, the advantages of pest control to society as a whole are probably larger than those to individual farmers or even all farmers. Also important are the long-run or dynamic benefits derived from pest control. Practically all actions taken in this respect have had positive and significant spillovers to similar problems. For example, the fight against the pests of the past century led to basic and applied scientific discoveries crucial to improving the knowledge needed to combat other problems. Moreover, in a number of cases the advances in agricultural sciences also had a direct bearing on improving human health. The disease and pest control methods perfected in the second half of the 1800s, such as improved chemical treatments, sophisticated breeding and grafting practices, and biological controls, have been used extensively ever since. Similarly, much of the legislation concerning plant protection, such as quarantine and inspections laws, and a great part of the agricultural research and administrative institutions have their origins in the second half of the 1800s. Both the body of legislation and the state institutions detailed in Table 8.2 have effectively contributed to limiting the introduction and spread of diseases in California and elsewhere.

TABLE 8.2 *Partial List of U.S. and California Efforts in Plant Protection*

Year	Law/Institution	Description
1870	First California plant pest control legislation	Various statues empowered counties to pay bounties for gophers and squirrels. In 1883, the California Political Code gave county boards of supervisors power to destroy gophers, squirrels, other wild animals, noxious weeds, and insects injurious to fruit or fruit trees, or vines, or vegetable or plant life.
1880	Board of State Viticultural Commissioners created	Supplemented University of California work in controlling grape pests and diseases with special emphasis on phylloxera. Remedy-oriented rather than research-oriented. The university was responsible for experimental and research work.
1881	California passed the first U.S. law granting plant quarantine authority	Enlarged the duties and powers of the Board of State Viticultural Commissioners and authorized the appointment of a State Viticultural Health Officer empowered to restrain import of plants that might be diseased.
1881	County Boards of Horticultural Commissioners created by County Boards of Supervisors	The county boards were empowered to inspect properties upon complaint and to require treatment of insect infestations. By 1882 county boards had been appointed in 21 counties.
1882	University of California offered its first course in economic entomology	
1883	State Board of Horticultural Commissioners created	The board was empowered with authority to issue regulations to prevent the spread of orchard pests and to appoint an "Inspector of Fruit Pests" and "quarantine guardians" as enforcement officers.
1885	First explicit legislative authority to inspect incoming interstate and foreign shipments	Authorized the state "inspector of fruit pests or quarantine guardian" to inspect fruit packages, trees, etc. brought into the state from other states or from a foreign country.

Year	Law/Institution	Description
1886	First county plant quarantine ordinance	Ventura County was the first county that prohibited transportation within the county of anything infected with scales, bugs, or other injurious insects. Other counties followed and, by 1912, at least 20 counties had enacted several ordinances against the entry of pests.
1890	Maritime inspection of cargoes of foreign vessels initiated	
1899	California State Quarantine Law	Required the holding and inspection of incoming shipments of potential pest carriers and the disposal of infestations to the satisfaction of a state quarantine officer or quarantine guardian of the district or county. Labeling of shipments was required, hosts of certain peach diseases were embargoed from infested areas, and importation of certain pest mammals was prohibited.
1903	State Board of Horticulture replaced by the State Commissioner of Horticulture	Empowered to promulgate interstate and intrastate quarantines.
1905 Mar. 3	*Insect Pest Act*	Prohibited the importation and interstate transportation of live insects injurious to plants.
1905	First California quarantine order	Issued in response to the citrus whitefly of Florida.
1907	Southern California Pathological Laboratory at Whittier established	Carried out research studies on plant diseases and insect problems in Southern California.
1910 Apr. 26	*National Insecticide Act*	
1912 Aug. 12	*Federal Plant Quarantine Act*	Prevented the importation of infested and diseased plants.
1912	*Federal Horticultural Board* created	Enforced the Plant Quarantine Act.

(*continued*)

TABLE 8.2 (*continued*)

Year	Law/Institution	Description
1912	Citrus Experiment Station and Graduate School of Tropical Agriculture at Riverside established	Superseded the Southern California Pathological Laboratory. Had strong divisions of Entomology and Plant Pathology.
1912	Work started at the University Farm at Davis	Carried out entomology and plant pathology research for the university.
1912	Agricultural Extension's County Farm Advisor service developed	
1915	*Terminal inspection of plants in U.S. post offices began*	
1919	*Western Plant Quarantine Board created*	
1919	State Department of Agriculture created	Assumed some of the duties of the State Commissioner of Horticulture.
1919	*Federal Quarantine Law No. 37*	Regulated the movement of plants and plant products.
1920	*Federal Quarantine Law No. 43*	Quarantined against the European corn borer.
1921	California border inspection of incoming motor traffic initiated	Stations established on the roads coming from Nevada and Arizona. The original purpose was to prevent the introduction of the alfalfa weevil. By 1963, 18 stations were in operation on all major highways entering from Oregon, Nevada, and Arizona.
1924	*Quarantine on grapes from Spain*	Prevented the introduction of Mediterranean fruit fly.
1925	*National Plant Quarantine Board organized*	
1926	*Federal Bulb Quarantine*	
1928	*Plant Quarantine and Control Administration created*	Superseded the Federal Horticultural Board in its task of inspection of imports of nursery stock and other plants and prevention of plant pests.

Note: Federal programs in italics.
Sources: Compiled from Weber, *Plant Quarantine and Control Administration*, 1–90; Essig, "Fifty Years of Entomological Progress," 40; Smith et al., "Protecting Plants from their Enemies," 239–315; Ryan and University of California (System), *Plant Quarantines in California*, 4–11.

The efforts to combat injurious insects and diseases in California were built on earlier innovations in the understanding and control of disease. By the 1850s American agricultural leaders, including entomological and horticultural groups, were developing institutional structures that would provide the foundation for education, research, and collective action. In the 1840s, Solon Robinson and others organized the National Agricultural Society with the objective of directing the Smithsonian bequest to agricultural research. In the 1850s Marshall P. Wilder organized the U.S. Agricultural Society to lobby for the establishment of land-grant colleges and the creation of a department of agriculture. The Morrill Act, which granted land to the states for agricultural and industrial colleges, was passed in 1862. By the early 1870s agricultural entomology courses were being offered in a number of colleges throughout the United States.

In California, important institutional structures began emerging shortly after statehood was granted. Among the early institutions created were the State Agricultural Society and the California Academy of Sciences (organized in 1853). Both of these bodies promoted discussion and the exchange of information but were ill equipped to perform basic and applied research and outreach. In 1868 the University of California and the College of Agriculture were established to help fill this void. One of the college's early leaders, Eugene Hilgard, proved to be a man of enormous vision, talent, and energy. Trained in Germany as a biochemist and soil scientist, Hilgard established the policy that faculty have both research and extension responsibilities, and he took the lead in setting up experiment stations and a publication program aimed at communicating directly with farmers.[78] Much of the technical and research work on plant pathology, which would lead to major breakthroughs in plant protection, was undertaken at the university. Gradually other state boards and institutions designed to deal with particular problems came into existence. One of the most important and active boards was the Board of State Viticultural Commissioners, which worked to disseminate information on phylloxera and supported research on Pierce's disease. But its legacy is tarnished in part by a long squabble with Hilgard and other university scientists.

Quarantine and inspection laws provided another important tool in the arsenal to control pests and diseases. In this California was a pioneer, enacting its first quarantine legislation in 1881. Enacting quarantines

[78] Eugene Hilgard earned his PhD in organic chemistry at the University of Heidelberg.

and other pest control measures was highly controversial, and it engendered the opposition of nonfarm interests, nurserymen, and many horticulturalists jealous of their right to do as they wished on their private property. The early state and county acts went unfunded and more than once the courts overturned legislation or limited enforcement. After 1881, county-appointed horticultural commissioners "could order disinfection of infested property or declare infested orchards a public nuisance if recalcitrant owners refused to take action against the pests." But in practice little action was taken because of bureaucratic constraints and financial shortcomings. By 1885 the legislature had stiffened state horticultural pest control laws, and by 1891 the institutional framework was secure enough that Los Angeles County succeeded in destroying 325,000 recently imported pest-infested orange trees.[79] The legacy of these early efforts is still with us today – even the casual tourist entering the state by car encounters the state agricultural inspection stations designed to block pests and diseases that might hitchhike a ride into the state's fields. For most states it would be nearly impossible to stop the migration of pests and diseases from neighboring states, but California's long coast to the west and mountains and deserts to the north, east, and south offer natural barriers to migrating insects and diseases. However, with improvements in transportation and the increased mobility of people and commodities, the challenge of preventing new infestations has become even more difficult. But all future efforts, be they biological, chemical, or administrative in nature, will be much easier to envision and implement because of the scientific and institutional foundations laid in the nineteenth and early twentieth centuries.

Perspectives

In this chapter we have only scratched the surface of the history of California agriculture. An analysis of the investments made to reshape the landscape adds significantly to our case that labor-intensive, land-saving investments were a dominant activity in one of the nation's highest wage states. All of the land-building activities that the induced innovation literature reserves for low-wage, labor-abundant economies were under way on a grand scale in California as farmers literally created their

[79] For an excellent treatment of the politics of California pest control legislation see Seftel, "Government Regulation," 379.

environment by leveling and clearing fields, controlling flood waters, moving surface water through extensive canal and ditch systems, and pioneering the use of agricultural pumps. By 1930 California farms had about 45,000 irrigation pumps, and by 1940 the state had nearly 5 million acres of irrigated farmland. To provide perspective, 5 million acres is equal to about 73 percent of all the irrigated land in Japan, the premier example of a country that followed the biological versus mechanical route of agricultural development.

As the evidence on agricultural pumps indicates, machines played a key role in land building. This reliance on machinery points to the interaction of mechanical and biological innovations. Farmers were every bit as interested in saving labor in their land-building enterprises as in their plowing or harvest activities. Californians invented Fresno scrapers (significantly improved land planes) for leveling their fields, and they employed giant steam-powered dredges (along with legions of immigrant laborers) for building thousands of miles of dikes, canals, and ditches. The symbiotic link tying biological and mechanical technologies together was particularly evident in the battles against diseases and insects. In addition to identifying chemicals that would address the threats and gaining knowledge about when best to apply those chemicals and in what doses, farmers also needed machines that would effectively spread the poisons. Santa Clara County, the center of the California deciduous fruit industry, also became a center for innovation of spraying and dusting equipment. In 1884 John Bean built a high-pressure, continuous-spray pump to combat SJS on his small almond orchard. Within a few years he had formed the Bean Spray Pump Company (which in the 1920s would evolve into the Food Machinery Corporation). Before Bean most sprayers had consisted of handheld squirt cans or hand pumps on a pickle barrel with a single hose. Within a decade Bean's larger porcelain-lined units could handle four hoses and generate twice the pressure that his early models did. Figure 8.5 shows a Bean spray rig. By the 1930s, working pressures had more than doubled again and steam- and gasoline-powered units could handle nine hoses. The general improvements in spraying and dusting equipment both saved labor and allowed for more effective distribution of chemicals.[80]

[80] Anderson and Roth, *Insecticides and Fungicides*, 214–17; Mason, *Spraying, Dusting and Fumigating of Plants*, 136; Advertisement, John Bean Mfg. Co., 1937, Bean files, Higgins Collection, Special Collections, UC Davis Libraries.

FIGURE 8.5. Bean spray pump circa 1915. *Source:* Courtesy of San Jose Public Library, California Room.

For all the biological investments, California's progress in dealing with the free rider problem and developing institutions for effective collective action perhaps stands above the others. The very existence of the array of legal–institutional structures created to combat diseases and pests testifies to the enormous individual and collective importance that farmers placed on biological issues. As a rule the mechanical innovations, which induced-innovation advocates have long asserted dominated agricultural productivity growth, did not require collective action. Public roads made many machines more productive by allowing them to serve larger areas. Informal agreements among neighboring farmers to share, rent, and hire machines also facilitated their diffusion.[81] But these were relatively simple transactions. The truly difficult challenges of marshalling entire communities to fight biological threats were often insurmountable, as with the utter failure of Texas to mount any effective collective response to the boll weevil invasion in the 1890s. USDA scientists rushed to southern Texas and recommended a wide cotton-free quarantine zone, but political paralysis reigned in Austin. California's experience of fighting a succession of invaders stands as a monument to what might have been possible elsewhere.

[81] Olmstead and Rhode, "Beyond the Threshold," 27–57, and "Reshaping the Landscape," 663–98.

The problems with the induced innovation story continue if we explore the post–World War II era. This is the period when California should have moved heavily into land-saving, biological technologies. Surely there was a continuation of biological investments, but there were no distinct breaks that set this period apart. Moreover, laborsaving mechanization continued as California led the world in the adoption of the mechanical cotton picker and the innovation of equipment to harvest a variety of specialty crops such as sugar beets, tomatoes, and numerous nuts and fruits.

9

More Crops, More Animals

Livestock and Feeds in the Farm Economy

The vast majority of agricultural history research has focused on crops. The relatively few discussions of livestock often focus on cattle and cowboys and stand apart from the mainstream of American economic history. Since 1925 the annual value of cash receipts from all livestock sales has almost always exceeded the cash receipts for all crops. As a recent example, in 2004 the value of cash receipts for crops was $117.8 billion, whereas the value of livestock receipts was $123.5 billion. Furthermore, the value of production of the most important animal products exceeded that of comparably ranked crops. The top five animal categories in 2004 were cattle and calves ($34.9 billion), milk ($27.5 billion), broilers ($20.4 billion), hogs ($13.1 billion), and eggs ($5.3 billion). By contrast, the top five crops in terms of value of production were corn ($24.4 billion), soybeans ($17.9 billion), hay ($12.2 billion), wheat ($7.3 billion), and cotton ($4.9 billion).[1] The three top-ranked crops were used primarily to feed animals, demonstrating the close link between the crop and animal sectors.

The gross value data in the preceding paragraph overstate the relative importance of animals because the value of animals includes the value of the crops they consumed. One must subtract the value of crops fed to animals from the gross value of animal products to obtain the value added of livestock production. For recent decades, such an adjustment would substantially lower the value of animals relative to that of crops. But animals also provide draft power and manure for the crop sector. These

[1] USDA, *Agricultural Statistics 2006*, Tables 9–24, 9–30, 9–43. The data for the value of livestock and products are preliminary.

contributions are not too important today, but in the age before the internal combustion engine, petroleum-based fuels, and synthetic fertilizers they were crucial. Thus, the relative size of the two sectors depended on the interrelationships linking the sectors and these changed significantly over time.

To reinforce our assertion that little attention has been given to animal products relative to the coverage of crops, note how much less has been written about the history of milk than of cotton. In recent years the cash sales of dairies have been about four times that of cotton farms, and in the 1930s, before the extensive mechanization of either activity, dairies employed about 50 percent more labor hours than cotton farms. Both industries experienced key technological changes, but far more has been written about the cotton gin and the mechanical harvester, which revolutionized cotton production, than about the creamery, refrigeration, and the mechanical milking machine, which transformed dairy operations. Both industries were closely associated with the broader political, social, and economic development of large regions. Finally, both activities were subject to devastating shocks, but the literature on the cotton boll weevil dwarfs that on the diseases and pests that threatened dairy herds even though, if left unchecked, several diseases had the potential of devastating cattle populations. One of these diseases, bovine tuberculosis, was killing thousands of Americans every year early in the twentieth century.[2] Scholarly fashions are hard to overcome.

This chapter begins our analysis of the biological changes that transformed animal husbandry and farm productivity in the roughly two hundred years before the advent of hybrid corn. Chapters 10, 11, and 12 continue this discussion by investigating animal breeding, dairy operations, and draft power. In addition to breeding, an array of geoclimatic factors, market forces, and technological changes revolutionized nineteenth-century animal production. Everything changed – the location of animal production, the methods of feeding and caring for animals, the relative importance of various breeds, and the ways animals were brought to market and processed. Animal husbandry in most regions was in a perpetual state of flux. The relation of animals to the crop sector evolved as the corn frontier advanced westward and northward and as better grasses, clovers, and legumes were introduced. Animal production depended on the fortunes of key crops – when crops failed, animal populations invariably

[2] Olmstead and Rhode, "Impossible Undertaking."

declined. As highlighted in Chapter 1, in the mid-1840s the potato crops of New England and New York began to succumb to the same rot that later devastated Ireland and northern Europe. The decline in potato production contributed to the regional fall in cattle and swine production in this corn-deficit region.[3]

The changes in the animal sector run counter to most accounts of American agricultural development, because successive generations of farmers adopted intensive systems of animal husbandry and cropping that were labor using instead of laborsaving. Mechanization played a supporting role, but the overwhelming impetus for the changes was nonmechanical. Moreover, in this age, mechanization depended on biological innovations such as the shift from oxen to horses and the breeding of horses better suited for pulling equipment. Thus, the pattern discussed in the California chapter was a national phenomenon: farmers were adopting laborsaving machines in specific activities, but at the same time they were substituting toward more labor-intensive activities.

It is important to remember that the indigenous peoples had few domesticated animals; thus, all livestock that make up the modern farm economy, with the exceptions of turkeys and dogs, were foreign to North America. When European colonists first came to settle what would become the United States, they brought along their livestock in the hope of replicating the lifestyle and productive relationships of their home countries. Their idea of a harmonious farm environment had its roots in European soil.[4] Due to the initial lack of farm capital, European settlers often left their livestock to forage in forests and swamps.[5] This led to the development of lean, tough critters that could defend themselves from predators. This changed with economic development as farmers (often repeatedly) redesigned their animals; substituted among species and breeds; restructured the environment in which the animals lived by building fences, barns, silos, and feedlots; planted pastures of foreign plants; expanded corn output; and eliminated predators, such as wolves and coyotes, and competitors, such as deer and buffalo. These labor-intensive investments fundamentally changed the farm economy.

[3] Leavitt, "Meat and Dairy Livestock," 38–9.

[4] Anderson, *Creatures of Empire*. Recent research shows that the first chickens introduced to the Americas came from Polynesia, not Europe. "First Chickens in Americas Were Brought from Polynesia," *New York Times* (5 June 2007).

[5] One of the valuable attributes of swine was their ability to kill snakes. Dohner, *Encyclopedia of Historical and Endangered Livestock*, 166.

TABLE 9.1. *Number of Animals on Farms (in thousands)*

	All Cattle	Cows Kept for Milk	Hogs	Stock Sheep	Horses	Mules
1850	18,379	6,385	30,354	21,723	4,337	559
1860	25,620	8,586	33,513	22,471	6,249	1,151
1870	23,821	8,935	25,135	28,478	7,145	1,125
1880	39,676	12,443	49,773	42,192	10,357	1,813
1890	57,649	16,512	57,427	40,876	15,266	2,252
1900	67,822	17,140	62,876	61,606	18,280	3,271
1910	61,950	20,633	58,206	52,525	19,849	4,218
1920	66,777	19,680	59,371	35,077	19,783	5,441
1930	64,036	20,124	56,319	57,014	13,523	5,383
1940	60,818	21,937	34,070	40,173	10,098	3,849
1950	76,920	21,233	55,789	31,406	5,409	2,204
1959	92,534	16,552	67,949	33,945	2,955	–

Source: Carter et al., *Historical Statistics*, Tables Da988–994.

Major Livestock Types

Livestock served numerous purposes for early American farmers. They provided work (carriage and draft), food (meat, fat, and milk), materials (fiber, horns, leather, and glue stuffs), and fertilizers (manure, bone meal, and blood). Animals also required feeding and care. The major types of livestock differed in herding tendencies, ability to fend for themselves, feeding requirements, and reproductive behavior. These characteristics were not completely fixed and were subject to selective pressures, both natural and artificial.[6] Understanding the roles and characteristics of different types of livestock a century or two ago provides a basis for understanding the movement from general-purpose to special-purpose animals and the patterns of regional specialization in livestock production. Table 9.1 gives a sense of the changing numbers of animals on farms.[7]

[6] For example, breeding efforts have been directed toward increasing litter size for pigs and twinning rates for sheep and to shortening the reproduction cycle.

[7] The census enumerated livestock on June 1 from 1860 to 1900, on April 15 in 1910, and on January 1 in 1920 and 1930. These changes create inconsistent counts, especially of young animals. Animal births normally peaked in the spring. Piglet births also had a secondary peak in the fall. Measuring the hog population was especially sensitive to the timing of the count. For both beef cattle and hogs, the population would peak in the summer and tended to decline as stock was slaughtered in the fall. Lambs were typically sold off the farm and slaughtered in spring. The census instructed its enumerators in 1870 to exclude spring-born animals, provided no direction in 1880 or 1890, and in

TABLE 9.2. *Representative Animal Reproduction Cycles and Feed Requirements*

Animal	Age of Puberty (months)	"Best" Age for Mating (months)	Gestation Period (months)	Litter Size	Litters per Year	Feed Units (Dairy = 1) Grain	Roughage
Horses	12–24	24–36	11–12	1	1	1.30	0.75
Beef cattle	4–18	20–27	9–9.5	1	1	0.16	0.88
Dairy cattle	4–18	18–24	9–9.5	1	1	1.00	1.00
Hogs	3–7	8–12	4	4–12	1–2	0.70	0.00
Sheep	6–10	18–20	5	1–2	1–2	0.02	0.20

Sources: Bailey, *Cyclopedia*, 30–1; Asdell, *Patterns of Mammalian Reproduction*; Jennings, "Consumption of Feed by Livestock," 836.

Our discussion will focus on cattle, horses, swine, and sheep. The physical stature of the animals in these major categories is so different that simply counting the number in each group can be misleading. One can use the relative feed requirements (shown in the last two columns of Table 9.2) to compare the number of animals in feed-equivalent units. The feed units, which come from U.S. Department of Agriculture (USDA) estimates for adult animal consumption in the mid-twentieth century, provide two pieces of information.[8] First, they indicate the amount of feed an individual animal of each species consumed relative to that of a dairy cow. By this measure, sheep are much "smaller" than dairy or beef cattle. Second, the numbers offer a sense of what proportion of feed was in the form of grain, which humans could eat; and in the form of roughage, which humans could not eat. Thus, the numbers show that horses consumed more grain than dairy cattle but less roughage. These weights and the numbers of stock appearing in Table 9.1 make it clear that bovines were the most important type of livestock (as measured in feed-equivalent units).

In early America, cattle served the greatest range of functions, providing draft power, meat, and dairy products. Initially, multipurpose animals were the most common. Cattle possessed one great advantage for early American farmers because, like sheep and goats, they were ruminants that could eat vegetation that humans could not consume directly.[9] Cattle

1900 circulated a schedule segregating the population by age. U.S. Bureau of Agricultural Economics, *Livestock on Farms*, 3–25.

8 Jennings, "Consumption of Feed by Livestock."

9 As draft stock, oxen could work for about eight hours if given a lengthy break to allow grazing. Dohner, *Encyclopedia of Historical and Endangered Livestock*, 210.

possessed a herding instinct and generally preferred to congregate at night; their large size, horns, and hooves offered some protection against predators; and they had relatively long natural lifetimes, stretching into the mid-teens and occasionally into the thirties. However, they reproduced relatively slowly. Contrasting dairy cows with swine highlights the difference in breeding potential. After reaching breeding maturity in a little less than two years, a dairy cow could deliver one calf a year until about twelve years of age. Due to swine's earlier maturation, shorter gestation period, and larger litter size, topped off by the possibility of bearing two litters per year, a sow (and her offspring) could produce more than a hundred female piglets in the time it took a dairy cow to produce two female calves.

It has frequently been noted that livestock were a preferred product on the agricultural frontier because animals could walk themselves to the market – this was especially true of cattle.[10] Cattle raising extended over a wide and open range, in part due to the animal's mobility. However, in the era before mechanical refrigeration, dairying tended to be located in cool environments close to markets. The rapid growth of bacteria inhibited the production of milk and milk products in warm environments. Cheese production was pushed into even cooler climates than butter. An additional factor influencing the location of dairying was the availability of succulent grasses, which benefited from cool, moist summer climes.[11]

Horses, asses, and mules were kept principally for work. They grew too slowly to be good meat producers, converted grass to meat less efficiently than cattle, and took twice as long to reach maturity.[12] Mares reached reproductive age at about two years and could produce about two foals every three years.[13] Also, horses live longer than cattle, swine, or sheep.[14] The horse was considered superior to cattle for draft power because its front end is heavier (by roughly 1.5-fold) than its rear end. This provided "an advantage in inertial motion" relative to oxen.[15] Because horses were

[10] U.S. Industrial Commission, *Report on the Distribution of Farm Products*, 226–7.

[11] Huntington, "Distribution of Domestic Animals."

[12] Ibid., 152.

[13] The fecundity of mares was especially uncertain. Mule breeding, which involved mating a jackass with a mare, was even more troublesome. Horse mares were often afraid of jackasses and in turn jackasses, unless kept from birth with horses, preferred to mate jennets, females of their own kind. Professional mule breeders resorted to using "teaser stallions" to determine whether the mare was receptive and ready for the jack. Bradley, *Missouri Mule*, 3.

[14] Nowak, *Walker's Mammals of the World*, 1017, 1057, 1157, 1234.

[15] Smil, "Horse Power," 125.

kept to provide work daily, they were located in close proximity to the crops they tended.

Farmers kept swine primarily for their meat and lard. As captured in the expression "eating like a pig," hogs were noted as voracious omnivores, but this may overstate the range of their diet. Hogs were not ruminants. Each had only one stomach, making feeding exclusively on grasses problematic. Instead, swine competed with humans for food but also ate roots, acorns, garbage, and feces. Ellsworth Huntington noted that, given their variability, "some kind" of hog was "adapted to almost every kind of environment." Unlike other major farm animals, swine are not covered by protective hairs and lack the ability to sweat, so they have difficulty surviving in hot climates (above 98 degrees Fahrenheit) when exposed to direct sunlight. This limited their domain somewhat. As a rule swine were concentrated in areas of high human density, including cities.[16] Hogs also grow rapidly relative to other farm animals, gaining weight in a shorter time and with less feed. A piglet's weight can increase as much as 5,000 percent in the six months after birth.[17] Swine can convert up to 35 percent of the energy they consume into body weight, whereas sheep and cattle retain only 10 to 15 percent.[18] Hogs can live upwards of fifteen years, but after the development of improved breeds and intensive feeding systems those intended for slaughter seldom lived more than a season or two.

Compared with horses and cattle, swine reproduced extremely rapidly. Sows reached sexual maturity in six to eight months. The gestation period was about four months – "three months, three weeks, and three days" in the vernacular – and produced litters averaging six to eight piglets. Under the right conditions, a sow could produce two litters per year. Swine were harder to drive long distances because they have shorter legs than cattle or most sheep and were more troublesome and lost weight quickly on the trail.[19] These factors drew production closer to the point of consumption although meatpacking methods partially offset this transport problem.

[16] Huntington, "Distribution of Domestic Animals," 163.

[17] Gade, "Hogs," 537.

[18] Power, *Planting Corn Belt Culture*, 155, citing Towne and Wentworth, *Pigs*, 7. See also Bogue, *From Prairie to Corn Belt*, 104–5.

[19] Cronon, *Nature's Metropolis*, 225–6, citing Clemen, *American Livestock and Meat*, 58–9; Burnett, "Hog Raising and Hog Driving," 99; Boucher, *History of Westmoreland County*, 253–64; Nimmo, *Report in Regard to the Range*, 113; Henlein, "Cattle Driving," 92; U.S. Census Bureau. 10th Census 1880, *Report on the Productions of Agriculture*, 990. But the Census, 974, states that "the distance traveled each day is 12 or 15 miles, according to grass and water," and "a herd traveling with calves cannot make 12 mile per day."

In early America, preventing fresh meat from spoiling was difficult. The traditional curing methods of salting and smoking produced a palatable pork but were less effective with beef. Hot temperatures were a problem, so hog packing typically began after the first frost and extended to the last thaw.[20] In addition to reducing spoilage, farmers began slaughtering in late autumn because animal feed became scarce in the winter months. One indication of the relative importance of these effects appears in the timing of cattle and swine packing. Both activities benefited from the cold, but frost did more damage to the grasses on which cattle fed than to the corn which fattened swine; hence, the "season for the slaughtering and packing of cattle always began before that of hogs."[21] In the late nineteenth and early twentieth centuries, the spread of refrigeration and silos relaxed these constraints.[22]

Sheep were kept to provide wool, meat, and milk. Compared with Europeans, Americans put far greater emphasis on wool production than in raising sheep for meat or dairying. Like cattle and goats, sheep are ruminants. Because they use their lips (rather than just their tongues) to gather vegetation, sheep can feed on shorter grass than cattle. Sheep also are willing to graze on steep slopes where most other livestock, except goats, fear to go. Sheep reproduce relatively quickly, with an average gestation period of five months. They have a stronger herd instinct than cattle and are somewhat easier to drive, at least over short distances. Of the four major types of animals, sheep were the most vulnerable to predators and thus needed the protection of shepherds. Because shepherds were a fixed cost, this encouraged the keeping of large flocks.[23] Sheep raised for wool production tended to be located in sparsely populated fringe areas and in broken terrain beyond the reaches of crop cultivation. In the age before mechanical refrigeration, sheep intended for the commercial meat market were raised near urban areas because mutton was harder to preserve than pork or beef. For similar climatic reasons, mutton production was relatively less common in the South.

Table 9.3 shows the changing centers of livestock production over the period 1840–1910. In 1840, sheep production started the farthest east of any livestock species and by 1910 it was the farthest west. Sheep were also

[20] Cronon, *Nature's Metropolis*, 225; Clemen, *American Livestock and Meat Industry*, 110–34; and Edminster, "Meat Packing and Slaughtering," 242–9.

[21] Clemen, *American Livestock and Meat Industry*, 110.

[22] Coburn, *Swine in America*, 22–4.

[23] Huntington, "Distribution of Domestic Animals," 167.

TABLE 9.3. *Geographic Center of Animal Stock, 1840–1910*

	Latitude		Longitude		Latitude		Longitude	
	Mean	Median	Mean	Median	Mean	Median	Mean	Median
	Sheep				Swine			
1840	40.62	41.21	78.67	77.77	37.68	37.88	83.44	84.20
1850	39.85	40.36	81.77	81.46	36.93	36.98	85.07	85.40
1860	39.29	40.12	85.53	83.04	37.20	37.70	86.47	86.14
1870	39.61	40.20	88.59	84.61	37.89	38.48	87.60	87.29
1880	38.96	39.86	93.04	87.81	38.54	39.44	88.84	89.08
1890	39.47	40.20	94.69	90.23	39.01	39.82	90.17	90.52
1900	40.80	41.17	100.02	104.83	39.12	39.95	90.57	90.92
1910	40.69	41.01	100.32	104.77	38.82	39.78	90.76	91.11
	Cattle				Dairy			
1840	38.35	39.08	81.55	81.66	NA	NA	NA	NA
1850	37.75	38.57	83.80	83.25	38.59	39.74	82.12	81.88
1860	37.50	38.57	87.35	86.07	38.73	39.93	84.65	83.63
1870	38.00	39.42	87.42	87.02	39.52	40.48	84.76	84.08
1880	38.60	39.78	89.27	89.13	39.51	40.42	86.25	86.07
1890	38.80	39.81	92.17	92.45	39.69	40.59	88.25	88.70
1900	38.76	39.78	93.47	94.07	40.00	40.82	88.39	88.72
1910	38.86	39.79	93.73	93.75	39.87	40.75	89.47	89.47
	Horses and Mules							
1840	38.47	38.99	82.29	82.69				
1850	38.15	38.83	83.80	84.13				
1860	38.14	38.91	86.13	85.72				
1870	38.88	39.67	87.30	87.07				
1880	38.85	39.72	88.98	88.66				
1890	39.33	40.00	91.31	90.70				
1900	39.16	39.77	92.21	91.83				
1910	39.20	39.72	92.86	92.68				

Source: Derived using data from Haines and Inter-university Consortium for Political and Social Research, *Historical, Demographic, Economic, and Social Data.*

the most northern animal stock with dairy cows a close second. Swine production started the farthest west in 1840 and moved the least.

In the age before the widespread use of feed concentrates and commercial fertilizers, the location of livestock production was closely linked to that of crop production. Feed crops were a leading input for animal production; manure was a principal means of replenishing the fertility of the soil; and both feed crops and manure were too bulky to bear long-distance shipping. Farmers in many regions found wisdom in the old saying, "more crops, more animals; more animals, more manure; more manure, more crops." The major types of livestock differed in the quantity and quality of manure they produced. Cattle ranked highest in the daily volume of manure produced per head, followed by horses, swine, sheep, and poultry. Inspired by the pioneering work of the great European scientists Justin von Liebig, John Lawes, and Joseph Gilbert, agricultural chemists unlocked the secrets of animal excrement over the second half of the nineteenth century. They demonstrated systematically how the chemical composition and value of manure as fertilizer varied with the type of animal, its age and activities, and its diet. One benefit of high-protein meal, it was learned, was that it yielded more nitrogen-rich manure.

Agricultural scientists also explored how the effects of different manures on plant growth depended systematically on the crop, soil, and season of application. As a rule, poultry manure, which had a high nitrogen content, was of the highest value. Manure from sheep and horses ranked next. All three products were dry and decomposed quickly, releasing their nutrients to the soil. Sheep manure contained more nitrogen and phosphoric acid but less potash than horse manure and was considered a quick-acting concentrated fertilizer. Pig excrement was highly variable as well as slow to decompose. Given animal husbandry practices, swine dung and urine were difficult to collect. Cow manure contained the least plant food per ton and released its nutrients slowly. Cattle urine was of potentially greater value than dung, especially as a source of nitrogen, if carefully collected and promptly applied; but in standard practice, much of its fertilizer value was lost due to drainage, leaching, fermentation, and volatilization. Even under the best conditions, soil scientists found that 15 to 20 percent of the nitrogen would be lost.

Advances in Feeds

In the late nineteenth century, agricultural textbooks and farm journals frequently published the results of feeding experiments conducted by American and European animal scientists. These technical studies sought to determine the effects that different rations of various feeds had on an animal's weight gain, its ability to perform work, and its reproductive behavior. Among the texts popularizing this research were Elliot W. Stewart's *Feeding Animals: A Practical Work upon the Laws of Animal Growth,* which went through seven editions between 1883 and 1895, and

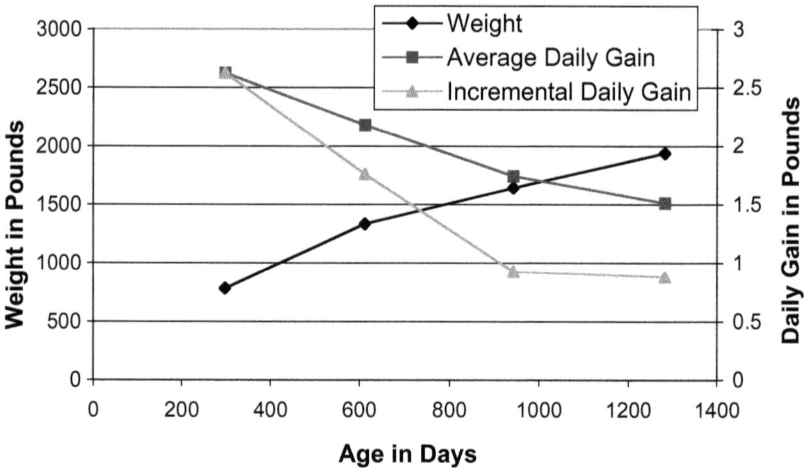

FIGURE 9.1. Relationship between age and weight among cattle at American Fat Stock Shows. *Source*: Stewart, *Feeding Animals* (1888), 529–31.

the classic text of William A. Henry and (later) Frank B. Morrison, *Feeds and Feeding*, which went through twenty-two editions over the period 1898–1956.[24]

Some of the principles, such as the notion that the weight gain of animals decreased with age, were fairly straightforward and became widely understood. As an example, Stewart's empirical study of 460 cattle examined the relationship between an animal's age and weight, with a special emphasis on the incremental daily gains. Figure 9.1 summarizes his findings. Stewart concluded that the "mathematical proposition, that... every additional pound put upon a young animal costs more in food than the previous pound" demonstrated "the importance of *early maturity*."[25]

As evidence of biological innovation, the vast majority of the grasses, grains, beans, vetches, and legumes that feed American farm animals today were, like the animals they sustained, introduced to North America from other continents. Corn was the major exception and the verdict is still out on a few forage crops. For example, it is likely, but not certain, that American symbols such as Kentucky Bluegrass and Timothy grass were imports. But there is no doubt about the foreign origins of important feeds such as alfalfa, soybeans, red and white clovers, dozens of important grasses, chickpeas, and many more. Moreover, most feed crops, even

[24] Stewart, *Feeding Animals* (1883).
[25] Stewart, *Feeding Animals* (1888), 529–31.

those native to North America, were transplanted and adapted to flourish in a wider range of conditions.[26]

The decades immediately before and after 1900 witnessed path-breaking advances in the understanding of food and feeds that led to the study of nutrition as a science. It is worth noting that, as with the germ theory of disease, working with animal models was crucial for progress in this area.[27] Through most of the nineteenth century, physiologists considered foods to be composed of only a handful of macronutrients: proteins, fats, carbohydrates, "ash" (mineral residuals), and water. In 1890, Christian Eijkman, a Dutch physician and student of microbiologist Robert Koch, found that chickens that were fed on a diet of polished rice developed an avian disease similar to beriberi. Eijkman, together with Gerrit Grijns, later (1900) identified a water-soluble factor in rice bran that prevented this polyneuritis in the birds. The factor was initially named vitamine, "an amine vital for life."[28] As the role of the nutrient was placed into a broader context, it was renamed thiamin or B1. In 1907, Norwegian researchers Alex Holst and Theodor Frölish used guinea pigs to identify scurvy in an animal model and within a few years scientists at the Lister Institute in London isolated a factor in lemons, later known as vitamin C, that acted as a preventive. These findings led Casimir Funk, a Polish biochemist at the Lister Institute, to propose in 1912 his general theory of *vitamins* as micronutrients, which when available in sufficient quantities prevent specific deficiency syndromes. Besides beriberi and scurvy, human diet–related syndromes included rickets, night blindness, and pellagra. Guided by the vitamin concept and newly developed experimental techniques using purified diets, by the end of the 1910s cutting-edge researchers, including Elmer McCollum, identified vitamin A, a fat-soluble factor that prevented night blindness, and vitamin D, a heat-stable factor that combated rickets. In 1919 biochemist Harry Steenbock of Wisconsin noted that the presence of vitamin A in plants was related to their yellow pigment.

These discoveries, generated by the global scientific community, soon had important effects on research in the United States. In 1917, the newly formed National Research Council impanelled an Agriculture Committee

[26] For classic treatments of grasses see Edwards, "Settlement of Grasslands"; Piper and Bort, "Early Agricultural History of Timothy"; and Carrier and Bort, "History of Kentucky Blue Grass."

[27] Combs, *Vitamins*, 16–17.

[28] Ibid., 11, 17–18; Gubler, "Thiamine," 235.

to help incorporate the latest scientific information into wartime efforts to increase the food supply. One target area was animal nutrition. The experimental advances combined with wartime shortages of supplemental proteins induced a search for methods of optimizing animal nutrition. An animal feeding subcommittee, chaired by Henry P. Armsby of Pennsylvania State University, began a series of cooperative experimental investigations to determine the protein requirements of cattle. Following the publication of these results in the early 1920s, the study group, renamed the Committee of Animal Nutrition in 1928, turned their attention to the requirements of swine and other livestock.[29]

We can trace what was known about feeds and feeding by comparing the contents of the leading book on the subject over the course of four decades. The aforementioned W. A. Henry of the University of Wisconsin published the first edition of his influential *Feeds and Feeding: A Handbook for the Student and Stockman* in 1898. The book was "entirely rewritten" in 1910 and again in 1915. At this later date F. B. Morrison, also of the University of Wisconsin (and later of Cornell), became a coauthor. In 1923 Morrison released the eighteenth edition, noting that "exceedingly important discoveries have been made concerning the science and practice of live stock feeding." As with previous editions, the new release summarized the results of numerous "feeding trials that have been conducted at the experiment stations" and related the practical experiences of "leading stockman."[30] In 1936 Morrison released the twentieth edition. At that time he boasted that "in no field of agriculture has more progress been made in recent years than in livestock feeding and nutrition."[31] Taken together these books offer something akin to a fossil record of what was known and taught about the science and practice of animal feeding between 1898 and 1940. Clearly much that was written in any of these volumes would be dated by later discoveries, but the advances in knowledge over these four decades are undeniable.

Over time the discussion of feeds became more strongly based on controlled experiments and an understanding of chemistry and nutrition. A closer examination of the treatment of cottonseed meal will illustrate this point. In his first edition Henry noted that "the utilization of cotton seed

[29] Ullrey, "Landmark and Historic Contributions," Ch. 1. Key publications, National Research Council (U.S.), *Plan for Cooperative Experiments*, and *Growth and Recommended Nutrient Allowances for Cattle*, and *Recommended Nutrient Allowance for Swine*.

[30] Henry and Morrison, *Feeds and Feeding*, iii.

[31] Morrison, *Feeds and Feeding*, 1.

and its products as food for both man and beast is an excellent example of what science has accomplished for the advancement of agriculture." He was correct. One of the most significant biological improvements in the period between the Civil War and World War II was the increased utilization of cottonseed. The extensive commercial uses of cottonseed oil, meal (or cake), and hulls (the latter two are now important animal feeds) were all innovations of the late nineteenth and early twentieth centuries.

Farm accounts show that many antebellum planters used cottonseed as a green manure, spreading it on corn fields or in orchards, and that some fed it to animals. There were problems with both uses. Subsequent research has shown that using the seed for fertilizer increased the acidity of the soil. Many southern soils were already acidic, so adding cottonseed was generally not wise. In addition, planters soon discovered that pigs fed on cottonseed often died. In general cattle showed no ill effects, but the typical plantation had few cattle to consume the seed. Thus, most seed, apart from about 15 percent saved for planting, was wasted. Worse, much was dumped into rivers, creating major environmental problems. After 1866, southerners began to turn this waste product into a valuable by-product. Figure 4.3 in Chapter 4 charts the rising fraction of U.S. cottonseed crushed for commercial purposes from 1872 to 1939. As the figure indicates, by the 1920s roughly 80 percent of U.S. cottonseed was being processed. At this time, farm revenues from cottonseed comprised about 13 percent of the combined earnings from seed and lint.[32]

Henry focused on feeding seed, and its meal and hull by-products, to animals. In 1898 he reported on experiments that tested various feed combinations on cattle, horses, sheep, and swine. He noted that "the practice of fattening steers exclusively on cotton-seed hulls and cotton-seed meal was begun in the South in 1883. The business has grown so that it is estimated that 400,000 cattle were fattened at the oil mills of the South for the season of 1893–94, besides large numbers of sheep." Work done at the Maine experiment station showed that substituting cottonseed meal for corn meal led to noticeable increases in milk output. Work at numerous other experiment stations yielded contradictory results on the effect of cotton feeds on milk production and quality. Experiments on feeding seed and meal to swine confirmed what farmers had long known – the vast majority of the test animals died. But nobody understood why they died or why cattle fed the same feed thrived. It was found that, for some

[32] Henry, *Feeds and Feeding* (1898), 154; Moloney, "Cottonseed," 431–63; Wrenn, *Cinderella of the New South*; Lamborn, *Cottonseed Products*, 16–22.

reason, boiling the seed substantially reduced the mortality rate of swine to about 25 percent. To add to the puzzle, calves fed cottonseed meal usually died. "All attempts to determine the poisonous principle ... have thus far proved futile, and the matter is still a mystery."[33]

Morrison's 1936 treatment of cottonseed was much more sophisticated. Most important, the "poisonous principle" was now understood. Cottonseed contains the poison gossypol, which is lethal to swine but not adult ruminants. It was the immature digestive systems of calves that made them vulnerable. Morrison also detailed other aspects of cottonseed chemistry. Cottonseed meal was exceptionally rich in protein, and by 1936 the meal was graded for its quality, including texture, fiber, fat, and protein content. The best grades contained more than 41 percent protein. Morrison advised on the calcium and phosphorus qualities of the seed by-products and noted deficiencies of vitamins D and A. He went into considerable detail about the effects of gossypol on various species and observed that "in the ordinary cooking of cottonseed kernels in the process of oil manufacture, most of the gossypol is converted into a substance (called d-gossypol, or bound gossypol) which has much less toxic properties." He reported the results of numerous studies, which showed that, as long as cattle received sufficient amounts of calcium and vitamins from other sources, they could be fed large quantities of cottonseed meal. Numerous studies determined that cottonseed meal could be fed to swine so long as the rations did not exceed 9 percent of the diet.[34]

The increasing importance of cottonseed by-products was part of a larger story as farmers came to rely increasingly on commercial feeds. Just as there were backward linkages tying farmers with the agricultural equipment industry, there were linkages tying farmers with enterprises producing scientific feeds. Henry devoted little attention to commercial feeds in his 1898 edition, but over the next few decades a new industry emerged. In 1936 Morrison noted that "the manufacture of commercial mixed feeds has become a very important industry in the United States. In 1929 there were ... 750 establishments manufacturing prepared feeds for animals and fowls," with a total output value of $400 million. To provide

[33] Henry, *Feeds and Feeding* (1898), 154–9.

[34] Morrison, *Feeds and Feeding*, 360–7, 452, 545, 599, 668–71, 757, 884–6. The advances in cottonseed-based animal feeds were part of a larger story as chemists developed industrial and human uses for the oil. Important breakthroughs include David Wesson's 1899 innovation in processing cottonseed oil, and Procter and Gamble's release of hydrogenated cottonseed oil (Crisco) in 1911. ConAgra Foods, *Company History*; Forristal, *Rise and Fall of Crisco*.

perspective, the total value of "agricultural machinery, and attachments and parts" manufactured in 1929 was $278 million.[35]

Other biological innovations, such as ensilage, alfalfa, and soybeans, further revolutionized animal feeding. We discuss ensilage in Chapter 11, but the important point to note is that silo construction only began in 1873 and that by the 1920s there were about one-half million silos in the United States. Alfalfa, or Lucerne, is a nitrogen-fixing legume and an important source of hay. A comparison of Henry's and Morrison's treatments between 1899 and 1929 shows a marked increase in both the attention given to alfalfa and the understanding of its feed value. Over three decades alfalfa acreage had increased more than fivefold, accounting for about one-third of all tame hay harvested in 1929. Its relative value was much greater because alfalfa "provides over twice as much digestible protein per acre as clover hay, [and] about five times as much as hay from timothy and mixed timothy-and-clover." By 1936 there was a thriving market for commercial alfalfa meal.

The spread of alfalfa is a further testament to biological research and innovation. Alfalfa was introduced to Europe from southwestern Asia. Spanish conquistadors brought it to the Americas and numerous immigrant groups carried it to the thirteen colonies. Both George Washington and Thomas Jefferson unsuccessfully experimented with alfalfa.[36] Alfalfa was introduced to California from Chile in the early 1850s and was spread to the western states, where it thrived.[37] The varieties of alfalfa first grown in the United States could not survive harsh winters and decades of attempts to grow it in the northern states met with failure. As late as 1898, Henry observed that alfalfa was not suitable for the more frigid states, but, as noted in Chapter 1, Wendelin Grimm's efforts changed this situation. According to many accounts, Grimm introduced seeds from Germany to Minnesota in 1858. Most of his crop died of winterkill, but a few plants survived. Year after year Grimm saved the seeds of the more winter-hardy plants until he had developed a strain suitable for cold climates. Grimm raised alfalfa far beyond its known area of production for about fifty years before his crop was "discovered" by plant breeders in 1900.[38] Tests proved its hardiness and extension publicity led to its rapid dissemination in the northern states.

[35] U.S. Census Bureau. 15th Census 1930, *Agriculture*, 1096.

[36] Russelle, "Alfalfa."

[37] U.S. Census Bureau. 12th Census 1900, *Census Reports*, Vol. 6, *Agriculture*, pt. 2, *Crops and Irrigation*, 203.

[38] For variants of this standard account, see Brand, "Acclimation of an Alfalfa Variety," 891–2; Westgate, "Another Explaination," 184–8; and Wedin, *Improvements*.

Parallel efforts by the USDA accelerated the northern expansion of alfalfa production. Concerned by the heavy alfalfa losses to disease, drought, and winterkill, Secretary of Agriculture James Wilson sent horticulturist Niels Hansen in search of new varieties. Hansen traveled over two thousand miles by wagon and sleigh across Eastern Europe, Siberia, and Central Asia and obtained eighteen thousand pounds of seed from both frigid regions and deserts. Hansen subsequently tested his seeds across a wide range of geoclimatic conditions in the United States. In 1898–99 a harsh winter killed roughly half of the crop in Nebraska, Colorado, and Wyoming, but Hansen's Turkestan variety flourished. Even at the South Dakota experiment station where the "common alfalfa" was killed, his new variety was unaffected. Hansen made two subsequent trips, returning with superior varieties from Kashmir and Siberia. As with his previous imports these were tested in the early 1900s throughout the North Central states.[39] These projects led to a significant regional shift in production in the first half of the nineteenth century. By a similar process the crop was acclimated for the southern states, but in this case the foundation seed evidently came from Peru (via Spain).[40] These research efforts helped make alfalfa the nation's most important hay crop, accounting for more than 58 percent of the value of all hay harvested in 2005.

In 1898 Henry devoted less than a page to "soja beans." He clearly considered it an experimental crop suitable for the South but "too tender to be generally useful in at the North." This view changed dramatically over the next four decades, and in 1936 Morrison noted that "175,000 tons of soybean oil meal were produced in the United States" in 1934. In this edition he made more than thirty references to soybeans and their by-products that covered dozens of pages. His analysis drew on research conducted at experiment stations across the country.[41]

In 2005 soybeans (for beans) were second only to corn in the value of crop production and more than two and one-half times the value of cotton lint and seed production. Soybeans also made up about 90 percent of the nation's total oilseed production. Given its importance, relatively little has been written about the crop's history in this country.[42] The introduction

39 Russelle, "Alfalfa."

40 Henry, *Feeds and Feeding* (1898), 203–9; Morrison, *Feeds and Feeding*, 251–8; Edwards and Russell, "Wendelin Grimm and Alfalfa," 21–33; Tysdal and Westover, "Alfalfa Improvement," 1122–53; USDA, *Agricultural Statistics 2006*, ix–18, ix–19; Stewart, *Alfalfa-Growing*.

41 Henry, *Feeds and Feeding* (1898), 161–2, 209; Morrison, *Feeds and Feeding*, 369–73.

42 USDA, *Agricultural Statistics 2006*, ix–18, ix–19.

and diffusion of soybeans in the United States and Western Europe was crucially dependent on scientific research and, as with so many other crops, demonstrates the importance of the international flow of biological information and material. In 1765, Samuel Bowen introduced soybeans into Georgia from China via London. Bowen experimented with soybeans on his farm near Savannah and in 1767 received a royal patent to manufacture soy sauce. As was common for the age, Bowen shared seeds for the common good, and a packet he sent to the American Philosophical Society in Philadelphia was likely the first introduction of soybeans into the northern colonies.[43] Benjamin Franklin independently introduced soybeans from France in 1770, and there are references to experimental plots in Massachusetts in 1829 and 1831. Admiral Perry brought two varieties from Japan in 1854, which the Commissioner of Patents distributed to farmers.

In 1878, Professor G. H. Cook of New Jersey imported soybean seeds from the Bavarian Agricultural Station and others obtained seeds from eminent Viennese plant scientist Friedrich Haberlandt. By this time, scientists across Europe were experimenting with the crop in an effort to find varieties suitable for particular regions. The North Carolina experiment station initiated experiments with soybeans in 1882, and most other state experiment stations soon followed suit. In 1898 the USDA started to introduce new varieties from across Asia. At this time there were "not more than eight varieties, with limited adaptation to soil and climate, grown in the United States." In 1907 the USDA began work with soybean hybridization. Between 1907 and 1935 the U.S. soybean acreage increased from less than 50,000 to about 5.5 million acres. Although soybeans were relatively new as a commercial crop, by 1935 America was the world's second largest producer with an output of about 40 million bushels of seed, well behind the leading producer, China (including Manchuria), but about double that of Korea and three times that of Japan. By 1939 American production had doubled to more than 90 million bushels. At this time the principal use of soybeans in the United States was to feed animals. Soybeans were preserved as silage or hay, cut and fed green, and used as pasturage for sheep and hogs.[44]

Researchers paved the way for the expansion of U.S. production in the 1920s and 1930s. For animal forage and pasture, "the development of such varieties as Virginia, Laredo, Otootan, Wisconsin Black,

[43] Hymowitz and Shurtleff, "Debunking Soybean Myths," 473–6.

[44] Morse and Cartter, "Improvement in Soybeans," 1155.

Manchu, Wilson-Five, Kingwa, Peking, and Ebony by selection from introductions has been the principal factor in the increased use and acreage." Researchers also focused on over a thousand selections and varieties for their oil content and their suitability for different regions. Other experiments identified varieties suitable for various industrial purposes and examined protein content, flavor, disease resistance, adaptability to different climates and soils, and more.[45] In 1935 soybeans were an important crop and were about to become even more so, thanks in large part to biological innovation.

The new feeds were especially important for the South. "A large part of the livestock history of the South after 1819 is tied up with the search for a suitable grass or forage plant to supplement corn. The most valuable northern artificial grasses, though quite frequently experimented with, could with difficulty withstand the summer heat, especially south of Virginia. Clover, which had been so successfully used in Virginia, suffered from all like difficulty, despite many successful experiments on a small scale, and little was grown."[46] In 1971 the USDA published a report analyzing one hundred native southern grasses, some of which grew over a wide range, but none had the value of the more popular northern grasses and clovers. Southern newspapers and journals gave considerable attention to experiments with foreign grasses and legumes. The cowpea (imported from Africa) was an important find, but it had drawbacks. The plant ran to vines on rich soils, and it gained a reputation as an animal killer. In any case, cowpeas emerged as an important forage crop and green manure by the 1850s and by the early twentieth century represented the region's most important legume.[47]

Most efforts to introduce feed crops into the South were unsuccessful, and some were downright disastrous. Widespread experiments in the antebellum years with a number of tropical grasses, including crab grass, gamma grass, guinea grass, Johnsongrass, and Bermuda grass, were all disappointing. In fact, the last three of these grasses have the ignominious distinction of being included on a list of the world's ten worst weeds.

[45] Ibid.; Horvath, *Soybean Industry*, 1–2; Piper and Morse, *Soybean*, 39–54; Markley and Goss, *Soybean Chemistry*, 2.

[46] Leavitt, "Meat and Dairy Livestock," 184.

[47] Cowpeas are also know as black-eyed peas. Leavitt, "Meat and Dairy Livestock," 9, 183–8; Leithead, Yarlett, and Shiflet, "100 Native Forage Grasses"; Henry and Morrison, *Feeds and Feeding*, 237. As an example of the attention given the cowpea by the southern press, Edmund Ruffin wrote several essays extolling its value as a feed and fertilizer. Boyd, "Edmund Ruffin." In 1848 a correspondent to the *Southern Cultivator* (January 1848) observed that "[t]he pea must take the place of the clover plant in the South, where the latter does not flourish."

Bermuda grass seemed to hold the greatest promise, but its tendency to take over fields and the difficulty of eradicating it were major drawbacks. In spite of these problems it did become an important cattle feed in the Deep South. Johnsongrass, which now thrives in most states, was probably the greatest offender and is one of the nation's most economically costly weeds.[48]

"Johnsongrass . . . is a warm-season grass that forms large colonies from stout, finger-sized, much-branched rhizomes. A medium-sized plant can produce up to 60 feet of rhizomes in a single year. From these rhizomes emerge a thicket of unbranched, pinkie-sized stems that grow 5 to 6 feet tall and are topped with a foot-long open panicle of flowers that are followed by purplish seeds. A medium-sized clump will produce two pounds of seed." Under normal conditions the grass was an excellent feed that rivaled Timothy in protein content and feeding value. However, under stress (from insects, drought, or frost) it produced prussic acid (hydrocyanic acid), which was often fatal to livestock. Far more damaging, Johnsongrass took over cropland, dramatically lowering yields and requiring massive labor inputs to keep it at bay. Much as the Hessian fly changed the economic fortunes of wheat farmers, the introduction of Johnsongrass and Bermuda grass in the antebellum era significantly dampened the economic prospects of cotton and corn growers.

Johnsongrass was imported to the United States from the Middle East. It may have been cultivated in Mississippi before 1830 and probably reached Georgia and South Carolina by 1835. John H. Means, who would become governor of South Carolina in 1855, was closely associated with one of these initial introductions, and many early accounts referred to the weed as "Means grass." He was one of the early advocates of the grass and one of its first victims. According to one account, Means received the Johnsongrass seeds circa 1835 from an agent whom he had sent to the Ottoman Empire to help educate the Turks in the art of growing cotton. The costly impact of the grass was not immediately evident and in the early 1840s others continued to plant it.

By the late 1840s experimenters were questioning whether the value of Johnsongrass as a forage crop justified the costs it imposed on farmers. For example, in 1848 the *Southern Cultivator* published several letters from correspondents as to the grass's merits and hazards. "B. M." of the University of Alabama wrote that two years earlier he had recommended Means grass, but was now "convinced that it is a very serious pest" and

[48] Leavitt, "Meat and Dairy Livestock," 185; Henry and Morrison, *Feeds and Feeding*, 213. Holm, "Weeds Problems in Developing Countries," 113–8.

dubbed it "mean grass." He described a grass that spread rapidly and was all but impossible to eradicate. Despite this condemnation, others, desperate for a fodder crop that "might assist in freeing us from the *disgraceful tax* we now pay to the Eastern States for hay," continued to plant the luxurious grass. By 1850 Means had no doubt that the plant was a menace. He considered selling his plantation in order to move to Louisiana, but "'the big grass has inspired such a terror that no one will even look at it [his plantation].' He continued, 'When the grass runs me off, then I must seek a home in the West.'" The grass was eventually renamed for Colonel William Johnson of Selma, Alabama. Johnson had obtained the seeds from Means in the early 1840s and distributed them in his vicinity. From these many epicenters and perhaps with fresh introductions from abroad, the weed spread throughout the cotton belt and then on to an even broader domain.[49]

The spread of Johnsongrass represents one of many cases when conscious efforts toward biological innovation substantially increased the costs of production for many crops and lowered agricultural efficiency. An understanding of the changing grass environment bears on a broad set of issues dealing with the seasonal demand for labor in the South. Researchers who use labor requirements drawn from the early twentieth century to impute requirements for the antebellum years do not appreciate the effect of the spread of new grass varieties. Fields overrun with Johnsongrass would not yield a crop, so extended periods of rain, which kept workers out of the fields, could be disastrous. The grass would overtop commercial crops such as cotton or corn, thereby blocking the sunlight, while robbing the plants of nutrients and water. Thus, in addition to increasing labor inputs the grasses sapped yields. Before the recent advent of improved herbicides, grasses required constant attention. Given that the nearest substitute for herbicides was a worker with a hoe, it is highly problematic to treat herbicides as primarily "land-saving" innovations.

Conclusion

This chapter set the stage for our analysis of animal breeding, the dairy industry, and draft power. Animals occupied an important and growing

[49] Interview by the authors with UC Davis weed ecologist Tom Lanini, 19 January 2007; *Southern Cultivator* (August 1848), 118, and (December 1848), 180; McWhorter, "Introduction and Spread of Johnsongrass," 496–500; Phares, *Farmer's Book of Grasses*, 118–23; Mitich, "Colonel Johnson's Grass," 112–13.

niche in American agriculture which evolved as the nation matured. The animal and crop sectors grew in tandem – each dependent on the other. But the expansion involved much more than the simple replication of what was already present. The development of a greater variety of high-quality feeds, which allowed farmers to sustain larger and higher quality animal populations in pasture- and pen-fed environments, was an important achievement of the pre–World War II era. This was fundamentally a labor-intensive activity because farmers created pastures and built silos and barns. Scientists also made great strides in understanding feeds and animal nutrition. The new feeds were, with the exception of corn, foreign to North America and their adoption required considerable experimentation – trial and error and scientific plant breeding programs both played a role. By 1800 foreign grasses and clovers were becoming a mainstay of northern pastures. The subsequent spread of cottonseed products, alfalfa, and soybeans revolutionized animal feeding. The fact that American farmers in 1929 were spending considerably more on commercial feeds than on agricultural equipment should give even the most ardent proponent of the "mechanization was all that mattered" hypothesis reason to pause and take stock.

10

Defining and Redesigning America's Livestock

The imprints of biological forces on the animal sector are even more evident than those on crops. Eighteenth-century paintings of livestock illustrate this point. Even if the artists exaggerated the attributes of their subjects, the animals of 1750 bore little physical resemblance to their 1940 descendants on America's farms. By this date horses, cattle, hogs, sheep, and poultry were far more productive and manageable in the tamer and more confined environments they inhabited. In general, specialized breeds had replaced general-purpose animals.[1]

Although they came to share a common conceptual basis in the science of genetics, animal and plant breeding differed in several respects. Plant breeders typically worked with large populations, whereas animal breeders worked with individuals or small groups. The contribution of the male in animal breeding was more obvious and identifying the father was easier. Although plant breeders were always on the lookout for exceptional rogue plants – recall Fife's Hard Red wheat – they generally focused on the population averages of a small number of quantitative characteristics and relied on random recombination and mutation to generate diversity. Animal breeders, on the other hand, tracked many qualitative and quantitative characteristics, developed elaborate pedigrees, and deliberately intervened in the mating process. The fecundity of animals was very limited compared to that of plants. Breeding could not proceed until female animals reached sufficient maturity and, even then, they produced relatively few offspring. By way of contrast, most plants produced thousands

[1] Apparently some artists intentionally misrepresented animals to conform to fashion and the demands of their patrons. Ritvo, *Animal Estate*, 59–60.

of viable seeds each year.[2] In addition, many plants possessed asexual pathways of propagation and could fertilize themselves, which allowed for the retention of desirable combinations of characteristics. Plant breeders also had a crucial cost advantage. The high cost of animals led many professional breeders to adhere to established methods and avoid the risks of radical experiments.[3]

Animal breeding was more advanced than plant breeding in the era before the rediscovery of Mendel's principles in 1900, but plant breeding benefited more from this breakthrough. Donald F. Jones, a pioneering breeder of hybrid corn, noted in 1925 that "animal breeding has been brought to a higher plane of development than plant breeding, and there are several good reasons for this. From the first, sex in animals has been obvious." In addition, livestock breeders realized early on "the importance of the pedigree." Jones concluded that "in comparison with animal breeding, some of the practices in plant improvement are two hundred years behind the times."[4] But "the new knowledge of genetics has contributed little" to the breeding of animals, which "proceeds in much the same way" as it had since classical times.[5] Dutch animal breeder A. L. Hagedoorn concurred, noting that the rediscovery of Mendel's principles caused a "veritable revolution in plant-breeding technique" whereas the breeding of animals, other than poultry, remained "remarkably speculative and economically wasteful."[6]

Differences between animal and plant breeding appeared not only in the efforts of professional breeders but also in the practices of everyday farmers. Although many American farmers might plant their fields with seeds selected without much care from a previous year's harvest, far fewer left the mating of farm animals to chance. With few exceptions, such as farmers who raised razorback hogs or Texas Longhorn cattle, animal owners intervened in the reproduction process. Moreover, when it came to breeding, the typical farm was not self-sufficient – it relied on services supplied by neighbors and the marketplace. Even when the

[2] Winters, *Animal Breeding*, 62, 281; Sanders, *History of the Percheron Horse*, 341.

[3] Hammond, *Farm Animals*, 127; Warren, *Elements of Agriculture*, 23; USDA, "Livestock Breeding at the Crossroads," 831–41.

[4] Jones, *Genetics in Plant and Animal Improvement*, 1–12.

[5] Ibid., 486–7. As early as 1904, George M. Rommel wrote that "the breeders of plants have passed the breeders of animals... and to-day possess a better insight into the principles underlying their science." U.S. Bureau of Animal Industry, "Plan for the Improvement of American Breeding Stock," 316. Webber, "Effect of Research in Genetics," 125–6.

[6] Hagedoorn, *Animal Breeding*, 24, 27. Also see Gilmore, *Dairy Cattle Breeding*, 571–2.

farmers involved were not professional breeders, the matching of mares and stallions, of cows and bulls, of sows and boars, and of ewes and rams was largely an intentional act.

The reason was simple. Bulls, stallions, boars, and rams were difficult to control. This led farmers to geld most male animals, creating steers, geldings, barrows, and wethers. Castration made males more docile and manageable and allowed males and females of a given species to feed and graze together without trouble. Castration hastened fattening and improved the flavor and texture of the meat of animals intended for slaughter. As a result, keeping animals for stud was relatively rare and breeding was a specialized activity. The prevalence of castration – typically performed before sexual maturity – implied conscious selection, the intentional culling of "inferior" stock.[7] These breeding choices represented an ongoing biological intervention that shaped the very nature of animal husbandry.

There is little information available about stallions, bulls, boars, and rams prior to the 1920 *Census of Agriculture*. In that year, only 1.5 percent of U.S. farms reported having stallions whereas 73 percent reported keeping horses. Nationally, there were only 129,000 stallions (two years old and older), in contrast to the more than 8 million geldings and 9 million mares (in the same age category). In all of New England there were well under 3,000 breeding males. The market price demonstrated the value of a horse's breeding potential. The price of geldings in 1920 was slightly less than that of mares and only one-fifth that of stallions. Data on the dairy sector in 1920 paint a similar picture. Dairy bulls (one year and older) accounted for only 2.5 percent of dairy cattle and their ownership again was relatively uncommon. Whereas more than 71 percent of U.S. farms reported having dairy cattle in 1920, fewer than 11 percent reported the presence of dairy bulls. Yet dairy bulls were more valuable per head – by an average premium of 12 percent – than the milk cows they serviced. Nationally, the frequency of boars roughly equaled that for dairy bulls with 11.6 percent of U.S. farms reporting "boars for breeding." But in the key hog-producing midwestern states, one-quarter to one-half of farms kept breeding boars. These boars were worth about 30 percent more than gilts and breeding sows.[8]

7 Castration was done early, before first trial mating. The timing of castration varied; if too early, the loss of hormones could harm normal development; if too late, dangerous infections could result. Jennings, *Cattle and Their Diseases*, 316; Rice, *Breeding and Improvement*, 34.

8 U.S. Census Bureau. 14th Census 1920, 534, 560–4, 590–4.

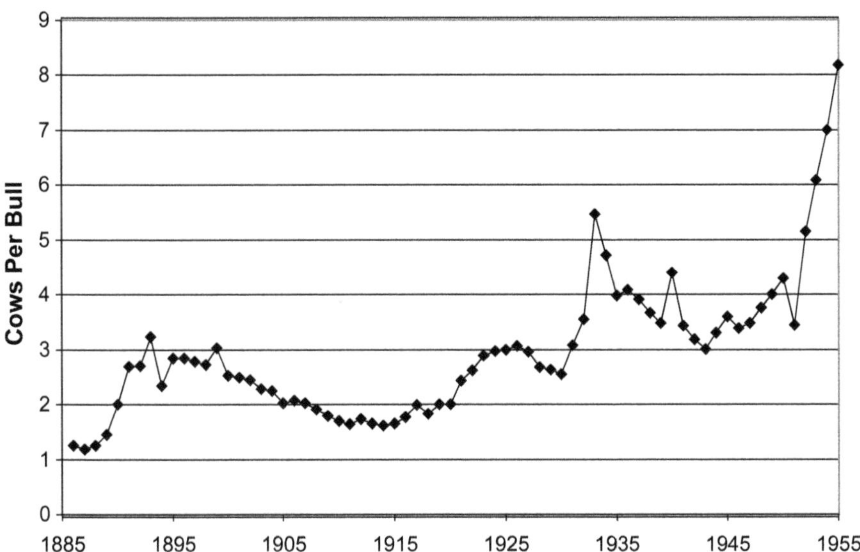

FIGURE 10.1. Number of registered Holstein cows per bull, 1886–1955. *Source:* Holstein–Friesian Association of America, *Herd-Book*, 1530–3.

The ratio of males to females of a given type of animal varied with market conditions. Figure 10.1 illustrates this point for Holstein–Friesians by graphing the share of bulls in all registered animals from 1886 to 1955.[9] The ratio of cows per registered bull rose from about 1.3 in the late 1880s to 7.3 in the mid-1950s. Although the rapid spread of artificial insemination after 1945 contributed to an increase in the number of dams served by a single bull, it is clear that the upward trend began well before this innovation. Earlier transportation improvements, which increased the effective service area of a favored animal, also contributed to the reduction of breeding males.

Among purebred horses, males outnumbered females by two to one around 1900. These stallions typically covered numerous mares, both grades and purebreds, each season. There was an active mating market with specialized professional stallioners offering studs for public service under a range of options. These included a single "leap," a full season of service, and the guarantee of a foal. Around 1912 the cost of a single service averaged $13 nationally, which was worth about nine days of

[9] Holstein–Friesian Association of America, *Herd-Book*, 1530–3.

farm labor at prevailing wages.[10] The attentions of top-quality boars, bulls, and rams also garnered high prices.

In the nineteenth century, Americans imported prized animals from around the world. In many early cases, the animals were smuggled out in contravention of the exporting country's prohibitions. In the mercantilist era, a country's improved livestock, like its laborsaving machines and the mechanics who could build them, were national assets to be protected. To encourage innovation, in 1793 Congress made animals imported for breeding duty free whereas a high tariff was applied to other stock.

North American animal breeders also imported the philosophy and practices that guided European breeding efforts. Many followed the purebred concept developed by Robert Bakewell of Leicestershire, England, in the 1760s and 1770s. Bakewell, known as the "father of animal breeding," advanced three guiding principles: set specific goals for desired traits; "like begets like"; and breed "best with best," regardless of the relationship between the female and male. By disregarding the contemporary abhorrence of incestuous mating, he sought pure breeding lines yielding uniform offspring.[11] Previously, breeders commonly had focused on outbreeding, that is, on mating females of the native stock with alien males possessing outstanding features.[12] Bakewell's purported success with line breeding to produce sheep and cattle with fixed desirable characteristics set the purebred movement on its way. Many of the goals set for breeding seem misguided to modern eyes. Bakewell emphasized specific aspects of appearance and "symmetry of form," which he intuitively correlated with productivity.[13] For his followers, pedigree and conformity to the type's ideal or "show ring" qualities often appear to have mattered more than performance.

The ability of inbred animals to produce offspring of relatively fixed characteristics gave breeders far more control than was previously possible. Farmers breeding from the common herd frequently complained of deterioration of the stock as desirable traits or combinations of traits were lost.[14] Mating close relatives alleviated such problems, as long as

[10] USDA, "Cost of Raising Horses," 28.

[11] Sanders, *Short-Horn Cattle*; Winters, *Animal Breeding*, 175–6.

[12] Marshall, *Breeding Farm Animals*, 17–20. Russell, *Like Engend'ring Like*, convincingly argued that popular accounts have overstated both Bakewell's originality and his actual achievements.

[13] Winters, *Animal Breeding*, 176; Marshall, *Breeding Farm Animals*, 35–6.

[14] It is also asserted that the better animals were marketed first, contributing to the decline in quality. Richard Parkinson complained in *Tour in America*, 297: "The method adopted

the extent of inbreeding was not too extreme, for example, as long as the harmful combinations of recessive genes did not accumulate. Attention was increasingly devoted to the concept of formal "breeds," which are classically defined as "animals that, through selection and breeding, have come to resemble one another and pass those traits uniformly to their offspring."[15] Twentieth-century animal scientists have found this concept elusive, more a matter of common consent among the breeders than of scientific precision.[16]

Animal breeding involved solving the puzzles of genetics, which provided interested farmers with an avenue to show off their abilities and acumen. In the late eighteenth and early nineteenth centuries, an astounding range of famous Americans, including George Washington, Thomas Jefferson, James Madison, Timothy Pinkering, Robert Livingston, Henry Clay, John Calhoun, Daniel Webster, and Nicholas Biddle, were prominent animal breeders. The long-standing participation of these elites suggests that this interest was more than a mere fad. After the 1850s, the role of public figures in breeding declined and the movement became more democratic, though not truly populist – it still required means and attention beyond what most dirt farmers could muster.

The second half of the nineteenth century saw the establishment of numerous formal breed associations. These organizations devised official breed standards and scorecards, tracked and published pedigrees in register books, and sponsored livestock shows and contests. They enabled individual farmers to come together to realize economies of scale in breeding. Table 10.1 summarizes the date of formation of the major breed associations for cattle, swine, and sheep in the United States. In the late

with the sheep by breeders is this: the butcher picks all that he likes out of the flock; and the remainder are kept to breed from; so that, instead of improving, the breed is more likely to degenerate."

15 Oklahoma State University, *Breeds of Livestock*. A related common term, the animal's "type," remained more vaguely defined and cut across and within breeds. "Type" generally referred to "a special purpose or form of an animal." Plumb, *Beginnings in Animal Husbandry*, 115. Horses were broadly categorized into draft, coach, light harness, or pony types; cattle into beef, dairy, or general-purpose types; sheep into mutton or wool types; and hogs into lard and bacon types. But there were other uses of "type." Depending on an animal's disposition or temperament, it could be called a "hot blood" type; depending on its conformation, it could be of the "smooth" type; and so on.

16 Animals were generally categorized as purebreds, grades, and scrubs. A scrub was an animal without any pedigree. If a scrub was mated with a purebred, the offspring was called a 1/2 breed. If that animal was mated with a purebred of the same line, the offspring was a 3/4 grade. Continuing the same process of "upgrading" resulted in a 7/8 grade and then a 15/16 grade.

TABLE 10.1. *U.S. Purebred Livestock Associations for Cattle, Sheep, and Swine Formed Prior to 1945*

Cattle	Date	Sheep	Date	Swine	Date
Angus	1883	Cheviot	1891	Berkshire	1875
Ayrshire	1863/1875	Columbia	1941	Chester White	1893/1913
Brahman	1924	Corriedale	1916	Duroc	1883
Brown Swiss	1880	Cotswold	1878	Hampshire	1893
Devon	1918	Delaine Merino	1885	Hereford	1934
Dexter	1912	Dorset	1891/1898	Poland China	1878
Dutch Belted	1880/1886	Hampshire	1889	Spotted	1914
Galloway	1882	Karakul	1929	Tamworth	1897
Guernsey	1877	Leicester	1888	Yorkshire	1893
Hereford (Horned)	1881	Lincoln	1891		
Holstein	1871/1885	Merino	1876		
Jersey	1868	Oxford	1882		
Milking Shorthorns	1912	Rambouillet	1889		
Polled Hereford	1900	Romney	1912		
Polled Shorthorn	1889	Shropshire	1884		
Red Poll	1883	Southdown	1882		
Shorthorn	1846/1882	Suffolk	1892/1929		
Sussex	1890	Tunis	1896		

Sources: Bixby et al., *Taking Stock*, 44–5, 88–101; and Briggs, *Modern Breeds of Livestock*.

nineteenth century virtually all of these organizations were headquartered in the Northeast and Midwest.

Britain was the world leader in animal breeding, but the United States was not far behind.[17] The formation of cattle breed associations in the United States lagged the British by about twelve years (for sheep the lag was only about two years). These small gaps, combined with American innovations in swine breeding, suggest that leading breeders in the United States were acting in parallel with their cousins across the Atlantic. In both countries, breed associations were formed as voluntary private enterprises to provide reliable information to buyers. Some were for-profit businesses and the ownership of herd registers were bought and sold. The U.S. federal government did little to recognize breed associations except

[17] There is clear evidence of improvement: "The weight of ordinary English cattle and sheep at slaughter nearly doubled in the course of the eighteenth century." Ritvo, *Animal Estate*, 69.

when regulating animal imports. The American and British practices differed from those in continental Europe and Canada, where governmental agencies controlled the herd and stud books.[18]

The initial formation of breed associations in Britain was a direct response to buyers in the Americas. The first herdbook in Britain was Coate's Shorthorn registry, published in 1822. At the time many breeders opposed revealing their animals' pedigrees, which they considered trade secrets. But disclosure widened the purebred export market. As one of many examples of the American influence, when Felix Renick traveled to Britain in the 1830s to purchase cattle for a consortium of investors, he opted to invest exclusively in animals with public pedigrees. Only Shorthorns fit this bill in the 1830s.[19]

Purebreds typically differed little from the local stock at their place of origin but were quite distinct from the animals in receiving countries. Importers in the United States, Canada, and Argentina became more obsessed with an animal's blueblood pedigree and its place in a closed herd book than did the elite breeders in aristocratic Europe. The international demand for certification to minimize fraud led European breeders to distinguish more sharply between registered animals and those without long pedigrees. As the U.S. Department of Agriculture (USDA) *Yearbook* of 1936 noted, "the emphasis on purity of breeding has naturally been greater in most importing lands than in most exporting lands."[20]

Although many breeders adopted the purebred concept, mule breeders pursued the advantages of hybridization.[21] Mules were the product of mating a donkey stallion (a jack or ass) and a horse mare, and they possessed qualities many considered superior to either parent. George Washington popularized mule breeding after he received gifts of asses from the King of Spain and the Marquis de Lafayette in the 1780s. Prior to Washington's endeavor, asses and mules were "practically unknown" to American farmers. By 1860, there were more than one million mules in the United States.[22]

There were many entrenched beliefs about mammalian genetics. Among the most important was telegony, an idea espoused by such

[18] Derry, *Bred for Perfection*, 34–9.

[19] Sanders, *Short-Horn Cattle*, 199–202, quoted in Derry, *Bred for Perfection*, 23.

[20] USDA, "Livestock Breeding at the Crossroads," 834. See also Lush, *Animal Breeding Plans*, 27–33; Rommel, "Government Encouragement of Imported Breeds," 155.

[21] See Williams and Jackson, "Improving Horses and Mules," 929–46; Ashton, *Jack Stock and Mules*.

[22] Howard, *Horse in America*, 93, 96.

luminaries as Charles Darwin, Herbert Spencer, and Louis Agassiz. It held that the offspring of a female was not the product of a single father, but was influenced by the whole series of the mother's mates, especially her first. While a given sire was responsible for the impregnating event, it was thought that the sires responsible for previous impregnations also influenced later offspring.[23] Telegony was part of a complex set of notions. One held that a sire could become "infected" when servicing a female with the result that his offspring with subsequent females would bear a resemblance to her. A related belief held that the female herself would take on a likeness to the male with whom she had mated. Thus, a mediocre milking cow could experience increased production if mated to a sire of a high-production breed. Another idea popular with breeders was "saturation," which held that, as a male was mated successively with a given female, their offspring would increasingly resemble him. Early breeders also spoke of the "prepotency" of one parent, often the male that had "superior power . . . in determining the character of the offspring."[24]

Recall that the effects of genetics are subtle. An animal's phenotype (its outward appearance and performance) is not a direct manifestation of its genotype (its breeding value). Nor is the offspring's genotype the "blending" of the parents' genotypes. In addition, the progeny may show characteristics of the grandparents that neither parent appears to possess. This skipping of the generations is known as "atavism" or "reversion."[25] Offspring could "nick," showing characteristics considered superior to either parental line. Diseases and environmental forces also influenced the phenotype. It is not entirely surprising that informed breeders continued to believe in phenomena such as telegony, acquired characteristics, and even maternal impressions well into the twentieth century.[26]

[23] Darwin, *Variations of Plants*, 435. See Ewart, "Principles of Breeding," 132–4, for the general concept of infection, and for infection of the male, 137. Winters, *Animal Breeding*, 159–62; Mumford, *Breeding of Animals*, 166–7, 171; Rice, *Breeding and Improvement*, 141.

[24] Winters, *Animal Breeding*, 124; Sinnott and Dunn, *Principles of Genetics*, 365.

[25] Sinnott and Dunn, *Principles of Genetics*, 144. A related idea was "reversion," which was the reappearance of ancestral characteristics after several generations. See also Ewart, "Principles of Breeding," 135–6.

[26] Winters, *Animal Breeding*, 153, noted that in 1925, "there is a rather widespread belief among breeders that acquired characters are transmitted." Winters, 162, further noted that "maternal impression has been one of the most persistent" of the beliefs that he judged "mere fantasies." Under this belief, "the developing fetus may become impressed by any unusual sights or experiences encountered by its dam, and will bear resemblance to these impressions. The story is that McCombie of Tillyfour [the Scottish Aberdeen–Angus breeder of the early nineteenth century] had erected a high black board fence around his

Experiences with mule breeding cast doubts on telegony and associated phenomena. In controlled breeding trials, mares were bred to a series of jacks and stallions. If the makeup of a colt was influenced by many sires, as telegony suggested, then the product of a stallion and a mare that had previously mated with a donkey should have some "mulish" characteristics, but this did not occur. University of Missouri Experiment Station experiments showed that none of the horse offspring from mares that had previously produced mule foals gave "visible evidence of the existence of telegony."[27] By 1900 many breeders were doubting telegony and related beliefs, and by 1930 it had become a disregarded myth, except among dog breeders.[28] As the 1936 USDA *Yearbook* observed, one of the chief contributions of modern genetics to animal breeding was explaining such puzzles as why "identical pedigrees need not mean identical heredity" and "dispelling superstitions."[29]

The 1936 *Yearbook* was particularly critical of breed definitions and associations for promoting unscientific standards and drawing attention away from "functional abilities." The USDA opined that "livestock breeding may be at a turning point" and that "while the methods and practices of the past have accomplished a great deal, giving us the fine breeds of livestock we have today, yet these methods and practices have taken us about as far as they can." Jay L. Lush, whose prints were on this article, was leading the scientific changes in animal husbandry. Lush would become known as the "father of modern animal breeding"; one of his major contributions was to apply statistical analysis to genetics.[30]

Much had been learned about mammalian reproduction before 1940 and this knowledge, along with a long record of breeding (albeit not perfectly informed by science), had a discernable impact on American livestock. Breeders had transformed the American livestock sector by replacing general-purpose, often feral animals with well-defined, special-purpose breeds. By 1940 almost all the major breeds of sheep, swine, and cattle found in America today had been established. Moreover, the geographical niches best suited for specific species and breeds were reasonably

breeding paddock, believing that by doing so he was more likely to get black calves." See also Ewart, "Principles of Breeding," 126–7; Rice, *Breeding and Improvement*, 141.

27 Mumford, *Breeding of Animals*, 174. Also Mumford, "Some of the Principles of Animal-Breeding," 28–53.

28 Gilmore, *Dairy Cattle Breeding*, 72; Mumford, *Breeding of Animals*, 169–74; Sinnott and Dunn, *Principles of Genetics*, 363–4; Rabaud, "Telegony," 389–99.

29 USDA, "Livestock Breeding at the Crossroads," 842.

30 Ibid., 831–5; Chapman, "Jay Laurence Lush," 277–305.

well understood, and regional specialization in both animals and feeding practices was well under way. Our discussion of sheep, swine, and cattle will demonstrate the economic significance of these changes.

Sheep

Sheep proved ill suited to the early colonial environment because of the threat from wolves and dogs. The sheep imported to Jamestown in 1609 "were largely destroyed by wolves" – an experience destined to be repeated many times on the shifting frontier.[31] Large colonial flocks were first confined to preserves such as Nantucket Island in Massachusetts, and elsewhere farmers often joined together to share the cost of a shepherd and fencing to protect their sheep and other animals. In New England the town common served this purpose.[32] Farmers might raise a few sheep for home consumption of the mutton and wool, but there was no significant export market for these goods.

Colonial sheep were typically scrawny, long-legged, large-boned animals that produced one or two pounds of coarse wool. "The flesh was generally very strong in taste,...so that some persons were unable to eat it."[33] But by the end of the eighteenth century, numerous improvers were breeding higher quality animals. George Washington was the best known of these early breeders. He kept abreast of Bakewell's work on sheep and corresponded on this issue with Arthur Young. By 1789 he had roughly 800 sheep that averaged over more than five pounds of high-quality fleece. George Washington Parke Curtis, the President's grandson, continued the work. Mating a Persian ram with sheep of English descent, Curtis bred "long-wooled sheep" that also yielded fleece in excess of five pounds.[34]

Americans began importing superior sheep from around the world, often in breach of the laws of the exporting countries. Spanish prohibitions on exporting Merino sheep are widely known, but "stringent English laws" also "prevented the American agriculturist from participating in the great improvement made in the English sheep from 1750 to 1810."[35] Spanish Merinos, prized for their high yields of long fine wool,

[31] Carman, Heath, and Minto, *Special Report*, 21, 25.

[32] Thompson, *History of Livestock Raising*, 74; Carman, Heath, and Minto, *Special Report*, 11–94.

[33] Carman, Heath, and Minto, *Special Report*, 11–94.

[34] Ibid., 58–62.

[35] Ibid., 87–8.

provided the first significant infusion of new blood. In 1793, William Foster of Boston introduced the first Merinos to the United States after smuggling three of the animals out of southern Spain. Foster's exploits came to naught because a friend, unaware of their pedigree, slaughtered them for mutton. In 1801 E. I. du Pont imported the ram Don Pedro to his Hudson River estate "to the great advantage of breeders" in the area. Du Pont moved the ram to his farm near Wilmington, Delaware in 1805 and continued his practice of allowing his neighbors free stud services. Don Pedro was kept very busy. By 1814 twenty-one local farmers possessed 746 full-blood Merinos, and 2,317 mixed bloods, mostly his offspring.[36] In 1801 Seth Adams imported a pair of Merinos via France to Dorchester, Massachusetts, where he began breeding full-blooded offspring. In 1807 he moved his flock to Ohio, setting the stage for improvements in the West. While serving as the American minister to France in 1802, Robert Livingston shipped two pairs of Merinos to his Hudson Valley farm. He built his flock by mating his Merinos with other purebloods. To generate interest in improved sheep, Livingston held annual public shearings. "Livingston Merinos" became foundation stock across New England and New York, with some of his rams commanding $1,000. Another diplomat, David Humphrey, returned from his post as minister to Spain in 1802 with ninety-one Merinos to stock his farm in Derby, Connecticut. Beginning in 1806, James Caldwell of New Jersey invested some $40,000 to build one of the largest flocks of Spanish Merinos in the country.[37]

This trickle of imports soon turned into a flood. The period of "hothouse industrialization" between Jefferson's Embargo of 1807 and the end of the War of 1812 set off a Merino boom as American farmers sought to supply the country's infant woolen industry. At the same time, war and political unrest in Iberia eroded the export barriers. Following the French invasion, the American consul to Portugal, William Jarvis, obtained 200 Escurials, one of the most prized Merino types, in late 1809. Thomas Jefferson and James Madison were among the recipients of these sheep. In 1810–11 Jarvis took advantage of renewed hostilities to purchase at least 3,400 Spanish Merinos, which he shipped to buyers

[36] Ibid., 134–5. In 1541 the Spanish introduced Merinos to North America with a shipment to Mexico. The quality of the descendents of these animals rapidly deteriorated. International Bureau of the American Republics, *Mexico*, 24.

[37] Carman, Heath, and Minto, *Special Report*, 136–48, 168–79, 217. Livingston, *Essay on Sheep*; Loehr, "Influence of English Agriculture," 5, 15; Vaughan, *Breeds of Live Stock*, 293–6; Woodward, "Sheep at $1,000 a Head," 14.

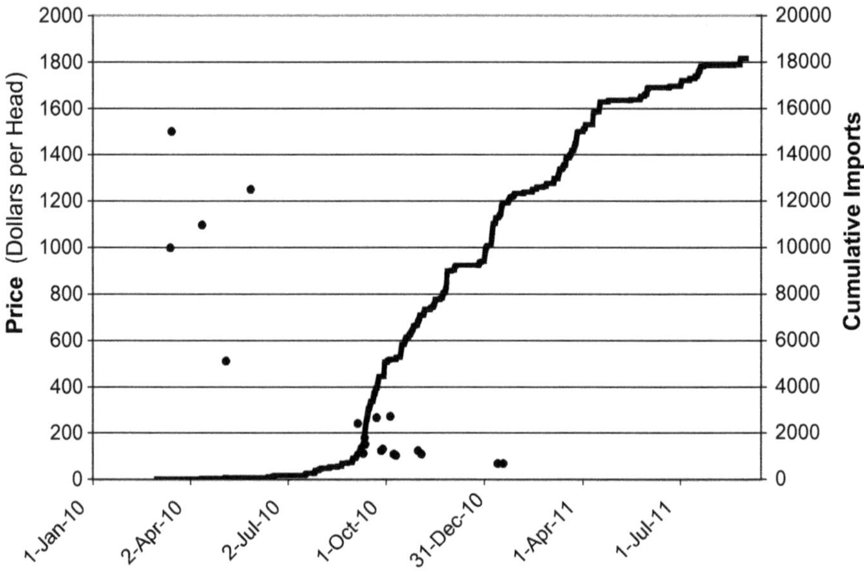

FIGURE 10.2. Merino imports and prices, 1810–1811. *Source:* Compiled from Carman, Heath, and Minto, *Special Report*, 167–213.

from Virginia to Massachusetts. Cash-strapped Spanish officials allowed Jarvis's agents the pick of the flocks.[38]

Numerous other importers also seized the moment. Between April 1810 and October 1811, 180 ships arrived in the United States carrying almost 20,000 Merinos. Figure 10.2 charts the path of imports and observations on prices. At the start of the boom Merino rams sold for more than $1,000 a head. With the surge in imports, prices dropped as low as $100 in early 1811. With the cessation of hostilities and the return of woolen imports from the English, the Merino mania abruptly came to an end. This experience was sobering but did not prevent subsequent booms and busts in the markets for breeding stock.

In the United States the federal government played little or no role in the introduction and propagation of sheep (unlike in France, Britain, and Sweden, where Merinos imported during the war were placed in government-run breeding farms). States and agricultural societies did offer bounties to promote the first importations, but in general the importation and breeding fell to private individuals, including ambassadors

[38] Carman, Heath, and Minto, *Special Report*, 172–5, 182–90; Vaughan, *Breeds of Live Stock*, 294–6.

acting on their own accounts.[39] In 1822 Samuel Henshaw of Boston imported the first Saxony Merinos, an exceptionally fine-wooled type descended from Spanish Merinos and introduced into Saxony in 1765. This set off a brief Saxon craze with 3,500 imported by 1830. Prices crashed after the breed proved too fragile for American conditions. There was a diffusion of Merinos across the Northeast, with high-grade Merinos or Saxons probably making up half of the region's sheep by 1840.[40] A further indication of the diffusion of Merino and Saxony blood comes from a report on the New York wool market dated 26 March 1850. About 85 percent of the more than 300,000 pounds of wool sold came from sheep whose pedigree was three-fourths or more Merino and/or Saxon.[41]

American breeders centered in Vermont vastly improved the Merino breed. Edwin Hammond of Middlebury, Vermont, became the nation's foremost Merino breeder. Hammond concentrated on inbreeding a domestic line developed since 1813 by Stephen Atwood of Connecticut. In 1856 Hammond bred his most famous ram, Sweepstakes, who at his peak weighed 140 pounds and yielded a 27-pound fleece of 2 1/2-inch wool. "In twelve years' time the Hammonds had nearly doubled the weight of fleece and percentage of fleece to live weight... of their sheep."[42] These qualities diffused widely as Vermont Merinos became a favorite for breeders from Virginia to Michigan.

By the 1850s top American Merinos were already yielding fleeces many times the weight of those of the original Spanish sheep. Generations of careful breeding had recast the Spanish Merino into a heavier, shorter legged, broader chested animal, with a far higher wool-to-body weight ratio. The wool-to-body weight ratio for the best rams rose from about 6 percent in 1812 to 15 percent in 1844, to 21 percent in 1865, and to 36 percent in 1880. This increase meant that farmers were getting much more wool per unit of feed.[43]

As a crude metric, in 1800 the average clip in the United States was probably about one to two pounds of coarse wool. The Census of 1860 suggested that average national wool yields were approaching four

[39] Vaughan, *Breeds of Live Stock*, 294–6. Cole, "Agricultural Crazes," 622–39; Carman, Heath, and Minto, *Special Report*, 165, 171–200, 222–31.

[40] Carman, Heath, and Minto, *Special Report*, 145, 225–39; Leavitt, "Meat and Dairy Livestock," 49; Woodward, "Sheep at $1,000 a Head," 14–15.

[41] "The Great Wool Sale in New York," *Cincinnati Price Current* (26 March 1850), 115.

[42] Vaughan, *Breeds of Live Stock*, 293–301.

[43] Ibid., 297–301.

FIGURE 10.3. Spanish Merino ram, 1810. *Source:* Courtesy of the Vermont Historical Society.

pounds per fleece. Between 1850 and 1860 wool yields increased by 18 percent in the New England states and by 27 percent in the mid-Atlantic states. Census data for 1869 indicate an average fleece exceeded four pounds.[44] Even after the center of production moved into less intensive open-range conditions, wool yields continued to climb. The Census of 1900 reported that "in the fifty years since the first agricultural census, the average production of wool per sheep has increased approximately 140 per cent."[45] In addition, there were significant quality improvements that yielded longer and finer fiber. At this time the average clip was 6.7 pounds. By 1940 the national yield averaged over 8 pounds. The steady climb in wool yields was due primarily to the improved genetic makeup of American sheep. Figures 10.3 and 10.4 offer a sense of the changing appearance of Merinos as breeders practiced their craft.

44 Carman, Heath, and Minto, *Special Report*, 140–6; Lebergott, *Americans*, 166; Carter et al., *Historical Statistics*, Tables Da980–982, Da761–762; U.S. Census Bureau. 8th Census 1860, *Agriculture*, cxxiii.

45 U.S. Census Bureau. 12th Census 1900, *Census Reports*, Vol. 5, ccxvi.

FIGURE 10.4. Vermont sheep breeder's 1906 business card. *Source:* Courtesy of the Vermont Historical Society.

Table 10.2, which shows the prevalence of sheep breeds in 1920, indicates that Merinos (including Rambouillets) remained popular into the early twentieth century but they faced new competitors, especially in the Northeast and Midwest, as farmers experimented with improved meat and dual-purpose breeds.[46] All of the breeds listed in Table 10.2 were postcolonial introductions (with the possible exceptions of "other" and "nondescript").[47]

The diffusion of meat breeds depended on imports from the British Isles – once again smugglers led the way. Around 1805 Captain Beanes of New Jersey smuggled out of England some full blood New Leicesters,

[46] Rambouillets were a derivative of Spanish Merinos that had been selectively bred on a state farm in France since 1786. Livingston imported the first Rambouillets into the United States in 1802. Vaughan, *Breeds of Live Stock*, 304–34.

[47] The 1890 census of agriculture reported that 16.7 million (47 percent) of the nation's 35.9 million sheep were Merinos of grade one-half to full blood; 7.4 million (22 percent) were English breeds of grade one-half to full blood. The remainder was reported as "all others." U.S. Census Bureau. 11th Census 1890, *Report*, 236.

TABLE 10.2. *Regional Distribution of Sheep by Breed, 1920 (in percent)*

	North Atlantic	East North Central	West North Central	South Atlantic	South Central	Far West	United States
Cheviot	1.4	0.6	0.3	0.9	0.1	0.1	0.3
Cotswold	3.6	4.4	8.0	3.1	2.9	9.5	7.2
Dorset	3.5	1.1	0.6	4.0	0.4	0.2	0.7
Hampshire	6.7	5.7	6.4	7.0	7.0	5.8	6.1
Leicester	1.1	0.8	0.3	0.2	0.1	0.1	0.3
Lincoln	1.0	2.1	1.4	0.7	1.2	6.0	3.8
Merino	15.8	18.8	9.9	6.0	18.1	35.0	25.4
Oxford Down	3.6	4.7	2.9	0.9	0.8	1.0	1.9
Rambouillet	1.9	3.1	6.2	1.1	15.4	19.6	13.3
Shropshire	32.8	41.6	48.7	28.0	15.9	12.1	23.2
Southdown	14.0	6.4	4.5	25.6	17.2	1.7	6.1
Tunis	0.1	0.1	0.1	0	0.2	0	0.1
Other	2.3	4.5	4.6	5.0	6.9	2.1	3.5
Nondescript	12.2	6.1	6.1	17.5	13.8	6.8	8.1

Note: Based on a survey of livestock reporters. Grades and scrubs are included in the breed in which the type predominates.
Source: USDA, *Yearbook 1920*, 747.

which he sold to New Jersey breeder George Farmer. Southdowns, Hampshires, Cotswolds, Cheviots, Lincolnshires, Oxford Downs, Dorsets, and Shropshires were all imported in the nineteenth century. As noted in Table 10.2, Shropshires, which arrived in 1855, were the most prevalent of these British introductions. Shropshires were a dual-purpose wool and meat breed that prospered in variable climates and in a variety of pastures. Their heavy coat of "close, oily wool" protected them from sleet and snow. (See Figure 10.5.) By 1920, the breed composed over 40 percent of the sheep in the North Central region and rivaled Merinos as the nation's most popular breed.[48] Meat producers relied much more on crossbreeding than wool producers who continued to increase Merino and other fine-wool sheep pedigrees.[49]

There were significant regional swings in the ratio of meat breeds to wool breeds as farmers responded to changes in demand and discovered breeds more suited to their locales. In the age before refrigerated

[48] Woodward, "Sheep at $1,000 a Head," 14; Carman, Heath, and Minto, *Special Report*, 23; Vaughan, *Breeds of Live Stock*, 336, 354, 366, 382, 403.

[49] U.S. Tariff Commission, *Wool-Growing Industry*, 21; Spencer et al., "Sheep Industry," 220–310.

FIGURE 10.5. Shropshire ram on the USDA Beltsville, Maryland farm, 1933. *Source:* Courtesy of the Department of Special Collections, National Agricultural Library, Manuscript Collection 251.

transport, farmers distant from urban markets stuck with wool breeds whereas those near markets shifted to meat breeds. But there were other factors in the decision because "mutton sheep ... required good grass pastures, [and] some grain and vegetables for fattening." Western competition and new opportunities in dairying contributed to a decline in the New England sheep population by more than 50 percent between 1840 and 1860. The remaining sheep were predominately meat breeds of English origin.[50]

The shift in sheep production to the western states created new challenges. By 1920 two-thirds of America's wool clip came from sheep in the Pacific and Rocky Mountain states. The history of the sheep industry in California offers an insight into the learning and improvement efforts under way across the West. There may have been a million Mexican sheep in California as early as 1825. These flocks of "Chaurros," were "small, course, bare-bellied, and hairy" creatures similar to those

[50] Carman, Heath, and Minto, *Special Report*, 239–41; Leavitt, "Attempts to Improve Cattle," 52.

throughout the Spanish (later Mexican) territory.[51] In the early 1850s several pioneers, including Col. W. W. Hollister, drove flocks from the Midwest. Pureblood Spanish Merino rams arrived a few years later. In 1863 "the Mexican sheep produces on an average two pounds of wool per year, worth from five to seven cents per pound;... the half-Merino six pounds, worth from eighteen to twenty-four cents." By this time the Golden State was already a leading wool producer. California ranked second in 1870 and was the nation's leading supplier in 1890. By 1920 the Mexican stock had disappeared and about 75 percent of California sheep were high-grade Merinos.[52]

The Census of 1880 detailed the rapid change in bloodlines. The leading ranchers managed flocks of thousands of sheep which fed on rangelands and typically employed less than one shepherd per thousand animals. Although the land-to-labor ratio was very high, there were no signs of mechanization, but there was clear evidence of rapid biological innovation. All ranchers owned high-grade Merino rams to upgrade their flocks, with about one ram for every fifty to one hundred ewes. There was a clear sense of the pros and cons of competing breeds. The breeders' primary objective was to increase wool production. One rancher noted that "my aim is a merino with fine, long staple, without too much grease, and having few or no wrinkles; a full, round body, and straight, short leg. Ewes should weigh 100 pounds, bucks 150 pounds.... We can only do this by breeding up to the 'Spanish merino family.'" Of the ranchers surveyed, the average clip ranged between 6 and 10 pounds, with many remarking that their average had increased about 20 percent in the past five years.[53]

The transformation of the sheep industry clearly demonstrates the importance of biological innovation. The 200 years before 1940 saw a complete redesign of the physical makeup of sheep. The excitement that early imports generated in the agricultural press, the extraordinary prices paid for purebloods, and the rapid diffusion of new breeds all testify to a vibrant sector. The importation of European sheep was only the beginning of the improvement process because American farmers repeatedly refined

[51] U.S. Census Bureau. 12th Census 1900, *Census Reports*, v. 5, ccxiii. The Census noted that the 1850 count vastly understated the number of sheep. The 1860 Census did show over one million. U.S. Census Bureau. 10th Census 1880, *Report*, 82; Wing, *Sheep Farming*, 216.

[52] U.S. Census Bureau. 10th Census 1880, *Report*, 39–40, 82; Hittel, *Resources of California*, 232.

[53] U.S. Census Bureau. 10th Census 1880, *Report*, 81–90.

the new breeds to increase wool and meat yields. The movement of sheep onto new arid western lands further challenged ranchers. Similar changes revolutionized hog farming.

Swine

Antebellum breeders made even greater improvements in swine than in sheep or indeed any other major livestock. Swine genotypes were relatively plastic, and the animal's great fecundity allowed the rapid transmittal of improved bloodlines. Hog breeders had great flexibility, with each sow producing five to ten females a year that offered a wide range of body types.[54] A breeder could concentrate on the top 10 or 20 percent of his sows to develop desirable characteristics. Features such as size, color, disposition, and the length and strength of the animal's legs were easy to observe. Once farmers started paying attention to weight gain per unit of feed, they could also breed more efficient animals.

Descriptions of hogs in seventeenth-century New England would have been familiar to farmers in the South and parts of the West in the mid-nineteenth century. "There were no choice breeds. Most of them were . . . of razor-back build, speedy runners, as might be inferred from the fact that they ranged the woods in a half-wild state, and the boars were quite capable, with their huge tusks, of taking care of any wolves that might attack them."[55] Swift, long-legged swine (similar to those shown in Figure 10.6) were a frontier livestock because of their ability to forage on mast and roots and to fend off predators. Before the advent of railroads, these long-legged pigs were also easier to drive to distant markets. As forests receded so did swine production (unless the climate and soil were favorable for corn).[56]

On the eve of the Civil War, northeastern hogs differed markedly from the swine of 1800. By the 1840s Americans had imported swine from nearly every region of Europe and from Africa, Asia, and South America. This diverse pool of germplasm gave American breeders great opportunities to adapt for specific traits. Chinese swine were an important source of improvement in early America. In the 1790s, George Washington owned hogs imported from China, and such swine were "common" in Massachusetts by 1800. However, Chinese animals did not gain wide favor

[54] Sims and Johnson, *Animals in the American Economy*, 50–67; Bogue, *From Prairie to Corn Belt*, 104; Leavitt, "Meat and Dairy Livestock," 232.

[55] Thompson, *History of Livestock Raising*, 25.

[56] Leavitt, "Meat and Dairy Livestock," 232–3.

FIGURE 10.6. These feral hogs in Texas resemble older generations of razorbacks. *Source:* Rick Taylor, *The Feral Hog in Texas*, 5, photograph copyright Dan Klepper, Courtesy of the Klepper Archive, www.edanklepper.com.

as a separate breed because they were too small and their fat was concentrated in a thick layer directly under the skin. Crossing Chinese hogs with American or European stock yielded animals with more desirable traits. By the early 1800s scores of northern European breeds, including Byfields, Bedfords, Berkshires, and Irish Graziers, had reached the United States.[57]

Economic development encouraged a change in breeds. In New England the decline in woodland mast and the increasing competition from western states spurred farmers to reduce the number of swine to what could be supported on farm wastes. The new diet eliminated the need for tough, wily beasts. The end of foraging also allowed farmers to control mating. Razorbacks were on the wane in New England by 1819 and were seldom seen in New York after 1840. Ohio, Kentucky, Indiana, and parts of the Southeast also witnessed rapid changes in swine breeds in the antebellum period. Ohio's Miami and Scioto valleys emerged as important

[57] Bogue, *From Prairie to Corn Belt*, 105–7; Shepard, *Hog in America*, 219–62; Leavitt, "Meat and Dairy Livestock," 219–62.

centers of corn-fed swine production, but farther west the ferocious "pike nose" razorback remained the dominant breed.[58] Selective breeding and changes in feed and market preferences resulted in dramatic changes in hog size and rearing practices. Eastern farmers increasingly favored early maturing animals that would rapidly and efficiently convert feed to pork, thus stimulating fresh importations in the 1840s of smaller English breeds.

There emerged a strong association between swine production and corn production. The availability of better feeds encouraged swine production, and the development of improved swine stimulated crop production. In 1840, six of the seven most important corn-producing states were also among the most important swine-producing states. Swine appealed to western farmers as a means of "transporting" corn to market. Foreign grasses and clovers suitable for feeding swine and cattle also gained favor.[59]

There is a large and weighty literature on the average size of hogs. The 162 hogs that John Pynchon of Springfield Massachusetts marketed between 1662 and 1683 averaged 169 pounds.[60] The heaviest tipped the scales at 182 pounds. Pynchon's hogs had "arched and narrow backs, were excellent runners, and good to fight bears and wolves." A century later an account of the same region recorded the average weight of a lot of 50 hogs at 201 pounds. This is hardly solid evidence that hogs were getting larger, but Deacon E. Hunt of Northampton did gain notoriety in 1774 for being the first farmer in the vicinity to kill a hog weighing over 300 pounds. Hogs of this size would later be common. Charles T. Leavitt reported that in 1825 the weight of New England hogs "of the same age" had increased by 30 to 50 percent in "recent years." This impressionistic account is consistent with Winifred Rothenberg's archival research that showed "a dramatic increase in live weights after 1800" in Massachusetts. The average for the years before 1800 was 164.5 pounds; the average for the period 1800–16 was 287.3 pounds.[61] In the Shaker community in New Lebanon, New York, the average weight of hogs at

[58] Leavitt, "Meat and Diary Livestock," 233–7; Inter-university Consortium for Political and Social Research, *Historical, Demographic, Economic and Social Data*.

[59] Leavitt, "Meat and Diary Livestock," 31, 72–5, 84, 116–17; Thompson, *History of Livestock Raising*, 146–7.

[60] This weight is presumably a gross weight, implying a net or dressed weight of about 135 pounds. Judd, *History of Hadley*, 378–9; Russell, *Long, Deep Furrow*, 153.

[61] Leavitt, "Meat and Dairy Livestock," 233–8; Rothenberg, *From Market-Places*, 101–8, 228–9. Russell makes references to several hogs over 600 pounds in New England in the early 1800s. Russell, *Long, Deep Furrow*, 287. We wish to thank Winifred Rothenberg for sharing her data.

the time of slaughter went from 239 pounds in 1792 to 534 pounds in 1829.[62] Farther west, hogs were smaller but increasing in size.[63]

The South did not experience as marked an upswing in hog weights, no doubt reflecting less breed improvement and care. Hogs in colonial Chesapeake were relatively small, with a gross weight of about 156 pounds over the period 1678–1820.[64] Antebellum plantation records for 11,212 hogs show that the average weight at time of slaughter was 146 pounds, with only a modest upward trend between 1840 and 1860.[65] Well into the twentieth century, southern hog farming typically was conducted on an entirely different basis than northern operations. In 1922 the USDA reported that southern hogs "are still given free range of the woods and are obliged to shift largely for themselves. Breeding is often a matter of chance, and the hogs are rounded up for butchering as needed."[66]

This evidence of hog weights understates the changes taking place because the age of slaughter was generally declining. An analysis of long-bone fusion and tooth wear provides hard evidence of the age of slaughter in the Chesapeake region. For the period 1660–1700, only 27 percent of hogs killed were less than one year old, 62 percent were over two years of age, and 45 percent were over three. After 1750 the norm in this region was to slaughter hogs at eighteen to twenty-four months. Literary accounts suggest that eastern farmers in the early nineteenth century slaughtered their hogs at about eighteen months, and in the South and West farmers slaughtered their razorbacks at eighteen to thirty-six months. By the 1850s many northeastern farmers were slaughtering their hogs after just eight to ten months, but in Virginia the norm was ten to fifteen months. This was the lower bound for the South. At the other extreme, Louisiana farmers slaughtered their hogs at about two years of age. One way to control for age is to examine data for spring pigs that

62 We thank John Murray for supplying us with copies of the Shaker documents. Western Reserve Historical Society, Cleveland, OH, Shaker Collection, items V: B–63–V: B–68, "Domestic Journal of Domestic Occurrences," reel 32.

63 Cuff, "Weighty Issue Revisited," 55–74.

64 Walsh, "Consumer Behavior," 247–8.

65 Hilliard, "Pork in the Ante-Bellum South," 465–6. See also Genovese, *Political Economy*, 115. Gallman, "Self-Sufficiency in the Cotton Economy," 15–16, reports a mean weight of 144 pounds for a sample of 2,115 hogs slaughtered on southern plantations in the late 1850s and early 1860s. But he argues these are likely slaughter weights, implying live weights of 192 pounds.

66 At this time the USDA reckoned that the most profitable hogs were "those that attain market weights of 175 to 225 pounds at 6 to 9 months of age." USDA, *Yearbook 1922*, 206, 197.

TABLE 10.3. *Regional Distribution of Hogs by Breed, 1920 (in percent)*

	North Atlantic	East North Central	West North Central	South Atlantic	South Central	Far West	United States
Berkshire	28.3	6.4	3.5	18.2	10.6	19.1	9.2
Cheshire	2.1	0.3	0.3	0.2	0.2	0.3	0.3
Chester White	35	15.3	11.7	3.7	3.1	7.6	10.7
Duroc Jersey	8.4	33.3	40.9	27.2	34.4	31.3	34.2
Hampshire	0.8	3.3	4.4	6.8	2.8	2.2	3.9
Yorkshire	2.4	0.8	0.7	0.4	0.3	0.3	0.6
Poland China	10.2	32.1	32.4	15.3	26.7	28.8	27.9
Tamworth	0.3	0.7	0.6	0.9	0.9	0.8	0.7
Razorback	0.2	0.2	0.3	14.4	10.3	1.4	4.2
Other	3.2	3.4	2	3.8	2.5	2.5	2.7
Nondescript	9.1	4.2	3.2	9.1	8.2	5.7	5.6

Note: Based on a survey of livestock reporters. Grades and scrubs are included in the breed in which the type predominates. Breeds in bold are bacon types.
Source: USDA, *Yearbook 1920*, 755.

were slaughtered by their tenth month. In 1799 the pigs slaughtered by the New Lebanon Shaker community averaged about 158 pounds; thirty years later such pigs averaged 267 pounds. A decline in the age of marketing is consistent with the changes in feeding and breeds.[67]

Table 10.3 provides evidence on the regional distribution of hogs by breed in the United States circa 1920. Four "lard" types – Duroc Jersey, Poland China, Chester White, and Berkshire – constituted more than 80 percent of the nation's swine. "Bacon" types, such as Tamworths and Yorkshires, were much less common.[68] The large regional differences in the prevalence of breeds reflected the geographic specialization that arose from a long process of experimentation. In the north Atlantic states, Berkshires and Chester Whites predominated. In the north central states, Duroc Jerseys and Poland China hogs vied for leadership, with Chester Whites holding third place. In the South, the patterns were similar but

[67] Leavitt, "Meat and Dairy Livestock," 237–8; Walsh, "Consumer Behavior," 247; Walsh, Martin, and Bowen, *Provisioning Early American Towns*; Walsh, *Feeding the Eighteenth-Century Town Folk*. The age of slaughter likely declined in most southern states between the periods 1840–50 and 1851–60. Hutchinson and Williamson, "Self-Sufficiency of the Antebellum South," 591–612; Domestic Journal of Domestic Occurrences, V: B-63-V: B-68, Reel 32, The Shaker Manscript Collection, Western Reserve Historical Society.

[68] Russell, "Breeds of Swine," 1–19. Hammond, *Farm Animals*, 96–9, observed that lard or pork types tended to mature earlier than bacon types.

Chester Whites were less prevalent and Berkshires more common. Even at this late date, razorbacks represented 10 to 15 percent of southern stock and 4.2 percent of hogs nationally. By way of contrast, purebreds comprised about 3.5 percent of American swine.[69]

Three of the top four breeds were distinctive American creations and the one exception – Berkshires – was largely redesigned in the United States. The experience with Berkshires highlights the process of biological learning. John Brentnall of New Jersey first imported this type from south central England in 1823. Beginning in the late 1830s, the breed gained considerable popularity – they matured quickly, were efficient converters of offal to pork, and yielded high-quality hams. But pork packers demanded larger and fatter animals, western farmers desired swine with strong legs to herd to market, and southerners needed animals that thrived in their hot climate. Berkshires fell short on all these accounts, so breeders went to work.[70] "To meet the demand for large hogs, fresh importations were made of the largest Berkshires that could be found in England," most notably the award-winning boar, Windsor Castle, imported in 1841. At the 1842 New York State Fair, Berkshires swept ten of the top awards. Vaughan notes that "American breeders have remolded" Berkshires to meet American conditions. The most prominent improver was N. H. Gentry of Sedalia, Missouri. Beginning in the mid-1870s with a few choice Berkshires from Canada, Gentry built a prized herd by relying largely on inbreeding. By the late 1940s "nearly all American-bred Berkshires" traced their origins to Gentry's stock.[71]

Chester Whites were one of the first distinctively American breeds. These large, white, coarse-haired hogs were first developed in southeastern Pennsylvania around 1818. The standard account credits James Jeffries with founding the line by importing a pair of swine from Bedfordshire, England. The animals and their descendents were interbred with Pennsylvania stock as well as with Byfield, Irish Grazier, and Normandy hogs. In the 1850s, Chester Whites spread to New York and the Midwest, reaching a "high-water mark" around 1870 when one authority claimed that they were "the most popular and extensively known breed of pigs in the United States."[72]

[69] U.S. Census Bureau. 14th Census 1920, v. 5, 633.

[70] Sims and Johnson, *Animals in the American Economy*, 56; Vaughan, *Breeds of Live Stock*, 487; Leavitt, "Meat and Dairy Livestock," 241–2; McDonald and McWhiney, "South from Self-Sufficiency to Peonage," 1095–118.

[71] Vaughan, *Breeds of Live Stock*, 487–90.

[72] Sanders, *Breeds of Live Stock*, 459; Shepard, *Hog in America*, 16; Vaughan, *Breeds of Live Stock*, 549; Ewing, *Southern Pork Production*, 56–7.

Another American breed, the Poland China, became the mainstay of the late-nineteenth-century swine industry. There are many stories of this breed's origins. The standard rendition has John Wallace, a trustee of the Shaker commune in Union Village, Ohio, importing four Big China swine from Pennsylvania in 1816. By the 1840s, crosses of these animals with other types resulted in the popular "Warren County" or "Shaker" hogs. Breeders then mated Warren County sows with boars of several imported breeds, including Berkshires and Irish Graziers, to yield the Poland China breed. D. M. Magie of Ohio laid claim to founding the breed, but many others were working on parallel paths. The National Swine Breeders' Convention officially bestowed the name Poland China on the breed in 1872 (no Polish swine contributed to the bloodline). In the post–Civil War era breeders repeatedly remade the breed to meet changing fancies, market conditions, and regional needs. According to Donald F. Malin, "no breed of swine has undergone such radical type changes as the Poland China."[73]

In the 1890s and 1900s, Poland China breeders increasingly devoted their energies to producing a "short-legged, small, compact-bodied hog popularly known as the 'hot blood.'" The ideals of the type included having a particular look to the ears and black coloration with six white points – four white feet and white on the tips of the nose and tail. As breeders focused on these "show ring" attributes, economic efficiency suffered. The sows were "neither prolific nor very good sucklers" and the type did not gain weight economically in feeding operations.[74] The boar Chief Perfection 2d 42559, bred by B. L. Gosick of Iowa in 1896, largely defined the bloodline. Many regarded this animal as the breed's greatest sire.[75] Chief Perfection received considerable notoriety in July 1903 when he was sold to a syndicate of midwestern breeders for $34,625 ($763,000 in 2005 purchasing power).[76] Some pig! The price was partially justified when his offspring captured both first and second place at the 1904 St. Louis World's Fair.

[73] Bogue, *From Prairie to Corn Belt*, 107–8; Shepard, *Hog in America*, 255–61; Malin, *Evolution of Breeds*, 163–7; Vaughan, *Breeds of Live Stock*, 500–28; *History and Biographical Cyclopaedia*, 132–6; Jones, *History of Agriculture*, 135–7; Davis, *History of the Poland China Breed*, 84–6.

[74] Russell, "Breeds of Swine," 8; Vaughan, *Breeds of Live Stock*, 507; Lush and Anderson, "Genetic History of Poland-China Swine, Pt. 1," 149–56, and "Pt. 2," 219–24.

[75] Plumb, *Types and Breeds of Farm Animals*, 495; Malin, *Evolution of Breeds*, 164–5.

[76] "More Poland-China Booming," *Breeder's Gazette* (19 August 1903), 271, which asserts the new owners intended to charge a service fee of $100 per sow. Daniels, *Twentieth Century History*, 724.

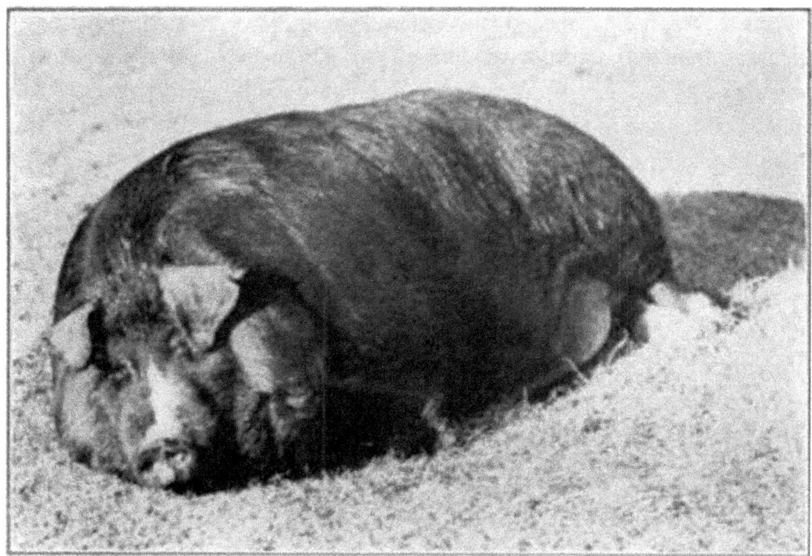

FIGURE 10.7. Four-hundred-pound Poland China barrow at 9 1/2 months. *Source:* Coburn, *Swine in America,* photo opposite p. 9.

As the divergence between the show qualities and meat-producing efficiency of Poland China hogs widened, practical farmers increasingly shifted to other breeds, most notably Duroc–Jerseys. In 1908, the "hot-blood" boom collapsed.[77] Competition with vegetable fats reduced the demand for lard, leading Poland China breeders to develop larger hogs with higher muscle-to-fat ratios. Breeders accomplished, and widely diffused, these changes within a decade, illustrating the rapidity of biological innovation.[78] By the early 1920s, small compact animals had become a rarity as "Big Type" Poland Chinas rose to prominence in the breed.[79] The four hundred-pound barrow shown in Figure 10.7 was a far cry from the razorbacks that populated farms a hundred years earlier.

By the 1920s, Duroc–Jerseys had become America's most popular breed, and this distinctive red hog remains important to this day. The most plausible story of the breed's origin traces the foundation stock of the Red Duroc line to the 1830s. Breeders crossed Red Berkshires with hogs of the Duroc strain developed in New York around 1823 by Isaac Frink. Breeding had a noticeable impact. "When the Duroc was

[77] Russell et al., "Hog Production and Marketing," 196; Malin, *Evolution of Breeds*, 165.

[78] USDA, *Yearbook 1922*, 229.

[79] Russell, "Breeds of Swine," 5, 8.

first introduced to cross with native swine . . . they added fifty to one hundred pounds to the dressed weight of a nine month old pig."[80] The early Durocs likely descended from a number of breeds, including red hogs that originated in Guinea. Other accounts highlight the importance of Henry Clay's importation of red Spanish hogs in 1837 and Daniel Webster's purchase of red swine from Portugal in 1852. By the time of the Civil War, two main types of red hogs existed in America – the small-boned Duroc and the large Jersey Red.[81]

Jersey swine were an oddity of technological history that satisfied an obsession for large animals. In fact, "prodigious weight, regardless of the amount of the feed required or the cost of producing it, to say nothing of economy of handling, seems to have been the object of New Jersey farmers." How fat were they? One hog, owned by Benjamin Rogers of Mansfield County, New Jersey, tipped the scales at 1,611 pounds. Such animals could barely walk and had to be moved by block and tackle. Clark Pettit of Salem County noted that breeders had "a great spirit of rivalry" to see who could produce the largest hogs at a given age. Pettit was among the country's most prominent hog breeders, shipping Jersey Reds to almost every state in the Union after 1865.[82] Midwestern breeders crossed these large red swine with Durocs to create the Duroc–Jersey breed.

The Duroc–Jersey was elevated to breed status at the 1872 National Swine Breeders' Convention. In 1883 the American Duroc–Jersey Swine Breeders' Association was founded in Chicago. As with other swine, breeders redesigned the basic features of the Duroc–Jersey breed. For several decades breed standards called for a medium-sized, early maturing hog, but around 1905 breeders focused on building larger hogs.[83]

Most breeds, if fed enough and allowed to live long enough, could grow quite large, and even in the colonial period there were accounts of hogs weighing more than five hundred pounds. Many later breeds were capable of exceeding one thousand pounds. But especially after the rise of pasture and pen feeding, farmers increasingly focused on the ability of animals to rapidly convert feed to pork and lard. The decline in the

[80] Evans, *History of the Duroc*, 20.

[81] Plumb, *Types and Breeds of Farm Animals*, 505–6; Malin, *Evolution of Breeds*, 211–12; Vaughan, *Breeds of Live Stock*, 529–31; Evans and Evans, *Story of Durocs*, 20.

[82] Woodward, "Jersey Red Hog," 14; Evans, *History of the Duroc*, 12; Vaughan, *Breeds of Live Stock*, 529.

[83] Malin, *Evolution of Breeds*, 211–12; Plumb, *Types and Breeds of Farm Animals*, 513–14; Vaughan, *Breeds of Live Stock*, 529–35; Evans and Evans, *Story of Durocs*, 15–22.

age of slaughter as market weights were increasing suggests that farmers were providing more ample rations, but swine were also becoming more efficient converters of feed. Data on the length of swine intestines provide strong evidence that economic efficiency was improving – yes, the length of intestines. In 1900, W. A. Henry of the University of Wisconsin noted that, given its physiology (specifically the length of its intestines), "the modern hog can digest his food more thoroughly than his ancestors, and he can eat a larger quantity of food in a given time."[84] The differences in the intestines of wild and domesticated animals had fascinated Charles Darwin.[85] In 1942 James Westfall Thompson elaborated on the efforts to design a more efficient animal:

> Structurally, the creation of the modern hog has been that of making an animal which would put flesh on the sides and quarters, instead of running to bone and a big head. Physiologically, the effort of the breeder has been to elongate the intestine of the hog so as to enable him to transmute into meat and fat the corn he consumes. According to naturalists the average length of the intestine of the wild boar compared with his body, is 10 to 1. In the ordinary domestic hog this proportion is 13.5 to 1; in the Siamese and Chinese hog it is 16 to 1; and in the Berkshires, Poland Chinas, Chester Whites, and Duroc–Jerseys it probably is 18 to 1 because of the many generations of careful feeding and selection.[86]

The physical makeup of swine digestive systems was evolving.

Longer intestines, along with related changes in intestinal surfaces and enzyme production, enhanced the digestion and absorption processes. Genetic changes also interacted with environmental factors. Modern hogs could digest a wider variety of foods and more efficiently convert inputs to outputs, especially in a high-input regimen. Domesticated breeds could consume many times the quantity of nutrients per day as could earlier breeds, and they gained weight much faster. A more rapid turnover saved on the capital cost.[87] The environmental changes in feeding and penning reinforced the effects of selective breeding.[88]

[84] Henry, *Feeds and Feeding* (1900), 543. Henry was conducting research on swine feeding and intestinal length at least ten years earlier. See Henry, "Experiments in Pig Feeding," 31–3.

[85] Darwin, *Variation of Animals*, 1:77, 2:293. The original edition was published in 1868. Darwin relied on Georges Cuvier's work.

[86] Thompson, *History of Livestock Raising*, 136.

[87] A number of scientists contributed to our understanding of animal reproduction. Interviews with C. Christopher Calvert, G. Eric Bradford, Trish Berger, Kirk Klasing, David W. Hird, Austin Lewis, and William H. Karasov conducted August–October 2006.

[88] The improvement in hog feed conversion efficiency could have been due to changes in breeding, the environment, and feeding. Austin Lewis suspected that the majority of the

As market conditions changed in the mid-twentieth century, so did hogs. Innovations in the crop sector (most notably the growth of vegetable oils), resulting from the rapid spread of soybean cultivation and the development of corn oil and hydrogenated cottonseed oil (Crisco), led to a pronounced movement away from lard to meat swine.[89] Changes in the hog population were facilitated by a national market with leading breeders shipping animals to distant states. This was similar to the market that developed for agricultural machinery, complete with specialized producers who advertised nationally, describing their animals' pedigrees, price lists, and testimonials from satisfied buyers.[90]

Changes in American hog production had costs as well as benefits. After an initial phase of importing hog types from around the world and diversifying the gene pool in the United States, there was a long history of breed extinction. Even on this issue, the period before 1940 has the dubious distinction of setting the pace. As just noted, changes in human eating habits and the growth of vegetable oils doomed many lard-type hogs. Most important, the mere process of selective crossbreeding, which was essential to hog improvement, led to the demise of many breeds. As farmers developed what they considered superior crossbreeds they ignored the older standards. The Bedford was among the first breeds to disappear sometime before 1870. A number of other, once prominent breeds, including the Byfield, Irish Grazier, Big China, Jersey Red, Duroc, Guinea, and Spanish Red also died out in the nineteenth century. In the twentieth century several others including the Suffolk, Cheshire, and Essex disappeared from American farms.[91]

Instead of finding different hogs to fit different environmental niches, farmers in the late twentieth century created controlled and uniform hog-raising environments to be stocked with a relatively homogenous selection of high-performing animals. More than half of the fifteen breeds of swine

credit was due to the development of new genotypes (that is, to breeding). However, to capture the full potential of the improved breeds required an increase in nutrition and a good environment.

89 The vegetable shorting industry, of which Procter & Gamble's Crisco was the pioneering product, depended on the hydrogenation process patented by American chemist William Normann in 1903. For the introduction of Crisco, see Forristal, *Rise and Fall of Crisco*. Also, J.M. Smucker Company, *Crisco*; tfX, *What are Trans Fats*.

90 For example, see the "Hog Circular" advertising "Premium Chester White Pigs" issued by N. P. Boyer of Parkersburg, Pennsylvania, circa 1867, Folder 196, Box 9, Burwell Family Papers, Coll. 112, Southern History Collection, University of North Carolina Libraries.

91 Kalk, "Extinction of Historic Breeds," 1–4.

listed in the USDA *Yearbook* for 1930 have disappeared. Today, four breeds – Hampshires, Landrace, Durocs, and Yorkshires – represent 87 percent of the total American purebred hog population. With the loss of genetic diversity comes the risk of the loss of disease resistance and the ability to adapt to changing circumstances. More primitive hogs, such as the razorback, likely had relatively strong immune systems and their loss was not an unvarnished blessing. Besides creating significant negative externalities on water supplies and air quality, the recent growth of mega-hog farms with thousands of animals has accentuated the concern about the loss of biodiversity. These large-scale and high-density production methods would not have been possible without developments in medical sciences and in particular the use of antibiotics.

Cattle

Bovines serve many purposes, and breeders devoted considerable effort to develop specialized breeds from stock originally used for work, milk, and meat. Our discussion of cattle reflects these functional divisions. This section focuses primarily on developments in beef breeds. The following chapter analyzes the development of dairy stock, and Chapter 12 deals with the changing role of cattle as beasts of burden. In all of these areas there were significant advances that improved farm productivity.

European settlers introduced a wide variety of cattle into North America, reflecting their home country's agricultural and pastoral heritage. The Spanish transported cattle originally from Andalusia and the Canary Islands into Florida around the 1520s; the Pilgrims imported the ruby-red Devons from southwestern England in 1623; the French settling Quebec came with the small black cattle from Brittany and Normandy that founded the Canadienne line; the Danes brought their large, yellow cows to the middle colonies; and the Dutch imported the smaller black-and-white and red-and-white cattle to New York (New Holland) from their native land.[92] In an age before formal breed associations and definitions, these animals were distinguished principally by conformation, color, and the size of their horns – long horns, short horns, or polled (no horns). Early Americans commonly followed the practice of referring to

[92] Today, 57.6 percent of registered beef cattle in the United States are of British breeds, 33.1 percent are of continental origin, 7.2 percent are Indian influenced, 1.7 percent are Spanish influenced, and 0.4 percent are African influenced. Bixby et al., *Taking Stock*, 46.

TABLE 10.4. *Regional Distribution of Cattle by Breed, 1920 (in percent)*

	North Atlantic	East North Central	West North Central	South Atlantic	South Central	Far West	United States
Aberdeen Angus	0.3	3.6	6.9	2.9	3.1	0.8	3.6
Ayrshire	4.8	0.4	0.3	0.3	0.0	0.2	0.6
Brown Swiss	0.4	0.7	0.2	0.1	0.0	0.1	0.3
Devon	0.4	0.1	0.1	0.7	0.3	0.4	0.3
Dutch Belted	0.4	0.1	0.1	0.0	0.7	0.0	0.2
Galloway	0.1	0.7	1.8	0.2	0.3	0.3	0.8
Guernsey	8.8	6.3	1.9	3.6	0.4	1.3	2.9
Hereford	1.7	7.8	24.2	9.1	22.5	40.9	21.0
Holstein	54.0	29.3	8.7	9.0	5.2	14.9	16.2
Jersey	14.3	15.3	4.8	29.5	23.8	8.8	14.0
Polled Durham	0.1	1.8	1.9	0.4	2.4	0.9	1.5
Red Polled	0.2	1.7	3.3	1.6	5.4	0.5	2.6
Short Horn Durham	5.4	23.3	36.2	9.5	15.9	21.9	22.6
Other	2.2	2.1	2.3	6.8	4.9	1.8	3.1
Nondescript	6.9	6.8	7.3	26.3	15.1	7.2	10.3

Note: Based on a survey of livestock reporters. Grades and scrubs are included in the breed in which the type predominates. Breeds in bold are dairy stock; all others are beef stock.
Source: USDA, *Yearbook 1920*, 731.

longhorns as "beef" breeds and shorthorns as "milk" breeds, but over time the categories were more sharply distinguished. Dairy cows became wedge-shaped creatures, with thin necks and shoulders, crooked backs, and high flanks. Beef cattle were rectangular, with thick necks, heavy shoulders, straight backs, and low flanks.[93]

Table 10.4 summarizes the distribution of cattle by breed in 1920. As with other animals, several strong patterns of regional specialization stand out. Holsteins predominated in the East, whereas Herefords were most prevalent in the West. Shorthorn Durhams clustered in the north central states, and nondescript cattle remained common in the South. At this time, about 3 percent of cattle were purebreds.

All of the purebreds and most of the breeds listed were postcolonial additions to North America. Their importation and development represented a major investment by the country's farmers. As with the invention and diffusion of agricultural machinery, a number of innovators played key roles in transforming American herds. Cattle-raising practices in

[93] Warren, *Elements of Agriculture*, 322–3.

colonial America differed substantially from those followed in England and most of Europe. Efforts in frontier areas to improve stock were often hindered by the absence of good feed, good shelter, and fences. "Seventeenth-century Chesapeake farmers built few barns, cowsheds, [and] stables." Colonists could not afford to provide feed to their cattle, choosing to let them roam in the marshes and pinelands. There was little hay for fodder, and cattle starved over the winter months. "Animals left to their own devices altered in appearance and behavior" and became smaller than those found in England.[94] In addition, the harsh conditions and inadequate diet slowed growth and delayed the age of slaughter. A study of bovine long bones showed that for the years 1660–1700 about 68 percent were over four years old.[95]

In 1783 John Holroyd Sheffield observed that the quality of American beef was generally poor and that in "the back part of the Carolinas and Georgia great herds of cattle are bred very small and lean; they run wild in the woods." The agricultural press repeatedly noted that higher quality animals could barely survive, much less thrive without complementary inputs.[96] New England farmers had to provide more care for their cattle in part because of the harsher winters. In all regions farmers tried to improve the situation by importing English grasses and clovers, but the flow of improved breeds was limited. According to G. A. Bowling, "the initial mass of importations of cattle from Europe into the North American colonies ceased about 1640. From that date to the American Revolution the cattle needs of the colonies were taken care of through inter-colonial trade, or through trade with the Spanish colonies in the Western Hemisphere." By this account, most imports occurred well before the start of the cattle improvement movement in Britain, and colonial farmers would have largely depended on selecting among domestic stock to upgrade their herds. Given the early deterioration in the quality of colonial cattle during the initial phases of settlement, there was little prospect for significant improvement. However, following the Revolutionary War there was an increase in the transatlantic cattle trade. The long process of overcoming the influence of the cattle bred in the first

94 Anderson, *Creatures of Empire*, 109, 117.

95 Walsh, Martin, and Bowen, *Provisioning Early American Towns*, 34–6.

96 Sheffield, *Observations on the Commerce of the American States*, 26. Another hypothesis for the small size of American cattle is that early colonists deliberately transported small animals to save space on crowded ships. Oakes, "Ticklish Business," 195–212; Henlein, *Cattle Kingdom*, 13; Anderson, *Creatures of Empire*, 109–22; Walsh, Martin, and Bowen, *Provisioning Early American Towns*.

250 years of settlement was under way. This story is fundamentally a tale of biological investment and learning.[97]

The first "blooded" cattle imported on a large scale from Britain to the United States were Devons, Bakewell Longhorns, and most notably Shorthorns. Shorthorns originated in northeast England and were sometimes referred to as Durhams or Teeswater cattle. The first recorded transatlantic shipment of these breeds was to Virginia in 1783. Among the leading importers was Harry Dorsey Gough of Perry Hall, Maryland.[98] In 1796, he offered for sale nine bulls crossed "from the famous Mr. Bakewell's breed and the Lincolnshire." Thereafter, Gough's name (sometimes spelled Goff) appeared in close association with Bakewell's. Other Chesapeake breeders, including John S. Skinner of Baltimore and Henry Miller of Augusta County, crossed native stock with larger, red cattle imported from England.[99]

Matthew Patton of Rockingham County, Virginia, introduced improved cattle into Kentucky, thereby setting the stage for the Bluegrass State's emergence as an important cattle region. In 1785 several of his sons drove cattle of the Gough and Miller lines across the Appalachians. In 1790 Matthew followed with several more animals, apparently Bakewell Longhorns. In 1795 Patton imported into Kentucky two more animals of the Gough type. The bull, Mars, the heifer, Venus, and their offspring became known as the "Patton stock" and provided an important part of the founding breed stock in the West.[100] By 1817, "Patton stock was said to have increased the weight of 4-year old bullocks in Kentucky 25 to 30 percent." Matthew's son, John, moved some Patton cattle to Ross County, Ohio, around 1799. When John Patton died in 1803 his neighbors, Felix and George Renick, purchased the herd. The

97 Bowling, "Introduction of Cattle," 151–2.

98 The most detailed and creditable account regarding imports of the early 1780s appears in Cabell and Swem, "Some Fragments," 168; Bevan, "Perry Hall," 40–1; Rice, "Importations of Cattle," 35–47.

99 "Rural Economy," *Farmers' Register* (15 August, 1798), 76; "Encouragement to Farmers to Improve Their Stock," *Philadelphia Gazette and Universal Daily Advertiser* (23 September 1796), 2; "The Beeves," *American Farmer* (31 March 1820), 7; "Agriculture," *American Farmer* (7 July 1820), 113; "Bulls and Cattle," *Republican Star and General Advertiser* (17 December 1822), 1.

100 Allen, *History of the Short-horn Cattle*, 581–2; Rice, "Importations of Cattle," 35–7. There is considerable confusion regarding the relationship of Gough and Miller. Many sources have claimed they were associates in the cattle trade, but our examination of archival and census materials finds no evidence of a partnership. See the *Breeders' Gazette* (20 December 1923), 856, and (17 January 1924), 72, for an alternative to the standard account.

Renicks became leading suppliers of improved stock in the Ohio River Valley.

Imports of Shorthorns increased in the aftermath of the War of 1812 and again during the boom of the early 1830s. The activities of Lewis Sanders of Grass Hills, Kentucky, illustrate the international flow of both information and animals as well as the initiative of leading stockmen. For many years Sanders perused English agricultural publications that described the advances in cattle breeding. In 1816, he instructed his British agent "to select a competent judge of cattle" to travel the length and breadth of England. "The agent, J. C. Etches, selected two pairs of Teeswater cattle, which were prized for their excellent forms, early maturity, and milking qualities; two pairs of the longhorns; ... and two pairs of the Holderness breed."[101] The total cost of these twelve animals delivered in Kentucky was $6,699. Numerous other farmers employed agents to procure, insure, and ship British cattle. Sanders, and others who acquired some of his cattle, soon offered the bulls for public service, with one bull, Tecumseh, fetching the princely sum of ten dollars per cow.

The hiring out of prized bulls was a common practice. The celebrated bull, Coelebs, earned its owner, Samuel Jacques of Charleston, Massachusetts, $3,800 for servicing about four hundred cows. John Hare Powell was one of many breeders to gain a national reputation. He first imported Shorthorns in 1822 and by 1825 boasted at least twenty on his spread near Philadelphia. Over the next fifteen years, the offspring of Powell's original herd stocked farms from Maine to Georgia with animals selling for $300 a head. Powell regularly hired out his bulls: "One bull often brought in $500 a year at $8 to $10 per cow." In the late 1830s the noted English breeder Jonas Whitaker sent seventy to eighty Shorthorns to the United States for public sales held on Powell's estate.[102]

As with crop improvement and the betterment of other animal species, there was often voluntary communal support for cattle improvement. State and county fairs encouraged farmers to show off their animals and offered prizes for those judged the best. State agricultural societies subsidized the importation of purebred cattle and at times purchased animals collectively. They also exchanged information with, and sold and donated purebred animals to, other state agricultural societies. The Massachusetts Society for Promoting Agriculture encouraged farmers to

[101] The press of the day was fascinated with such activities. For an example, see *American Farmer* (29 December 1820); Rice, "Importations of Cattle."

[102] Rice, "Importations of Cattle," 38–9; Lemmer, "Spread of Improved Cattle," 80–2.

FIGURE 10.8. Thomas Bates' Cleveland Lad (3407). Cattle breeders throughout the United States imported Bates' shorthorns. *Source:* Sanders, *Shorthorn Cattle*, 88.

import purebloods, imported others on its own account, and kept high-grade Shorthorns and Devons on the society farm and on the farms of leading members. Unlike their counterparts who raised sheep, swine, and horses, early American cattle improvers devoted little effort to developing a distinctive national bovine breed.[103] Instead, they chiefly imported improved animals from Britain.

The Renick brothers took the lead in organizing fifty shareholders in the "Ohio Company for Importing English Cattle" in 1833.[104] In 1834, the company imported nineteen carefully selected British Shorthorns, bred by Thomas Bates, J. Clark, and others. (Figure 10.8 depicts a Bates Shorthorn bull from the 1830s.) The company acquired additional English stock in 1835 and 1836. Collectively, these animals provided the foundation for many of the Shorthorn "families which are well known in this country." Investors in Kentucky, Illinois, and elsewhere copied the Ohio model by forming animal-improvement associations.[105] In late

[103] Brinkman, "Historical Geography of Improved Cattle," 62–6.

[104] Lemmer, "Spread of Improved Cattle," 82, 87; Henlein, *Cattle Kingdom*, 75–9.

[105] Briggs, *Modern Breeds of Livestock*, 35; Bailey, *Cyclopedia*, 372; Rice, "Importations of Cattle," 43.

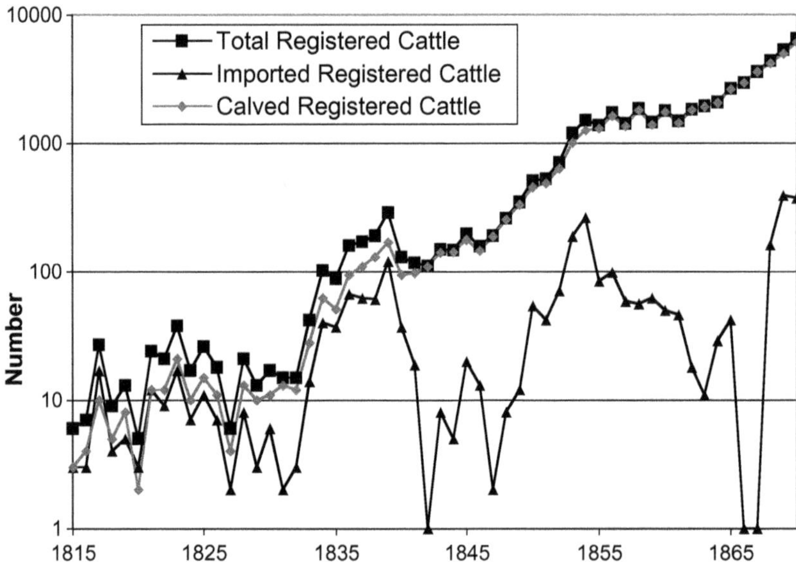

FIGURE 10.9. Number of newly registered purebred animals. *Source:* Brinkman, "Historical Geography of Improved Cattle," 45–8.

1836 and 1837, the Ohio Company sold off its livestock at public auction at prices averaging $880 per head. These sales generated revenue of more than $50,000 on the roughly $10,000 invested a few years before.[106] Lower prices in the 1840s slowed the Shorthorn improvement movement until the revival of imports in the 1850s and 1860s. Figure 10.9 traces the waves in imports of registered purebreds, most of which were Shorthorns. It also graphs the number of domestically born registered purebred calves descended from these imports.

A series of booms, first for Bates-bred cattle, then for animals from the Duchess line, and finally for Scotch stock, ensued. Prices for Shorthorns of the "right" pedigree regularly exceeded $10,000 per head, and in 1873 an Englishman paid a record price of $40,600 for the American Shorthorn, 8th Duchess of Geneva. In the mid-1870s, fashions changed and the price of purebred Shorthorns fell significantly. Fresh British imports of Hereford and Angus cattle began to outshine Shorthorns at stock shows. Shorthorns were thinly fleshed and slow to mature compared to the new arrivals.

[106] Derry, *Bred for Perfection*, 22–4.

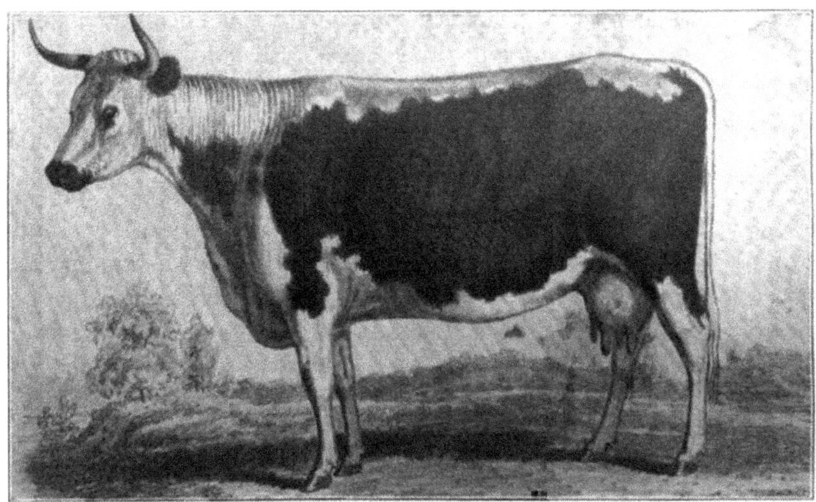

FIGURE 10.10. Jessica. Owned by Col. Lewis Sanders of Kentucky. This cow was a direct descendent of Henry Clay's importations. *Source:* Sanders, *Story of the Herefords*, 269.

As its name suggests, the Hereford breed originated in Herefordshire in southwest England. This breed of large, solid, red cattle had been largely perfected by 1815. Although it is likely that a few Herefords arrived in America earlier, Henry Clay is often credited with introducing the breed in 1817 to his Ashland farm in the Bluegrass region of Kentucky. These cattle became part of the wave of improved western stock known as the "Seventeens." They were outcasts to the purebred movement because they represented high-quality animals without well-documented pedigrees. (Figure 10.10 depicts a descendent of one of Clay's animals.)

Seven years after Clay's initiative, the Massachusetts Society for Promoting Agriculture imported a breeding pair for local use. Erastus Corning and W. H. Southam of upstate New York were also notable importers and breeders of Herefords. Between 1848 and 1886, about 3,700 registered Herefords entered the United States, establishing the foundation for what would become one of the most prevalent breeds of beef cattle. Corning and Southam provided breeding stock to farmers across the nation. Daniel Webster was among their most famous customers, but another buyer, Edward Wells of Johnstown, New York, deserves special mention because of his experiments. Wells "purchased eight to ten head and for several years bred them on the same farm with purebred Shorthorns in order to compare the feeding and grazing qualities of the two breeds."

Herefords were noted for tolerance of drought, summer heat, and winter cold, for their early maturation, and for the ability to spread out over the grassland and walk large distances to find food and water. These characteristics made the breed particularly suited to the American range. Briggs observed that "in many areas of the West, cattle and Herefords are practically synonymous."[107]

Angus cattle were another late addition to North America. This polled strain was developed in northern Scotland over the late eighteenth and early nineteenth centuries. The first Angus imported to North America arrived in Canada around 1860. A Scotsman, George Grant, imported four bulls to Kansas in 1873, and the firm of Anderson and Findlay of Lake Forest, Illinois, introduced the first registered animals in 1878. The popularity of the short-legged, cylindrically shaped breed grew because, as it is today, the Angus was marked as a "butcher's beast," producing high-quality meat.[108]

The progress of American beef followed a rather straightforward path as the industry moved inland from the eastern seaboard. Most of the new breeds that would come to dominate the American landscape entered the United States from Britain in the century following the Revolutionary War. The South played an important role in this process as cattle from the Carolinas and Virginia moved into the Ohio Valley.[109] However, the South did not participate in the mainstream of the herd improvement movement during the nineteenth century, in large part because many improved animals shipped into the South fell victim to tick fever. But before the Civil War, some southerners pursued paths more suited to their region by experimenting with Zebu or humped cattle, including the Brahman type. In 1835, Campbell R. Bryce of Columbia, South Carolina, imported two Indian bulls and four cows from Egypt. In 1849, James Bolton Davis of Fairfield County, South Carolina, brought stock directly from India. He had long reasoned "that all cattle brought from a Northern to our Southern climate must necessarily degenerate to the peculiarities of our location, and that it would be easier to improve cattle already acclimated, *or imported animals from a still warmer region*." Much like plant breeders, he paid close attention to latitude, altitude, and climate in placing his animals. He crossed his Brahmans "upon Ayrshire,

[107] Briggs, *Modern Breeds of Livestock*, 115; Rice, "Importations of Cattle," 40–2; Lemmer, "Spread of Improved Cattle," 87–8.

[108] Briggs, *Modern Breeds of Livestock*, 129, 139–40, 147–8.

[109] Jones, "Beef Cattle Industry," 168–94, 287–319; Jones, *History of Agriculture*, 77–119, 175–6.

Devon and Durham breeds, as well as upon our common cattle." He purportedly sold his half-breeds for $1,000 a pair. Reports of Davis' activities were widely disseminated in the press. Another early introduction occurred in 1854, when the British crown gave two bulls to Richard Barrow, a St. Francisville, Louisiana planter, in recognition of his services in teaching cotton and sugar cane cultivation to colonial administrators. These animals provided the foundation for the "Barrow" type of the Gulf Coast. Despite advantages such as superior resistance to heat and disease, Zebus did not take hold in the United States to the extent that they did in Brazil.[110]

The South's major impact on the beef industry came out of Texas. The great American cattle frontier that emerged in Texas following 1865 had an entirely different history involving different problems, different cattle types, and dramatically different methods of production. The Texas model would have a profound influence on later cattle frontiers in the Rocky Mountain states. In the late 1860s, Texas emerged as a major cattle-exporting state. In 1866 about 260,000 Longhorns were driven north. After a decline in 1867 and 1868, Texas exports climbed to 350,000 in 1869, to 600,000 in 1871, and then averaged over 250,000 for the next fourteen years.[111] The legendary Texas Longhorns bore little resemblance to the cattle that populated the midwestern frontier. Longhorns were hardy beasts that were especially adapted to the hostile environment of the Southwest. They could survive and thrive despite poor pasture and limited water. Their hard hooves, long legs, and ability to maintain their weight on the trail made this breed ideal for the long drive. Over a relatively short period, they evolved a partial immunity to Texas tick fever and more generally had excellent immune systems, which helped them ward off diseases that plagued most other breeds. The bulls became exceptionally aggressive and the cows very fertile over a long reproductive lifespan. Longhorns successfully used their formidable horns and hooves to protect their young against wolves and coyotes rather than panicking and taking flight. These characteristics ensured a relatively high rate of natural increase.[112] Thus, their development was more a product

[110] *Pittsfield Sun* (15 February 1855), 4; Oklahoma State University, *Breeds of Livestock*.

[111] In the early 1840s, Texans drove cattle to New Orleans; in 1846 Edward Piper drove a thousand cattle to Ohio; and by 1856 Texas Longhorns had appeared in Chicago. But the major shipments north began after the war. Dale, *Range Cattle Industry*, 24, 59.

[112] Longhorns today exhibit considerable genetic variation and have relatively high levels of heterozygosity. Reflecting the animals' strong immune systems, breeders today administer far fewer vaccinations to Longhorns than to other breeds. Modern breeders

FIGURE 10.11. Modern Texas Longhorn. Descriptions of earlier Longhorns suggest a leaner and more boney animal. *Source:* Photograph courtesy of Dickinson Cattle Co. Inc., http://www.texaslonghorn.com.

of natural selection than artificial breeding; but their selection by Texas cattlemen as the breed of choice was clearly a conscious decision.

This distinctive type was built on a foundation of Spanish Criollo or Retinto cattle that had been introduced to Hispaniola in 1493. Spanish cattle had arrived in Mexico and Florida in 1521. Texas Longhorns most likely came from Criollo cattle that had advanced up the eastern coast of Mexico to Louisiana and southeastern Texas during the eighteenth and early nineteenth centuries. The extent and source of non-Criollo blood in the makeup of Longhorns is much less certain. There were numerous opportunities for Criollo cattle to mix with animals of northern European lineage, but surviving Longhorns of the type shown in Figure 10.11 have

note that the ferocious reputation Longhorns developed may be an exaggeration and more a product of their environment than an innate attribute. Longhorns in captivity are extremely docile, but it is likely that the near-wild animals on the open range were far more aggressive. Phone interview with David Hillis, Professor of Integrative Biology, University of Texas, Austin, 29 December 2006. Longhorn calves had relatively low birth weights which contributed to easy, live, and unassisted births, with relatively low losses of calves and mothers. Casey, *Genetic Advantages*.

FIGURE 10.12. Comparison of a Longhorn steer with an improved beef animal, circa 1930. *Source:* Mohler, "Science in the Livestock Industry," 508.

little Anglo influence.[113] USDA accounts disparagingly compared earlier generations of Longhorns with improved breeds (see Figure 10.12).

Longhorns spread to other states as northern ranchers stocked their ranges with Texas cattle. Shortly after the Civil War, Longhorns were becoming established in Kansas and Colorado, with the latter state boasting a million cattle by 1869. The first Texas breeding stock entered the Platte River district of Nebraska in 1869 and soon thrived in regions previously thought unsuitable for livestock. By 1871 there were roughly forty Texas-style ranches in southeastern Wyoming with 86,000 cattle. In the late 1870s and 1880s, with the advance of the railroad and the lessening threat of Indian raids, Longhorns spread into the Dakotas and much of Montana, Alberta, and Saskatchewan. "A period of less than fifteen years following the Civil War had witnessed the establishment of

[113] Jordan, *North American Cattle-Ranching*, 106, 147–53; Worcester, *Texas Longhorn*, 24; Rouse, *Criollo*, 21, 46, 74; Charles, "Cattle Raising in Spanish Florida," 116–24; Gard, *Chisholm Trail*, 7, 15; Dale, *Range Cattle Industry*, 24. Blood typing and genetic research has failed to uncover strong ties to Anglo stock. Murphey et al., "Blood Type Analyses," 231–4. David Hillis notes that other present-day Longhorn populations are genetically most similar to the cattle now found in Portugal and do not show a strong Anglo influence. Hillis interview.

every noteworthy foothold of the Texas system in the Great Plains, from New Mexico to Montana."[114]

The demise of the Longhorn era came about as rapidly as its advance and represents another example of rapid biological change. Just as farmers who pushed onto the Great Plains suffered from an inadequate knowledge of the weather and suitable cropping practices, so cattlemen misjudged the suitability of Longhorns and the extensive Texan method of production for much of the Great Plains, including northern Texas. "The system died of cold and drought."[115] Longhorn cattle had evolved from a breed that for centuries had lived in wet coastal marshlands. Even as the cattle and Texas system were expanding, ominous signs appeared when calf yields fell drastically as the cattle moved north. Yields in Kansas were only four-fifths of that common in central Texas, in Nebraska only two-thirds, and in the Dakotas about one-half. Longhorns did not reproduce as efficiently in colder climates. A series of hard winters, including those of 1871–72, 1884–85, 1886–87, and 1889–90, coupled with devastating droughts in parts of the expanded Longhorn domain in 1879, 1880, 1886, 1891, and 1892 spelled doom. During the winter of 1886–87, mortality was severe from Canada to Texas. Cattle died of exposure, starvation, and thirst by the hundreds of thousands as rivers and watering holes froze solid.

Noted animal scientist David Hillis has questioned whether Longhorns were inherently ill suited for the northern ranges, noting that today many Canadian ranchers raise Longhorns without apparent problems. Although it stands to reason that the denser and fatter Shorthorns and Herefords might do better in cold weather, we have found no reports of how these English breeds fared in the exceptionally harsh winters of the 1870s and 1880s that froze out the Longhorns. Claims that droughts hastened the end of the Longhorn era surely require qualification because Longhorns were noted for being relatively drought-hardy cattle.[116] The arguments about both cold and dry weather make more sense when tied to production and market factors.

Ecological problems encouraged cattle owners to move to more capital- and labor-intensive ranching systems that included the construction of

[114] Jordan, *North American Cattle-Ranching*, 225–7; Dale, *Range Cattle Industry*, 60–1, 70–114.

[115] Jordan, *North American Cattle-Ranching*, 236–7; Libecap and Hansen, "Rain Follows the Plow," 86–120. For more lengthy treatments of this issue see Webb, *Great Plains*, 319–84; Hargreaves, *Dry Farming*.

[116] Hillis interview.

fenced pastures to allow for winter feeding.[117] Grazing became inexorably tied to crop production, both in the West and in the corn belt states, where plains animals were shipped to be fattened on hay and grain. This growing division of labor linking cattle-rearing and cattle-finishing states hastened the demise of Longhorns because they were not as economical fatteners on corn as other breeds. In the 1870s ranchers on the Great Plains began importing purebred bulls, mostly Shorthorns, from the Midwest. Later Hereford bulls became the favorite. The improvement of cattle on the plains stimulated corn belt farmers to increase the quality of their cattle, both to supply the demand for breeding stock farther west and to better compete with western cattle. In 1883 the USDA asserted that the value of cattle in the United States had more than doubled due to advances in breeding.[118] Hand-in-hand with the increasing capital and labor intensity of production was a change in preferences as consumers opted for a more marbled cut of beef. The change in consumer demand, coupled with cheaper transportation, had a profound impact on cattle operations in the East as eastern cattlemen began sending young animals to the Midwest for fattening. Previously such animals would have been slaughtered for veal. This abrupt increase in the flow of cattle from east to west occurred in the early 1880s and had the negative side effect of significantly increasing the transfer of contagious bovine diseases to the Midwest. The extension of the railroad system further contributed to the rapid demise of Longhorns. The animals' horns, an asset on the open range, proved a drawback in the tight confines of the stockyards and railcars, where they were prone to hooking each other. Longhorns are very social animals with a well-established hierarchy within the herd. In the course of shipping, the introduction of new animals into a herd would incite the others to push and shove as they sought to establish where the new animals fit into the hierarchy; injuries occurred in the process.[119]

The Longhorn story was just a part of the larger pattern of changing rational land-use policy as the focus of the Great Plains economy evolved from buffalo hunting, to Longhorn production, to a mixed agriculture that could provide year-round feed and water to newly imported breeds. Although laborsaving mechanization would play a part in this evolutionary story, it was just a bit part. The more fundamental historical

[117] Jordan, *North American Cattle-Ranching*, 236–40; Dale, *Range Cattle Industry*, 84–5, 108–12.

[118] Dale, *Range Cattle Industry*, 157–66. See USDA, *Report of the Commissioner 1883*, 282.

[119] Hillis interview.

process was labor using as farmers inexorably moved to more intensive systems of agriculture.

Conclusion

The experience of roughly two hundred years of animal breeding summarized in this chapter reinforces our general findings for the crop sector regarding the importance of biological innovations in American agricultural productivity growth. Far from being single-mindedly focused on mechanization, farmers in all regions were working hard to improve their animals. Although some of these efforts went down blind alleys, many paid handsome dividends. There were major strides in the science of animal breeding and in the understanding of feeds that, by the twentieth century, were having a significant impact on practical farmers. Well before many of the intricacies of animal genetics were understood, farmers and professional breeders were redesigning the livestock on American farms. This was a global effort as American innovators imported and experimented with breeding stock from around the world.

The changes for sheep were spectacular; the average national wool yield per sheep increased at least fourfold from the colonial era to 1940. There were parallel increases in both the quality of wool and the quantity and quality of meat produced. These improvements were made possible by conscious and systematic breeding. Starting with the first highly publicized importation of Spanish Merinos, American farmers redesigned their flocks. Sheep breeders significantly improved upon the purebloods first imported – the average American Merino of 1940 produced several times the wool of the best sheep first smuggled out of Iberia. Although some of the most publicized early endeavors were carried out on the estates of famous Americans such as George Washington, E. I. du Pont, and Robert Livingston, by the mid-nineteenth century the breeding habit had become established even in frontier areas.

The changes for swine were even more rapid, in large part due to their greater fecundity. Between 1700 and 1940 just about everything changed, including the animals, the locus of production, feeding methods, and the on-farm environment. By the 1840s the vast majority of the genetic material needed for hog improvement was already in place.[120] The popular hogs of 1940 bore little resemblance to the razorbacks of the

[120] Eight of the nine breeds (or the bloodlines that would go into creating them) listed in Table 10.1 were already on American farms by 1840. The one exception was the

eighteenth century. The modern animals gained weight faster and more efficiently on the higher input diets and yielded different proportions of pork, fat, and bacon. As economic conditions changed, breeders changed hog characteristics. Many of the most successful breeds were distinct American creations.

Cattle followed the same basic path as that of sheep and hogs, beginning with the celebrated importation of high-priced foreign stock, active breeding programs, and the investment in better feeding and management systems. The makeup of beef cattle changed fairly rapidly, as reflected by the rise and fall of Longhorns; but even so, the impact of genetic change on the total bovine population was slower to materialize than the impact on hogs and sheep due to different reproduction, maturation, and slaughter regimes. The movement of cattle onto the western range also required fundamental changes in herding systems that drew heavily on Hispanic Texan rather than Anglo technologies. We will continue our analysis of cattle in the next chapter.

Animal breeding and husbandry required collective action. This took many forms – some were apparent in market transactions, some developed in voluntary institutions, and some imposed by government fiat. Communal purchasing of breeding stock; renting stud services; defining breeds; establishing breed associations, herdbooks, and advanced registries; and developing cooperatives to advance artificial insemination are notable examples. These associations helped reduce information costs, overcome capital constraints, and achieve economies of scale. In contrast to plant breeding and seed production, most commercial animal breeding remained a decentralized farm-level enterprise.

There were many institutional achievements that deserve mention. As Shawn Kantor and others have shown, the evolution of fencing laws over the nineteenth century redefined property rights in many states to reflect changing relative scarcities and to increase economic efficiency.[121]

Tamworth, which was introduced from England into Illinois in 1882. USDA, *Yearbook 1920*, 755; Shepard, *Hog in America*, 231–62; Vaughan, *Breeds of Live Stock*, 582.

[121] For an analysis of some of the legal and economic issues underlying the changing definition of property rights as economic conditions evolved, see Kantor, *Politics and Property Rights*.

11

Nature's Perfect Food

Inventing the Modern Dairy Industry

The dairy industry is an integral part of our story due to the sector's growing significance in the agricultural economy. The share of dairy production in U.S. farm output climbed from about 16 percent in 1900 to 30 percent in 1940. In 2004 the farm value of milk output was second only to the value of cattle and calves in the animal sector and was greater than the value of corn production. The changes in recent decades have been remarkable. Since 1940, the quantity of milk production has risen by almost 50 percent, even though the number of milk cows has declined by more than one-half. Annual milk output per cow soared from 4,624 pounds in 1940, to 9,751 pounds in 1970, to 18,201 pounds in 2000, growing almost 2.3 percent per annum. Since 1940, milk output per hour of dairy work has increased about eightfold.[1] These productivity increases have been accompanied by significant structural changes. The number of dairy operations declined from 650,000 in 1970 to 90,000 in the early 2000s; at the same time the average herd size rose from about twenty to about one hundred cows.

Science has helped propel the transformation of the dairy industry. Biomedical innovations, including bovine growth hormones, antibiotics, and breeding via artificial insemination, have contributed to the increase

[1] Man hours per hundredweight of milk went from 3.8 in 1910–14 to 0.2 in 1982–86. Carter et al., *Historical Statistics*, Tables Da1143–1171. An Economic Research Service study of "specialized dairy farms" in 1993 reported that 0.39 hours of labor were required per cwt – a ninefold increase in the hourly output per worker since 1950–54. Short, *Structure, Management, and Performance Characteristics*, 31. For output per cow see USDA, *Agricultural Statistics*, various years.

in output per cow.[2] New genetic engineering technologies promise to further increase output per cow. Farmers have also tinkered with the animal's environment, working on the principle that a "happy cow is a more productive cow." In the hot, arid environment of the Southwest, farmers have provided sun shades and microsprayers to keep the animals cool. The shift to "confinement dairies," where cows are kept in large holding pens rather than in pastures, has also led to higher yields. Taken together the innovations of the past half-century have been nothing short of spectacular.[3]

This chapter argues that the changes before 1940 rival the importance of those since 1940. Rather than being a stagnant period with few innovations, the nineteenth and early twentieth centuries witnessed crucial qualitative changes – indeed the very birth and definition of dairying as a distinct industry. As Henry Alvord, Chief of the Dairy Division of the Bureau of Animal Industry, noted in 1900, "no branch of agriculture in the United States has made greater progress than dairying during the nineteenth century."[4] Reflecting similar sentiments, T. R. Pirtle, an authority on the industry, dubbed the period 1870–1920 dairying's era of "Great Development."[5] There was good reason for Alvord and Pirtle's enthusiasm. In 1800 dairying was, at best, a haphazard sideline for all but a few farmers. Milk came from cows held principally for their beef and draft power. These mongrel "native" multipurpose cattle often wasted away and even perished during the winter. Low-quality butter and cheese were produced on the farm in crude and unsanitary conditions. The nineteenth century witnessed great change with the introduction and improvement of specialized dairy breeds – cows kept principally for their milk. The century also saw the spread of year-round feeding and milking. These

[2] The bovine hormone, which is a genetically engineered copy of hormones produced naturally by cows, is alternatively referred to as rBGH (recombinant bovine growth hormones) and rBST (recombinant bovine somatotropin). Cows injected daily with rBST generally increase milk output by 10 to 20 percent, but they probably account for no more than a 2 to 4 percent increase in national output. Short, *Characteristics and Production Costs*, 4; Kingsnorth, "Bovine Growth Hormones," 266–9. Another controversial change on the horizon is the use of cloned cows and bulls as breeding stock. Pollack and Martin, "F.D.A. Says Food from Cloned Animals Is Safe," *New York Times* (29 December 2006).

[3] U.S. Economic Research Service, *Dairy*; Graves and Fohrman, "Superior Germ Plasm in Dairy Herds," 997–1142; USDA, *Yearbook 1943–1947*, 160–75; Ensminger, *Animal Science*, 68–99; Schertz et al., *Another Revolution in U.S. Farming*, 85–256.

[4] Alvord, "Dairy Development," 381.

[5] Pirtle, *History of the Dairy Industry*, 7.

changes required that farmers create pastures, increase fodder production, and construct silos and barns – all labor-intensive activities. Many recent yield-increasing advances, such as improvements in herd quality and better feeding systems, are continuations of these past developments. As in the recent era, scientific research opened new horizons as medical and public policy breakthroughs allowed dairying to spread to the South and made milk a safe and reliable product. Before 1920 commercial milk was anything but "Nature's Perfect Food."[6]

When all of these factors are considered, the history of the dairy industry stands the induced innovation story on its head. The period before 1940 was, on balance, an era of labor-intensive investments and revolutionary biological innovations that transformed the nature of the industry. These advances set the stage for many recent improvements. The attention given to dairying in the agricultural press and the emergence of specialized dairy journals offer pointed testimony that American farmers were keenly interested in learning about biological innovations.[7] It was not until after World War II that the industry's most important laborsaving innovation, the milking machine, diffused widely.

Output per Cow

There are many conflicting estimates of milk yields per cow for the second half of the nineteenth century. Fred Bateman has done the most thorough work, but even his numbers should not be treated as "sacred cows." The difficulties arise mainly from inconsistencies in census reporting. The 1850 and 1860 Censuses published data only on butter and cheese production. The 1870 Census added "fluid milk sold" to butter and cheese factories. The 1880 Census treated production similarly, excluding fluid milk consumed on farms or sold to urban consumers. The 1890 Census provided fuller coverage but it was not until 1910 that comprehensive data were collected.[8]

Table 11.1 reports milk yield estimates for the period 1850–1920 from Alvord and Pirtle; from Frederick Strauss and Louis H. Bean, and Marvin W. Towne and Wayne D. Rasmussen; and two sets from Bateman between 1850 and 1910. The table also reports yields for the early

[6] For a more critical analysis of the changes, see DuPuis, *Nature's Perfect Food*.

[7] Schlebecker and Hopkins, *History of Dairy Journalism*.

[8] The trend in milk yields per cow is indicative of progress in dairying but is potentially misleading. Ideally, one would want evidence on the value of output relative to the costs of production, but cost data over the long run are not available.

TABLE 11.1. *Average Annual Milk Yield per Dairy Cow, 1850–1940*

Year	Alvord–Pirtle	Bateman 1968	Bateman 1962	Strauss–Bean & Towne–Rasmussen	Voelker & Agricultural Statistics
1850	1,436	2,371	1,839	1,879	
1860	1,505	2,559	1,904	1,922	
1870	1,772	2,670	2,139	1,970	
1880	2,004	2,797	2,382	2,475	
1890	2,709	3,050	2,777	2,604	
1900	3,646	3,352	3,384	3,384	
1910	3,113	3,570	3,521	3,200	3,759
1920	3,627				4,008
1930					4,508
1940					4,624

Notes: Voelker's series has the virtue of matching those published in *Agricultural Statistics* beginning in 1920.
Sources: Pirtle, *Handbook of Dairy Statistics* (1928), 3; Alvord, "Statistics of the Dairy," 52; Bateman, "Economic Analysis," 8, 47, and "Improvement," 263, and "Marketable Surplus," 347; Strauss and Bean, "Gross Farm Income," 91; Towne and Rasmussen, "Farm Gross Product," 288–9; USDA, *Agricultural Statistics 1942*, 462. Data for 1910 are from Voelker, "History of Dairy Recordkeeping."

twentieth century from D. E. Voelker and the U.S. Department of Agriculture (USDA). The Appendix at the end of the chapter compares the series. Alvord and Pirtle suggest that milk yields per cow increased by 1.3 percent per year between 1850 and 1920. At the other extreme, Bateman's 1968 estimate indicates a rise of 0.7 percent a year. Even Bateman's lower estimate of the rate of growth is respectable although it is well below the post-1940 rate of 2.3 percent. Most of the difference in the growth rates in these pre-1910 series is due to the different estimates for 1850.[9]

We are confident that yields increased before 1850. Alvord and Pirtle also held this opinion. Elinor Oakes makes a convincing case that improvements began before 1800. She details investments in care and feeding and improved breeding in commercially oriented pockets near northeastern cities. The efforts to drain land and build improved pastures with foreign grasses and clovers were especially impressive.[10] Our review of *Early American Imprints* and *Early American Newspapers* buttresses

[9] Ontario data show an increase on par with what took place in the United States. Between 1883 and 1920 yields increased from 2,800 to 4,400 pounds. Ankli and Millar, "Ontario Agriculture in Transition," 211–12.

[10] Oakes, "Ticklish Business," 195–8.

Oakes' identification of early advances.[11] There were very few substantive accounts of dairying in the seventeenth century. By the early national period articles increasingly conveyed a spirit of experimentation and improvement. Newspapers were replete with stories of wonder cows, some producing more than 9,000 pounds in a season. Around 1750 a large dairy with seventy-three cows in Narragansett, Rhode Island, averaged about 3,757 pounds per cow for one year.[12] These would have been exceptional animals because the Massachusetts Society for Promoting Agriculture suggested that a "fair" amount for "ordinary cows" was about 2,500 pounds of milk per year in 1800. Percy Bidwell and John Falconer reckoned that in 1830 output per cow in Massachusetts averaged about 3,225 pounds.[13]

The story appears similar in the middle states. Rare mid-eighteenth-century evidence of two superior New Jersey herds, feeding on excellent pasture, shows that the yield was about 2,400 pounds per cow (allowing for milk consumed on the farm). This was probably exceptional for the age.[14] In comments on conditions in Maryland around 1800, noted agriculturalist Richard Parkinson observed that "a good cow, properly fed, will give sixteen quarts of milk in twenty four hours; but that is not common. For want of care in feeding and milking, they frequently give only one quart in twenty-four hours. It is not uncommon for a cow to go unmilked for two or three days." Farmers frequently stored milk in the cows because "milk is hard to keep from the frost in winter; and the heat in summer very soon turns it sour; and so they milk it as they want it." Infrequent and irregular milkings "caused the cows to decline in their milk very much."[15] One quart per day would have yielded about 500 pounds a year assuming a milking season of about 250 days.

As an 1840 benchmark, Stevenson Fletcher asserted that dairy cows in Pennsylvania averaged about 2,500 pounds of milk per year.[16] Joan M.

[11] *Early American Imprints, Series 1 and 2*; and *Early American Newspapers.*

[12] Schumaker, *Northern Farmer and His Markets*, 19–20. To arrive at 3,757 we converted the butter and cheese produced to milk.

[13] Bidwell and Falconer, *History of Agriculture*, 229. Pirtle, *History of the Dairy Industry*, 27, converted the society's estimates of butter and cheese production to obtain the estimate of 2,500 pounds of milk.

[14] Woodward, *Ploughs and Politicks*, 325.

[15] Parkinson, *Tour in America*, 285–6. For the backward state of dairying in the South during the colonial period, see Gray, *History of Agriculture*, 204–6. For 1860, Bateman calculates an average southern yield of 1,247 pounds. Bateman, "Marketable Surplus," 347, and "Improvement in American Dairy Farming," 258.

[16] Fletcher, *Pennsylvania Agriculture 1840–1940*, 169.

Jensen maintains that "the search for a better milk cow preoccupied both gentlemen farmers, who read the latest books, and agricultural journals for advice, and country farmers, who gossiped, swapped, and bought their way to better herds. By the 1840s, many dairies in the Philadelphia hinterland had improved Durham shorthorns cows that gave as much as thirty-six quarts a day. In the late eighteenth century, even the most ambitious farmers had been satisfied with eight quarts a day."[17] Few cows would have averaged 36 quarts over a sustained period, but in any case yields at this date in the West and South were considerably below those in Massachusetts and Pennsylvania. As one indication, as late as 1850 the average yield in most southern states was still less than one thousand pounds a year.

Yields in frontier regions were always low. A narrative from Canada circa 1800 offers an example of frontier conditions: "We brought a cow with us, who gave us milk during the spring and summer; but owing to the wild garlic (a wild herb, common to our woods), on which she fed, her milk was scarcely palatable, and for want of shelter and food, she died the following winter." The problem of animals, and milk cows in particular, eating poisonous plants was common enough that contemporary newspapers regularly published folk remedies on how to resuscitate animals and revive foul milk.[18]

From these fragments, it is reasonable to assume that the national average yield per cow at the end of the eighteenth century was less than 1,500 pounds and possibly as low as 1,000 pounds.[19] This would imply that milk output per cow likely increased 3.1-fold and perhaps by as much as 4.6-fold between 1800 and 1940. The sources of these yield increases are out of sync with the induced innovation model's account of American agricultural growth. Bateman noted that, "before 1900 . . . there were no mechanical improvements in dairying even remotely comparable to, for example, the mechanical reaper. Thus most of the influence on the dairy production function and dairy efficiency had to originate from other (nonmechanical) sources, particularly improved breeds and feeding

[17] Jensen, *Loosening the Bonds*, 96.

[18] Traill, *Backwoods of Canada*, 276. For an example of the interest in dangerous plants, see "Effectual Method of Removing the Taste," *Green Mountain Patriot* (23 February 1803), 4.

[19] Evidence presented by Pirtle for 1916 output per cow in Siberia (1,192 pounds) and Chile (1,520 pounds) suggests that the estimates for colonial America are in rough accord with low-yielding producers more than one hundred years later. Pirtle, *Handbook of Dairy Statistics* (1928), 3; Bateman, "Labor Inputs and Productivity," 222–3.

and care techniques."[20] Improvements in feed and shelter are especially important in his account. Although we concur with Bateman's general emphasis, mechanical innovations did play a role. Most important, the diffusion of the mechanical mower helped increase the hay supply to carry cows over the winter months.[21] Improved feed and care were only two of the many changes that transformed the dairy industry.

The Development of Specialized Dairy Animals

Bateman downplayed the contributions of breed improvements before 1910, but by this date breeding was having an important impact on the dairy industry. Referring to the 1930s and 1940s, T. C. Byerly noted that "the production of milk per cow in the United States has increased by more than 1000 lb, or by about 20 percent in the past 20 years. This has been accomplished largely through the use of superior sires, especially progeny-tested sires and their sons."[22] Due to the relatively slow reproduction rate of cattle, developing superior milkers required decades of work.

Changes began in the Northeast and gradually diffused to other areas.[23] In 1800 there were few, if any, purebred dairy cows in the United States, but this situation soon changed with importations from Europe. By 1885 there were about 90,000 registered purebred dairy cattle in the United States and by 1895 the number had grown to about 273,000. These animals had a significant impact on herd quality far beyond their actual number. By the turn of the past century, "their blood is so generally diffused that half-breeds or higher grades are very numerous wherever cows are kept for dairy purposes. Therefore, although pure-bred animals form less than 2 per cent of the working dairy herds, their influence is so

[20] Bateman, "Improvement in American Dairy Farming," 255–6. Output per cow increased in spite of countervailing factors that pushed down yields. As an example, after its introduction to the United States in 1887, the horn fly became a serious cattle pest, adversely affecting weight gain and milk production. Bolton, Butler, and Carlson, "Mating Stimulant Pheromone of the Horn Fly," 951–64.

[21] The mower, like the reaper, began its commercial spread in the 1840s and by 1860 was widely used in dairy regions. Olmstead and Rhode, "Beyond the Threshold," 27–57.

[22] Byerly, "Role of Genetics," 328. Bateman further develops this view in his book with Jeremy Atack. Atack and Bateman, *To Their Own Soil*, 147. For the counterview see Leavitt, "Attempts to Improve Cattle," 51–67; Lemmer, "Spread of Improved Cattle," 79–93; McGaw, "Specialization and American Agricultural Innovation," 134–49; Prentice, *American Dairy Cattle*.

[23] Bidwell and Falconer, *History of Agriculture*, 132.

great that it is probable the average dairy cow in the United States at the close of the century will carry nearly 50 per cent of improved blood. The breeding and quality of this average cow, and consequently her productiveness and profit, have thus been steadily advanced."[24] By the 1920s the evidence comes into sharper focus, with the number of dairy purebreds reaching about 900,000 animals, and virtually all dairy cows were classified as grade animals of the purebred stocks. By 1930 the number of registered purebreds had increased to almost 1.3 million animals. In 1926, the 3 percent of the dairy population that were purebreds accounted for 10 percent of the milk output. Pirtle further noted that "within a few decades the production of milk has been raised from about 10,000 pounds for the highest cow to more than 37,000 pounds in a year."[25] Clearly the enormous breeding advances that occurred after 1940 were built on the foundation laid over the previous century. In fact, the post–World War II achievement was, to a large extent, the diffusion of the characteristics of the best producers of 1940. An examination of the development of the leading dairy breeds reinforces this point.

Holstein–Friesians are the premier modern dairy cow. They represented nearly one-half of all dairy cattle in 1920, and by 1975 the share was over 85 percent. This dominance is largely a result of their high milk production. The familiar black-and-white breed originated in the northern provinces of the Netherlands. A few Holsteins were introduced to North America by the Holland Land Company (1795) and by William Jarvis (1810), whose success as a Merino importer was discussed in Chapter 10. But these early introductions did not have a lasting impact. Winthrop W. Chenery of Belmont, Massachusetts, established the first permanent herd in the 1850s. After a Holstein cow purchased off a Dutch ship proved a successful producer, Chenery imported a bull and two cows in 1857 and four more cows in 1859. Due to a contagious pleuropneumonia outbreak in 1859, the state of Massachusetts condemned and slaughtered all of Chenery's cows, but he persevered and rebuilt his herd. Several New York dairy operators, including Gerritt Miller, the Smiths–Powell firm, and Henry Stevens, helped popularize the Dutch breed. Competing Holstein and Friesian breed associations were established in the 1870s and 1880s (see Table 11.2). The two main groups unified in 1885 to form

[24] Alvord, "Dairy Development," 392. Houck, *Bureau of Animal Industry*, 187; Nystrom, "Dairy Cattle Breeds," 3; Edwards, "Europe's Contribution," 72–84; Bowling, "Introduction of Cattle," 129–54.

[25] Pirtle, *History of the Dairy Industry*, 33–5.

TABLE 11.2. *Purebred and Associated Grade Dairy Cattle*

Breed	Date First Imported	Founding of Breed Association	Established: Advanced Register	Established: Herd Improvement Register	Number of Registered Purebreds 1885	1895	1903	1920	Grade Number 1920
Holstein–Friesian	1857	1871	1885	1928	21,138	187,750	115,009[b]	528,621	10,500,000
Jersey	1815[a]	1868	1882	1928	51,000	150,000	237,680	231,834	9,300,000
Red Polled	1847	1883	1908		nd	4,408	30,300	30,000	1,800,000
Ayrshire	By 1837	1863	1902	1925	12,867	18,750	25,541	30,509	400,000
Brown–Swiss	1869	1880	1911		nd	1,930	3,258	8,283	nd
Dutch Belted	1838	1886	1912		nd	971	1,524	5,900	150,000
Guernsey	1815[a]	1877	1901		4,947	12,547	24,337	79,440	1,933,000[c]
TOTAL (excl. Red Polled)					89,952	202,948	407,349	916,602	22,283,000

[a] Alderney.
[b] Holstein only; the comparable number in 1895 is 90,325.
[c] Guernsey includes purebred and grades.

Notes: Red Polled cattle were considered a dual-purpose breed for milk and beef and are thus excluded from the total. Advanced Registers included individual animals who achieved high performance levels. Herd Improvement Registers applied DHIA-like tests (see text following) to entire herds of purebred cows. See Peters, *Livestock Production*, 112, 125–26, 135, 145, 152, 160, 170; and Gilmore, *Dairy Cattle Breeding*, 372–85.

Sources: Alvord, "Statistics of the Dairy," 67; Pirtle, *History of the Dairy Industry*, 35–56, 166, and *Handbook of Dairy Statistics* (1928), 21–6; Houck, *Bureau of Animal Industry*, 187, listed "registered purebreds." Alvord, "Dairy Development," 381–403, estimates that there were roughly "200,000 to 300,000" purebreds in 1890, noting that not all were registered. Alvord reports the breed association estimates of the number of "living purebreds," which was only 60 to 70 percent of the number "registered." This is because associations' registers were not adjusted to account for mortality.

the Holstein–Friesian Association. Directly after this change, importing practically ceased.[26] Thus, nearly all the Holstein "blood" now present in the United States was introduced by 1886.

Holsteins became dominant only after a lengthy "battle of the breeds." Among the early contenders were Milking Shorthorns, descendents of the dual-purpose Durham cattle prominent in the early nineteenth century. Breeders on each side of the Atlantic took Durhams in different directions: those in Scotland developed the beef-producing qualities whereas American farmers sought to perfect a dual-purpose animal. U.S. farmers established the Milking Shorthorn Cattle Club of America in 1912.[27]

Other contenders in the battle of the breeds included Guernseys and Jerseys, two distinct lines of cattle that originated in the Channel Islands. Each gained prominence in the late nineteenth century when their exceptional ability to produce butterfat was highly valued. Early U.S. sources tend to lump together all Channel Island imports, making the separate histories of Guernseys and Jerseys difficult to decipher. A further problem is that many of these animals were simply recorded as "ship's cows." We know that in 1815 Maurice and William Wurts began importing and distributing brown Alderney cattle (sometimes referred to as Jerseys). In 1818 James Creighton of Baltimore imported an Alderney bull and by 1820 was selling other cattle that he had "purchased on the Islands of Guernsey and Alderney" in local auctions. Among the better known importers was the financier Nicholas Biddle, who obtained three cows from Guernsey for his Andalusia, Pennsylvania farm. In the 1850s, R. L. Coit of New Jersey, Charles H. Fisher of Philadelphia, and James P. Swain of Bronx, New York, imported blooded Guernseys. At about the same time John A. Taintor of Connecticut and Daniel Webster of Massachusetts imported Jerseys. Advocates of these two breeds formed the American Jersey Cattle Club in 1868 and the American Guersney Cattle Club in 1877.[28]

Other competing breeds included the Ayrshire, Brown Swiss, Dutch Belted, and Red Polleds. The histories of these breeds detail the first importations, the activities of early improvers such as P. T. Barnum, and the organization of breed associations.[29] Although each had its devotees,

[26] Briggs and Briggs, *Modern Breeds of Livestock*, 232, 244; Prescott and Price, *Holstein–Freisian History*, 1–7.

[27] Briggs and Briggs, *Modern Breeds of Livestock*, 217–19.

[28] Vaughan, *Breeds of Live Stock*, 185–93; *American Farmer* (2 June, 23 June, 15 July 1820), 79, 101, 113; Pirtle, *History of the Dairy Industry*, 45–9.

[29] Vaughan, *Breeds of Live Stock*, 211–12.

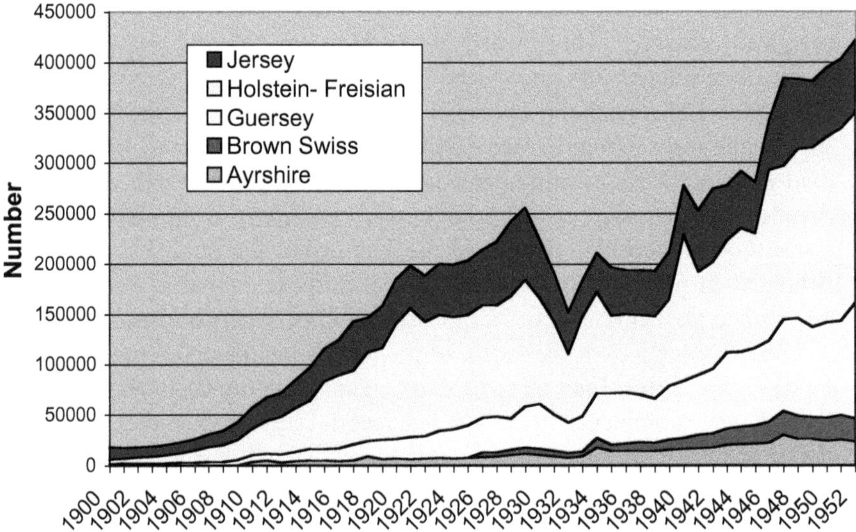

FIGURE 11.1. Number of purebred dairy cows registered each year, 1900–52. *Sources:* USDA, *Agricultural Statistics 1930*, 901; Sarbaugh, "Dairy Cattle Breeds."

these breeds represented only a small fraction of registered purebreds. Figure 11.1 graphs the number of dairy cattle on the books of the five major breeding associations for each year from 1900 to 1952. One notable trend was the rising dominance of Holsteins, which produced large volumes of milk with relatively low butterfat ratios, and the shifts away from Jerseys and Guernseys, which yielded less but richer milk. Driving this change was a shift in consumer taste away from butter and cream toward milk with lower fat content.

By the 1880s all of the important breeds had associations that maintained herd registries, sponsored contests, devised scorecards, and generally promoted the breeds. Although these organizations performed valuable services, knowledgeable observers justifiably complained that the associations excessively emphasized characteristics and lineages that were unrelated to productive merits. In its 1936 *Yearbook* the USDA argued that breed definitions and associations promoted unscientific standards and drew attention away from "functional abilities."[30] Jay L. Lush, the "father of modern animal breeding" and advocate of combining modern genetics with statistical analysis, was especially critical of the closed

[30] USDA, "Livestock Breeding at the Crossroads," 831–5.

and backward-looking view of breed associations and their emphasis on pedigree rather than performance.[31]

In spite of these criticisms, the breed associations did provide some valuable services. In 1882 the American Jersey Cattle Club began conducting seven-day butterfat tests. In 1885 the Dutch Friesian (Holstein) Association created an "Advanced Registry," which set milk output standards and kept "blue books" that recorded individual cow production.[32] Advanced registries that listed high-performing animals were subsequently established for Guernseys (1901), Ayrshires (1902), and Jerseys and Red Polleds (1908). Surviving data clearly document the advance in record performances. The 1902 Guernsey champion produced 11,623 pounds of milk, and the 1915 champion yielded 24,008 pounds – an average annual increase of more than 950 pounds for the highest producer in this breed. The Holstein–Freisian cows that "made history" in the 1920s also vastly outproduced the "best of the best" of earlier decades. In 1870 Dowager, owned by Gerrit S. Miller of Peterboro, New York, gained acclaim by producing 12,681 pounds of milk. By 1924, 54 cows exceeded 30,000 pounds of milk and the top Holstein–Freisian yielded 37,465 pounds.[33] These data on champion producers offer a sense of the potential impact of breeding, because the "highest cows" in the 1870s were pampered with abundant feed and good care. A comparison of Figures 11.2 and 11.3 offers a sense of the change in the physical conformation of milk cows.

Improvements in milk testing facilitated the modernization of the dairy industry. In 1890 Stephen Babcock, a biochemist at the University of Wisconsin, developed a revolutionary butterfat test (see Figure 11.4). The relatively simple technology, which employed sulfuric acid, a calibrated bottle, and a small hand-powered centrifuge, allowed the butterfat content of milk to be measured quickly and cheaply. A reliable butterfat test was a technology ripe for invention. Between 1888 and 1890, researchers at several experiment stations devised and published reports on six competing chemical tests, and Swiss chemist N. Gerber independently invented a test using sulfuric acid shortly after 1890.[34] Babcock's

31 Chapman, "Jay Laurence Lush," 277–305.

32 Briggs and Briggs, *Modern Breeds of Livestock*, 238; Vaughan, *Breeds of Live Stock*, 160, 166, 193, 217.

33 Prescott and Price, *Holstein–Freisian History*, 11–19, 60; Prescott and Schall, *Holstein–Freisian History*, 18–19; Pirtle, *History of the Dairy Industry*, 47–51. The data of Pirtle differs slightly from that found by Prescott and Price.

34 Wing, *Milk and Its Products*, 44–7.

FIGURE 11.2. Representation of a Jersey cow from the 1800–30 era. The cow depicted here bears little resemblance to the elite stock of just a few decades later. *Source:* New York State Agricultural Society, *Transactions 1850*, 321.

FIGURE 11.3. The Holstein Friesian Clothilde owned by Smiths & Powell Co. produced a record 26,021 pounds of milk in 1886. *Source:* Photograph courtesy of Holstein Association USA, Inc.

FIGURE 11.4. Stephen Babcock with an electric centrifuge butterfat tester circa 1926. The early models were crudely constructed and driven by a hand crank. *Source:* Courtesy of the Wisconsin Historical Society.

test won the day, replacing the old, unreliable method of churning a milk sample to estimate the fat content.[35] In this era when skimmed milk was for calves or pigs, not people, butterfat was milk's most valued attribute. The ability to monitor quality reduced the free rider problem associated with selling milk based on weight alone. As Wisconsin Governor William Dempster Hoard purportedly said, the Babcock test "made more

[35] Judkins, *Principles of Dairying*, 59.

dairymen honest than the Bible ever made."[36] Under the old incentive structure, dairymen stood to gain by adding water to the milk or skimming off the rich cream. Indeed, "watering" and "cream skimming" are enshrined in economic terminology on information failures. The Babcock test lessened the static incentives to "cheat" and gave farmers stronger dynamic incentives to adopt better practices and breeds.[37] Farmers could now identify and cull cows that produced milk with a low fat content. In the words of S. J. Van Kuren, "It stimulated a marked increase in dairying everywhere, because it turned on the searchlight of profit-vs.-loss on the business of feeding and milking dairy cows."[38]

Responding to the new incentives, farmers across the country began forming local "cow testing associations" to select those genetic lines that yielded the most productive offspring. Within ten years of the founding of the world's first dairy herd improvement association in Denmark, Danish immigrant Helmer Rabild helped found the first U.S. association in Newaygo County, Michigan, in 1906. The idea took off. By 1909, associations existed in Vermont, Iowa, California, Wisconsin, and Nebraska. This movement promoted extensive recordkeeping of both feed consumption and milk and butterfat output. The realization that genetic material from males also contributed to dairy performance led to the formation of cooperative bull testing associations by 1908. Bull associations were designed to provide the services of proven bulls at a reasonable cost and to prevent the slaughter of good bulls.

The USDA extension service encouraged the spread of the cow testing movement and in 1924 developed national guidelines. Shortly thereafter the testing program was rechristened Dairy Herd Improvement Association (DHIA). By 1926, there were 777 DHIAs involving more than 327,000 cows on 19,000 farms, as well as 225 bull associations.[39] Figure 11.5 graphs the course of the two movements from 1906 to 1960. The associations' efforts led to rapid results. In 1926 the average "association" cow produced 7,200 pounds of milk and 282 pounds of butterfat

[36] *New York Times* (16 June 1930), 14. Citing the *New England Farmer* 15 (1836), 66, Charles Townsend Leavitt, "Meat and Dairy Livestock," 70, noted the "universal practice of watering milk 25 percent to 50 percent."

[37] Lampard, *Rise of the Dairy Industry*, 153–62, 197–204.

[38] Van Kuren, "Dairy Machinery," 80; Pirtle, *History of the Dairy Industry*, 78–80.

[39] Pirtle, *History of the Dairy Industry*, 31–2, 35–56. Herd improvement associations are still an important feature in American dairying. U.S. Agricultural Research Service, *Summary of DHI Participation*.

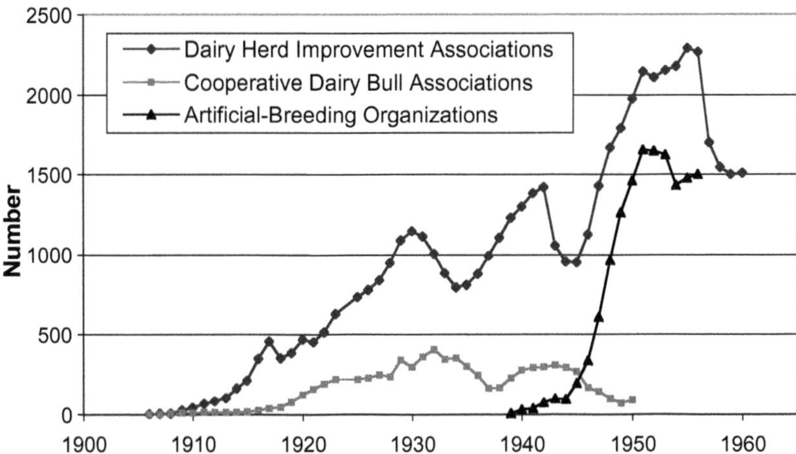

FIGURE 11.5. Dairy Herd Improvement Associations and related breeding associations. *Sources:* USDA, *Agricultural Statistics* (various years); Voelker, "History of Dairy Recordkeeping."

compared with 4,368 and 175 pounds, respectively, for all cows nationally.[40] In 1925 purebred dairy groups bought into the herd improvement movement, initiating the random testing of entire herds and improving recordkeeping of production.[41] By the mid-1930s, purebred associations and the DHIAs were pursuing nearly identical paths, and breed association leaders were touting the DHIA system as "the greatest single move ever made for the constructive improvement of the dairy cattle of the United States."[42]

Among the most significant changes after 1940 was the perfection and diffusion of artificial insemination (AI). AI dates to Lazzaro Spallanzani's experiments with dogs beginning in Italy in 1784 and to E. I. Ivanoff's work with cattle in Russia in 1899. Experiments in several countries helped make AI practicable for farmers. Better methods of harvesting and implanting sperm, and later the use of frozen sperm, increased the amount of sperm collected and impregnation rates while lowering the percent of sperm wasted. In addition, institutional advances helped lower

[40] Parker, "Cow-Testing Associations a Factor in Low-Cost Dairying," 280. There is an obvious causality problem here, but many other sources detail how local programs in fact improved productivity.

[41] Peters, *Livestock Production*, 112, 125–6, 135, 145, 152, 160, 170; Gilmore, *Dairy Cattle Breeding*, 372–85.

[42] Elwood, Lewis, and Struble, *Changes in Technology and Labor Requirements*, 17.

FIGURE 11.6. Sir Pietertje Ormsby Mercedes. *Source:* Courtesy of the Holstein Association USA, Inc.

costs. The first cooperative dairy AI program dates to Denmark in 1936. The cooperative approach spread to New York the same year and by 1940 similar efforts were also under way in New Jersey, Wisconsin, and Minnesota. As Figure 11.5 shows, AI organizations took off nationally after 1945.

AI vastly increased the impact of sires known to produce high-milk-yielding daughters. With traditional methods a bull could mount about fifty dams a year, generating twelve to thirteen females. In 1924 the Holstein–Freisian Association listed the top sires by the number of daughters with yearly milking records. The champion bull, Sir Pietertje Ormsby Mercedes, stood atop the list, with fifty-six such daughters over his lifetime (see Figure 11.6).

By comparison, with AI a bull's influence could be multiplied a thousandfold, siring 5,000 to 20,000 daughters per year.[43] This progression of supply-side changes set the stage for the acceleration in breeding activity after 1940. "Artificial insemination has helped greatly in spreading the superior germ plasm of superior sires. The number of cows bred

[43] Sir Mercedes probably had other female offspring not included in a recordkeeping program. Pirtle, *History of the Dairy Industry*, 52; Foote, *Artificial Insemination to Cloning*, 6–8.

by artificial insemination has increased from about 10,000 in 1939 to more than 3 1/2 million in 1951."[44]

Specialization and Structural Changes

Since the beginnings of agriculture, farmers have struggled to obtain sufficient feed to carry livestock over the winter.[45] Even cows saved from slaughter were not fed enough to maintain milk flows. "Medieval cows usually lactated no longer than four months, and probably some of the more poorly attended colonial cows gave milk for only slightly longer periods of time."[46] This changed in the nineteenth century. Between 1850 and 1910, the average milking season increased by one-quarter, from roughly 237 days to 300 days in 1910.[47] The standard milking year is now about 305 days, so there has been very little change since 1910.

Several demand and supply shifts motivated farmers to provide better feed for their cows. On the supply side, in addition to mowing machines, the introduction of ensilage extended the milking season by up to ten weeks. Most accounts credit Fred L. Hatch of McHenry County, Illinois, with building North America's first tower silo in 1873. The technology was an immediate success, spreading to the dairy regions of the northern states and Canada by the early 1880s.[48] In 1889 Herbert Myrick boasted to Congress that "no innovation in American agriculture has ever made more rapid progress than the system of preserving green fodder in silos. The improvements which have been made in this system are also characteristic of the American genius for bettering and simplifying the inventions of others. The ensilage system, as a practical reality, is hardly ten years old in the United States; yet . . . there are upward of 7,000 silos in the

[44] Byerly, "Role of Genetics," 323–31.

[45] A USDA survey published in 1875 found that the cost of wintering dairy cattle averaged $17.10 per head nationally – slightly more than two-thirds of the value per head. Costs were highest in the Northeast and declined to the west and south. USDA, "Wintering Farm-Animals," 84–90; U.S. Bureau of Agricultural Economics, *Livestock on Farms*.

[46] Jensen, *Loosening the Bonds*, 96. Farther north the feeding constraints were more severe. In 1880 the lactation of most cows in Quebec tapered off after only six months. U.S. Dept. of State, "Cattle-Raising in Quebec," 572–3.

[47] Bateman, "Improvement in American Dairy Farming," 266–71; cows are not milked year-round because they need time to dry out before "freshening," that is, giving birth and resuming lactation.

[48] USDA, "Silos and Ensilage," 1–70; Carrier, "History of the Silo," 175–82; Meeks, *Time and Change in Vermont*, 164; Fletcher, *Pennsylvania Agriculture, 1840–1940*, 179. Howard Russell disputes Hatch's claim to erecting the first American silo and dates the achievement to Vermont circa 1869. Russell, *Long, Deep Furrow*, 441.

FIGURE 11.7. Wood plank silo in Oconto County, Wisconsin. By the end of the nineteenth century, wood, stone, and tile silos were a common sight in dairy regions. *Source:* Courtesy of Rita Neustifter, Oconto County WIGenWeb Coordinator and Webmaster.

country."[49] He went on to explain how in many areas Americans had cut silo costs in half by building them out of wood (see Figure 11.7). In 1918 there were 404,000 silos in the United States with a capacity of over 30 million tons.[50] By 1936, 62 percent of dairy operations in the USDA's "Eastern Dairy" region and 85 percent in the "Western Dairy" region reported having silos.[51] As noted in Chapter 9, farmers improved pastures by introducing better grasses, clovers, alfalfa, and sorghums and by increasingly adopting commercial feeds.[52]

Changes in the forces of demand transformed the markets faced by farmers. Among the most revolutionary structural changes was the development of the "factory system" that shifted butter and cheese production off the farm. The first cheese factory in the United States dates to Jesse

[49] Myrick, "Improved Silo and Ensilage," 743.

[50] Pirtle, *Handbook of Dairy Statistics* (1922), 17.

[51] Elwood, Lewis, and Struble, *Changes in Technology and Labor Requirements*, 28.

[52] Pirtle, *History of the Dairy Industry*, 67.

Williams of Oneida County, New York, in 1851. Williams was a skilled cheesemaker who started buying milk from neighbors so that he could specialize in what he did best. Alanson Slaughter, of Orange County, New York, is generally credited with building the first creamery or butter factory in 1861.[53] As with many popular testaments to "history's first," these claims are contested, but what matters for our argument is that the application of the factory system to dairying was new at mid-century and rapidly diffusing.[54] As the census of agriculture for 1860 noted, "The 'cheese factory' system . . . was introduced a few years ago" to allow "farmers with a few cows, to avoid the expense of the necessary building, and to introduce the best apparatus for the manufacture of cheese, unite to send their milk every morning to a certain point, where it is converted into cheese, and each farmer receives his proportion (or the money received for it) according to the quantity of milk he has furnished. . . . At the factory a competent person is employed to attend to the business, and the cheese is made on the most approved principles."[55] A direct result of the factory system was an improvement in the quality of cheese and butter and an increase in the international market for American products.[56]

Another change was more subtle. By 1890 some dairymen altered the breeding period so that their cows dropped calves in September and October instead of April, which had been the custom. These cows were well fed and maintained their milk flow over the winter and spring when they were put onto pasture. This allowed for butter production at a lucrative time when most other animals were dry. In addition, during the winter it was easier to keep the milk fresh and make quality butter. As an added advantage the cows went dry in the summer, when flies and

53 Durand, "Migration of Cheese Manufacture," 271–2; Alvord, "Dairy Development," 384–6. Just as whiskey and hogs embodied a method of shipping corn, butter and cheese provided a way to ship milk. In addition to being less perishable than milk, 100 pounds of milk were reduced to about 10 pounds of cheese and 5 pounds of butter. Durand, 262.

54 As an example, a "Mr. Norton" operated a cheese factory ca. 1844. U.S. Patent Office, *Report of the Commissioner 1845*, 989–90. Also see Pirtle, *History of the Dairy Industry*, 154–5; and Larson et al., "Dairy Industry," 311. The first cheese factory in Canada was built in Oxford County, Ontario, in 1864 by Harry Farrington, who carried the concept from New York. U.S. Dept. of State, "Cattle and Dairy Farming in Ontario," 544.

55 U.S. Census Bureau. 8th Census 1860, *Agriculture*, lxxxvi.

56 Ibid., lxxxv. In the early twentieth century considerable quantities of butter were still made on the farm. In 1921, 22.4 percent of all milk produced went to making creamery butter, and 13.8 percent went to the manufacture of butter on the farm. Larson et al., "Dairy Industry," 293.

other insects that tormented cows were at their worst. Dairymen who produced for the fresh milk market also adapted breeding so that some cows dropped every month, in order to ensure a year-round milk supply. These calving/milking systems would not have been possible without advances in ensilage and the production and storage of other feeds.[57]

Several additional developments ushered in the era of increasing specialization. Among the most important were improvements in transportation and the resulting changes in market structure. Milk and milk products were shipped on railroads from an early date, but the advent of refrigeration following the Civil War meant that perishable products could be transported greater distances without compromising quality. Refrigerated rail cars were first used in 1867 to ship dairy products into the Chicago and New York markets. Refrigeration "made it possible for large cities to reach further into the country for whole milk and by doing so they pushed the butter and cheese factories back from the great markets."[58] As an example, the Boston milkshed extended from 65 miles in 1870 to 275 miles in 1910.[59] The market for fresh products was further extended with the introduction of large, commercial, cold storage facilities around 1900 and of motor trucks in 1911. The development of improved refrigeration systems contributed to growth in the production and popularity of ice cream.

Additional innovations revolutionized the way milk was marketed. Gail Borden's 1856 invention of condensed (evaporated) milk was a major step in protecting infants in an age before good refrigeration and pasteurization. By the early twentieth century canned milk made up a considerable share of total fluid milk consumption. The invention of malted milk (1883) and several methods to make powered milk (1901–04) further assisted preservation. Advances in handling fresh milk included the introduction of glass-lined vats around 1910 and of improved cleansing solutions for washing dairy equipment. Pirtle observed that the development of "washing powder was essential to the dairy industry with its multiplicity of utensils and equipment which must be kept free from bacteria of all kinds."[60] By 1940 almost all milk sold in urban America was

[57] Wing, "Dairy Industry," 580.

[58] Pirtle, *History of the Dairy Industry*, 152–3.

[59] Bateman, "Improvement in American Dairy Farming," 259; also see Atack and Bateman, *To Their Own Soil*, 149.

[60] Pirtle, *History of the Dairy Industry*, 148–58. U.S. Federal Trade Commission, *Report*, 28.

pasteurized and most was distributed in sealed glass bottles. The days of itinerant peddlers ladling milk into consumers' containers were on the wane.

Labor Inputs

Dairy farming was traditionally a very labor-intensive activity. Prior to the World War II era, the makeup of the dairy labor force underwent several important changes. The expansion of market-oriented dairying in the early nineteenth century created economic opportunities, especially for women. Milking cows and making butter and cheese were almost exclusively women's work. Dairying ranked with textile manufacturing as creators of jobs for northern women in the antebellum years. As milk output increased and cow lactation periods lengthened, dairying provided more than just seasonal employment. Making butter was arduous. After milking the cows, women had to carry the milk to the milkhouse, pour it into skimming pans, skim the cream, churn it into butter, knead out the buttermilk and knead in salt or another preservative, pack the butter into molds to form bricks, and then pack the bricks into pails or boxes. In addition utensils had to be cleaned. Even at this early date, many women knew that boiling utensils reduced the chances of producing foul butter. This was one of many nonmechanical principles that gained favor without any deep scientific understanding.[61]

Although labor-intensive biological innovations accounted for most of the increase in milk yields, there was a stream of laborsaving innovations in butter production, many aimed at producing a better churn. "From 1802 to 1849, the Office of Patents issued 244 patents on machinery related to butter making, 86 percent of them for churns." The new technologies had a significant impact. During the colonial period women often churned butter using a bowl and spoon. Dasher churns were common in some areas by the 1770s. "A simple dasher churn often held only a gallon of cream, produced one or two pounds of butter at a time, and took a woman three hours to churn." This was a significant improvement over using a spoon and bowl. "By 1850, farm women could choose from a variety of improved churns that held up to fifty gallons of cream and produced 100 to 150 pounds of butter in as little as one and one–half hours."[62] The larger churns were powered by dogs, sheep, or calves

[61] Jensen, *Loosening the Bonds*, 92–113.
[62] Jensen, "Butter Making and Economic Development," 820.

walking on a treadmill. The advent of the creamery changed the gender division of labor by bringing more men into butter and cheese production. In addition, the pace of patenting accelerated with 1,360 butter-related patents issued between 1850 and 1873.[63]

As cheese and butter making moved from the farm to the factory, creameries adopted a number of biological and mechanical innovations. Among the most important was the continuous centrifugal cream separator invented by Carl Gustav Patrik De Laval of Sweden in 1878. The early machines could process about three hundred pounds of milk a day. This was an enormous advance over the earlier process of pouring milk into shallow pans and, over a two- to three-day period (in which milk often soured), letting gravity separate the cream. Two Danish immigrants brought the first separator to the United States in 1882. Once again we see an example that runs counter to the predictions of the induced innovation model because the first practical mechanical separator was invented in a country that was a major exporter of labor.[64]

Subsequent inventive efforts focused on mechanizing milking. Between 1872 and 1909, 127 milking machine patents were issued in the United States.[65] Practical success with such devices proved elusive despite incentives to save labor. The period of dairying's "Great Development" saw an intensification of effort. Between 1850 and 1910, the estimated total labor input per milk cow nearly doubled, from 77 to 141 hours.[66] The yield of milk per cow did not increase proportionately (at least by Bateman's reckoning), so the average amount of milk per hour of work fell. One hour of labor yielded about 30.8 pounds of milk in 1850 but only 25.4 pounds in 1910. The output per hour in the milking operation increased over this period (from 40.2 pounds to 47.6), but more hours were devoted to sanitation and the year-round care and feeding of the stock. Overall, technical change in the dairy industry was labor using.

Before World War I, a dairy farmer had to work ten times as long as a modern farmer to produce a gallon of milk. As an indication of the dairy industry's demand for labor, around 1940, dairying employed over one and one-half times as much labor as was devoted to growing cotton.

[63] Jensen, *Loosening the Bonds*, 104–5.

[64] Christensen, "First Cream Separator," 338; Edwards, "Europe's Contribution," 78–80; Jenson, *Loosening the Bonds*, 98–9; Fussell, *English Dairy Farmer*, 160–202.

[65] Bateman, "Labor Inputs and Productivity," 211.

[66] Ibid., 208–9. Bateman makes no adjustment for the gender division of labor. Using the standard assumption that males had higher opportunity costs than women would further increase the growth in labor inputs relative to the early nineteenth century.

When first introduced early in the twentieth century, pulsating suction milking machines promised to save about one-fifth of the annual labor expended per cow. As with the cream separator, Europeans were even more active than Americans in developing this laborsaving innovation. The machines spread slowly at first in the United States due to the structure of dairy farming, the lack of electricity, improper sanitary practices, and "hard times" on the farm. In 1910, about 12,000 farms possessed milking machines. This was less than one of every four hundred farms that reported keeping dairy cows. By 1940, the number of farms with milking machines had reached 190,000. Nonetheless, on the eve of the Second World War, perhaps 90 percent of all cows were milked by hand. Up to the 1950s the labor savings generated by the diffusion of milking machines and other mechanical technologies was roughly offset by the extra time needed to meet new sanitary standards.[67] In the postwar period diffusion accelerated, with one-half of the cows milked mechanically by 1950 and nearly all dairy cows in commercial operations by the mid-1960s. By 1965, some 500,000 farms reported milking machines. The spread of milking machines was only the most visible part of a larger mechanical revolution on dairy farms.[68]

Milk Yesterday and Today

In 1956 the *Journal of Dairy Science* published a special edition commemorating the fiftieth anniversary of the founding of the American Dairy Science Association. Several dairy scientists recounted the technological advances that in their view had revolutionized their industry. Some of the advances were theoretical insights, others were general-purpose technologies, and others were solutions to specific problems. We have already touched on some advances, including the Babcock test, the invention of ice cream and condensed milk, refrigeration systems, AI, and the development of effective cleansing solutions. A full list would include the application of Mendelian genetics along with the discovery of amino acids and vitamins A, D, and K. Surprisingly none of the articles in this golden jubilee edition addressed what were arguably the most important advances. Prior to 1940, scientists and public officials, along with (and

[67] Witzel, "Development of Dairy Farm Engineering," 777–82. Because cows cooperated in the production and release of milk and were prone to mastitis – an output-reducing inflammation of the mammary glands – perfecting a mechanical method of milking posed significant technological challenges.

[68] Forste and Frick, "Dairy," 119–47.

sometimes in spite of) farmers, redesigned milk. This is one of the most important stories in American agricultural history.

Our discussion of the changes in yields per cow and the transformation of the dairy industry only hinted at the cleaning up of the milk supply. Just as the cars, dentistry, and eye surgery of today bear little resemblance to the products delivered in 1900, so it is with milk. But in this instance almost all of the key quality-related innovations occurred and were widely diffused before 1940. Milk in 1900 could kill you. Milk in 1940 was of similar quality as the milk now available in the twenty-first century. Accounting for quality changes, advances in the fifty years between 1890 and 1940 rival those of the past five decades. These quality changes were spawned by the germ theory of disease. This represents an early case where breakthroughs in basic science had a fundamental impact on agricultural productivity.

Since the mid-eighteenth century, physicians had suspected that consuming milk could be dangerous to one's health. Milk was an excellent medium for the multiplication and transmission of the bacteria that caused typhoid fever, paratyphoid fever, diphtheria, streptococcus (septic) sore throat, scarlet fever, milk fever, undulant fever (brucellosis), dysentery and other gastrointestinal infections (including many caused by *Escherichia coli* due to fecal contamination), tuberculosis, and many other diseases.

By the early twentieth century a growing number of public health advocates became alarmed about milk safety. Armed with a new understanding of the germ theory of disease and with mounting epidemiological evidence, these crusaders left a rich documentation of the "Milk Problem." Around 1900 contaminated milk caused infections that resulted in tens of thousands of deaths per year in the United States. Most of the victims were children. By 1940 most of the portals of diseased milk had been closed as a result of two distinct movements. The first involved campaigns to impose sanitary and health standards to clean up dairies, identify and cull sick cows, improve milk handling processes, prevent milk adulteration, and the like. The goal was to prevent disease-causing organisms from entering and multiplying in the milk supply. Closely coupled with the testing program was the highly controversial drive to mandate milk pasteurization to destroy those harmful bacteria and viruses that evaded the cleanup efforts. Commercially pasteurized milk, which was virtually nonexistent in 1900, comprised about 98 percent of the milk supply in major cities by 1936.[69]

[69] Fuchs and Frank, "Milk Supplies and Their Control," 29.

Recent work by Samuel Preston and Michael Haines shows that gastrointestinal diseases accounted for nearly 25 percent of all infant deaths in the period 1899–1900. Although problems with water sanitation accounted for many of these untimely deaths, cow's milk was also heavily implicated. In the summer months, when it was harder to keep milk fresh, diarrhea-related deaths soared. In addition there were numerous studies comparing the death rates of breast-fed and bottle-fed babies. The results were stark. Preston and Haines reported the results of a Children's Bureau study of eight cities between 1911 and 1915: "death rates among those [infants] not breastfed were 3–4 times higher than among the breastfed."[70] Across Europe similar studies showed that babies fed cow's milk were dying from intestinal problems at about five to ten times the rate of breast-fed infants.[71] The records of New York City's Randall's Island Hospital bolster the indictment of cow's milk. For the years 1895–97 the city admitted 3,609 children, of which 1,509 or about 42 percent died. In 1898, the facility began serving pasteurized milk and, between 1898 and 1904, only 1,349 (22 percent) of the 6,200 children admitted died. A contemporary investigation asserted that there were "no other changes" besides pasteurization to account for the dramatic drop.[72]

Although the most serious threat was to infants and young children who were not breast-fed, adults were by no means immune to the danger of contaminated milk. Recall that Nancy Hanks Lincoln, Abraham Lincoln's mother, suffered a painful death from contracting milk sickness (also known as the trembles) in 1818. This was a terrifying and often fatal disease that frequently afflicted entire communities. It resulted from consuming the milk of cows that had eaten white snakeroot, a plant common in the Midwest and parts of the South.[73] The threat of milk sickness represented the tip of the iceberg. As of 1908 epidemiologists had traced about 500 epidemics of typhoid fever, diphtheria, and scarlet fever in the United States to specific dairies or milk handlers. There were undoubtedly many more that could not be positively linked to the milk supply.[74]

70 Preston and Haines, *Fatal Years*, 27. The data pertain to the Death Registration Area.

71 Eager, "Morbidity and Mortality Statistics," 229–42.

72 Ibid., 229–42. Also see Rosenau, "Pasteurization," 612.

73 Beck, "Facts Relative to a Disease," 315–19; Roadhouse and Henderson, *Market Milk Industry*, 71–2; U.S. National Parks Service, *Lincoln Notebook*. Farmers learned of the association between white snake root and the sickness in the first half of the nineteenth century and began taking precautions.

74 Kerr, "Certified Milk and Infants' Milk Depots," 565–88. A later U.S. Public Health Survey for the years 1923–46 showed another 974 milkborne disease outbreaks in the United States, documenting 40,972 cases and 804 deaths. Roadhouse and Henderson, *Market Milk Industry*, 58.

E. coli poisoning was also a serious problem given the prevalence of fecal material in milk.

To address these problems most cities eventually established milk regulations with commissioners (often doctors of medicine) to inspect and regulate the sanitary conditions of dairies and milk handlers. Milk commissioners developed detailed scorecards much like those used for inspecting restaurants. Some cities began testing milk samples in order to head off problems and track down the source of contaminated products. When epidemics did break out, public health officials were often able to identify the cause and impose cleanup measures to limit the damage. The USDA, state experiment stations, and the agricultural colleges also got into the act with a long list of publications and educational efforts to instruct farmers on the newly understood dangers and on how to clean up their premises and procedures. This involved many measures that seem obvious today, such as cleaning up dairy water supplies, washing the cows' udders and the milkers' hands before milking, requiring milk workers to be free of communicable diseases, controlling insects and rodents, taking precautions against introducing foreign matter into the milk (in particular the use of covered pails designed to keep out cow droppings), and many more. Even in the absence of more sophisticated efforts such as pasteurization, dairy sanitation campaigns, coupled with the introduction of milk depots to provide clean milk to the urban poor (especially in the summer months), had a dramatic impact on infant mortality. As an example, initiatives in Rochester, New York, purportedly decreased the infant mortality rate from 33 to 15 percent.[75]

The combined effect of cleaner dairies, pasteurization, condensed milk, and more sanitary handling and storage systems on reducing infant diarrhea represents one of the major public health achievements of the past two hundred years. However, the single most ambitious public health initiative to improve the milk supply was the aggressive federal–state campaign to eradicate bovine tuberculosis. Tuberculosis (TB) was the leading cause of death in the United States, taking about 148,000 human lives in 1900 – accounting for about one out of every nine deaths.[76] Roughly 10 percent of TB sufferers had contracted the bovine form of the disease,

[75] Rosenau, "Pasteurization," 591–628.

[76] U.S. Census Bureau, "Tuberculosis in the United States," 516, and *Mortality Rates 1910–1920*, 16, 27. For a more complete analysis of the bovine TB eradication program see Olmstead and Rhode, "Impossible Undertaking," 734–72, and "Tuberculous Cattle Trust," 929–63.

most commonly from contaminated milk. Thus, around 1900 bovine TB was responsible for the deaths of about 15,000 Americans every year. Death represented only a fraction of the disease's cost because countless others were permanently crippled or lingered in pain as they wasted away.

Bovine TB was an insidious disease because apparently healthy animals could be both infected and contagious.[77] Rates of TB infection tended to be higher among closely confined cattle than in free-range animals. The prevalence of the disease also increased with an animal's age. As a result, bovine TB was much more common in dairy herds and purebred stock and, more generally, in the "advanced" agricultural regions of the country.[78] In 1890, Robert Koch developed tuberculin, which made it possible to detect TB in animals before they developed visible symptoms. Early tuberculin tests showed that the extent of the infection was far more widespread than had been suspected. Around 1915 about 10 percent of the nation's dairy cattle were infected with bovine TB, and the infection rate was growing at an alarming pace. Left unchecked the incidence of the disease would have approached European rates, which were several times that prevailing in the United States.[79] Efforts to improve herds paradoxically contributed to the spread of bovine TB (and other diseases) by increasing the trade in animals. The rate of infection of purebreds was two to three times the rate for ordinary dairy stock.

The eradication campaign was a massive endeavor that touched virtually every farm in the country as the federal government, in cooperation with state and local governments, systematically tested and retested cattle, destroying those that tested positively. Between 1917 and 1940 veterinarians administered roughly 232 million tuberculin tests and ordered the destruction of about 3.8 million cattle (from a population that averaged 66.4 million animals over this period). The vast majority of condemned animals were valuable dairy cows or breeding stock.

The eradication program was an enormous success. By 1941 every county in the United States was officially accredited as being free of bovine TB (that is, with an infection rate below 0.5 percent). Although

[77] National Research Council (U.S.), *Livestock Disease Eradication*, 13.

[78] Myers, *Man's Greatest Victory over Tuberculosis*, 264, 267–8, 309, 323.

[79] Ibid., 115; U.S. Bureau of Animal Industry, *Special Report on Diseases of Cattle 1916*, 409, 416–7, 422, and *1912*, 417; Mitchell, "Animal Diseases and Our Food Supply," 168; *1912*, 417; Myers and Steele, *Bovine Tuberculosis Control*, 256–7, 280–1. For a discussion of the controversy in combating bovine TB in Europe see Orland, "Cow's Milk and Human Disease," 179–202.

the costs of the eradication program were huge, the savings to farmers and meatpackers alone (resulting from increases in animal productivity and a decline in the number of condemned carcasses at slaughterhouses) exceeded the costs by at least a ratio of ten to one. However, the real savings were not to be found in the farm sector. As with the campaign to clean up dairies, the spillover effect on human health was the main story. By 1940, before effective chemotherapy was available, new cases of bovine TB in humans had become a rarity. The eradication program, coupled with the spread of milk pasteurization, prevented at least 25,000 U.S. TB deaths per year in the period immediately preceding World War II. In addition, the TB program created a model used to eliminate brucellosis (ungulate fever) as well as other diseases. While the United States was making one of the largest government-directed biological investments in the history of American agriculture, the advanced European nations did little, resulting in hundreds of thousands of needless deaths. As in so many other cases noted in this book, the United States was a world leader in adopting biological technologies.

Conclusion

The history of the dairy industry conforms nicely to our general thesis and reinforces our findings for both staple and specialty crops and for animals. The pre-1940 era saw the development of modern dairying, with year-round feeding and housing of cows in barns during the winter season. The transition involved an increase in the division of labor, with the emergence of the cheese factory and butter creamery. The manufacture of dairy products, once the job of farm women, increasingly became the task of male factory workers. With transportation improvements and refrigeration came significant shifts in market areas, further stimulating specialization. The period also witnessed the emergence of specialized dairy breeds in the United States. The multipurpose animals of the early nineteenth century gave way to animals that were increasingly efficient in converting feed into milk. Given the relatively slow reproduction rate of cattle and the large reservoir of inferior stock, this was a slow process. In 1889 H. H. Wing conceded that most American dairy cows were still "natives or scrubs" and lamented that many, "perhaps even a majority, of our dairymen make no attempt to increase the product of their cows." But the transition to better breeds had made considerable headway, because he also pointed out that "in the United States [there] are more Jerseys than on the Island of Jersey, more Holstein–Friesians than in Holland,

and more Short Horns than in England."[80] By the twentieth century, contemporaries were describing the then state-of-the-art cows as "milk producing machines."[81]

The development of the Babcock butterfat test accelerated the transition to more efficient milk producers by allowing farmers to better assess which animals were high fat producers. This test also gave farmers an increased incentive to improve their herds because milk buyers could now grade milk more easily. Recordkeeping took on new importance and both purebred and testing associations emerged to encourage the improvement of dairy stock. Almost all of these changes rested on biological innovations and many were labor-using rather than laborsaving. Over the period 1850–1910, on-farm labor inputs per cow and per unit of milk rose substantially. Most likely there were offsetting gains in labor productivity in creamery operations when they were moved from the farm to the factory. The most important laborsaving mechanical innovation in dairying, the milking machine, did not diffuse widely until after World War II. From the end of the colonial era to 1940, milk yields per cow likely increased threefold. This was not as fast as the rate of growth of yields in the post–World War II era, but the more recent gains were based on the institutional and biological foundations built over the long haul. Moreover, adjusting for quality changes shifts the calculus dramatically. All of the scientific and institutional advances that transformed milk from a mass killer of humans to "Nature's Perfect Food" were in place by 1940.

Appendix: Milk Yield Estimates, 1850–1940

The earliest estimates are from USDA experts Alvord and Pirtle, who reported in gallons and pounds, respectively. The two series match assuming an average of 8.6 pounds per gallon. Both series are misleading because they exclude fluid milk production for the early years.[82] Bateman's 1968 data are the most accepted although they are not fully documented. The data from his 1962 master's thesis are better documented but differ from his 1968 *Journal of Economic History* data. The 1962 estimates for the period 1850–80 are based on butter and cheese production and assume that 1 pound of butter required 21 pounds of milk, 1 pound of cheese required 10 pounds of milk, and the combined

[80] Wing, "Dairy Industry," 580.
[81] Kirkland, *History of American Life*, 185.
[82] Alvord, "Statistics of the Dairy," 58.

total of milk devoted to manufactured dairy products comprised a constant fraction (0.67) of total milk production. These conversion ratios differ significantly from those used by Alvord and others. As an example, the Census of 1900 used conversion ratios of about 30.8 pounds of milk per pound of butter and 11.4 pounds of milk per pound of cheese. If correct, these Census ratios imply that Bateman's 1962 estimates significantly underestimated yields for the early years.[83] In addition, the constant two-thirds share does not match that used by Bean and Strauss. Their numbers, in a similar manner, fail to capture the effect of changing market conditions on these production relationships. Bean and Strauss note that "total milk production was extrapolated from 1899 [back] to 1869 on the basis of the actual relationship after 1899" to processed dairy production.[84] An additional difficulty with the yield estimates is that they do not indicate butterfat production, which was more relevant than pounds or gallons for some purposes. The butterfat content of milk likely increased over the nineteenth century.

Towne and Rasmussen accepted the Strauss–Bean numbers and estimating techniques for the period 1869–1909 and extrapolated them back to the early nineteenth century. Bateman (1962), Strauss and Bean, and Towne and Rasmussen all rely on conversion ratios (where 1 pound of butter required 21 pounds of fluid milk and 1 pound of cheese required 10 pounds of milk) from Vail, "Production and Consumption," 56. These conversion ratios, drawn from the period 1924–40, differ from those assumed by Alvord and Pirtle (30 pounds for butter and 11 pounds for cheese) taken from the turn-of-the-century censuses. Increases in the butterfat content of milk likely account for some of the differences.

[83] Bateman was clearly aware that the use of 0.66 to represent a constant manufactured to fluid milk ratio had problems because of changes in the industry over time. For example see Bateman, "Marketable Surplus," 347. The butterfat content of milk probably rose between 1850 and 1900 suggesting the need for an even greater conversion ratio for the earlier years.

[84] Strauss and Bean, "Gross Farm Income," 91. To add another fly to the ointment, it is possible that many observers have overstated milk output data by adding the raw milk needed for cheese and butter production. Much cheese appears to have been made from skim milk, so the same milk that was used to make butter was used to make cheese after the cream was separated.

12

Draft Power

The long view of American agricultural development shows that mechanical and biological technologies were not entirely separable and that the strict identification of land saving with biological innovations and labor-saving with mechanization is problematic. Examining the evolving use of horses and mules drives these points home. Not only were crops a key intermediate input into the production of draft power, draft animals in turn were an intermediate input of crop production. Equines were the biological technologies used to drive the early farm machines, thereby making the adoption of biological and mechanical technologies closely intertwined. In the 1850s, for example, the use of the mechanical reaper and the use of horses were complementary; thus, using one technique increased the productivity of the other. The combined effects of these intermediate inputs and their complementary relationships can lead to paradoxical results. As one example, land abundance in antebellum America stimulated mechanization by lowering the cost of draft power. The cost of feeding a horse relative to the cost of hiring a worker in the United States was one-sixth that in Britain. The farm tractor provides another paradoxical case because one of the most important mechanical innovations in all of agricultural history was a major land-saving technology.[1]

[1] Christensen, "Land Abundance," 315. The weekly cost of feed in the United States in 1850 was $1.03 and the cost of wages was $3.80, whereas in Britain the sums were 15s and 9.58s, respectively. These relationships also had important effects on the endogeneity of horse costs. See "The Tractor as a Land-Saving Innovation" later in this chapter, and Olmstead and Rhode, "Reshaping the Landscape," 663–98.

TABLE 12.1. *Shares of Oxen, Horses, and Mules in Farm Draft Power*

	Oxen (%)	Mule (%)	Horse (%)
1850	25.8	8.4	65.7
1860	22.9	12.0	65.1
1870	13.8	11.7	75.5
1880	7.5	13.8	78.7
1890	6.1	12.3	81.7

Notes: The Census ceased collecting data on working oxen after 1890. The numbers may not sum to 100 due to rounding. Our analysis ignores the trivial number of "asses" reported in 1890. Note that movements in relative shares do not always capture movements in absolute numbers. The total of oxen on farms peaked at 2.2 million in 1860, and the number is slightly higher in 1890 than in 1880, despite the relative decline.

Sources: U.S. Census Bureau. 7th Census 1850, *Agriculture*, 8th Census 1860, *Agriculture*, 9th Census 1870, *Statistics of Wealth and Industry*, 10th Census 1880, *Report on the Productions of Agriculture*, and 11th Census 1890, *Report on the Statistics of Agriculture.*

Over the course of the nineteenth century, farmers and regions became increasingly specialized and thus increasingly dependent on others for inputs, including machines, seeds, and stud services. The changes often involved introducing entirely new crops and livestock, thereby dramatically shifting preexisting production patterns. But few of the changes left as large a footprint on American farms as the wholesale shift in draft power.

Transition from Oxen to Equines

The Census of Agriculture began collecting data on the number of working oxen and horses and mules in 1850, but by this date the transition away from oxen was already well under way. Narrative evidence makes it clear that the predominant mode of draft power on American farms in the colonial and early national periods was oxen; mules were largely unknown and horses were used mainly for transport. As the Census data in Table 12.1 indicate, by 1850 the share of oxen in total horsepower capacity had already fallen to about one-quarter. The oxen share was down to about 6 percent in 1890 and was miniscule by 1925.[2]

Agriculturalists have long debated the relative advantages of the different sources of animal power.[3] Horses were noted to work twice as fast and require fewer breaks than oxen, thus saving human labor, but were also more expensive. Horses cost twice as much per head as oxen, consumed twice the feed, were more prone to illness and injury, and their harnesses were more costly than oxen yokes.[4] Because of differences in hooves, equines provided greater traction than oxen under icy conditions but poorer traction in muddy conditions. The speed of horses on good roads and their ability to trot was especially valued. Horses were also more suited to use with reapers and other new farm machines. The relative advantages of equines grew as an area modernized and as farm wages rose.

Cattle, however, possessed a number of characteristics that made them highly desirable for specific draft power uses, especially by settlers. Hence, the classic picture of the westward-bound Conestoga wagon pulled by a pair of yoked oxen. Rather than requiring a daily grain ration, cattle could live on native grasses and occasionally skip meals. In addition to providing draft power, cattle supplemented the settlers' diet by supplying milk and meat. Cattle were better able to swim across rivers and were less prone to appropriation by raiding Indians. Oxen walked about two miles per hour slower than equines but possessed greater endurance and, according to some accounts, could travel a greater distance in one day. Oxen provided steady, if slower, draft power and proved to be superior in heavy-duty tasks such as breaking new soils, pulling stumps, and hauling logs.

[2] According to Kinsman, "Appraisal of Power Used on Farms," 9, U.S. farms possessed 200,000 oxen in 1924, less than one-hundredth of the more than 20 million horses and mules.

[3] As an example, in 1819 the *American Farmer* published a series of articles titled "On the Comparative Utility of Oxen and Horses," (17, 24, and 31 December), which favored the continued use of oxen. Articles publicizing the reports of the "Working Oxen" committee of the Massachusetts Society for Promoting Agriculture had a similar bias. A more balanced assessment appeared in "Horses versus Oxen," *Pittsfield Sun* (24 September 1850), 4.

[4] Allen, *Domestic Animals*, 190–2; International Museum of the Horse, *Draft Animals in Early America*. "Horses and Oxen," *Middlesex Gazette* (18 October 1826), 1, noted that "horses have 261 kinds of diseases, and oxen only 47." Also see "Oxen Vs. Horses," *Barre Gazette* (22 August 1851), 4; "Shall We Use Horse or Oxen?" *Pittsfield Sun* (22 April 1858), 4.

TABLE 12.2. *Dollar Value per Head of Working Age Animals in 1869 by Region*

	Ox	Horse	Mule	Milk Cow
Mid-Atlantic	84	167	166	51
East North Central	70	122	124	33
West North Central	56	131	137	39
South Atlantic	44	116	154	33
East South Central	42	130	157	33
West South Central	26	88	137	22
Mountain	65	141	164	54
Pacific	67	131	149	63

Note: No data were reported for New England.
Source: Young, *Special Report on Immigration.*

Working oxen were typically less valuable per head than horses or mules but more valuable than milk cows. Table 12.2 displays data on the representative value per head for a working ox, horse, mule, and dairy cow by region circa 1870.[5] In the North, one horse of working age was worth about the same as two oxen; in the South, it was closer to a one-to-three ratio. Mules were even more valuable in that region.

Figure 12.1 maps the changing share of oxen in farm draft power in 1850 and 1890 by county. It is clear from the 1850 map that oxen predominated in the New England region and upper Midwest.[6] They were also common on the frontier and the newly settled areas of the South. Horses (and mules) predominated in the middle latitudes. Between 1850 and 1890, the regions where equines prevailed expanded across the Midwest and later to the West and South. The remaining oxen in the post-bellum period tended to be clustered on the frontier and in logging areas. By 1890, oxen remained important only at the extremes, in the upper Great Lakes and the lower South. The shift in sources of animal power had profound effects on the farm sector by changing the composition

[5] The values come from state-level data reported in Young, *Special Report on Immigration.* The regional value per head for each type of stock was derived by weighting the state-level data by the number of animals as reported in U.S. Census Bureau. 9th Census 1870, *Statistics of the Wealth and Industry.* Note that the *Special Report* lists the price for a pair of working oxen.

[6] According to Jones, "Horse and Mule Industry in Ohio," 77–8, farmers of New England extraction who settled in the Buckeye State possessed a greater cultural preference for oxen than those from Pennsylvania, Virginia, or Kentucky. This predisposition for the older mode of draft power stands in direct contrast to the usual characterization of Yankee migrants to the Midwest as progressive and those from the upper South as traditionalists.

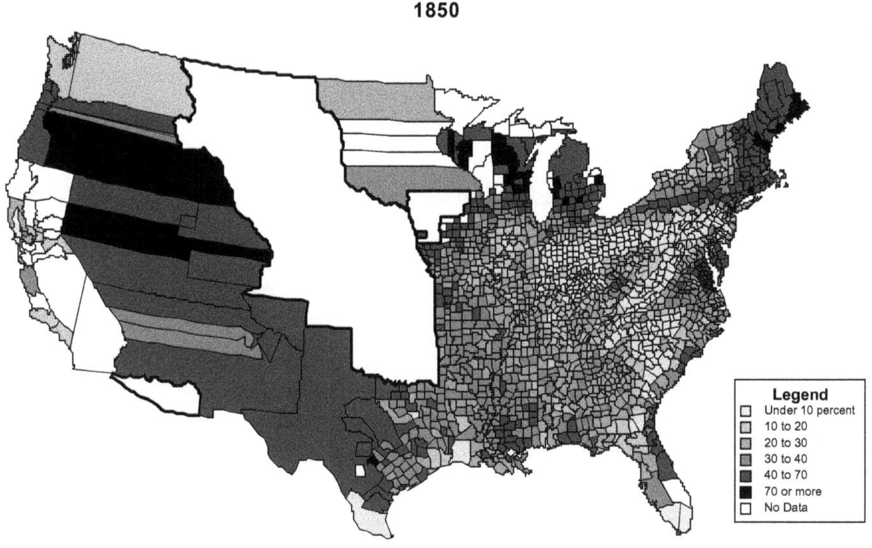

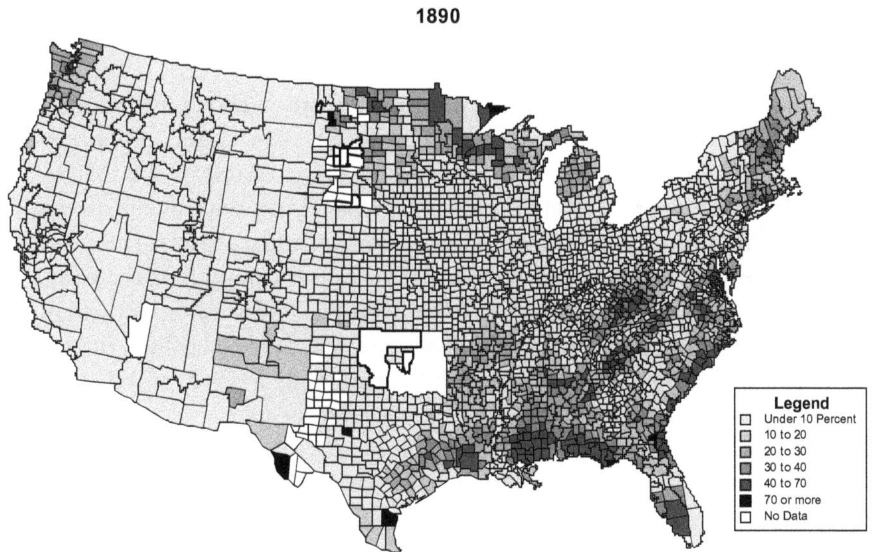

FIGURE 12.1. Share of oxen in draft power, 1850 and 1890. *Source:* Derived using data from Haines and Inter-university Consortium for Political and Social Research, *Historical, Demographic, Economic, and Social Data*.

of crop output toward hay and oats; by shifting the livestock pool from multipurpose meat–milk–power bovines to a portfolio of specialized beef cattle, milk cows, and draft horses; and most critically, by facilitating the adoption of farm machinery.

Adoption of the Mule

As with the contest between oxen and horses, agriculturalists long debated the choice between horses and mules. Advocates of the mule saw the hybrids as "uniting most of the good qualities of the horse and ox."[7] Mules worked slightly slower than horses but ate less, withstood heat better, were smarter, were more even-tempered, matured earlier, and suffered less from abuse and neglect. Obviously mules could not be used for breeding purposes, but then neither could geldings. Mules typically sold for higher prices than horses – the per-head value of mules averaged 18 percent higher than that of horses over the period 1867–1914. This reflected, in part, the greater difficulty of breeding the hybrid. Jacks and mares did not easily mix.[8]

The greater durability of mules was a benefit to draft animal owners who could not closely monitor how their animals were driven and kept, as these tasks were often in the hands of hired laborers, slaves, or sharecroppers. As early as the 1780s, Johann David Schöpf observed that mules were "perfectly adapted for the American economy" because "scant attention and bad feed" sufficed.[9] In his 1856 classic, *A Journey in the Seaboard Slave States*, Frederick Law Olmsted relates the following exchange with a Virginia planter: "When I ask why mules are so universally substituted for horses on the farm, the first reason given, and confessedly the most conclusive one, is that horses cannot bear the treatment that they always must get from negroes; horses are always soon foundered or crippled by them, while mules will bear cudgelling, or lose a meal or two now and then, and not be materially injured, and they do not take cold or get sick, if neglected or overworked."[10]

In his magisterial survey of antebellum southern agriculture, L. C. Gray observed that, after 1800, "[i]t gradually became a widely held opinion

7 Lamb, *Mule in Southern Agriculture*, 27, is citing Davie, *Address Delivered*, an 1818 report of the Agricultural Society of Pendleton, South Carolina, as quoted in the *American Farmer* (5 October 1821), 220–1; Warder, "Mule Raising," 180–90.

8 Trowbridge, "Mule Production," 21; Lamb, *Mule in Southern Agriculture*, 26–9.

9 Lamb, *Mule in Southern Agriculture*, 4, cites Schöpf, *Travels in the Confederation*, 2:48.

10 Olmsted, *Journey in the Seaboard Slave States*, 47; Marx, *Capital*, ch. 7, fn 17, reproduces Olmsted's account.

of Southern farmers and planters that mules were preferable to horses for field work. It was thought that mules... were much hardier under careless treatment by Negroes, less likely to become frightened... and would tread down less corn and cotton in cultivation."[11] The application of principal-agent theory lends further credence to the hypothesis that mules were more popular in the South because they were less susceptible to the aftereffects of abuse.[12]

As indicated in Table 12.1, most of the increase in the proportion of draft power provided by mules occurred in the 1850s. These national shares hide substantial regional disparities. Figure 12.2 maps the share of draft power provided by mules. Mules were most prevalent in the South and in parts of the frontier. Their prevalence in the South increased over time, reflecting a substitution away from both oxen and horses. The domain of mules was largely the black belt cotton regions where sharecropping was most common. Although mules were most prevalent in the South, horses remained the dominant source of power and outnumbered mules in 90 percent of southern counties.

The mules that toiled in the lower South were typically bred elsewhere, principally in the upper South. The location of mule breeding changed significantly over time. In the colonial period, breeding was centered in the Connecticut Valley before shifting to the Chesapeake. In the 1820s breeding moved to the Kentucky Bluegrass region – the great nursery of the mule – and the Nashville basin of Tennessee. Missouri, which received its first mules as part of the Sante Fe trade in the 1820s, emerged as a major breeding center in the 1870s, and mule breeding in Texas took off in the 1890s. After the Civil War, Kentucky specialized in producing jacks for other breeding areas; Tennessee focused on breeding light, "hot-blooded" black types for the cotton fields of the South; and Missouri concentrated on large, heavy-boned, more docile black or sorrel types for the corn belt and the urban draft market.[13]

Horse Breeding

Pedigree breeding first took hold in the raising of equines for racing. In this pursuit, Americans emulated their English brethren. Thoroughbreds, however, represented only a minute fraction of U.S. equines. Away from

[11] Gray, *History of Agriculture*, 48.

[12] Kauffman, "Why Was the Mule Used in Southern Agriculture," 336–51.

[13] Clark, "Live Stock Trade," 567–81; Bradley, *Missouri Mule*, 39, 93, 96; Anderson and Hooper, "American Jack Stock," 390–404.

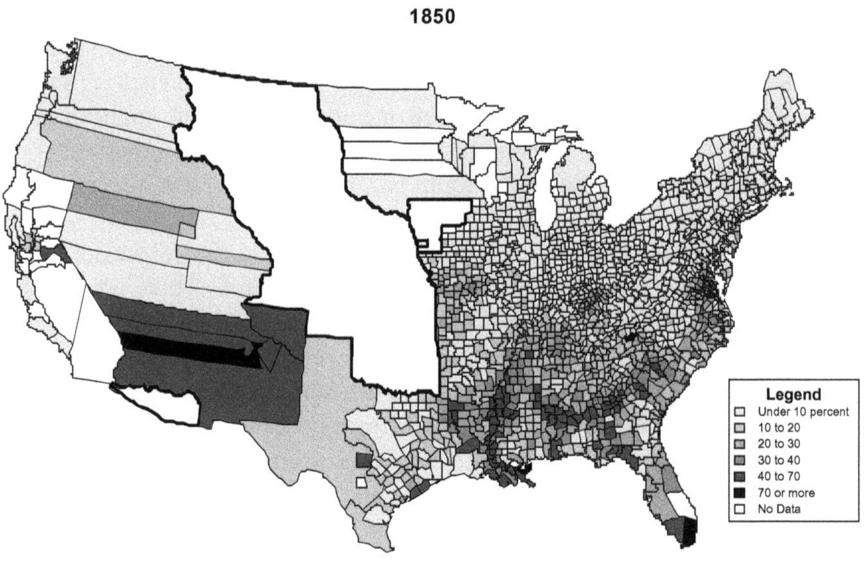

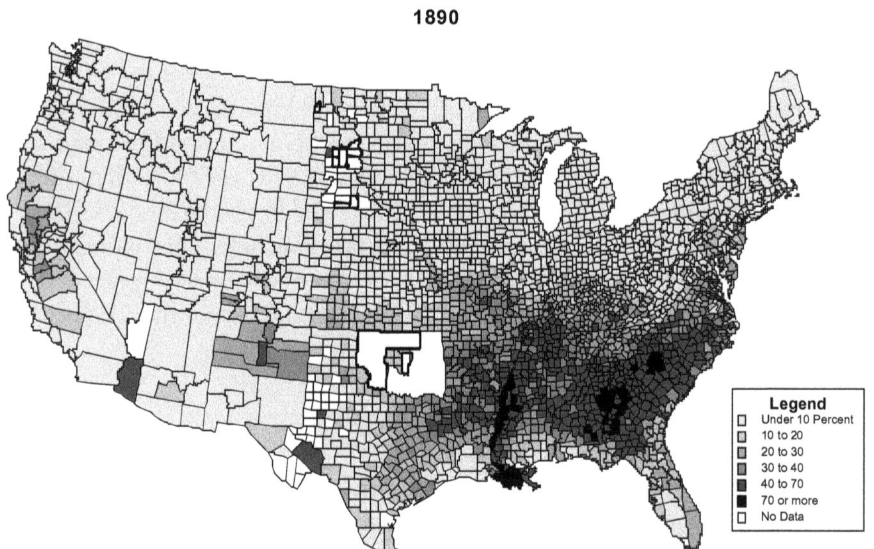

FIGURE 12.2. Share of mules in draft power. *Source:* Derived using data from Haines and Inter-university Consortium for Political and Social Research, *Historical, Demographic, Economic, and Social Data.*

this ennobled sphere, early breeders succeeded in developing a number of distinctive native lines: the Narragansett pacer, Conestoga, Morgan, Tennessee Walking Horse, Standardbred, Quarter Horse, American Saddle Horse, Florida Cracker, and Appaloosa.[14] In the late nineteenth century, Americans intensified their efforts to breed improved draft lines. Increasing numbers of heavy powerful horses were imported from northwestern Europe – Percherons from Normandy, Belgians from the Low Countries, Shires and Suffolk Punches from England, and Clydesdales from Scotland. Edward Harris of Moorestown, New Jersey, is credited with importing the first Percherons to the United States in 1839; Dr. A. G. Van Hoorebeke of Monmouth, Illinois, introduced the first Belgians in 1866; and Scottish immigrants to Canada brought the first Clydesdales. In addition to improving animal quality through breeding, farmers and ranchers became more adept at training their animals. This was especially true of cattlemen in the Southwest and California, who developed the highly dexterous cutting horses (and the skilled riders to utilize them). The animals and methods employed in the West represented a total break from those common to the East and Midwest.

Promoters of improved draft stock formed breed associations and maintained studbooks. Among these were the Norman–Percheron Association (1876), the American Clydesdale Association (1879), the American Shire Horse Association (1885), the American Association of Importers and Breeders for Belgian Draft Horses (1887), and the American Suffolk Horse Association (1911). In 1882, the *Breeder's Gazette* exulted: "The growth of the business of breeding draft-horses in the United States within the past ten years has been simply wonderful. . . . [It] has passed beyond the domain of accident in this country. We have our well-established breeds of heavy horses that reproduce themselves with as much certainty as any sort of livestock; and we have our pedigree records for draft-horses, that are kept with as much care as are any records of pedigrees for any other sort of farm stock."[15] Percherons were, by far, the favored breed. Around the peak of activity in 1920, there were 70,600 registered purebred Percherons but only 5,600 Shires, 5,000 Belgians, and 4,200 Clydesdales on America's farms.[16] These figures vastly understate the

[14] Thompson, *History of Livestock Raising*, 51, 63–4; Strohm, "Conestoga Horse," 175–80; Lush, *Animal Breeding Plans*, 34. The Appaloosa was developed by the Nez Perce from horses that escaped from the Spanish.

[15] "Growth of the Draft-Horse Interest," *Breeder's Gazette* (21 December 1882), 773.

[16] Mischka, *Percheron Horse in America*, 41–7; Plumb, *Types and Breeds of Farm Animals*, 98–146; Bogue, *From Prairie to Corn Belt*, 120–2.

influence of purebreds because many grade horses were bred up from the registered stock.

The regulation of horse breeding represents a government intervention into agriculture seldom addressed in the literature. Numerous states enacted laws to require registration of studs available for public service and to regulate the operation of the market for breeding services. Wisconsin pioneered this movement with a law effective 1 January 1906. Within a decade, twenty states, including virtually all of the Midwest, followed suit.[17] The state laws typically involved the licensing of breeding horses for public service, instituting inspection procedures to prevent the spread of communicable diseases, and standardized contract forms, especially with regard to liens on prospective colts. Given that a mare had only a moderate probability of conceiving when covered by a stallion, participants in breeding activity found it advantageous to condition payments on performance. The stallioner, on the other hand, desired guarantees that the owner of a mare which conceived and bore a colt would not hide the fact and claim that the mare had miscarried. The aim of the legislation was to eliminate "inferior" or scrub stallions from service. In 1916, of the nearly 56,000 stallions registered in the eighteen states reporting data, almost 60 percent were purebreds and only 8 percent were "mongrels" or crossbreeds.[18]

Federal government involvement in horse breeding never reached the level prevailing in Europe. For example, the French, Belgium, Austrian, and Prussian states had well-established national stud farms to improve their draft stock, and in Russia the Czar had a chief of the Royal Stud.[19] The U.S. federal government intervened in horse breeding relatively late, during the administration of Theodore Roosevelt. In 1906, the U.S. Department of Agriculture (USDA) established a breeding farm in Weybridge, Vermont, to provide foundation sires – stallions of the Morgan horse breed – for the U.S. Calvary.[20] The famous Morgan horse was originally bred by Justin Morgan in nearby Middlebury, Vermont. The Weybridge farm worked in conjunction with the Army's newly established (1908) Remount Service, which sought to improve the quality of

[17] U.S. Census Bureau. 14th Census 1920, v. 5, 534–8. With the exception of Oklahoma, no southern states participated in this movement. Glenn, "Stallion Legislation," 289–99.

[18] Glenn, "Stallion Legislation," 292–3.

[19] USDA, "Progress in Horse Breeding," 19–22; and Gay, *Productive Horse Husbandry*, 179–88. The French national stud farm, Les Haras nationaux, established by Colbert in 1665, exists to this day.

[20] USDA, *Yearbook 1912*, 43–4.

military horses. In 1921, the Army assumed full responsibility for the horse breeding program.[21]

The USDA engaged in another, more speculative, line of research during the Roosevelt–Taft years. The Division of Animal Husbandry of the U.S. Bureau of Animal Industry sought to replace the mule with a hybrid possessing "more quality and finish." The plan was to breed a number of Grevy's Zebra stallions (which Teddy Roosevelt received from the Emperor of Abyssina) to jennets to produce "Zebrules." The project was funded for nine years but not renewed when the Wilson administration assumed control in 1913.[22]

Mechanization in the Horse Epoch

In the first great period of farm mechanization (1840–1910) the adoption of new technologies was closely tied to the adoption of horses and mules. Indeed, the classroom definition of farm mechanization involved the adoption of machines that allowed farmers to substitute horsepower for human labor, thereby enabling workers to cultivate more acreage. To illustrate the relationship between machines and horses we matched a sample of 628 Illinois farmers, who had all purchased McCormick reapers, to the 1860 manuscript census of agriculture.[23] The reaper owners in our sample each possessed an average of 5.9 equines and 0.43 working oxen – twice the ratio of equines to bovines found in the same area of Illinois in the Bateman–Foust sample of the general population of farmers. Their sample included an unknown number of reaper owners, so the ratio would be more than twice as high if it was possible to contrast reaper owners with nonowners. The new technology was associated with the use of a faster draft mode – mechanization was proceeding hand-in-hand with complementary biological advances.

Table 12.3 carries the analysis forward by documenting changes in the national ratios of improved land per worker, horsepower per worker, and improved land per unit of horsepower over the second half of the nineteenth century. Improved acres per workers increased by about 75 percent between 1850 and 1900 whereas horsepower per worker increased by more than one-half. Notably, there was only a modest increase in the number of improved acres per unit of horsepower. We know of two

[21] Reese, "Breeding Horses for the United States Army," 341–56.

[22] Bradley, *Missouri Mule*, 13–14; USDA, *Yearbook 1909*, 61–2.

[23] Bateman and Foust, *Agricultural and Demographic Records*.

TABLE 12.3. *Trends in National Land, Power, and Labor Ratios, 1850–1900*

	Improved Acres per Worker	Horsepower per Worker	Improved Acres per Horsepower
1850	23.2	1.4	17.1
1860	26.2	1.5	17.0
1870	30.2	1.5	19.7
1880	34.8	1.6	21.6
1890	38.9	2.0	19.5
1900	40.4	2.1	19.4
Ratio:			
1900/1850	1.7	1.5	1.1

Notes: Horsepower sums the number of horses, mules, and working oxen of all ages. The labor force data are from Craig and Weiss and are the sum of male adults, 16 years and older; female adults, 16 years and older; and children. The use of slave data, which provide no age breakdown for 1850 and 1860, along with the desire to maintain consistency, necessitate the inclusion of the younger workers throughout.

Sources: U.S. Census Bureau. Census of Agriculture 1950, v. 2, 11, 361–2; Craig and Weiss, *Rural Agricultural Workforce*.

reasons, and can speculate as to a third, to explain why this ratio did not change more. First, although machine producers had an eye on saving labor they also improved designs to save on horsepower. Data from early reaper trials show that between 1852 and 1857 the average draft power needed to pull a machine fell by 37 percent.[24] Second, draft animals improved over time. Finally, it is likely that farmers became more adept at managing and feeding their animals, thereby increasing their animals' efficiency.

The Tractor Revolution

While mules and horses easily survived Roosevelt's Zebrule threat, they faced a far more serious challenger – the gasoline tractor. The gasoline tractors of the early 1900s were behemoths, patterned after the giant steam plows that preceded them. They were useful for plowing, harrowing, and belt work but not for cultivating fields of growing crops or for powering farm equipment in tow. Innovative advances between 1910 and

[24] Olmstead, "Mechanization of Reaping and Mowing," 349–50.

1940 vastly improved the machine's versatility and reduced its size, making it suited to a wider range of farms and tasks. At the same time, largely as a result of progress in the mass production industries, the tractor's operating performance greatly increased while its price fell.

Several key advances marked the otherwise gradual improvement in tractor design. The Bull (1913) was the first small and agile tractor, Henry Ford's popular Fordson (1917) was the first mass-produced entry, and the revolutionary McCormick–Deering Farmall (1924) was the first general-purpose tractor capable of cultivating growing row crops. The Farmall was one of the first to incorporate a power takeoff, which enabled it to transfer power directly to implements under tow. Allied innovations such as improved air and oil filters, stronger implements, pneumatic tires, and the Ferguson hydraulic three-point hitch system greatly increased the tractor's lifespan and usefulness. Seemingly small changes often yielded enormous returns in terms of cost, durability, and performance. As an example, rubber tires reduced vibrations, thereby extending machine life, enhanced the tractor's usefulness in hauling (a task previously done by horses), and increased drawbar efficiency in some applications by as much as 50 percent. The greater mobility afforded by rubber tires also allowed farmers to use a tractor on widely separated fields.[25]

We know of no attempt to formally measure the actual productivity effects of these improvements in tractor design, but knowledgeable observers placed great stock in their importance. As an example, Roy Bainer, an early adopter of the tractor in Kansas, noted that without improved air filters his machines lost power and needed a valve job within a year of entering service. Bainer went on to become a dean of the agricultural engineering profession. He and other agricultural engineers emphasized that the Ferguson hitch revolutionized the tractor's capabilities and efficiency in towing implements. The extraordinarily rapid diffusion of some of the changes lends credence to Bainer's view. The standard wheeled models dropped from about 92 percent of all tractors sold for domestic use in 1925 to about 4 percent in 1940. Conversely, general-purpose tractors, first introduced in 1924, made up 38 percent of sales in 1935 and 85 percent by 1940. In a similar fashion, within six years of Allis–Chalmers' 1932 introduction of pneumatic tires on its new

[25] Developments since World War II were largely limited to refining existing designs, increasing tractor size, and adding driver amenities. Fox, "Demand for Farm Tractors," 33. For the general evolution of the tractor, see Gray, "Development of the Agricultural Tractor"; Williams, *Fordson, Farmall, and Poppin' Johnny*, 85–128.

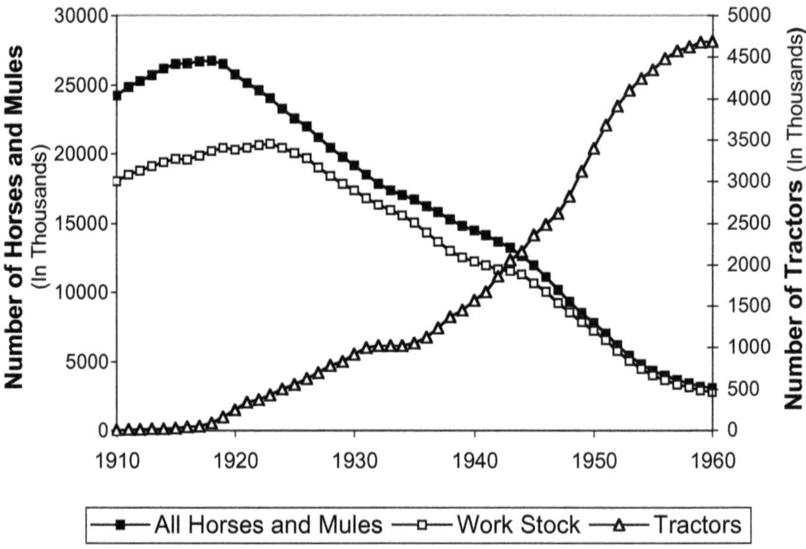

FIGURE 12.3. Number of draft animals and tractors in the United States, 1910–60. *Sources:* Brodell and Jennings, "Work Performed and Feed Utilized," 10; USDA, *Agricultural Statistics 1962*, 432; and Carter et al., *Historical Statistics*, Table Da623.

models, 95 percent of new tractors produced in the United States were "on rubber." By 1945 about 72 percent of all wheeled farm tractors had rubber tires.[26]

A summary picture of the replacement of the animal mode by the mechanical mode is offered in Figure 12.3, which charts the number of tractors and draft animals in the United States between 1910 and 1960. The total number of farm horses and mules reached its peak of 26.7 million head in 1918, and the number of workstock (animals age three and older) crested in 1923 at 20.7 million head, roughly the level that would have been required over the period 1920–60 to maintain the 1910 ratio of workstock to cropland harvested. After 1925 draft animal numbers steadily declined, falling below 3 million by 1960.[27] The stock of tractors expanded rapidly during World War I, rising to about one million

[26] McKibben and Griffin, *Changes in Farm Power and Equipment*, 13; U.S. Census Bureau. Census of Agriculture 1945, *Special Report*, 71; Roy Bainer, interviewed by authors, Davis, California, 1982–84.

[27] The population of horses and mules off farms began to decline well before the population on farms. There were about 3.45 million horses off the farm in 1910. This number fell to an estimated 2.13 million by 1920 and to 380,000 by 1925, according to C. L. Harlan

machines in 1929. A second growth spurt began in the late 1930s, leading the tractor stock to climb above 4.5 million units by 1960.

The diffusion of the tractor exhibited significant regional variation as indicated in Table 12.4, which presents the proportion of farms reporting tractors and draft animals. The Pacific and West North Central regions led the way with roughly 8 percent of farms in 1920 reporting tractors. The development of the general-purpose tractor in the mid-1920s quickened the pace of diffusion in the East North Central region and, to a lesser extent, in the three southern regions. All regions experienced a slowing of diffusion during the Great Depression and an acceleration during and immediately after World War II. The postwar spread of tractors was especially rapid in the South.[28]

The adoption and diffusion of the machine mode ultimately represented far more than the mere replacement of the animal mode. Rather it signaled a dramatic increase in horsepower capacity on the farm. Between 1910 and 1960 draft power on farms soared by over four and one-half times whereas cropland harvested remained roughly constant. Our estimate of the relative horsepower capacity (measured in terms of draw-bar power) supplied by tractors and workstock indicates that in 1920 tractors accounted for about 11 percent of national farm horsepower capacity, 40 percent in 1930, 64 percent in 1940, 88 percent in 1950, and 97 percent in 1960. Compared with the traditional measure of diffusion which relies on the percentage of farms reporting tractors, our alternative measure using power capacity shows that diffusion started earlier and grew faster and more smoothly.[29]

Scholars have long debated why both machine and animal modes of production coexisted for almost a half-century.[30] Researchers failed to recognize that many key variables in the tractor adoption decision were endogenous to the agricultural sector taken as a whole. The old technique was embodied in the horse, a durable capital good with an inelastic short-run supply and a price that could adjust to keep the animal mode competitive. Figure 12.4 graphs the movements in real prices (discounted by the gross national product [GNP] deflator) of medium-size tractors, mature horses (age two and older), and hay over the period 1910–60.

of the USDA Bureau of Agricultural Economics, as cited in Dewhurst, *Americas Needs and Resources*, 1103, 1108.

28 Brodell and Ewing, "Use of Tractor Power," 5–11; Carter et al., *Historical Statistics*, Tables Da661–666, Da983–986.

29 Sargen, *Tractorization*, 32–41, esp. 38.

30 See Olmstead and Rhode, "Reshaping the Landscape."

TABLE 12.4. *Percentage of Farms Reporting Tractors, Horses, or Mules, 1900–1969*

	U.S.	NENG	MALT	ENC	WNC	SALT	ESC	WSC	MTN	PAC
					Tractors					
1920	3.6	1.4	3.1	5.1	8.4	0.9	0.5	1.8	6.5	7.5
1925	7.4	4.6	11.3	13.5	13.7	2.6	1.3	2.9	7.7	13.4
1930	13.5	10.5	21.5	24.7	26.5	4.2	2.1	5.7	17.9	20.8
1935	nd	nd	nd	nd	nd	nd	nd	nd	nd	nd
1940	23.1	19.2	33.0	39.9	44.7	5.6	3.6	14.9	28.3	27.6
1945	34.2	31.1	49.7	56.4	61.2	11.1	7.7	25.3	44.3	38.4
1950	46.9	44.9	63.6	69.4	71.8	23.2	18.8	38.4	64.8	56.8
1954	60.1	60.2	77.4	82.6	83.3	38.6	31.8	50.0	74.6	67.3
1959	72.3	74.2	86.4	88.8	88.0	55.8	47.2	62.7	82.1	76.8
1964	76.6	74.8	88.1	90.3	90.4	63.2	56.3	67.4	83.4	73.8
1969	80.8	81.0	89.4	87.6	88.0	74.0	70.7	75.1	82.6	76.0
					Horses or Mules					
1925	84.2	74.0	81.6	87.8	92.2	79.9	81.2	85.9	89.2	65.5
1930	79.9	66.3	74.6	84.3	89.0	77.7	73.8	83.5	81.9	55.4
1935	73.4	54.3	70.4	77.9	83.8	68.0	70.1	76.1	78.1	48.9
1940	71.5	46.6	66.5	74.4	79.3	71.6	69.1	74.8	75.5	43.5
1945	63.8	37.2	56.0	61.8	77.1	65.3	64.2	63.5	69.2	35.5
1950	54.0	35.4	40.9	39.6	57.8	62.8	64.5	57.5	59.3	27.4
1954	37.6	26.0	23.9	20.9	32.8	49.3	50.8	42.9	49.6	21.8
1959	30.6	22.2	20.2	16.4	23.1	40.6	44.6	37.6	48.6	22.7
1964	nd	nd	nd	nd	nd	nd	nd	nd	nd	nd
1969	20.0	20.1	20.4	15.6	18.6	16.0	19.8	25.0	40.8	22.5
					Horses					
1900	79.0	84.8	88.6	91.6	93.1	55.5	62.6	78.7	91.8	87.3
1910	73.8	79.3	85.3	90.7	93.0	48.2	52.8	72.8	87.2	84.2
1920	73.0	81.4	87.1	91.2	93.6	46.1	53.6	69.1	90.8	78.2
1935	51.9	54.0	67.6	74.8	79.6	23.2	25.6	46.6	76.9	45.5
1940	51.6	46.4	63.4	71.3	75.4	24.7	29.8	50.0	74.6	40.6
1945	48.3	37.4	53.6	60.9	74.5	25.3	31.4	49.0	69.5	33.2
1950	39.4	35.1	39.4	38.1	55.8	26.4	31.6	45.5	58.7	26.5
					Mules					
1900	25.8	0.3	4.0	7.9	18.5	36.1	48.5	49.0	7.2	8.5
1910	29.4	0.5	4.6	9.1	21.1	41.2	50.4	52.5	8.9	8.1
1920	35.0	0.8	6.8	10.9	24.9	52.1	55.1	60.5	12.6	9.2
1935	33.1	0.5	7.1	9.8	17.4	50.4	57.8	55.8	8.8	7.0
1940	30.3	0.4	6.8	7.6	13.4	53.1	54.6	48.6	6.3	5.3
1945	25.4	0.3	4.7	5.2	10.2	47.0	50.3	36.7	4.4	3.2
1950	20.5	0.5	2.6	2.6	4.6	41.6	46.3	24.8	2.9	2.0

Sources: U.S. Census Bureau. Census of Agriculture 1950, v. 2, 382–91, and Census of Agriculture 1959, 214–17, 506–9, and Census of Agriculture 1969, 2:23, 4:14, 5:32.

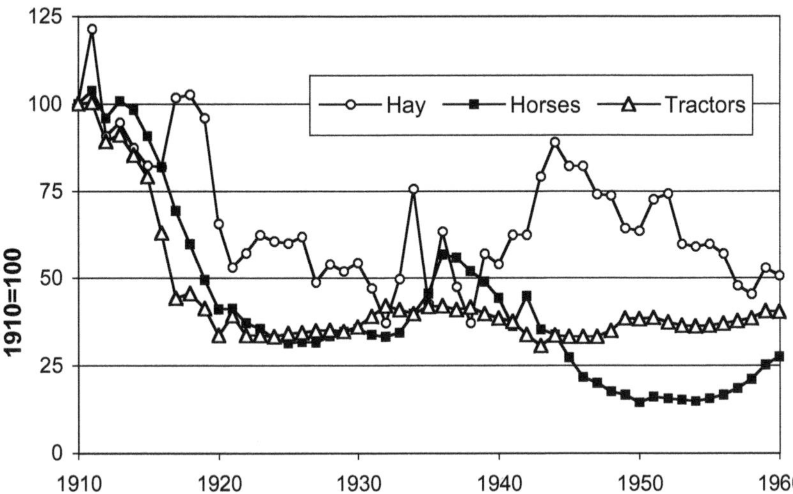

FIGURE 12.4. Real prices of hay, horses, and tractors. *Note:* All prices are deflated by a GNP deflator derived from Balke and Gordon, "Estimation," 84–5, before 1929, and the official series (U.S. Census Bureau, *Historical Statistics*, Table F5) thereafter. *Sources:* Prices for horses two years and older are from USDA, *Crops and Markets* 12, no. 2 (1935), 34, up to 1935, and spliced to Carter et al., *Historical Statistics*, Table Da984, thereafter. Hay prices are from Carter et al., *Historical Statistics*, Table Da735. The tractor price series is based on "medium tractors" from U.S. Bureau of Agricultural Economics, *Income Parity for Agriculture*, 50, up to 1939, and spliced to the series for 20–29-belt horsepower tractors from USDA, *Agricultural Prices 1950*, 34; USDA, *Agricultural Statistics 1956*, 468, and *1962*, 563.

Real tractor prices fell dramatically over the 1910s and early 1920s, but what is notable is that the fall in horse prices nearly kept pace. The real value of mature horses in 1925 was less than one-half that of their 1915 value. As well as lowering horse prices through direct competition, advances in mechanical technology and the diffusion of the tractor increased productivity and shifted out the supply of agricultural products. Given a downward-sloping demand function for agricultural products, this shift led to lower feed prices, thereby reducing the major expense of the animal mode. Thus, via its effects on both horse and feed prices, the tractor had the paradoxical effect of making its major rival more competitive in the short run. Many scholars have noted that cost differentials between the machine and animal modes were small. But rather than reflecting long-run conditions, the small difference in costs merely represented a transitional phase in the adjustment process. As in the debate over the profitability and viability of slavery, the real question answered by such static microeconomic comparisons is whether the market for durable capital goods (here, the horse) was working – it was.

To address the viability of the horse mode, we modeled the diffusion process as a capital replacement problem. The key is to understand how horse breeders responded to changing economic conditions. As we have shown, there was a well-functioning and sophisticated market for the provision of draft power well before the tractor was introduced, and horse breeders responded to market signals in a predictable fashion. For example, a decline in horse prices in the mid-1890s resulted in a marked reduction in breeding activity. Although there were substantial and persistent differences in the regional prices of draft animals (with prices rising as the distance from Missouri increased), prices in every region tended to move together over time. Finally, it is important to understand that the draft animal mode of power was not technologically stagnant. As a result of selective breeding, the average weight of work horses on American farms rose from 1,203 pounds per animal in 1917 to 1,340 pounds in 1943, increasing the pulling power of the horse.[31] Detailed estimates for 1912 of the cost of raising horses from birth to age three reveal that such costs were closely related to the price of mature horses and generally declined from east to west.[32] The cross-sectional data on rearing costs combined with information on input prices allowed us to construct two time-series estimates for the cost of rearing horses.[33] The key variable of interest is the ratio of the price of mature horses to the cost of rearing colts, which we will call Dobbin's q. Conceptually, it is analogous to "Tobin's q," which is defined as the ratio of an asset's market price to its reproduction costs. As is well known in the finance and investment literatures, if "Tobin's q" exceeds unity, then it is advantageous to invest in new capital equipment. A ratio of less than unity discourages new investment because it is cheaper for a firm to acquire additional capital through the purchase of existing assets.[34]

[31] Brodell and Jennings, "Work Performed and Feed Utilized," 18–19; Williams and Jackson, "Improving Horses and Mules," 929–46.

[32] USDA, "Cost of Raising Horses," 28.

[33] The first series uses a cross-sectional regression of rearing costs on a set of input prices to derive the horse-rearing cost function. The second is formed as a Laspeyres index from the average weights of the key inputs (grain, hay, labor, land, and interest rates) in overall costs and the relevant price changes over time.

[34] For an entry point into the investment literature, see Abel, "Consumption and Investment," 763–7.

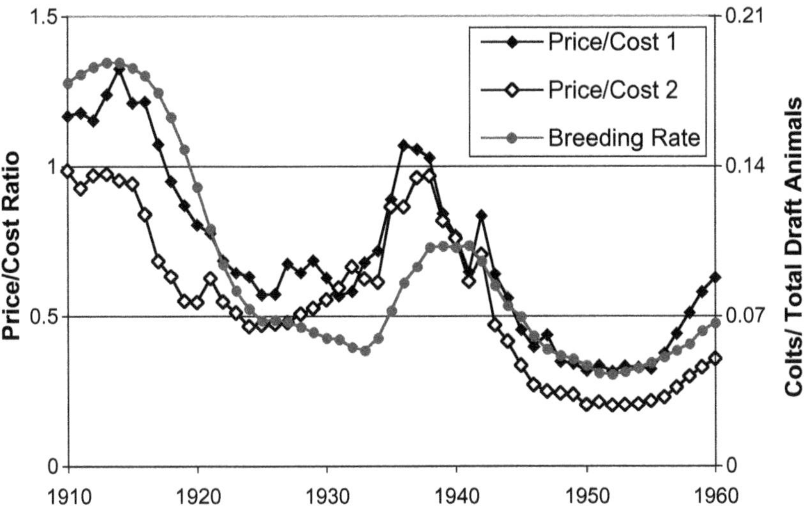

FIGURE 12.5. Dobbin's q and breeding rates. *Source:* Olmstead and Rhode, "Reshaping the Landscape," 686.

The trends in the price–reproduction cost ratio and the extent of horse and mule breeding, as measured by the ratio of colts under one year of age to total animal stocks, are displayed in Figure 12.5. The correspondence is striking. The breeding rate and price–cost ratios all moved sharply down over the early 1920s, recovered slightly in the mid-1930s, and then fell again after the late 1930s. The investment decisions of individual breeders helped ensure the ultimate victory of the tractor mode of production.

Another sign of the quick adjustment by horse breeders to the changing demand for animal power was the rapid reduction in the number of stallions and jacks licensed under the state registration laws discussed earlier. Figure 12.6 graphs the number of registered studs in six important breeding states – California, Illinois, Indiana, Iowa, Kansas, and Wisconsin – between the early 1910s and early 1940s.[35] Sporadic data available for Kentucky and Missouri paint a largely similar picture. Figure 12.6 reveals that by the mid-1920s the numbers of registered stallions and jacks were

[35] California Department of Agriculture, *Monthly Bulletin*, 725–7; Purdue University Agricultural Experiment Station, "Indiana Stallion Enrollment"; Illinois Department of Agriculture, *17th Annual Report*, 25, and *29th Annual Report*, 167; Iowa Department of Agriculture, *Year Book 1941*, 185; Kansas State Board of Agriculture, *Stallion Registry 1939*, 6; Wisconsin Department of Agriculture and Markets, *Biennial Report 1935–36*, 194–5, and *Biennial Report 1939–40*, 9; and U.S. Bureau of Agricultural Economics, "Horses, Mules, and Motor Vehicles," 9–10.

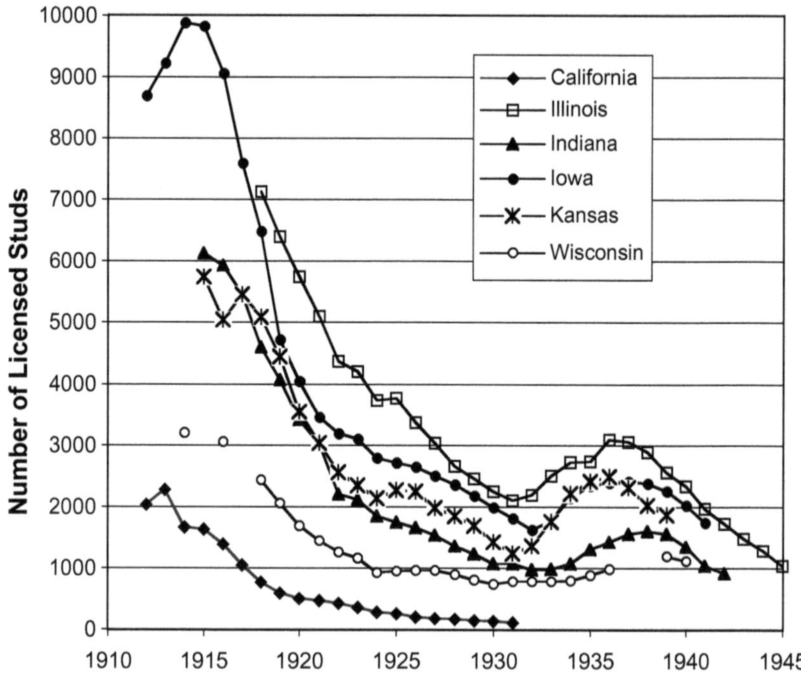

FIGURE 12.6. Stallion and jack registration in selected states. *Source:* Olmstead and Rhode, "Reshaping the Landscape," 688.

typically less than one-third of the 1915 levels. Given that registration fees were nominal in most states, the rapid and steady decline in licensing after World War I undoubtedly reflected the withdrawal of many breeders from the business. The small upsurge in licensing in the 1930s indicates that the trend was reversible when economic conditions changed.

The decline in breeding was sufficiently rapid that it raised alarms. By the mid-1920s, USDA authorities began to publish concerns about impending problems in articles such as, "Horse Production Falling Fast in US" and "Shortage of Work Animals in Sight." The 1931 USDA *Yearbook* followed suit. Under the headline "Horses and Mules Now Raised Are Much Fewer than the Replacement Needs," the *Yearbook* noted that breeders were annually producing only one-half of the one million horse colts and 300,000 mule colts needed to sustain the animal population.[36] Clearly the people whose livelihood depended on horse and

[36] USDA, *Yearbook 1926*, 437–9, *Yearbook 1931*, 328–31, and "Shortage of Work Animals in Sight," 108–9.

mule breeding saw the handwriting on the wall – the horseless age was near.

Opposition to the End of the Horse Epoch

Not every interested party was willing to accept the end of the horse era without a fight. The most important group opposing change was the Horse Association of America (HAA), formed in November 1919. The HAA championed the cause of livestock dealers, saddle manufacturers, farmers, breeders, and other business interests that had a financial or emotional interest in horses and mules. From a modest start the association rapidly emerged as a vigorous lobbying organization. By the end of 1920 it had acquired an extensive dues-paying membership and had raised about one million dollars "to aid and encourage the breeding, raising, and use of horses and mules."[37]

The HAA worked hard to get its message out. Over the period 1922–24, it distributed more than three million copies of its publications. Between 1920 and 1935, executive secretary Wayne Dinsmore traveled over 200,000 miles on behalf of the association.[38] The HAA initiated an extensive direct-mail campaign, which by 1923 was annually sending out more than 250,000 personally addressed letters. The association also disseminated its messages via radio and film. Its movie, "Horse Power in Action," was seen by 60,000 viewers in the years 1936–37.[39] The HAA sought to convince bankers, farmers, and breeders that the horseless age was not at hand. The association provided advice about how to better feed and care for horses to enhance their competitiveness. It also promoted the use of larger horses and big hitches that would allow farmers to harness teams two to three times more powerful than they had previously used. Figure 12.7 shows a 21 horse team at work in an Oregon wheat field.

The HAA promoted the advantages of homegrown power to the farm community. The HAA's appeals emphasized that reliance on horses kept money within the community whereas the use of tractors required an outflow of the cash required to purchase and operate the equipment. According to the HAA, farmers who adopted tractors were more exposed

37 A fuller treatment appears in Olmstead and Rhode, "Agricultural Mechanization Controversy," 35–53.

38 *Journal of the American Veterinary Medical Association* (October 1935), 408–9.

39 Horse Association of America [HAA], *Leaflet*, nos. 78, 105, 132; *Journal of the American Veterinary Medical Association* (October 1928), 728–31, and (October 1938), 406–8.

FIGURE 12.7. A combine harvester powered by 21 horses in The Dalles, Oregon, 1939. Western farmers, unlike their eastern brethren, learned to manage such large teams. *Source:* Photograph by Al Monner, negative 4282, courtesy of HistoricPhotoArchive.com.

to the vagaries of the market and thus were the worst credit risk.[40] More importantly, HAA's propagandists viewed the decline in horse use as the root cause of the farm crisis of the 1920s and 1930s. As the depression spread from the farm sector to engulf the entire nation, the HAA argued that "every man, woman, and child in America has been affected by the decrease in horses and mules." The connection between the decline in animals and the Depression is "direct and inescapable."[41]

The Tractor as a Land-Saving Innovation

Paradoxically, from the perspective of the induced innovation model, the spread of the tractor was a major land-saving innovation because it

[40] HAA, *Leaflet*, nos. 91, 106.

[41] HAA, *Leaflet*, no. 211. The HAA also argued that the adoption of motor vehicles in the nonfarm sector created havoc for farmers. We estimate that the spread of nonfarm motor vehicles freed an additional 54 million acres for the production of food and fiber for human use.

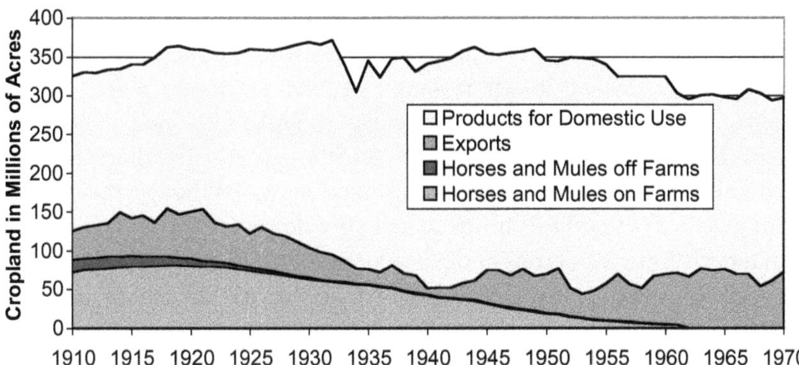

FIGURE 12.8. U.S. cropland harvested by use, 1910–70. *Source:* Carter et al., *Historical Statistics*, Tables Da661–666; USDA, *Agricultural Statistics 1962*, 537.

liberated the pasture and cropland previously used to feed horses and mules and increased the timeliness of farm work.[42] Given the crop yields prevailing over the late nineteenth and early twentieth centuries, a mature farm horse required about three acres of cropland for feed each year. In aggregate over the period 1880–1920, farm draft animals consumed the output of roughly 22 percent of all cropland harvested and draft animals in cities and mines consumed another 5 percent. Figure 12.8 charts the acreage of cropland required to feed America's horses and mules as well as the acreage used for export and other domestic purposes between 1910 and 1970.[43] The total cropland harvested was a little higher at the beginning of the period than it was at the end. Cropland used to feed horses and mules peaked in 1915 at about 93 million acres – 79 million for maintaining work animals on farms and 14 million for those off farms – and declined steadily thereafter. By 1930, the acres of cropland devoted to animal feed had declined to 65 million, with all but 2 million acres devoted to farm stock. Three decades later, only 5 million acres were needed. During the farm depression of the 1920s, the acreage released from feeding horses and mules exceeded the reduction of acreage devoted to the export markets.

[42] See Johnson, "Review of Vaclav Smil." For the important contribution of the tractor to the timeliness of operations in the upper Great Plains and corn belt, see, respectively, Sargen, *Tractorization*; and Grotewold, *Regional Changes*, 21–2.

[43] USDA, *Agricultural Statistics 1962*, 537. The data do not include pasture requirements. For a detailed breakdown of the effect on specific crops and on the farm surplus, see U.S. Census Bureau. 15th Census 1930, *Agriculture. The Farm Horse*, 49–77.

Linking the farm crisis to the tractor takes the form of the classic immiserating growth hypothesis: if demand is inelastic, productivity advances can lead to a decline in net revenue, making producers worse off as a group. Willard Cochrane's "Agricultural Treadmill" and the recent debate over biological technologies – such as genetically altered, high-productivity dairy cattle – support this hypothesis.[44] Whether the immiserating growth hypothesis holds in actuality depends on the elasticity of demand for American crops in the international markets. There is no consensus on this issue. However, it is worth noting that the federal acreage reduction schemes of the 1930s were premised on the assumption that demand was inelastic. The quantity of land removed from production by these programs approximately offset the acreage freed by the conversion to tractors and motor vehicles. For example, in 1934 the wheat, cotton, and corn programs contracted to remove 36 million acres from production. This was equal to the actual decline of cropland that had been devoted to feed horses and mules since the peak in 1915.[45]

Conclusion

Improved plows, wagons, reapers, threshers, and myriad other farm implements revolutionized farm life. Throughout the nineteenth and twentieth centuries the United States was a world leader in the invention and adoption of farm machinery. But none of these implements functioned without a complementary power supply, and, until the ascendancy of the tractor after World War I, animals supplied the vast majority of that power. Thus, mechanization went hand-in-hand with adding biological inputs to American farms.

Although draft power was deficient in the early colonial period, by the time of the Revolution, American farms already were noted for their intensive use of draft power. Land abundance lowered the price of feed, which in turn lowered the cost of draft power. Farmers didn't simply add more animals, they devoted great effort to continually modify their power supply to meet changing needs. New breeds of cattle, such as Devons, gained popularity in New England in large part due to their

[44] Cochrane, *Farm Prices, Myths and Realities*, 85–107; also see King, "Gasoline Engine," 64–78; and Bruce Ingersoll, "U.S. Dairy Board, Drug Makers Accused of Illegally Promoting Bovine Hormone," *Wall Street Journal* (30 November 1990), B4B.

[45] USDA, *Yearbook 1936*, 39, 44, 702; Carter et al., *Historical Statistics*, Tables Da661–666.

value as draft animals. As farming areas became more established and as new types of machinery appeared, such as reapers and mowers, there was a wholesale shift in the northern states from oxen to equines; in the South there was a significant movement to mules. These shifts represented enormous changes for farmers.

Equine breeding became a highly specialized activity, as evidenced by the relatively small number of stallions and the extensive markets for stud services. Moreover, and further reflecting the division of labor and regional specialization, breeding was increasingly concentrated in a few border states. Given that, in principle, anyone could breed horses and mules, this specialization signaled an increase in efficiency. The importation of superior draft animals from Europe, the emergence of purebred associations, and the diffusion of improved blood in the general equine population increased the quality of the stock. The change in the quality of equines more or less mirrored the changes that were transforming other farm animals. Between 1917 and 1943 the average size of draft horses increased by about 10 percent. This movement to larger and more powerful horses started much earlier, in the mid-nineteenth century. In a similar fashion, breeders perfected different types of mules for different markets. Clearly, the nineteenth-century mechanical revolution also required a biological revolution.

One of the hallmarks of the American experience over the nineteenth century was the westward march of commercial agriculture. This process symbolically ended in 1890 with the closing of the frontier. As newly settled areas matured, the acres of cropland harvested continued to increase until topping out in 1920. However, the tractor, by allowing farmers to convert roughly 180 million acres from draft animal feed to the production of food and fiber for human consumption, essentially continued the process of land augmentation for another four decades. The released cropland was roughly equal to two-thirds of the total cropland harvested in 1920 in the territory of the Louisiana Purchase.[46]

[46] USDA, *Agricultural Statistics 1962*, 537. In addition to increasing the effective stock of land, the tractor was also a land-augmenting innovation via its effect on increasing yields. With a tractor, farmers could accomplish their tasks in a more timely fashion and increase the extent of double cropping.

13

Tying It Together

We began this book with the story of the potato blight of the 1840s. As an indicator of the advances achieved over the following forty years, if the fungus had first appeared in the late 1880s it *could have been* held in check with Bordeaux Mixture – copper sulfate mixed with lime and water, developed by French botanist Alexis Millardet in 1883.[1] By 1885 word of Millardet's research was reaching scientists in Europe and North America – the world had its first effective fungicide. "Bordeaux Mixture was instantly welcomed by the wine growers of France and within twelve months was being eagerly adopted for the control of potato blight in North America thanks to Beverly Galloway who not only recognized its importance but made great efforts to ensure that the farming community did too."[2] In 1886, on the heels of Millardet's announcement, U.S. Commissioner of Agriculture Norman J. Colman distributed a circular on Bordeaux Mixture and other possible means of controlling downy mildew in vines. The USDA immediately began extensive experiments

[1] In October 1882, Millardet noticed that vines along a road in Medoc still had leaves, while those off the road had lost their leaves to downy mildew. He learned that farmers had sprayed the healthy plants with a mixture of copper sulfate and lime in the hope that the unappealing appearance of the grapes would deter theft. This led Millardet to conduct a series of experiments and to announce his discovery of Bordeaux Mixture in 1883. His better known writings of 1885 are usually taken as the point when the information became generally available. Many others discovered the power of copper sulfate and lime to retard fungal spores at about the same time. Johnson, "Early History of Copper Fungicides," 67–79; Millardet, *Discovery of Bordeaux Mixture*.

[2] Ayres, "Alexis Millardet," 23–6. In 1886 Galloway was a plant pathologist at the University of Missouri. He joined the USDA in 1887 and later became the founding Chief of the Bureau of Plant Industry.

with copper sprays and in 1887 began cooperative demonstrations with leading grape farmers. By 1890 more than five thousand grape growers were participating. Similar demonstration work was under way with potato farmers. In 1893 the USDA noted that more attention had been given to the practical application of Bordeaux Mixture in the United States than in any other nation, including France. By this time American scientists had demonstrated the compound's efficacy in treating seventeen plant diseases. Aiding the diffusion process were rapid improvements in spraying and dusting technologies. USDA scientists and the emerging agricultural chemical industry ensured that the miraculous fungicide rapidly became a standard treatment on potatoes, tomatoes, vines, and fruit trees. However, this rapid dispersion and acceptance did not take place in labor-abundant Ireland. Even preliminary experiments with Bordeaux Mixture in Ireland did not begin until 1891, and in 1894 the potato crop failed again. Diffusion was delayed due to flawed experiments, adulterated chemicals, and farmer indifference. Not until World War I was Bordeaux Mixture in general use in Ireland.[3] In the diffusion of this technology, as in so many other cases, income, government outreach programs, education, and linkages with industry appear to have swamped whatever effect factor scarcity may have had.

American farmers and their allies in universities, government, and industry were keenly interested in biological innovation, invested heavily in learning and applying biological technologies, and reaped considerable returns from those investments. Given the rapidly changing pest environment, to have done otherwise would have been disastrous. Table 13.1 offers an indication of the dynamic effects of biological innovation on increasing yields relative to a world without learning, circa 1940. The table lists the major crops, a few of the most destructive insects and diseases, and the estimated losses absent biological learning. We emphasize two points. First, the table, like our discussions throughout this book, only touches on a small fraction of the more troublesome insects and diseases. In 1889 C. V. Riley estimated that there were about twenty-five thousand *native* injurious insect species, of which more than seven thousand did enough damage to rank as "pests." There were more than one thousand that preyed on fruit trees, and over two hundred attacked cereals and grasses. In addition to native species, farmers had to contend

[3] Peterson and Campbell, "Beverly T. Galloway," 29–43; Large, *Advance of Fungi*, 40–3, 237–9, 261–71; Connell, "Potato in Ireland," 67; Galloway, "Plant Pathology," 54–6, and "Report of the Chief," 263–5.

TABLE 13.1. *Agriculture Without Biological Innovation to Combat Pests*

Crop	Insects	Diseases	Without Bio	With Bio
Wheat	Hessian fly, midge, chinch bug	Rusts, bunt	D	L–M
Corn	European corn borer		M	L
Cotton	Bollworm, cotton worm, boll weevil	Rots, seeding diseases, wilts	D	L–M
Vines	Phylloxera	Mildews, Pierce's disease	E	L
Oranges	Cottony cushion scale		D	L
Various Fruit	San Jose scale		D	L
Tobacco	Hornworms	Granville wilt, black shank, seed diseases	G	M–S

Note: E = commercial extinction (over 80% losses); D = disastrous (50–80% losses); G = grievous (30–50% losses); S = severe (20–30% losses); M = moderate (10–20% losses); L = low (less than 10% losses).

with introduced species, many far more injurious in the United States than in their home countries where birds, insects, and diseases kept them in check. Second, we are assuming a counterfactual world without biological learning, not the real world where actual losses were tempered.[4] In the absence of learning, events such as the Irish potato famine, the phylloxera epidemic, and the initial devastation caused by the Hessian fly, wheat midge, boll weevil, San Jose scale, cottony cushion scale, and so forth would have been far more common. The wheat specialists we consulted agreed that, if no precautions had been taken against the Hessian fly, farmers would not have had a wheat crop worth harvesting. Farmers growing many other crops would have met the same fate if they hadn't adapted to the ever-evolving threats of insects and diseases. Also, without quarantines, even more exotic pests and diseases would have attacked American crops and livestock.[5]

The estimated losses in the table are based on the testimony of plant scientists, field trials, and natural experiments. To give perspective to our estimates, around 1980 scientists estimated that, without pesticides, crop

4 Riley, "Injurious and Beneficial Insects," 603.

5 For a list of seventy-five foreign pests and diseases that the quarantine system helped keep out of the United States, see U.S. House Committee on Appropriations, *Agricultural Department Appropriation Bill 1966*, 127–30.

and livestock production in the United States would fall by 30 percent, and in the United Kingdom potato and cereal yields would plummet by more than 40 percent. If farmers were forced to rely on means of defense consistent with the state of biological knowledge at some arbitrary earlier date, such as 1840, the losses would have been correspondingly larger.[6] The table omits the effect of a worsening weed environment. Adding Johnsongrass, Bermuda grass, Russian thistle, and other foreign introductions to the table would further dampen farm prospects.

The table focuses on land yields, but as we have noted output per worker hour also would have suffered if crops were ravished. More farmers would have plowed, planted, and cultivated, only to see their crops destroyed. Picking a field with one-quarter bale of seed cotton demanded much more than one-fourth the effort of harvesting a field with one bale. Also recall that many of the investments that limited damage were very labor-intensive – hoeing weeds, hand picking insects, replanting vines, tearing out barberry bushes, repeatedly spraying and dusting, hand scrubbing trees with caustic solutions, and sterilizing seedbeds come to mind. The final column offers a crude estimate of actual losses in the first decades of the twentieth century, given biological innovations. The table does not consider the extra losses that would have occurred due to winterkill, drought, and heat without varieties of plants introduced to survive in more varied conditions.

The vintage of crop varieties grown in the 1910s and 1920s in the United States also reveals the pervasiveness of biological innovation. Edward Montgomery noted that three-quarters of the maize varieties available in 1913 had been developed since 1840. The USDA's first systematic survey of the distribution of wheat, conducted in 1919, showed that only 8 percent of acreage was planted in varieties in existence before 1840. Tobacco witnessed similar changes in its major types – varieties grown using specific cultural practices and processed using specific curing techniques. In 1929, only 16 percent of tobacco acreage was planted in types that were in standard use in 1840. Cotton varieties also experienced enormous turnover, genetic recombination, and, in many cases, extinction due to the onslaught of the boll weevil. If any cotton land

[6] Braunholtz, "Crop Protection," 22. Throughout this book we referenced a number of historical and modern sources to support our dire view of a world without biological learning. For current estimates of losses to insects and diseases absent control measures see Palm, "Estimated Crop Losses," 139–57; Smalley, "Estimated Losses Without Pesticides," 169–71; Schwartz and Klassen, "Estimate of Losses Caused by Insects," 15–77; James, "Estimated Losses," 79–84; Drummond et al., "Estimated Losses," 111–27.

in the 1910s and 1920s was planted with pre-1840 "varieties," it was only after decades of growing newer higher yield or easier-to-pick varieties that were abandoned after falling prey to the weevil. Among forage crops, many grasses and clovers, soybeans, alfalfa, and corn for ensilage were either totally new or in an experimental stage in 1840. The new varieties offered substantial value added, as with the differences between soft and hard wheats, and were often a necessary condition for the geographical spread of agriculture. The changes in animal breeds paralleled the changes in crops.

As a further indication of the importance of biological innovation in crop production, Table 13.2 offers a summary of the impact of foreign crop introductions on reshaping American agriculture. With the exception of corn, all major crops and most specialized crops benefited from the repeated infusion of new genetic material. Seeds reached U.S. shores by many means. Immigrants carried the finest seeds and cuttings from their native lands. The Mennonites who brought Turkey wheat to Kansas carefully selected their best seeds and one entrepreneur, Bernhard Warkentin, reportedly imported 25,000 bushels of seed from southern Russia. Throughout American history, leading farmers imported seeds and cuttings in a quest for superior crops. The search was often focused on solving a particular problem – finding a grass that would thrive in the South, or a variety resistant to a disease. Most experimenters freely shared their new products with their neighbors, sometimes with disastrous consequences – recall the introduction of Johnsongrass. State agricultural societies and the federal government were also in on the quest. Naval captains and American consuls regularly returned home with promising genetic materials, and the Commissioner of Patents mailed seeds to farmers across the country. In the late nineteenth and early twentieth centuries, the USDA's famous seed hunters imported thousands of varieties and scoured the world in search of predators to control pests. The importation of new materials was the first step. Plant breeders, be they farmers or trained scientists, carefully selected and crossed varieties to produce the plants that revolutionized crop production.

Animals also underwent enormous changes due to foreign introductions and domestic breeding. Table 13.3 provides a glimpse of a few of the more prominent changes. In every case the animals of 1940 bore little resemblance to those of 1800. The new specialized breeds gained weight faster and provided more and better services to their owners than the animals they replaced. Milk output per cow and sheep wool yields soared. The age at which hogs, steers, and sheep were slaughtered declined. By

TABLE 13.2. *Importance of Foreign Plant Introductions*

Wheat	In 1919, 80 percent of U.S. wheat acreage consisted of varieties that had not existed in North America before 1873, and less than 8 percent was planted in varieties dating earlier than 1840. New varieties were crucial in fighting pests and diseases and in allowing wheat cultivation to succeed in a variety of geoclimatic regions.
Corn	The genetic material used to remake American corn was present in North America at the time of European settlement.
Cotton	A succession of introductions from Mexico and Central America following 1806 revolutionized production. Introductions from North Africa and Asia were less important but contributed to the mix. When hybridized with earlier-maturing cottons and bred for local conditions the Mexican-type cottons such as Petit Gulf increased yields, provided resistance to diseases, and increased picking rates more than threefold. These varieties gave the United States a competitive advantage in the world cotton market. Following the boll weevil, fresh introductions helped reestablish cotton quality and became the foundation for production in the far West.
Tobacco	Tobacco predated European colonists. Tobacco was particularly malleable, so domestic breeders played an especially important role; but beginning with John Rolfe's initiatives fresh introductions invigorated American production. Later examples include the introduction of Havana in the 1870s and Sumatran in 1896 for cigar leaf production.
Vines	Although grapes existed in North America prior to European settlement, they did not produce quality wine. Hundreds of European varieties (*Vitis vinifera*) that became the foundation of the California industry were introduced after 1850. Most notable was Agoston Haraszthy's importation of about 100,000 vines (perhaps 300 varieties) following his expedition of 1861.
Fruit Trees	The introduction of the navel orange from Brazil is one of many stories of the establishment of commercial fruit production in the United States. In California, Spanish missionaries introduced most types of tree fruits around 1800, but later cohorts of European settlers supplemented those early introductions with the superior varieties that became the hallmark of the state's fruit industry.
Feeds	Numerous varieties of soybeans, alfalfa, and grasses that made up a major portion of feed crops were introduced and bred for different conditions.
Beef Cattle	Carcass weight increased while market age fell. Major breeds were introduced and improved to meet changing conditions. Longhorns spread across the Great Plains and then were replaced by Shorthorns and Herefords.

TABLE 13.3. *Changes in Animals*

Dairy Cows	Milk output per cow probably tripled in the period before 1940. The creamery was invented. Major breeds were introduced. Herd associations, advanced registries, and the butterfat test advanced productivity. The scientific and institutional foundations for artificial insemination were in place. Revolutionary sanitation changes cleaned up dairies, animals, and milk.
Draft Animals	Long before the introduction of tractors and trucks there were deep-seated shifts in the sources of farm power with the transition from oxen to equines. New draft breeds were introduced from Europe and many farmers learned to use larger teams. Mules which were novelties in the late colonial period gained important niches, particularly in the South. The division of labor between breeders and users of equines denoted an extensive market for breeding services with breeders possessing specialized skills and capital that were not evenly distributed across the farm population. This was true for many breeds of farm animals.
Sheep	With the introduction of Spanish Merinos and the later spread of improved American-bred Merinos the wool clip per animal increased four- to eightfold from the colonial era to 1940. The introduction of British breeds vastly improved meat output and quality.
Swine	Breeds were introduced from around the world and distinctive American breeds created. Razorbacks gave way to animals that gained weight faster and more efficiently on higher intensity diets. Average weight increased while market age fell.

1940 hogs and steers generally reached market age in less than half the time they did in 1800. In addition to breeding this reflected the deeper biological transformation that was built upon more intensive systems. Soybeans, corn, alfalfa, cottonseed products, and new grasses, along with processing and storage facilities, replaced or supplemented low-density grazing operations. In the broader world there is often a quantity–quality tradeoff; however, in spite of increases in output, the quality of many animal products improved significantly. Wool was finer and longer in 1940 than in 1800. Milk in 1940 posed little threat of transmitting a large number of diseases, but just a few decades earlier milk contributed to the deaths of tens of thousands of children every year. The taste of lamb became milder and generally more pleasing to the consumer. The fat-to-pork ratio of hogs increased and then decreased as breeders responded to market signals. Counter to this trend, breeding and feeding practices increased the fat content of beef, making a more tender (and less healthy) product, which consumers fancied. The "first mechanical revolution" also entailed a biological revolution whereby equines replaced oxen. Farmers also bred larger horses and learned to work with larger teams.

Table 13.4 provides an overview of the geographic shifts of the major American staple crops and the climatic challenges these movements entailed. The table reports statistics on the distribution of county-level production of tobacco, corn, cotton, and wheat in 1840 and 1910 by latitude, longitude, annual precipitation, annual temperature, and January temperature.[7] It includes data not only for the center (the median characterizing split placing 50 percent of production on either side) but also for the breakpoints of 10, 25, 75, and 90 percent. To illustrate how the data may be interpreted, the amount for annual precipitation for tobacco in 1840 in the 10 percent row is 38.4 inches, which means that 10 percent of the crop was produced in counties that received, in an average year, less than 38.4 inches and 90 percent was produced in counties that received more than that amount. The table shows that for tobacco there was relatively little change in any of the variables between 1840 and 1910.

The table shows that in 1840, 90 percent of all corn was grown east of 90.1 degrees west longitude and, thus, 10 percent was grown to the west of that longitude. In 1910 more than 50 percent of corn was grown farther west (90.4°W) than the most western lands, which accounted for only 10 percent of production in 1840. The westward shift for wheat was even greater; over 75 percent of the crop was grown farther west (89.6°) than the most westerly 10 percent in 1840 (87.0°). The western movement had dramatic consequences for the climatic conditions under which the crops were raised. In 1910, more than 50 percent of the corn crop was grown in places drier than the driest places that accounted for 10 percent of production in 1840. Wheat and cotton were moving into more arid

[7] The January temperature, for example, represents the average January temperature based on the county-level distribution of production in each census year weighted by the county's mean January temperatures over the period 1950–70. The annual temperature and precipitation statistics were calculated using similar procedures. Note that these measures do not capture temporary weather conditions or secular climate changes such as global warming. Rather they reflect the changing distribution of production, taking location-specific climatic conditions as constant.

TABLE 13.4. *Changing Geographical Distribution of Corn, Cotton, Tobacco, and Wheat, 1840–1910*

			Corn		Cotton		Tobacco		Wheat	
		Percent	1840	1910	1840	1910	1840	1910	1840	1910
Annual	Driest	10	36.9	26.5	45.4	33.2	38.4	37.6	33.2	16.6
Precipitation		25	39.6	31.9	48.5	43.1	41.6	41.6	36.0	19.3
(inches)	Median	50	43.9	36.2	51.8	47.7	42.8	43.4	39.0	26.2
		75	49.4	40.7	54.4	51.5	46.0	46.6	42.5	36.4
	Wettest	90	53.3	47.4	57.0	53.9	49.4	49.4	47.0	40.9
Annual	Coldest	10	50.9	47.2	60.9	60.8	54.8	50.1	47.8	40.1
Temperature		25	53.3	49.7	62.0	61.8	56.3	54.4	49.7	43.6
(°F)	Median	50	56.3	52.5	64.1	63.6	57.3	56.8	52.6	50.4
		75	60.0	56.4	65.9	65.4	58.2	58.2	55.3	54.1
	Warmest	90	63.3	61.6	66.8	66.9	59.0	59.8	58.7	56.7
January	Coldest	10	26.7	17.8	41.2	40.4	32.6	26.8	23.6	5.9
Temperature		25	30.1	22.4	44.1	42.4	34.9	32.4	24.2	12.0
(°F)	Median	50	35.1	27.3	46.2	44.7	36.5	35.1	30.1	25.8
		75	40.6	33.8	49.2	47.4	38.3	38.0	33.9	30.6
	Warmest	90	45.4	41.2	50.9	49.8	39.3	40.7	38.5	33.8
Latitude	South	10	33.3	34.8	30.9	31.2	36.3	36.0	36.2	37.7
Degrees		25	35.3	37.8	31.7	32.2	36.7	36.7	38.5	39.2
	Median	50	37.8	40.1	32.8	33.3	37.3	37.6	40.0	41.3
		75	39.5	41.6	34.0	34.5	38.0	38.7	41.2	46.1
	North	90	40.6	42.8	35.0	35.3	39.0	40.1	42.9	47.7
Longitude	East	10	76.7	82.6	81.1	80.7	77.0	77.0	76.0	82.6
Degrees		25	80.4	86.1	83.6	83.3	78.3	79.0	77.5	89.6
	Median	50	84.5	90.4	87.6	88.7	79.8	84.2	80.7	97.1
		75	87.3	95.2	91.1	95.5	87.1	86.5	84.2	99.3
	West	90	90.1	97.4	91.5	97.3	88.3	88.3	87.0	112.4
90–10 Split										
Annual Precipitation			16.5	20.9	11.7	20.7	11.1	11.8	13.8	24.3
Annual Temperature			12.4	14.4	5.9	6.1	4.2	9.7	10.9	16.6
January Temperature			18.7	23.4	9.7	9.4	6.7	13.9	14.9	27.9
Latitude			7.3	8.0	4.1	4.1	2.6	4.1	6.8	9.9
Longitude			13.4	14.8	10.4	16.6	11.3	11.3	11.0	29.8

lands as well.[8] Focusing on the driest 10 percent of production (top row) shows that the marginal corn lands received 10.4 fewer inches of rain in 1910 than in 1840, the marginal cotton lands received 12.2 fewer inches,

[8] For wheat production the table mutes the temperature and precipitation changes along with the northern shift because it excludes Canadian data.

and the marginal wheat lands received 16.6 fewer inches. For wheat this was exactly half of what the driest 10 percent received in 1840. Corn, cotton, and wheat production were also shifting to regions with colder winters, as measured by January temperatures. At any point in time corn occupied more temperate and moister zones than wheat, but by 1910 corn was typically grown in harsher, drier climates than had supported wheat in 1840.

These cross-sectional data are pertinent to the current debates on global warming, because they provide an indication of farmers' ability to respond to changing climatic conditions. The 90–10 splits, displayed at the bottom of the table, offer yet another perspective on the magnitude of the increasing climatic diversity in American agriculture. The data in this section are derived from the data displayed in the upper part of the table by subtracting the 10 percent entries for any category from the 90 percent entries. The range of growing conditions widened substantially for corn, cotton, and wheat between 1840 and 1910. As an example, in 1840 the January temperatures of the coldest fields accounting for 10 percent of wheat production and the warmest fields accounting for 10 percent of production differed by 14.9 degrees Farhenheit. In 1910 the difference was 27.9 degrees Farhenheit.

The movement of production into more arid regions with more variable climates was one of the hallmarks of American agricultural development. Biological innovation was a necessary condition for this westward march. Some of America's most distinguished historians, including Frederick Jackson Turner, Walter Prescott Webb, and their many disciples, explored the broader causes and consequences of the geographical expansion of agriculture. We believe that our quantitative analysis provides a new perspective on the magnitude of the challenges that farmers confronted.

The Yield Takeoff

We have argued that biological innovation was an integral part of American agriculture, starting with the first colonial settlements. Time and time again, new crops, more productive animals, and improved cultural systems were developed that revolutionized American farming. Recognizing these past changes does not mean that nothing distinctive took place around 1940. After a long period of stagnancy, yields per acre of a broad range of crops began to increase rapidly. The upsurge in yields coincides with the introduction of hybrid corn but is so widespread that a more general explanation is required. After all, the use of F2 (and F1) hybrid seed is specific to maize, sorghum, and a handful of other crops, but the yields of numerous additional crops also shot up. Some have asserted that the increasing use of synthetic nitrogen fertilizers, made possible by the Haber–Bosch process, explains the upward break in yields.

Vaclav Smil, G. J. Leigh, and Michael Pollan have all elevated Fritz Haber and Carl Bosch's nitrogen synthesis processes to high prominence. Smil, for example, claims that "without this synthesis about 2/5 of the world's population would not be around."[9] Haber began experimenting with ammonia in 1904 and, after a hit-and-miss start, gained the support of Badische Anilin– und Soda–Fabrik (BASF) in 1908. In July 1909 Haber sent a letter to the BASF directors describing his recent breakthrough in synthesizing ammonia. Led by Carl Bosch, who headed BASF's nitrogen fixation research, the company overcame a number of technical obstacles to translate Haber's experimental procedures into a commercial operation. BASF's first ammonia fertilizer plant went online 13 September 1913. Subsequent improvements in the production process dramatically increased the supply of nitrogen while lowering its price. These changes indeed revolutionized agriculture.[10]

To evaluate the hypothesis that the upsurge in yields across a broad range of crops had a common origin – one resulting from the increased use of synthetic nitrogen – we use county-level crop yield data to investigate the timing of the structural breaks. USDA data, which are available for an increasing number of states from 1910 on, allow us to control for the shifting geographic distribution of output.[11] The large number of observations per year raises confidence in the results. Our analysis asks, for each crop, when did the break in yields occur, that is, which break year produces the best statistical fit (the highest R-squared). We consider the period from 1910 to 1970. The results largely confirm the story of a common shift, but not one closely tied to the spread of nitrogen fertilizer. A more complex, nuanced story is required.

The results indicate that the break year for corn resulting in the best fit is 1934, around the period when hybrid corn takes off. But the best-fitting break year for wheat, rye, and barley, determined independently, is also 1934. For oats, 1933 and 1934 are tied as the break year. For hay, 1933 has the best fit (1934 is a candidate for cotton, but the highest R-squared

9 Smil, *Enriching the Earth*, xv. Also see Leigh, *World's Greatest Fix*; and Pollan, *Omnivore's Dilemma*.

10 Smil, *Enriching the Earth*, 60–107, and *Creating the Twentieth Century*, 189–94.

11 U.S. National Agricultural Statistics Service, *Quick Stats*.

for this crop occurs in 1950). There are no F2 hybrids for the small grain or hay crops introduced in that period. So far, so good, for those pushing for a general explanation of the yield takeoff. But is synthetic nitrogen the answer? No – not for explaining a break in 1933–34. In the early 1930s, the use of synthetic nitrogen, and indeed all fertilizers, decreased. Much of the nitrogen added to the soil in this period came from the increased cultivation of soybeans. The production of this crop expanded considerably thanks in large part to the activities of the USDA in importing and breeding improved varieties and then to the exclusion of soybeans from acreage limitation controls.

The timing of the break in 1934 suggests that forces other than fertilizer use were in play. Two forces immediately come to mind: the crop programs initiated during the New Deal that took more marginal lands out of production, and the onset of a period of better climate. The dust bowl years in the early and mid-1930s were noted for especially bad weather that suppressed yields. But the dry years and the dust storms also took low-yielding land out of production. The weather in the decades after the mid-1930s, due in part to increasing levels of the greenhouse gas CO_2, has been more conducive to plant growth.[12] Our account of one-variety cotton production showed that an increase in yields was closely associated with the introduction of improved varieties before the upswing in fertilizer use. Similar detailed studies of other crops could shed light on this issue.

Numerous studies in the plant science literature have sought to gauge the share of yield increases attributable to genetic improvement of the cultivated varieties. Typically these studies compared the yield of seeds of varieties released at different dates grown in the same experimental plots under identical environmental and agronomic regimes. Often the experimental design included a range of cultural practices – varying levels of planting densities, fertilizer application, and so on. The studies generally found that genetic improvement accounted for most of the long-run increase in yields.[13]

As one example, in the period 1977–79 Donald Duvick grew a series of U.S. maize varieties, ranging from the open-pollinated Reid (vintage 1930) to the double-cross hybrids of the 1940s through 1960s, to the single-cross hybrids of the 1970s, in the Midwest under the same experimental conditions. On the basis of these trials, Duvick attributed 71

[12] Evans, *Crop Evolution*, 373–9.

[13] Ibid., 296–306; Slafer, Satorre, and Andrade, "Increases in Grain Yield in Bread Wheat," 7–10.

to 89 percent of the increase in yields between 1930 and 1980 to genetic gains.[14] As another example, William Meredith, Jr., and Robert Bridge surveyed fifteen trials in which obsolete and modern varieties of upland cotton were grown, extending from Half-and-Half (1910) to Stoneville 826 (1978). They found that genetic improvement accounted for between 67 and 100 percent of the increase in yield.[15] Finally, T. S. Cox and colleagues tested 38 hard red winter wheats, ranging from Turkey (1874) and Kharkof (1900) to the latest introductions of the mid-1980s. They found that for each year of vintage (that is, moving from a variety introduced in one year to a variety introduced in the next year), yields increased an average of 16.2 kilograms per hectare.[16] This was greater than the actual annual gain in yields between 1890 and 1990 in Kansas, where such wheats were grown. Experiments with sorghum, soybeans, oats, and barley produced similarly large effects attributable to varietal change.

In most of these trials, the new varieties grew best under high-input agronomic regimes. The new seeds were bred to complement the use of synthetic fertilizers – for wheat, the plants were shorter in stature with stiffer stalks to support fuller heads of grain (and, hence, possessed a higher harvest index); for corn, the new varieties were more resistant to lodging. But it is crucial to note that the new varieties typically were superior even under older practices. These findings again confound accounts that place exclusive emphasis on the advent and increasing use of synthetic fertilizers in explaining the yield takeoff.[17] Other factors also argue against a monocausal nitrogen story. Animal yields shot up at about the same time as crop yields. While synthetic nitrogen contributed to the supply of more plentiful and cheaper feeds, numerous studies have shown that breeding was the dominant factor that explained the increases in animal productivity in the post–World War II era.

Collective Action, Spillovers, and Creating Public Goods

It is common to depict farmers in Europe, and in many other parts of the world, as working within village communities that governed production

[14] Duvick, "Genetic Contributions," 46. Russell, "Genetic Improvement of Maize Yields," 259, summarized the results of eleven U.S. studies. The fraction of the yield increases attributed to genetic gain varies between 33 and 94 percent, with a median contribution of 73 percent.

[15] Meredith and Bridge, "Genetic Contribution," 75–87.

[16] Cox et al., "Genetic Improvement," 756–60.

[17] Evans, *Crop Evolution*, 269–334.

relationships. By contrast, the stereotypical farmer in the United States is portrayed as highly individualistic and isolated from his neighbors. The differences have been exaggerated and, in many instances, American farmers were united by a farm research-policy establishment that was more innovative and more successful than those elsewhere.

Farmers banded together, often eliciting the aid of government, and perhaps more frequently responding to the visionary policies of government scientists and bureaucrats to fight insects, diseases, and weeds. Individual efforts would have been doomed to failure because of the free rider problem and the inability to capture the returns to research from creating public goods.[18] A farmer who employed the recommended crop rotation systems, cleaned his field, or sprayed to control insects created positive spillovers that would benefit a neighbor who did nothing. Conversely the neighbor's neglect created negative spillovers. Farmers understood this and subscribed to uniform policies enforced by the state. This took many forms, ranging from local to national prohibitions against planting barberry bushes which hosted rust spores, to collective cleanup efforts, to the creation of government agencies to quarantine the movement of crops and livestock. The common desire to control the spread of animal diseases led to legislation that codified long-standing common law granting governments the right to destroy private property without compensation. The development of compensation schemes required a clear understanding of farmer incentives. Payments had to be high enough to entice farmers to reveal the existence of a problem, but not so high as to generate moral hazards by encouraging farmers to neglect their animals.

From the mid-nineteenth century the federal government and many state governments developed increasingly sophisticated institutions to restrict the importation and trade of diseased animals. By the time of World War I, reasonably effective procedures were in place and the USDA's Bureau of Animal Industry had already achieved spectacular successes in controlling a number of disease outbreaks, including contagious pleuropneumonia and foot-and-mouth disease. National programs were under way to control hog cholera, tick fever, bovine tuberculosis, anthrax, and a host of other threats. The United States was a world leader in advancing the basic science and creating the instructional structures

[18] We introduced the concept of public goods in Chapter 8. The term impure public goods, referring to a large class of goods that are not entirely public but not purely private, better describes many of the creations of individual farmers and various levels of government. For a discussion of a large literature on this issue see Dalrymple, "Impure Public Goods," 71–89.

needed to control tick fever and bovine tuberculosis. These programs involved the cooperation of millions of farmers. The lack of resolve in Western Europe led to the loss of hundreds of thousands of lives to bovine tuberculosis well after the disease had been expelled from most American herds.

Farmers also banded together to protect the genetic "purity" of their seeds. An important example was the one-variety community movement, beginning in the 1920s, that transformed American cotton production. Other countries attempted to copy this model but with little success. As part of a larger effort to establish and enforce standards and grades in the marketplace, state governments created testing programs to certify seed quality for a wide range of crops. This often required the development of efficient and accurate tests to overcome "lemons" problems resulting from asymmetric information. Better certification systems extended to the animal sector where, for example, Stephen Babcock invented a cheap and effective method to determine the fat content of milk. Better systems also extended to community breeding associations with collective ownership of prized sires and extensive recordkeeping of progeny lines. The nineteenth and early twentieth centuries gave birth to a set of progressive private, state, and federal research and regulatory institutions to enforce food safety, to control plant and animal pests, and to help establish common grades and standards that increased the efficiency of agricultural markets.

When researchers cannot capture all the benefits of their creative endeavors, there will be an underinvestment (relative to a socially optimal level). Partly for this reason, societies create patent and copyright systems to give inventors temporary monopolies. But although agricultural machinery and chemicals were patentable, genetic materials were not. This changed in 1930 with the passage of the world's first plant patent act, which protected asexually reproduced plants. The enactment of the Plant Variety Protection Act in 1970 had to wait for sophisticated tests that could determine the difference between one seed and another. But well before plants received patent protection there was a plethora of private-sector inventive activity, where leading farmers and seed companies made significant contributions to plant improvement. State and federal agencies added to this brew. Animal breeders were at least as active, and many developed national markets for their creations. A large and important literature has identified inventions with patents. The absence of patent records for a large class of biological activities has led to the inference that little happened. However, a search of the press, farm journals, Patent Commission reports, and various state and federal reports suggests that innovators were making great strides in the introduction and creation of new and more productive types of plants and animals.

Induced Innovation

Throughout this book we have referred to the induced innovation hypothesis. The actual historical changes in factor prices – in particular the long-run increase in the price of land relative to the price of labor – should have induced biological innovations *if* one believed that biological innovations primarily saved land. However, the strict associations of mechanical innovations with laborsaving and of biological innovations with land saving are highly problematic. New cotton varieties vastly increased picking rates, and herbicides were substitutes for hoes and workers. The output or employment effect of many so-called laborsaving machines, of which the cotton gin was a notable example, was to increase the demand for labor. The tractor saved labor but was among the most land-saving innovations in American history. What is beyond dispute is that the sum of all non-mechanical farm investment dwarfed the investment in mechanization. We have analyzed only a subset of the nonmechanical investments associated both with crop and animal breeding and with the efforts to combat pests. Well before the world had heard of Cyrus McCormick, thousands of American farmers had invested $1,000 or more in a single purebred animal. By comparison, the price of early mowers and reapers was in the range of $100 to $200. We have made little mention of nonmechanical investments to create the farm infrastructure – clearing and leveling land, fencing, draining, irrigating, and the like. The induced innovation literature considers such investments as land saving and, thus, most discussion of them deals with labor-abundant economies. Accounting for these activities in the United States would dramatically tip the scale in favor of biological innovations. Robert Gallman's work provides empirical confirmation. In 1900, the value of land improvements represented more than one-half (54 percent) of all reproducible capital in American agriculture, whereas equipment made up less than 10 percent. The total value of investment in land improvements between 1840 and 1900 was over four times that invested in equipment.[19]

[19] Gallman, "United States Capital Stock," 165–213.

We have repeatedly referred to the interaction effects that link mechanical and biological innovations. Draft animals provided the major source of farm power until well after World War I, so most machines required complementary biological inputs. Various power sources and improved spraying machines helped apply chemicals. Although many trenches were dug via pick and shovel, American farmers increasingly relied on draft animals, steam engines, and then internal combustion engines to power the plows, dredges, and land planes that reshaped American farms. Power saws cut the wood for silos, barns, and fences; electric and diesel pumps moved water – the list is nearly endless. The closer the inspection the more confusing the standard paradigm of neatly associating one type of innovation with saving land and the other with saving labor.

The international agricultural development literature is fond of pointing to the United States, and other land-abundant countries such as Canada and Australia, as following a mechanical course of innovation, and to labor-abundant nations, such as Japan, the Philippines, Korea, and the like, as following a biological path (with Europe in between these extreme cases). The view is that both land-abundant and labor-abundant nations invented and adopted technologies that conformed to their relative factor scarcities (as opposed to changing factor scarcities). If labor-abundant nations experienced little mechanization, they may well have invested *relatively* more in biological technologies than the land-abundant countries. But this proposition says little about the absolute levels of research, investment, and returns to biological activities in the two polar sets of countries. The literature too often simply assumed that the advanced land-abundant nations neither invested in nor accomplished much significant biological innovation. This issue needs to be reexamined in the light of the actual American (Canadian and Australian) experience. Clearly there is much still to learn.

References

Abel, Andrew B. "Consumption and Investment." In *Handbook of Monetary Economics*, Vol. 2, edited by Benjamin M. Friedman and Frank H. Hahn, 725–78. New York: North–Holland, 1990.

Adusei, Edward O., and George W. Norton. "The Magnitude of Agricultural Maintenance Research in the USA." *Journal of Production Agriculture* 3, no. 1 (1990): 1–6.

Affleck, Thomas. "The Early Days of Cotton Growing in the South-West." *De Bow's Review* 10, no. 6 (1851): 668–9.

Agelasto, A. M., C. B. Doyle, G. S. Melroy, and O. C. Stine. "The Cotton Situation." In *U.S. Dept. of Agriculture Yearbook 1921*, 323–406. Washington, DC: GPO, 1922.

Agricultural Research Center – Hays. *Climatological Summary* [Web site] Hays, KS: The Center, 2007 [updated regularly]. Available from http://www.wkarc.org/Arch/arch.asp.

Agricultural Research Center – Hays. *Weather Extremes* [Web site] Hays, KS: The Center, 2007 [updated regularly]. Available from http://www.wkarc.org/Arch/arch.asp.

Aiken, Charles S. "The Evolution of Cotton Ginning in the Southern United States." *Geographical Review* 63, no. 2 (1973): 196–224.

Akerlof, George A. "The Market for 'Lemons': Quality Uncertainty and the Market Mechanism." *Quarterly Journal of Economics* 84, no. 3 (1970): 488–500.

Alexander, Donald Crichton. *The Arkansas Plantation, 1920–1942*. New Haven, CT: Yale University Press, 1943.

Allard, R. W. *Principles of Plant Breeding*. New York: John Wiley, 1960.

Allen, Lewis F. *History of the Short-horn Cattle: Their Origin, Progress and Present Condition*. Buffalo, NY: Author, 1872.

Allen, Richard Lam. *Domestic Animals*. New York: Orange Judd, 1848.

Alvord, Henry E. "Dairy Development in the United States." In *U.S. Dept. of Agriculture Yearbook 1899*, 381–403. Washington, DC: GPO, 1900.

Alvord, Henry E. "Statistics of the Dairy." *U.S. Bureau of Animal Industry Bulletin*, no. 55 (1903).

American Farmer [Baltimore, MD], Various issues and dates.

American Mammoth Ox. [Photograph] Beltsville, MD: Special Collections, National Agricultural Library. Available from http://www.nal.usda.gov/speccoll/collect/history/bigoxpg.htm.

American Mercury [Hartford, CT], Various issues and dates.

Anderson, Edgar, and William L. Brown. "The History of the Common Maize Varieties of the United States Corn Belt." *Agricultural History* 26, no. 1 (1952): 2–8.

Anderson, Edgar, and William L. Brown. "Origin of Corn Belt Maize and Its Genetic Significance." In *Heterosis: A Record of Researches Directed Toward Explaining and Utilizing the Vigor of Hybrids*, edited by John W. Gowen, 124–48. Ames, IA: Iowa State College Press, 1952.

Anderson, O. G., and F. C. Roth. *Insecticides and Fungicides, Spraying and Dusting Equipment*. New York: John Wiley, 1923.

Anderson, P. J. "Growing Tobacco in Connecticut." *Connecticut Agricultural Experiment Station Bulletin*, no. 564 (1952): 1–5.

Anderson, Terry L., and Robert Paul Thomas. "Economic Growth in the Seventeenth-Century Chesapeake." *Explorations in Economic History* 15, no. 4 (1978): 368–87.

Anderson, Virginia DeJohn. *Creatures of Empire: How Domestic Animals Transformed Early America*. Oxford: Oxford University Press, 2004.

Anderson, W. S., and J. J. Hooper. "American Jack Stock and Mule Production." *Kentucky Agricultural Experiment Station Bulletin*, no. 212 (1917).

Ankli, Robert E., and Wendy Millar. "Ontario Agriculture in Transition: The Switch from Wheat to Cheese." *Journal of Economic History* 42, no. 1 (1982): 207–15.

Arizona Producer [Phoenix, AZ], Various issues and dates.

Asdell, S. A. *Patterns of Mammalian Reproduction*. Ithaca, NY: Comstock, 1946.

Ashton, John. *Jack Stock and Mules in Missouri*. Columbia, MO: Curators of the University of Missouri, 1987.

Atack, Jeremy, and Fred Bateman. *To Their Own Soil: Agriculture in the Antebellum North*. Ames, IA: Iowa State University Press, 1987.

Atack, Jeremy, Fred Bateman, and William N. Parker. "The Farm, the Farmer, and the Market." In *The Cambridge Economic History of the United States*. Vol. 2, *The Long Nineteenth Century*, edited by Stanley L. Engerman and Robert E. Gallman. New York: Cambridge University Press, 2000.

Atack, Jeremy, and Peter Passell. *A New Economic View of American History from Colonial Times to 1940*. 2nd ed. New York: W.W. Norton, 1994.

Atkinson, Alfred, and M. L. Wilson. "Corn in Montana: History, Characteristics, and Adaptation." *Montana Agricultural Experiment Station Bulletin*, no. 107 (1914).

Atkinson, George F. "Diseases of Cotton." In *The Cotton Plant: Its History, Botany, Chemistry, Culture, Enemies, and Uses*, edited by Charles Dabney, 279–316. Washington, DC: GPO, 1896.

Axton, W. F. *Tobacco and Kentucky*. Lexington, KY: University of Kentucky Press, 1975.

Ayres, Peter G. "Alexis Millardet: France's Forgotten Mycologist." *Mycologist* 18, no. 1 (1994): 23–6.

Babcock, E. B. "Recent Progress in Plant Breeding." *Scientific Monthly* 49, no. 5 (1939): 393–400.

Bacon, Edmund. Letters, 1802–1820. Louisiana and Lower Mississippi Valley Collections. Louisiana State University Libraries, Baton Rouge, LA.

Bailey, Liberty H. *Cyclopedia of American Agriculture: A Popular Survey of Agricultural Conditions, Practices and Ideals in the United States and Canada*. 4th ed. Vol. 3, *Animals*. New York: Macmillan, 1907–1909.

Baker, Raymond. "Indian Corn and Its Culture." *Agricultural History* 48, no. 1 (1974): 94–7.

Baker, W. A., and W. G. Bradley. "The European Corn Borer." *U.S. Dept. of Agriculture Farmers' Bulletin*, no. 1548, rev. (1948).

Baker, W. A., and W. G. Bradley. "The European Corn Borer." *U.S. Dept. of Agriculture Farmers' Bulletin*, no. 1548, rev. (1941).

Baker, W. A., and W. G. Bradley. "The European Corn Borer." *U.S. Dept. of Agriculture Farmers' Bulletin*, no. 1548 (1935).

Balke, Nathan S., and Robert J. Gordon. "The Estimation of Prewar Gross National Product: Methodology and New Evidence." *Journal of Political Economy* 97, no. 1 (1989): 38–92.

Ball, Carleton R. "The History of American Wheat Improvement." *Agricultural History* 4, no. 2 (1930): 48–71.

Ball, Carleton R., and J. Allen Clark. "Experiments with Durum Wheat." *U.S. Dept. of Agriculture Bulletin*, no. 618 (1918).

Ballard, W. W., and C. B. Doyle. "Cotton-Seed Mixing Increased by Modern Gin Equipment." *U.S. Dept. of Agriculture Circular*, no. 205 (1922).

Ballinger, Roy A., and Clyde C. McWhorter. "Results Achieved by One-Variety Cotton Communities in Oklahoma." *Current Farm Economics* 7, Series 49, no. 4 (1934): 68–71.

Barre Gazette [Barre, MA], Various issues and dates.

Bateman, Fred. "An Economic Analysis of the American Dairy Agriculture During the Period of Western Expansion." MS Thesis, University of North Carolina, Chapel Hill, 1962.

Bateman, Fred. "Improvement in American Dairy Farming, 1850–1910: A Quantitative Analysis." *Journal of Economic History* 28, no. 2 (1968): 255–73.

Bateman, Fred. "Labor Inputs and Productivity in American Dairy Agriculture, 1850–1910." *Journal of Economic History* 29, no. 2 (1969): 206–29.

Bateman, Fred. "The 'Marketable Surplus' in Northern Dairy Farming: New Evidence by Size of Farm in 1860." *Agricultural History* 52, no. 3 (1978): 345–63.

Bateman, Fred, and James D. Foust. *Agricultural and Demographic Records for Rural Households in the North, 1860* [Computer file]. Ann Arbor, MI: Inter-university Consortium for Political and Social Research [producer and distributor], 1976. Available from http://www.icpsr.umich.edu [membership required].

Beck, Lewis C. "Facts Relative to a Disease Generally Known by the Name of Sick Stomach or Milk Sickness." *New York Medical and Physical Journal* 1, no. 3 (1822): 315–9.

Bell, A. A. "Diseases of Cotton." In *Cotton: Origin, History, Technology, and Production*, edited by C. Wayne Smith and J. Tom Cothren, 553–97. New York: John Wiley & Sons, 1999.

Bennett, Charles A. *Saw and Toothed Cotton Ginning Developments*. Dallas, TX: Texas Cotton Ginners' Journal and Cotton Gin and Oil Mill Press, 1960.

Betts, Edwin M. *Thomas Jefferson's Farm Book*. Princeton, NJ: n.p., 1953.

Betts, Ronald E. "The Green Card Pays Off." *U.S. Dept. of Agriculture Marketing Activities* 15, no. 8 (1952): 13–6.

Bevan, Edith Rossiter. "Perry Hall: Country Seat of the Gough and Carroll Families." *Maryland Historical Magazine* 45, no. 1 (1950): 33–46.

Bidwell, Percy Wells, and John I. Falconer. *History of Agriculture in the Northern United States, 1620–1860*. Washington, DC: Carnegie Institute of Washington, 1925.

Bixby, Donald, Carolyn J. Christman, Cynthia J. Ehrman, and D. Philip Sponenberg. *Taking Stock: The North American Livestock Census*. Blacksburg, VA: McDonal & Woodward, 1994.

Blair, R. E. "The Grade and Staple of California Cotton." *California Dept. of Agriculture Monthly Bulletin* 16, no. 12 (1927): 628–31.

Blanchard, Henry F. "Improvement of the Wheat Crop in California." *U.S. Dept. of Agriculture Bureau of Plant Industry Bulletin*, no. 178 (1910).

Bledsoe, R. P., and E. C. Westbrook. "History and Progress of the One-Variety Community Cotton Work in Georgia." *Commercial Fertilizer* 50, no. 5 (1935): 16–19.

Bogue, Allan G. *From Prairie to Cornbelt: Farming on the Illinois and Iowa Prairies in the Nineteenth Century*. Chicago: University of Chicago Press, 1963.

Bolton, Herbert T., Jerry F. Butler, and David A. Carlson. "A Mating Stimulant Pheromone of the Horn Fly, *Haematobia irritans* (L.). Demonstration of Biological Activity in Separated Cuticular Components." *Journal of Chemical Ecology* 6, no. 5 (1980): 951–64.

Boss, Andrew, L. B. Bassett, C. P. Bull, et al. "Seed Grain." In *Seventeenth Annual Report*, 363–79. Delano, MN: Minnesota Agricultural Experiment Station, 1909.

Boss, Andrew D., and George A. Pond. *Modern Farm Management: Principles and Practice*. Saint Paul, MN: Itasca Press, Webb Publishing, 1951.

Boucher, John N. *History of Westmoreland County, Pennsylvania*. Vol. 1. New York: Lewis Publishing, 1906.

Bowling, G. A. "The Introduction of Cattle into Colonial North America." *Journal of Dairy Science* 25, no. 2 (1942): 129–54.

Bowman, M. L., and B. W. Crossley. *Corn: Growing, Judging, Breeding, Feeding, Marketing*. Ames, IA: Authors, 1908.

Boyd, Candice, item ed. "Edmund Ruffin." In *Virtual American Biographies*, Stanley L. Klos, site ed. Carnegie, PA: Virtualology.com, c1999–2006. Available from http://virtualology.com/apedmundruffin/.

Bradley, Melvin. *The Missouri Mule: His Origin and Times*, Vol. 1. Columbia, MO: Curators of the University of Missouri, 1998.

Brand, Charles J. "The Acclimation of an Alfalfa Variety in Minnesota." *Science, NS* 28, no. 729 (1908): 891–2.

Braunholtz, J. T. "Crop Protection: The Role of the Chemical Industry in an Uncertain Future." *Philosophical Transactions of the Royal Society of London. Series B, Biological Sciences* 295, no. 1076 (1981): 19–34.

Breeder's Gazette [Chicago, IL], Various issues and dates.

Breen, T. H. *Tobacco Culture: The Mentality of the Great Tidewater Planters on the Eve of the Revolution*. Princeton, NJ: Princeton University Press, 1985.

Brewer, William H. "Report on the Cereal Production of the United States." In U.S. Census Bureau. 10th Census 1880. *Report on the Productions of Agriculture as Returned at the Tenth Census, June 1, 1880*, Vol. 3. Washington, DC: GPO, 1883.

Briggs, Hilton M. *Modern Breeds of Livestock*. New York: Macmillian, 1949.

Briggs, Hilton M., and Dinus M. Briggs. *Modern Breeds of Livestock*, 4th ed. New York: Macmillan, 1980.

Brinkman, Leonard W., Jr. "The Historical Geography of Improved Cattle in the United States to 1870." PhD Thesis, University of Wisconsin, 1964.

Brodell, A. P., and J. A. Ewing. "Use of Tractor Power, Animal Power, and Hand Methods in Crop Production." *U.S. Bureau of Agricultural Economics F.M.*, no. 69 (1948).

Brodell, A. P., and R. D. Jennings. "Work Performed and Feed Utilized by Horses and Mules." *U.S. Bureau of Agricultural Economics F.M.*, no. 44 (1944).

Brown, Harry Bates. *Cotton: History, Species, Varieties, Morphology, Breeding, Culture, Diseases, Marketing, and Uses*, 1st ed. New York: McGraw–Hill, 1927.

Brown, Harry Bates. *Cotton: History, Species, Varieties, Morphology, Breeding, Culture, Diseases, Marketing, and Uses*, 2nd ed. New York: McGraw–Hill, 1938.

Brown, Harry Bates, and Jacob Osborne Ware. *Cotton*, 3rd ed. New York: McGraw–Hill, 1958.

Brown, William L., and Edgar Anderson. "The Northern Flint Corns." *Annals of the Missouri Botanical Gardens* 34, no. 1 (1947).

Brubaker, Curt L., E. M. Bourland, and Jonathan F. Wendel. "The Origin and Domestication of Cotton." In *Cotton: Origin, History, Technology, and Production*, edited by C. Wayne Smith and J. Tom Cothren, 3–31. New York: John Wiley & Sons, 1999.

Buller, A. H. Reginald. *Essays on Wheat*. New York: Macmillan, 1919.

Bulow, Jeremy I. "Durable-Goods Monopolists." *Journal of Political Economy* 90, no. 2 (1982): 314–32.

Bulow, Jeremy I. "An Economic Theory of Planned Obsolescence." *Quarterly Journal of Economics* 101, no. 4 (1986): 729–49.

Burges, Austin Earle. "Break This Vicious Circle Which Shuts You Out from Cotton Seed Sales." *Southern Seedsman* 1, no. 1 (1938).

Burnett, Edmund Cody. "Hog Raising and Hog Driving in the Region of the French Broad River." *Agricultural History* 20, no. 2 (1946): 86–103.

Burns, J. "A Pioneer Fruit Region." *Overland Monthly, 2nd Series* 12, no. 67 (1888).

Burwell Family. Papers, 1745–1977. Southern Historical Collection. University of North Carolina, Chapel Hill, NC.

Bushnell, William R., and Alan P. Roelfs, eds. *The Cereal Rusts*. Orlando, FL: Academic Press, 1984.

Butterfield, H. M. *History of Deciduous Fruits in California*. Sacramento, CA: Inland Press, 1938.

Byerly, T. C. "Role of Genetics in Adapting Animals to Meet Changing Requirements of Human Food." *Scientific Monthly* 70, no. 5 (1954): 323–31.

Cabell, N. F., and E. G. Swem. "Some Fragments of an Intended Report on the Post Revolutionary History of Agriculture in Virginia." *William and Mary Quarterly* 26, no. 3 (1918): 145–68.

Caffrey, D. J., and L. H. Worthley. "The European Corn Borer and Its Control." *U.S. Dept. of Agriculture Farmers' Bulletin*, no. 1294 (1922).

California Department of Agriculture. *Monthly Bulletin* (December 1931).

California State Board of Agriculture. *Statistical Report* (Various issues and dates).

California Historical Landmark, No. 929: Site of Propagation of the Thompson Seedless Grape. Sacramento, CA: California Environmental Resources Evaluation System.

California State Agricultural Society. *Transactions 1899*. Sacramento, CA: State Printer, 1900.

Calvert, Benedict Leonard. "Letter of Benedict Leonard Calvert to the Lord Proprietary, October 26, 1729." In *Proceedings of the Council of Maryland, 1698–1731*, Vol. 25, 601–10. Baltimore, MD: Archives of Maryland Online, 2006.

Camp, Wofford B. "Cotton Culture in the San Joaquin Valley in California." *U.S. Dept. of Agriculture Circular*, no. 164 (1921).

Camp, Wofford B. *Cotton, Irrigation, and the AAA: Transcript of Interviews Conducted 1962–1966 by Willa Klug Baum*. Berkeley, CA: Regional Oral History Office, Bancroft Library, University of California, 1971.

Campbell, C. Lee, and David L. Long. "The Campaign to Eradicate the Common Barberry in the United States." In *Stem Rust of Wheat: From Ancient Enemy to Modern Foe*, edited by Paul D. Peterson, 16–50. St. Paul, MN: APS Press, 2001.

Campbell, John D. "Comparisons of One-Variety with Multiple-Variety Cotton and Related Economic Considerations." *Southern Cooperative Series Bulletin*, no. 41 (1954): 7–33.

Campbell, John D. "One-Variety Cotton in Oklahoma." *Oklahoma Agricultural Experiment Station Bulletin*, no. B-386 (1952).

Carleton, Mark Alfred. "The Basis for the Improvement of American Wheats." *U.S. Dept. of Agriculture Division of Vegetable Physiology and Pathology Bulletin*, no. 24 (1900).

Carleton, Mark Alfred. "Cereal Rusts of the United States: A Physiological Investigation." *U.S. Dept. of Agriculture Division of Vegetable Physiology and Pathology Bulletin*, no. 16 (1899).

Carleton, Mark Alfred. "Hard Wheats Winning Their Way." In *U.S. Dept. of Agriculture Yearbook 1914*, 391–420. Washington, DC: GPO, 1915.

Carlson, Laurie Winn. *William J. Spillman and the Birth of Agricultural Economics*. Columbia, MO: University of Missouri Press, 2005.

Carman, Ezra A., H. A. Heath, and John Minto. *Special Report on the History and Present Condition of the Sheep Industry in the United States*. Washington, DC: U.S. Bureau of Animal Industry; GPO, 1892.

Carosso, Vincent P. *The California Wine Industry: A Study of the Formative Years*. Berkeley, CA: University of California Press, 1951.

Carr, Lois G. "Diversification in the Colonial Chesapeake: Somerset County, Maryland, in Comparative Perspective." In *Colonial Chesapeake Society*, edited by Lois G. Carr, Philip D. Morgan, and Jean B. Russo, 342–88. Chapel Hill, NC: University of North Carolina Press, 1988.

Carr, Lois G., and Russell R. Menard. "Land, Labor, and Economies of Scale in Early Maryland: Some Limits to Growth in the Chesapeake System of Husbandry." *Journal of Economic History* 49, no. 2 (1989): 407–18.

Carrier, Lyman. *The Beginnings of Agriculture in America*. New York: McGraw–Hill, 1923.

Carrier, Lyman. "The History of the Silo." *Journal of the American Society of Agronomy* 12, no. 5 (1920): 175–82.

Carrier, Lyman, and Katherine S. Bort. "The History of Kentucky Blue Grass and White Clover in the United States." *Journal of the American Society of Agronomy* 8, no. 4 (1916): 256–66.

Carter, Susan B., Scott Sigmund Gartner, Michael R. Haines, Alan L. Olmstead, Richard Sutch, and Gavin Wright. *Historical Statistics of the United States: Earliest Times to the Present*. Millennial ed., Vol. 4, Part D, *Economic Sectors*. New York: Cambridge University Press, 2006.

Casey, Michael. *Genetic Advantages of Texas Longhorn Cattle*. Nicasio, CA: Fairlea Longhorn Ranch [cited 12 June 2007]. Available from http://www.fairlealonghorns.com/Articles/genetic.html.

Castonguay, Stéphane. "Naturalizing Federalism: Insect Outbreaks and the Centralization of Entomological Research in Canada, 1884–1914." *Canadian Historical Review* 85, no. 1 (2004): 1–34.

Cates, H. R. "The Weed Problem in American Agriculture." In *U.S. Dept. of Agriculture Yearbook 1917*, 205–15. Washington, DC: GPO, 1918.

Ceci, Lynn. "Fish Fertilizer: A Native North American Practice?" *Science, NS* 188, no. 4138 (1975): 26–30.

Chaplin, Joyce E. *An Anxious Pursuit: Agricultural Innovation and Modernity in the Lower South, 1730–1815*. Chapel Hill, NC: University of North Carolina Press, 1993.

Chapman, Arthur B. "Jay Laurence Lush." *Biographical Memories* 57 (1987): 277–305.

Charles, Arnade. "Cattle Raising in Spanish Florida, 1513–1761." *Agricultural History* 35, no. 3 (1961): 116–24.

Cherry, John P., and Harry R. Leffler. "Seed." In *Cotton*, edited by R. J. Kohel and C. F. Lewis, 512–70. Madison, WI: American Society of Agronomy, 1984.

Chester, K. Starr. "Plant Disease Losses: Their Appraisal and Interpretation." *Plant Disease Reporter*. Supplement 193 (1950): 189–362.

Christensen, Paul P. "Land Abundance and Cheap Horsepower in the Mechanization of the Antebellum United States Economy." *Explorations in Economic History* 18, no. 4 (1981): 309–29.

Christensen, Thomas P. "The First Cream Separator." *Hoard's Dairyman* 84 (1939): 338.

Christenson, L. Peter. "Raisin Grape Varieties." In *Raisin Production Manual*, edited by L. Peter Christenson, Ch. 6, 38–47. Berkeley, CA: University of California Agriculture and Natural Resources, 2000.

Christidis, Basil G., and George J. Harrison. *Cotton Growing Problems*. New York: McGraw–Hill, 1955.

Cincinnati Price Current [Cincinnati, OH], Various issues and dates.

Clairborne, J. F. H. *Mississippi, as a Province, Territory, and State with Biographical Notices of Eminent Citizens*, Vol. 1. Jackson, MS: Power and Barksdale, 1880.

Clark, Gregory. "Why Isn't the Whole World Developed? Lessons from the Cotton Mills." *Journal of Economic History* 47, no. 1 (March 1987): 141–73.

Clark, J. Allen, and John H. Martin. "Varietal Experiments with Hard Red Winter Wheats in the Dry Areas of the Western United States." *U.S. Dept. of Agriculture Bulletin*, no. 1276 (1925).

Clark, J. Allen, John H. Martin, and Carleton R. Ball. "Classification of American Wheat Varieties." *U.S. Dept. of Agriculture Bulletin*, no. 1074 (1922).

Clark, J. Allen, John H. Martin, and Ralph W. Smith. "Varietal Experiments with Spring Wheat on the Northern Great Plains." *U.S. Dept. of Agriculture Bulletin*, no. 878 (1920).

Clark, J. Allen, and Karl S. Quisenberry. "Distribution of the Varieties and Classes of Wheat in the United States in 1929." *U.S. Dept. of Agriculture Circular*, no. 283 (1933).

Clark, Thomas D. "Live Stock Trade Between Kentucky and the South, 1840–1860." *Register of the Kentucky State Historical Society* 27 (1929): 567–81.

Clay, Clarence Albert. *A History of Maine Agriculture, 1604–1860*, University of Maine Studies, 2nd Series, no. 68. Orono, ME: University Press, 1954.

Clayton, John. "A Continuation of Mr. John Clayton's Account of Virginia." *Philosophical Transactions* 17, no. 205 (1693): 941–8.

Clemen, Rudolf Alexander. *The American Livestock and Meat Industry*. New York: Ronal Press, 1923.

Clinton, Elizabeth S. "The Origin and Spread of Maize (Zea mays) in New England." In *Histories of Maize: Multidisciplinary Approaches to the Prehistory, Linguistics, Biogeography, Domestication, and Evolution of Maize*, edited by John E. Staller, Robert H. Tykot, and Bruce F. Benz, 539–48. Amsterdam; Boston: Elsevier Academic Press, 2006.

Clinton, George Perkins. "Russian Thistle and Some Plants That Are Mistaken for It." *Illinois Agricultural Experiment Station Bulletin*, no. 39 (1895): 87–118.

Cobb, Cully A. *The Cotton Section of the Agricultural Adjustment Administration, 1933–1937: Oral History Transcript of Interview Conducted by Willa*

Klug Baum in October, 1966. Berkeley, CA: Regional Oral History Office, Bancroft Library, University of California, 1968.

Coburn, Foster Dwight. *Swine in America: A Text-Book for the Breeder, Feeder & Student*. New York: Orange Judd, 1912.

Cochrane, Willard W. *The Development of American Agriculture: A Historical Analysis*. Minneapolis: University of Minnesota Press, 1979.

Cochrane, Willard W. *Farm Prices, Myths and Realities*. Minneapolis: University of Minnesota Press, 1958.

Coclanis, Peter A. "David R. Coker, Pedigreed Seeds and the Limits of Agribusiness in Early-Twentieth-Century South Carolina." *Business and Economic History* 28, no. 2 (1999): 105–14.

Coker's Pedigreed Seed Company Catalog. Durham, NC: Rare Book, Manuscript, and Special Collections Library, Duke University, Spring 1917, Spring 1918, Spring 1927.

Colby, Charles C. "The California Raisin Industry: A Study in Geographic Interpretation." *Annals of the Association of American Geographers* 14, no. 2 (1924): 49–108.

Cole, Arthur H. "Agricultural Crazes: A Neglected Chapter in American Economic History." *American Economic Review* 16, no. 4 (1926): 622–39.

Coleman, J. Winston. Kentuckiana Collection. Transylvania University Library Special Collections, Lexington, KY.

Collings, Gilbeart H. *Production of Cotton*. New York: John Wiley, 1926.

Collins, Charles W. *Ohio, an Atlas*. Madison, WI: American Printing & Publishing, 1975.

Colwick, Rex F., and E. B. Williamson. "Harvesting to Maintain Efficiency and to Protect Quality." In *Advances in Production and Utilization of Quality Cotton: Principles and Practices*, edited by Fred C. Elliot, Marvin Hoover, and Walter K. Porter, Jr., 434–66. Ames, IA: Iowa State University Press, 1966.

Combs, Gerald F., Jr. *The Vitamins: Fundamental Aspects in Nutrition and Health*, 2nd ed. San Diego, CA: Academic Press, 1998.

Comstock, J. Henry. *Report upon Cotton Insects*. Washington, DC: GPO, 1879.

ConAgra Foods. *Company History, 1861–1919* [Web page]. Omaha, NE: ConAgra Foods [cited 5 July 2007]. Available from http://www.conagrafoodscompany.com/corporate/aboutus/company_history_timeline.jsp.

Connell, K. H. "The Potato in Ireland." *Past and Present*, no. 23 (1962): 57–71.

Constantine, John H., Julian M. Alston, and Vincent H. Smith. "Economic Impacts of the California One-Variety Cotton Law." *Journal of Political Economy* 102, no. 5 (October 1994): 951–74.

Cook, O. F. "Cotton Improvement on a Community Basis." In *U.S. Dept. of Agriculture Yearbook 1911*, 397–410. Washington, DC: GPO, 1912.

Cook, O. F. "Local Adjustment of Cotton Varieties." *U.S. Bureau of Plant Industry Bulletin*, no. 159 (1909).

Cook, O. F. "One-Variety Cotton Communities." *U.S. Dept. of Agriculture Bulletin*, no. 1111 (1922).

Cook, O. F., and C. B. Doyle. "One-Variety Community Plan Shows Numerous Practical Advantages." In *U.S. Dept. of Agriculture Yearbook 1933*, 132–8. Washington, DC: GPO, 1934.

Cook, O. F., and R. D. Martin. "Community Cotton Production." *U.S. Dept. of Agriculture Farmers' Bulletin*, no. 1384 (1924).

Cook, R. James, and Roger J. Veseth. *Wheat Health Management*. St. Paul, MN: APS Press, 1991.

Corn Refiners Association. *A Brief History of the Corn Refining Industry* [Web page]. Washington, DC: The Association, c2002 [cited 3 July 2007]. Available from http://www.corn.org/web/history.htm.

Coruthers, John Milton. "One-Variety Cotton Communities." PhD Diss., Cornell University, 1934.

Cox, A. B. "Cotton Classing and Standardization." In *Cotton Production, Marketing, and Utilization*, edited by W. B. Andrews, 318–36. State College, MS: W. B. Andrews, 1950.

Cox, T. S., J. P. Shroyer, Liu Ben-Hui, R. G. Sears, and T. J. Martin. "Genetic Improvement in Agronomic Traits of Hard Red Winter Wheat Cultivars from 1919–1987." *Crop Science* 28, no. 5 (1988): 756–60.

Crabb, A. Richard. *The Hybrid-Corn Makers: Prophets of Plenty*. New Brunswick, NJ: Rutgers University Press, 1947.

Craig, Lee A., Michael R. Haines, and Thomas Weiss. *Development, Health, Nutrition, and Mortality: The Case of the 'Antebellum Puzzle' in the United States*, NBER Working Paper Series on Historical Factors in Long Run Growth, Historical Paper 130. Cambridge, MA: National Bureau of Economic Research, 2000.

Craig, Lee A., Michael R. Haines, and Thomas Weiss. *U.S. Censuses of Agriculture, by County, 1840–1880* [Computer file]. Raleigh, NC: Unpublished files graciously provided by the authors, Department of Economics, North Carolina State University, 2000.

Craig, Lee A., and Thomas Weiss. "Agricultural Productivity Growth During the Decade of the Civil War." *Journal of Economic History* 53, no. 3 (1993): 527–48.

Craig, Lee A., and Thomas Weiss. *Rural Agricultural Workforce by County, 1800 to 1900* [Online database]. Oxford, OH: EH.Net, Miami University [distributor], 1998. Available from http://eh.net/databases/agriculture/.

Craven, Avery O. *Soil Exhaustion as a Factor in the Agricultural History of Virginia and Maryland*, Vol. 13, no. 1, University of Illinois Studies in the Social Sciences. Urbana, IL: University of Illinois, 1926.

Crawford, G. L. "Point Buying of Cotton Versus Buying on Quality Basis: Proceedings of the 31st Annual Convention of the Association of Southern Agricultural Workers, February 5–7, 1930." Jackson, MS: The Association, 1930.

Cronon, William. *Nature's Metropolis: Chicago and the Great West*. New York: W. W. Norton, 1991.

Cuff, Timothy. "A Weighty Issue Revisited: New Evidence on Commercial Swine Weights and Pork Production in Mid-Nineteenth Century America." *Agricultural History* 66, no. 4 (1992): 55–74.

The Cultivator [Albany, NY], Various issues and dates.

Dahms, R. G. "Insects Attacking Wheat." In *Wheat Improvement*, edited by Karl S. Quisenberry and L. P. Reitz, 411–43. Madison, WI: American Society of Agronomy, 1967.

Dale, Edward Everett. *The Range Cattle Industry*. Norman, OK: University of Oklahoma Press, 1930.

Dalrymple, Dana G. "Changes in Wheat Varieties and Yields in the United States, 1919–1984." *Agricultural History* 62, no. 4 (1988): 20–36.

Dalrymple, Dana G. "Impure Public Goods and Agricultural Research: Toward a Blend of Theory and Practice." *Quarterly Journal of International Agriculture* 45, no. 1 (2005): 71–89.

Danhof, Clarence H. *Changes in Agriculture: The Northern United States, 1820–1870*. Cambridge, MA: Harvard University Press, 1969.

Daniels, E. D. *A Twentieth Century History and Biographical Record of LaPorte County, Indiana*. Chicago: Lewis Publishing, 1904.

Darst, W. H. "Cotton-Seed Production." In *Seed Production and Marketing*, edited by Joseph F. Cox and George E. Starr, 190–212. New York: John Wiley, 1927.

Darwin, Charles. *The Variation of Animals and Plants Under Domestication*. Vol. 1 and 2. New York: D. Appleton, 1896.

Darwin, Charles. *Variations of Plants and Animals Under Domestication*. 2nd ed. Vol. 1. New York: D. Appleton, 1883.

Davenport, E., and Wilber John Fraser. "Corn Experiments 1895." *University of Illinois Agricultural Experiment Station Bulletin*, no. 42 (1896): 163–80.

Davie, William R. *An Address Delivered Before the South Carolina Agricultural Society at Their Anniversary Meeting Held in Columbia on the 8th of December 1818*. Columbia, SC: Telescope Press, 1819.

Davis, J. J. "The European Corn Borer: Past, Present, and Future." *Journal of Economic Entomology* 28, no. 2 (1935): 324–33.

Davis, Joseph Ray. *History of the Poland China Breed of Swine*, Vol. 1. Omaha, Poland China History Association, 1921.

Day, Richard. "The Economics of Technological Change and the Demise of the Sharecropper." *American Economic Review* 57, no. 3 (1967): 427–49.

de Kruif, Paul. *Hunger Fighters*. New York: Harcourt, Brace & World, 1928.

Delay, Peter J., ed. *History of Yuba and Sutter Counties, California*. Los Angeles: Historical Record Company, 1924.

Delta and Pine Land Company. Records, 1886–1982. Manuscript Collections. Mississippi State University Libraries, Starkville, MS.

"Delta & Pine Land Co." *Fortune* 15, no. 3 (1937): 125–32.

Derry, Margaret E. *Bred for Perfection: Shorthorn Cattle, Collies, and Arabian Horses Since 1800*. Baltimore, MD: Johns Hopkins University Press, 2003.

Dethier, V. G. *Man's Plague? Insects and Agriculture*. Princeton, NJ: Darwin Press, 1976.

Dewey, Lyster H. "Canada Thistle." *U.S. Division of Botany Circular*, no. 27 (1900).

Dewhurst, J. Frederic, and Associates. *America's Needs and Resources: A New Survey*. New York: Twentieth Century Fund, 1955.

Doebley, John, Jonathan D. Wendel, J. S. C. Smith, Charles W. Stuber, and Major M. Goodman. "The Origin of Cornbelt Maize: The Isozyme Evidence." *Economic Botany* 42, no. 1 (1988): 120–31.

Dohner, Janet Vorwald. *The Encyclopedia of Historical and Endangered Livestock and Poultry Breeds*. New Haven, CT: Yale University Press, 2001.

Dondlinger, Peter Tracy. *The Book of Wheat: An Economic History and Practical Manual of the Wheat Industry*. New York: Orange Judd, 1908.

Donnell, E. J. *Chronological and Statistical History of Cotton*. Reprint of 1872 ed. Wilmington, DE: Scholarly Resources, 1973.

Doutt, R. L. "Vice, Virtue, and the Vedelia." *Bulletin of the Entomological Society of America* 4, no. 4 (1958): 119–23.

Doyle, C. B. "Cotton Growing in One-Variety Communities." In *U.S. Dept. of Agriculture Yearbook 1926*, 263–7. Washington, DC: GPO, 1927.

Doyle, C. B. "Multiplicity of Varieties Handicaps Improvement in the American Cotton Crop." In *U.S. Dept. of Agriculture Yearbook 1933*, 107–14. Washington, DC: GPO, 1934.

Doyle, C. B. "One-Variety Cotton Communities Provide for Superior Cotton Fabrics." *Research Achievement Sheet*, no. 45(P) (1945).

Doyle, C. B., G. S. Meloy, and O. C. Stine. "The Cotton Situation." In *U.S. Dept. of Agriculture Yearbook 1921*. Washington, DC: GPO, 1922.

Dreyer, Peter. *A Gardener Touched with Genius: The Life of Luther Burbank*. Rev. ed. Berkeley, CA: University of California Press, 1985.

Drummond, R. O., G. Lambert, H. E. Smalley, and C. E. Terrill. "Estimated Losses of Livestock to Pests." In *CRC Handbook of Pest Management in Agriculture*, Vol. 1, edited by David Pimentel, 111–27. Boca Raton, FL: CRC Press, 1981.

Duggar, J. F. "Descriptions and Classification of Varieties of American Upland Cotton." *Alabama Experiment Station Bulletin*, no. 149 (1907).

Dunbar, William. *Life, Letters, and Papers of William Dunbar: of Elgin, Morayshire, Scotland, and Natchez, Mississippi: Pioneer Scientist of the Southern United States*, compiled by Eron Rowland. Jackson, MS: Press of the Mississippi Historical Society, 1930.

DuPuis, E. Melanie. *Nature's Perfect Food: How Milk Became America's Drink*. New York: New York University Press, 2002.

Durand, Loyal, Jr. "The Migration of Cheese Manufacture in the United States." *Annals of the Association of American Geographers* 42, no. 4 (1952): 263–82.

Duvick, Donald N. "Genetic Contributions to Yield Gains of U.S. Hybrid Maize, 1930 to 1980." In *Genetic Contributions to Yield Gains of Five Major Crop Plants*, edited by W. R. Fehr, 15–47. Madison, WI: CSSA, 1984.

Eager, J. M. "Morbidity and Mortality Statistics as Influenced by Milk." In *Milk and Its Relation to the Public Health*, 229–42. Washington, DC: GPO, 1908.

Early American Imprints, Series 1: Evans, 1639–1800. [Online database]: New Canaan, CT: Readex; Worcester, MA: American Antiquarian Society, 2002–. Available from http://infoweb.newsbank.com/ [Subscription required].

Early American Imprints, Series 2: Shaw–Shoemaker, 1801–1819. [Online database]: New Canaan, CT: Readex; Worcester, MA: American Antiquarian Society, 2004–. Available from http://infoweb.newsbank.com/ [Subscription required].

Early American Newspapers, 1690–1876. [Online database]: New Canaan, CT: Readex; Worcester, MA: American Antiquarian Society, 2004–. Available from http://infoweb.newsbank.com/ [Subscription required].

East, E. M., and D. F. Jones. "Round Tip Tobacco – A Plant 'Made to Order': From Specifications Drawn by Manufacturers and Consumers of Cigars, and the Growers of Tobacco, a New Plant is Grown to Satisfy the Demands of Commerce." *Journal of Heredity* 12, no. 2 (1921): 51–6.

Edminster, Lynn Ramsay. "Meat Packing and Slaughtering: History and American Developments." In *Encyclopedia of the Social Sciences*, Vol. 19, edited by Edwin R. A. Seligman, 242–9. New York: Macmillan, 1933.

Edmonds, James E. "Around the Clock on the South's Largest Cotton Plantation." *Cotton Trade Journal* 37, no. 17 (1957): 40–3, 80.

Edwards, Everett E. "Europe's Contribution to the American Dairy Industry." *Journal of Economic History* 9, *Supplement: The Tasks of Economic History* (1949): 78–84.

Edwards, Everett E. "The Settlement of Grasslands." In *U.S. Dept. of Agriculture Yearbook 1948*, 16–25. Washington, DC: GPO, 1948.

Edwards, Everett E., and Horace H. Russell. "Wendelin Grimm and Alfalfa." *Minnesota History* 19 (1938): 21–33.

Elliot, Fred C., Marvin Hoover, and Walter K. Porter, Jr. *Advances in Production and Utilization of Quality Cotton*. Ames, IA: Iowa University Press, 1968.

Elliott, E. E., and C. W. Lawrence. "Some New Hybrid Wheats." *Washington Agricultural Experiment Station Popular Bulletin* 9 (1908).

Elwood, Robert B., Lloyd E. Arnold, D. Clarence Schmutz, and Eugene G. McKibben. *Changes in Technology and Labor Requirements in Crop Production: Wheat and Oats*, Works Projects Administration. National Research Project. Studies of Changing Techniques and Employment in Agriculture. Report A-10. Philadelphia: WPA, 1939.

Elwood, Robert B., Arthur A. Lewis, and Ronald A. Struble. *Changes in Technology and Labor Requirements in Livestock Production: Dairying*, Works Projects Administration. National Research Project. Studies of Changing Techniques and Employment in Agriculture. Report A-14. Philadelphia, PA: WPA, 1941.

Emerson, William D. *History and Incidents of Indian Corn and Its Culture*. Cincinnati, OH: Wrightson, 1878.

Ensminger, M. E. *Animal Science*, 7th ed. Danville, IL: Interstate Printers and Publishers, 1977.

Essig, E. O. "Fifty Years of Entomological Progress, Part 4, 1919 to 1929." *Journal of Economic Entomology* 33, no. 1 (1940): 30–58.

Evans, B. R., and George G. Evans. *The Story of Durocs: The Truly American Breed of Swine*. Peoria, IL: United Duroc Record Association, 1946.

Evans, L. T. *Crop Evolution, Adaptation and Yield*. New York: Cambridge University Press, 1993.

Evans, Robert J. *History of the Duroc: A Short History of the Duroc Jersey Breed of Swine*. Chicago: James J. Doty Publishing, 1918.

Ewart, J. Cossar. "The Principles of Breeding and the Origin of Domesticated Breeds of Animals." In *Report of the Bureau of Animal Industry for the Year 1910*, 125–86. Washington, DC: GPO, 1912.

Ewing, P. V. *Southern Pork Production*. New York: Orange Judd, 1918.

Farmers' Register [Chambersburg, PA], Various issues and dates.

Farrell, Richard T. "Advice to Farmers: The Content of Agricultural Newspapers, 1860–1910." *Agricultural History* 51, no. 1 (1977): 209–41.

Faught, William A. "Cotton Price Relationships in Farmers' Local Markets." *Southern Cooperative Series Bulletin*, no. 51 (1957).

"Field Experiments with Corn 1888." *Illinois Agricultural Experiment Station Bulletin*, no. 4 (1889): 37–87.

"Field Experiments with Corn 1889." *Illinois Agricultural Experiment Station Bulletin*, no. 8 (1890): 215–45.

"Field Experiments with Corn 1890." *Illinois Agricultural Experiment Station Bulletin*, no. 13 (1891): 389–404.

Fisher, Franklin M., and Peter Temin. "Regional Specialization and the Supply of Wheat in the United States, 1867–1914." *Review of Economics and Statistics* 52, no. 2 (1970): 134–49.

Fitzgerald, Deborah. *The Business of Breeding: Hybrid Corn in Illinois, 1890–1940*. Ithaca, NY: Cornell University Press, 1990.

Fletcher, Stevenson Whitcomb. *Pennsylvania Agriculture and Country Life, 1640–1840*. Harrisburg, PA: Pennsylvania Historical and Museum Commission, 1950.

Fletcher, Stevenson Whitcomb. *Pennsylvania Agriculture and Country Life, 1840–1940*. Harrisburg, PA: Pennsylvania Historical and Museum Commission, 1955.

Flint, Charles L. "Progress in Agriculture." In *Eighty Years' Progress of the United States*, 19–102. Hartford, CT: L. Stebbins, 1865.

Flint, Timothy. *A Condensed Geography and History of the Western States, or, the Mississippi Valley*, Vol. 1, American Culture Series, 384:10. Cincinnati, OH: E. H. Flint, 1828.

Flower, K. C. "Field Practices." In *Tobacco: Production, Chemistry, and Technology*, edited by D. Layten Davis and Mark T. Nielsen, 4C, 76–103. Oxford; Maiden, MA: Blackwell Science, 1999.

Foley, John P., ed. *The Jeffersonian Cyclopedia*. New York: Funk & Wagnalls, 1900.

Foote, Robert H. *Artificial Insemination to Cloning: Tracing 50 Years of Research* [Online document]. Ithaca, NY: Cornell University, 1998. Available from http://dspace.library.cornell.edu/bitstream/1813/3661/3/AI_to_Cloning_online.pdf.

Forristal, Linda Joyce. *The Rise and Fall of Crisco* [Web page]. S.l.: Mother Linda's Olde World Cafe and Travel Emporium, [cited 5 July 2007]. Available from http://www.motherlindas.com/crisco.htm.

Forste, Robert H., and George E. Frick. "Dairy." In *Another Revolution in U.S. Farming?*, edited by Lyle P. Schertz and others. Washington, DC: U.S. Department of Agriculture; GPO, 1979.

Fortenberry, William H. "The Story of Cotton." *U.S. Dept. of Agriculture Marketing Bulletin*, no. 37 (1967).

Fox, Austin. "The Demand for Farm Tractors in the United States: A Regression Analysis." *U.S. Dept. of Agriculture Economic Report*, no. 103 (1966).
Freeman, E. M., and E. C. Stakman. "Smuts of Grain Crops." *University of Minnesota Agricultural Experiment Station Bulletin*, no. 122 (1911).
Fryxell, P. A. *The Natural History of the Cotton Tribe: (Malvaceae, Tribe Gossypieae)*, 1st ed. College Station, TX: Texas A&M University Press, 1979.
Fuchs, A. W., and L. C. Frank. "Milk Supplies and Their Control in American Urban Communities of over 1,000 Population in 1936." *Public Health Bulletin*, no. 245 (1939).
Fussell, G. E. *The English Dairy Farmer: 1500–1900*. London: Frank Cass & Co., 1966.
Gade, Daniel W. "Hogs." In *The Cambridge World History of Food*, edited by Kenneth F. Kiple and Kriemhild Coneé Ornelas, Entry II.G.13, 536–42. New York: Cambridge University Press, 2000.
Gage, Charles G. "Historical Factors Affecting American Tobacco Types and Uses and the Evolution of the Auction Market." *Agricultural History* 2, no. 1 (1937): 43–57.
Gains, R. C. "The Boll Weevil." In *U.S. Dept. of Agriculture Yearbook 1952*, 501–4. Washington, DC: GPO, 1952.
Galinat, Walton C. "Domestication and Diffusion of Maize." In *Prehistoric Food Production in North America*, edited by Richard I. Ford, 245–82. Ann Arbor, MI: Museum of Anthropology, University of Michigan, 1985.
Gallman, Robert E. "Self-Sufficiency in the Cotton Economy of the Antebellum South." *Agricultural History* 44, no. 1 (1970): 5–23.
Gallman, Robert E. "The United States Capital Stock in the Nineteenth Century." In *Long-Term Factors in American Economic Growth*, edited by Stanley L. Engerman and Robert E. Gallman, 165–213. Chicago, IL: University of Chicago Press, 1986.
Galloway, B. T. "Plant Pathology: A Review of the Development of the Science in the United States." *Agricultural History* 2, no. 2 (1928): 49–60.
Galloway, B. T. "Report of the Chief of the Division of Vegetable Pathology." In *Report of the Secretary of Agriculture 1893*, 245–76. Washington, DC: GPO, 1894.
Galloway, Beverly Thomas. Papers. Series 7, Photographs. U.S. National Agricultural Library Manuscript Collections, Beltsville, MD.
Gard, Wayne. *The Chisholm Trail*. Norman, OK: University of Oklahoma Press, 1954.
Gardner, M. W., and W. B. Hewitt. *Pierce's Disease of the Grapevine: The Anaheim Disease and the California Wine Disease*. Berkeley, CA: University of California Press, 1974.
Garner, W. W. *Production of Tobacco*. Philadelphia: Blakiston, 1946.
Garner, W. W. "Tobacco Culture." *U.S. Dept. of Agriculture Farmers' Bulletin*, no. 571 (1949).
Garner, W. W., and H. A. Allard. "Effect of the Relative Length of Day and Night and Other Factors of the Environment on Growth and Reproduction in Plants." *Journal of Agricultural Research* 18, no. 11 (1920): 553–606.

Garner, W. W., H. A. Allard, and E. E. Clayton. "Superior Germ Plasm in Tobacco." In *U.S. Dept. of Agriculture Yearbook 1936*, 785–830. Washington, DC: GPO, 1936.

Garner, W. W., E. G. Moss, H. S. Yohe, F. B. Wilkinson, and O. C. Stine. "History and Status of Tobacco Culture." In *U.S. Dept. of Agriculture Yearbook 1922*, 395–459. Washington, DC: GPO, 1923.

Garside, Alston Hill. *Cotton Goes to Market*. New York: Frederick A. Stokes, 1935.

Gay, Carl W. *Productive Horse Husbandry*, 3rd ed. Philadelphia: Lippincott, 1920.

Genovese, Eugene D. *The Political Economy of Slavery: Studies in the Economy and Society of the Slave South*. New York: Vintage Books, 1967.

Giebelhaus, August W. "Farming for Fuel: The Alcohol Motor Fuel Movement of the 1930s." *Agricultural History* 54, no. 1 (1980): 173–84.

Giesen, James C. "The South's Greatest Enemy: The Cotton Boll Weevil and Its Lost Revolution, 1892–1930." PhD Diss., University of Georgia, 2004.

Gilbert, W. W. "The Root-Rot of Tobacco Caused by Thielavia Basicola." *U.S. Bureau of Plant Industry Bulletin*, no. 158 (1909).

Gilmore, Lester O. *Dairy Cattle Breeding*. Chicago: J. B. Lippincott, 1952.

Glenn, Charles C. "Stallion Legislation and the Horse-Breeding Industry." In *U.S. Dept. of Agriculture Yearbook 1916*, 289–99. Washington, DC: GPO, 1917.

Goodman, Lowell R., and R. Jerry Eidem. *The Atlas of North Dakota*. Fargo, ND: North Dakota Studies, 1976.

Goodman, Major M., and William L. Brown. "Races of Corn." In *Corn and Corn Improvement*, edited by G. F. Sprague and J. W. Dudley, 33–80. Madison, WI: American Society of Agronomy, 1988.

Graebner, L. A. "Economic History of the Fillmore Citrus Protection District." PhD Diss., University of California, Riverside, 1982.

Graves, R. R., and M. H. Fohrman. "Superior Germ Plasm in Dairy Herds." In *U.S. Dept. of Agriculture Yearbook 1936*, 997–1142. Washington, DC: GPO, 1936.

Gray, Lewis Cecil. *History of Agriculture in the Southern United States to 1860*. Reprint ed. 2 vols., Carnegie Institution of Washington Publication, no. 430. New York: Peter Smith, 1941.

Gray, R. B. "Development of the Agricultural Tractor in the United States." *U.S. Dept. of Agriculture Information Series*, no. 107 (1954).

The Green Mountain Patriot [Peacham, VT], Various issues and dates.

Griliches, Zvi. "Agriculture: Productivity and Technology." In *International Encyclopedia of the Social Sciences,* Vol. 1, edited by David L. Sills, 241–5. New York: Macmillan and Free Press, 1968.

Griliches, Zvi. "Hybrid Corn and the Economics of Innovation." *Science* 132, no. 3422 (1960): 275–80.

Grotewold, Andreas. *Regional Changes in Corn Production in the United States from 1909 to 1949*. University of Chicago Department of Geography Research Paper no. 40. Chicago: University of Chicago, 1955.

Gubler, Clark J. "Thiamine." In *Handbook of Vitamins*, edited by L. J. Machlin, 233–81. New York: Marcel Dekker, 1991.

Hackleman, J. C., and W. O. Scott. *A History of Seed Certification in the United States and Canada, Appendix 1*. Raleigh, NC: Association of Official Seed Certification Agencies, 1990.

Haeussler, G. J. "Insects as Destroyers." In *U.S. Dept. of Agriculture Yearbook 1952*, 141–6. Washington, DC: GPO, 1953.

Hagedoorn, A. L. *Animal Breeding*. London: Crosby Lockwood and Son, 1939.

Hahn, Barbara. "Making Tobacco Bright: Institutions, Information, and Industrialization in the Creation of an Agricultural Commodity, 1617–1937." PhD Diss., University of North Carolina, 2006.

Haines, Michael R., and Inter-university Consortium for Political and Social Research. *Historical, Demographic, Economic, and Social Data: The United States, 1790–2000* [Computer file]. Hamilton, NY: Colgate University; Ann Arbor, MI: ICPSR, 2004. Available from http://www.icpsr.umich.edu/cocoon/ICPSR/STUDY/02896.xml [Membership required].

Hairr, John. *Harnett County: A History*. Charleston, SC: Arcadia Publishing, 2002.

Hamilton, Laura M. "Stem Rust in the Spring Wheat Area in 1878." *Minnesota History* 20, no. 2 (1939): 156–64.

Hammond, John. *Farm Animals: Their Breeding, Growth, and Inheritance*. New York: Longmans, Green & Co., 1940.

Hancock, N. I. "A New Method of Delinting Cottonseed With Sulfuric Acid." *University of Tennessee Agricultural Experiment Station Circular*, no. 61 (1938): 1–2.

Handy, R. B. "History and General Statistics of Cotton." In *The Cotton Plant: Its History, Botany, Chemistry, Culture, Enemies, and Uses*, edited by Charles W. Dabney, 17–66. Washington, DC: GPO, 1896.

Haney, Philip B., Gary Herzog, and Phillip M. Roberts. "Boll Weevil Eradication in Georgia." In *Boll Weevil Eradication in the United States through 1999*, edited by Willard A. Dickerson, Anthony L. Brashear, James T. Brumley, Frank L. Carter, William Greenstette, and F. Aubrey Harris, 259–90. Memphis, TN: Cotton Foundation, 2001.

Haney, Philip B., W. J. Lewis, and W. R. Lambert. "Cotton Production and the Boll Weevil in Georgia: History, Cost of Control, and Benefits of Eradication." *Georgia Agricultural Experiment Stations Research Bulletin*, no. 428 (1996).

Hardeman, Nicholas P. *Shucks, Shocks, and Hominy Blocks: Corn as a Way of Life in Pioneer America*. Baton Rouge, LA: Louisiana State University Press, 1981.

Hargreaves, Mary. *Dry Farming in the Northern Great Plains: Years of Readjustment, 1925–1990*. Lawrence, KS: University of Kansas Press, 1993.

Hargrett Library. Broadside Collection, 1850–1859. Hargrett Rare Book & Manuscript Library. University of Georgia Libraries, Athens, GA.

Harnetty, Peter. *Imperialism and Free Trade: Lancashire and India in the Mid–Nineteenth Century*. Vancouver: University of British Columbia Press, 1972.

Harrison, George J. *A History of Cotton in California*. Typed manuscript, 1950.

Hartman, Hudson T., and Dale E. Kester. *Plant Propagation: Principles and Practices*, 2nd ed. Englewood Cliffs, NJ: Prentice Hall, 1968.

Hayami, Yujiro, and Vernon Ruttan. *Agricultural Development: An International Perspective*. Rev. and exp. ed. Baltimore, MD: Johns Hopkins University Press, 1985.

Hayden, Ada. "Distribution and Reproduction of Canada Thistle in Iowa." *American Journal of Botany* 21, no. 7 (1934): 355–73.

Hays, W. M. *Breeding Animals and Plants*. St. Anthony Park, MN: Farm Students' Review, 1904.

Head, Robert B. *Management and Control of Boll Weevils* [Web page]. Starkville, MS: Mississippi State University, 1992. Available from http://msucares.com/pubs/publications/p1830.htm.

Headlee, Thomas J., and James W. McColloch. "The Chinch Bug." *Kansas Agricultural Experiment Station Bulletin*, no. 191 (1913).

Headlee, Thomas J., and J. B. Parker. "The Hessian Fly." *Kansas Agricultural Experiment Station Bulletin*, no. 188 (1913): 83–138.

Hecht, Reuben W., and Glen T. Barton. "Gains in Productivity of Farm Labor." *U.S. Dept. of Agriculture Technical Bulletin*, no. 1020 (1950).

Hecht, Reuben W., and Keith R. Vice. "Labor Used for Field Crops." *U.S. Dept. of Agriculture Statistical Bulletin*, no. 144 (1954).

Hedrick, Ulysses Prentiss. *A History of Agriculture in the State of New York*. New York: New York Agricultural Society, 1933.

Helms, Douglas. "Just Looking for a Home: The Cotton Boll Weevil and the South." PhD Diss., Florida State University, 1977.

Helms, Douglas. "Revision and Revolution: Changing Cultural Control Practices for the Cotton Boll Weevil." *Agricultural History* 54, no. 1 (1980): 108–25.

Helms, Douglas. "Technological Methods for Boll Weevil Control." *Agricultural History* 53, no. 1 (1979): 286–99.

Henlein, Paul C. "Cattle Driving from the Ohio Country, 1800–1850." *Agricultural History* 28, no. 2 (1954): 83–94.

Henlein, Paul C. *Cattle Kingdom in the Ohio Valley, 1783–1860*. Lexington, KY: University of Kentucky Press, 1959.

Henry, W. A. "Experiments in Pig Feeding." In *Annual Report of the Agricultural Experiment Station of the University of Wisconsin, for the Year Ending June 30, 1889*, 6–41. Madison, WI: Democrat Printing, 1889.

Henry, W. A. *Feeds and Feeding: A Handbook for the Student and Stockman*, 2nd ed. Madison, WI: Author, 1900.

Henry, W. A. *Feeds and Feeding: A Hand-Book for the Student and Stockman*. Madison, WI: Author, 1898.

Henry, W. A., and F. B. Morrison. *Feeds and Feeding*, 18th ed. Madison, WI: The Henry–Morrison Co., 1923.

Herndon, G. Melvin. *William Tatham and the Culture of Tobacco, Including a Facsimile Reprint of an Historical and Practical Essay on the Culture and Commerce of Tobacco, by William Tatham*. Coral Gables, FL: University of Miami Press, 1969.

Heylin, Henry Brougham. *Buyers and Sellers in the Cotton Trade*. London: Charles Griffin & Company, 1913.

Hibbard, Benjamin Horace. "The History of Agriculture in Dane County Wisconsin: A Thesis Submitted for the Degree of Doctor of Philosophy, University of Wisconsin, 1902." *Bulletin of the University of Wisconsin*, no. 101 (1904): 67–214.

Higgins, Floyd Halleck. Collection, 1796–1982. Special Collections. UC Davis Libraries, Davis, CA.

High Plains Regional Climate Center. *Dickinson Experiment Station, ND (322188) Period of Record Monthly Climate Summary* [Web page]. Lincoln, NE: The Center, 2006 [updated occasionally]. Available from http://www.hprcc.unl.edu/cgi-bin/cli_perl_lib/cliRECtM.pl?nd2188.

High Plains Regional Climate Center. *Hays 1 S, KS (143527) Period of Record Monthly Climate Summary* [Web page]. Lincoln, NE: The Center, 2006 [updated occasionally]. Available from http://www.hprcc.unl.edu/cgi-bin/cli_perl_lib/cliRECtM.pl?ks3527.

Hilgard, Eugene W. *Report on Cotton Production in the United States, Also Embracing Agricultural and Physico-Geographical Descriptions of the Several Cotton States and of California,* Part 1, *Mississippi Valley and Southwestern States*. Washington, DC: GPO, 1884.

Hilliard, Sam B. "Pork in the Ante-Bellum South: The Geography of Self-Sufficiency." *Annals of the Association of American Geographers* 59, no. 3 (1969): 461–80.

Hinds, W. E. "Facing the Boll Weevil Problem in Alabama." *Alabama Polytechnic Institute, Alabama Agricultural Experiment Station Bulletin*, no. 178 (1914): 87–99.

A History and Biographical Cyclopaedia of Butler County Ohio, with Illustrations and Sketches of its Representative Men and Pioneers. Cincinnati, OH: Western Biographical Publishing, 1882.

Hittel, John S. *The Resources of California: Comprising Agriculture, Mining, Geography, Climate, Commerce, Etc. Etc. and the Past and Future Development of the State*. San Francisco, CA: A. Roman & Co., 1863.

Hodgson, Robert W. "The California Fruit Industry." *Economic Geography* 9, no. 4 (1933): 337–55.

Holley, Donald. *The Second Great Emancipation: The Mechanical Cotton Picker, Black Migration, and How They Shaped the Modern South*. Fayetteville, AR: University of Arkansas Press, 2000.

Holley, William C., and Lloyd E. Arnold. "Changes in Technology and Labor Requirements in Crop Production: Cotton." *National Research Project Report*, no. A–7 (1938).

Holm, LeRoy. "Weeds Problems in Developing Countries." *Weed Science* 17, no. 1 (1969): 113–8.

Holstein–Friesian Association of America. *Herd–Book*, Vol. 139. Brattleboro, VT: The Association, 1955.

Holt, Rackham. *George Washington Carver: An American Biography*. Garden City, NY: Doubleday Doran, 1943.

Hopper, Norman W., and Robert G. McDaniel. "The Cotton Seed." In *Cotton: Origin, History, Technology, and Production*, edited by C. Wayne Smith and J. Tom Cothren, 289–318. New York: John Wiley & Sons, 1999.

Horse Association of America. Leaflets. Special Collections. National Agricultural Library, Beltsville, MD.

Horton, J. R., and A. F. Satterthwait. "The Chinch Bug and Its Control." *U.S. Dept. of Agriculture Farmers' Bulletin*, no. 1223 (1922).

Horvath, A. A. *The Soybean Industry*. New York: Chemical Publishing, 1938.

Houck, Ulysses Grant. *The Bureau of Animal Industry of the United States Department of Agriculture: Its Establishment, Achievements and Current Activities*. Washington, DC: Author, 1924.

Houghton, Frederick L. *Holstein–Friesian Cattle: A History of the Breed and Its Development in America*. Brattleboro, VT: Holstein–Friesian Register, 1897.

Howard, L. O. "The Chinch Bug." *U.S. Dept. of Agriculture Division of Entomology Bulletin*, no. 17 (1922).

Howard, L. O. "Insects Affecting the Cotton Plant." *U.S. Dept. of Agriculture Farmers' Bulletin*, no. 47 (1896).

Howard, L. O. "The Insects Which Affect the Cotton Plant in the United States." In *The Cotton Plant: Its History, Botany, Chemistry, Culture, Enemies, and Uses*, edited by Charles Dabney, 317–50. Washington, DC: GPO, 1896.

Howard, L. O., and C. L. Marlatt. "The San Jose Scale: Its Occurrences in the United States with a Full Account of Its Life History and the Remedies to Be Used Against It." *U.S. Dept. of Agriculture Division of Entomology Bulletin, New Series*, no. 3 (1896).

Howard, Robert West. *The Horse in America*. Chicago: Follett Publishing, 1965.

Howell, L. D., and John S. Burgess, Jr. "Farm Prices of Cotton as Related to Its Grade and Staple Length in the United States, Seasons 1928–29 to 1932–33." *U.S. Dept. of Agriculture Technical Bulletin*, no. 493 (1936).

Howell, L. D., and Leonard J. Watson. "Cotton Prices in Relation to Cotton Classification Service and to Quality Improvement." *U.S. Dept. of Agriculture Technical Bulletin*, no. 699 (1939).

Hudson, John C. *Making the Corn Belt: A Geographic History of Middle-Western Agriculture*. Bloomington, IN: Indiana University Press, 1994.

Hughes, Jonathan R. T. *American Economic History*. Glenview, IL: Scott, Foresman, 1987.

Hunter, Brooke. "Creative Destruction: The Forgotten Legacy of the Hessian Fly." In *The Economy of Early America: Historical Perspectives and New Directions*, edited by Cathy Matson, 236–62. University Park, PA: Pennsylvania State University Press, 2006.

Hunter, W. D. "The Boll-Weevil Problem." *U.S. Dept. of Agriculture Farmers' Bulletin*, no. 1329 (1923).

Hunter, W. D. "Methods of Controlling the Boll Weevil." *U.S. Dept. of Agriculture Farmers' Bulletin*, no. 163 (1903): 1–15.

Huntington, Ellsworth. "The Distribution of Domestic Animals." *Economic Geography* 1, no. 2 (1925): 143–72.

Hutchens, T. W. "Tobacco Seed." In *Tobacco: Production, Chemistry, and Technology*, edited by D. Layten Davis and Mark T. Nielsen, 4A, 66–9. Oxford; Maiden, MA: Blackwell Science, 1999.

Hutchinson, John N. "Northern California from Haraszthy to the Beginnings of Prohibition." In *The University of California/Sotheby Book of California Wine*, edited by Doris Muscatine, Maynard A. Amerine, and Bob Thompson, 30–48. Berkeley, CA: University of California Press, 1984.

Hutchinson, William A., and Samuel H. Williamson. "The Self-Sufficiency of the Antebellum South: Estimates of the Food Supply." *Journal of Economic History* 31, no. 3 (1971): 591–612.

Hymowitz, T., and W. R. Shurtleff. "Debunking Soybean Myths and Legends in the Historical and Popular Literature." *Crop Science* 45, no. 2 (2005): 473–6.

Hyslop, J. A. *Losses Occasioned by Insects, Mites, and Ticks in the United States.* Washington, DC: U.S. Bureau of Entomology and Plant Quarantine, Division of Insect Pest Survey and Information, 1938.

Illinois Department of Agriculture. *17th Annual Report.* Springfield, IL: State Printing Office, 1934.

Illinois Department of Agriculture. *29th Annual Report.* Springfield, IL: State Printing Office, 1946.

International Bureau of the American Republics. *Mexico: Geographical Sketch, Natural Resource Laws, Economic Conditions, Actual Development, Prospects of Future Growth.* Washington, DC: GPO, 1904.

International Museum of the Horse. *Draft Animals in Early America* [Web page]. Lexington, KY: Kentucky Horse Park [cited 31 May 2007]. Available from http://www.kyhorsepark.com/museum/drafthorse.php?pageid=11.

Inter-university Consortium for Political and Social Research. *Historical, Demographic, Economic, and Social Data: The United States, 1790–1970* [Computer file]. Ann Arbor, MI: ICPSR [producer and distributor], 197? Available from http://www.icpsr.umich.edu [Membership required].

Iowa Department of Agriculture. *Year Book 1941.* Des Moines, IA: The Department, 1941.

Iowa State University Department of Entomology. *The European Corn Borer: Introduction* [Web page]. Ames, IA: The University, 1996, c2006 [cited 3 July 2007]. Available from http://www.ent.iastate.edu/pest/cornborer/intro/intro.html.

Iranzo, Susana, Alan L. Olmstead, and Paul W. Rhode. "Historical Perspectives on Exotic Pests and Diseases in California." In *Exotic Pests and Diseases of California*, edited by Daniel Sumner, 55–67. Ames, IA: Iowa State University Press, 2002.

J.M. Smucker. *Crisco: History and Timeline* [Web page]. Orrville, OH: J.M. Smucker Co., 2007 [6 July 2007]. Available from http://www.crisco.com/about/history/1911.asp.

James, W. Clive. "Estimated Losses of Crops from Plant Pathogens." In *CRC Handbook of Pest Management in Agriculture*, Vol. 1, edited by David Pimentel, 15–77. Boca Raton, FL: CRC Press, 1981.

Jennings, Ralph A. "Consumption of Feed by Livestock, 1909–47." *U.S. Dept. of Agriculture Circular*, no. 836 (1949).

Jennings, Robert. *Cattle and Their Diseases.* Philadelphia: John E. Potter, 1864.

Jensen, Joan M. "Butter Making and Economic Development in Mid-Atlantic America from 1750 to 1850." *Signs* 13, no. 4 (1988): 813–29.

Jensen, Joan M. *Loosening the Bonds: Mid-Atlantic Farm Women, 1750–1850.* New Haven, CT: Yale University Press, 1986.

Johnson, Aaron G., and James G. Dickson. "Wheat Scab and Its Control." *U.S. Dept. of Agriculture Farmers' Bulletin*, no. 1217 (1921).

Johnson, D. Gale. "Agriculture and the Wealth of Nations." *American Economic Review* 87, no. 2 (1997): 1–12.

Johnson, D. Gale. "Review of Vaclav Smil, 'Enriching the Earth: Fritz Haber, Carl Bosch, and the Transformation of World Food Production'." In *EH.NET*. Oxford, OH: EH.NET Economic History Services, November 2001.

Johnson, D. Gale, and Robert L. Gustafson. *Grain Yields and the American Food Supply: An Analysis of Yield Changes and Possibilities*. Chicago: University of Chicago Press, 1962.

Johnson, George Fiske. "The Early History of Copper Fungicides." *Agricultural History* 9, no. 2 (1935): 67–79.

Johnson, James. "Steam Sterilization of Soil for Tobacco and Other Crops." *U.S. Dept. of Agriculture Farmers' Bulletin*, no. 1629 (1930): 1–13.

Johnson, James. "Tobacco Diseases and Their Control." *U.S. Dept. of Agriculture Department Bulletin*, no. 1256 (1924): 1–56.

Johnson, Paul C. *Farm Inventions in the Making of America*. Des Moines, IA: Wallace–Homestead Book Co., 1976.

Johnson, W. H. *Cotton and Its Production*. London: Macmillan, 1926.

Johnston, C. O., and L. E. Melchers. "The Control of Sorghum Kernel Smut and the Effect of Seed Treatments on Vitality of Sorghum Seed." *Kansas Agricultural Experiment Station Technical Bulletin*, no. 22 (1928).

Jones, Donald F. *Genetics in Plant and Animal Improvement*. New York: Wiley, 1925.

Jones, Robert Leslie. "The Beef Cattle Industry in Ohio Prior to the Civil War." *Ohio History* 64, Pt. 1 (1955): 168–94, and Pt. 2 (1955): 287–319.

Jones, Robert Leslie. *A History of Agriculture in Ohio to 1880*. Kent, OH: Kent State University Press, 1983.

Jones, Robert Leslie. "The Horse and Mule Industry in Ohio to 1865." *Mississippi Valley Historical Review* 33, no. 1 (1946): 61–88.

Jordan, Terry G. *North American Cattle-Ranching Frontiers: Origins, Diffusion, and Differentiation*. Albuquerque, NM: University of New Mexico Press, 1993.

Journal of Economic Entomology (Various issues and dates).

Journal of the American Veterinary Medical Association (Various years).

Judd, Sylvester. *History of Hadley Including the Early History of Hatfield, South Hadley, Amherst and Granby, Massachusetts*. Northampton, MA: Metcalf, 1863.

Judkins, Henry F. *Principles of Dairying: Testing and Manufactures*. New York: Wiley, 1924.

Kalk, Bruce H. "The Extinction of Historic Breeds of Swine." *American Minor Breeds Conservancy Newsletter* 8, no. 3 (1991): 1–4.

Kansas State Board of Agriculture. *Stallion Registry 1939*. Topeka, KS: State Printer, 1940.

Kantor, Shawn. *Politics and Property Rights: The Closing of the Open Range in the Postbellum South*. Chicago: University of Chicago Press, 1998.

Karlinsky, Nahum. *California Dreaming: Ideology, Society, and Technology in the Citrus Industry of Palestine, 1890–1939*. Albany, NY: State University of New York Press, 2005.

Kauffman, Kyle. "Why Was the Mule Used in Southern Agriculture?: Empirical Evidence of Principal-Agent Solutions." *Explorations in Economic History* 30, no. 3 (1993): 336–51.

Kemmerer, Donald L. "The Pre–Civil War South's Leading Crop, Corn." *Agricultural History* 23 (1949): 236–9.

Kerr, John W. "Certified Milk and Infants' Milk Depots." In *Milk and Its Relation to the Public Health*, 565–88. Washington, DC: GPO, 1908.

Kiesselbach, T. A. "Winter Wheat Investigations." *Nebraska Agricultural Experiment Station Research Bulletin*, no. 31 (1925).

Killebrew, J. B. *Report on the Culture and Curing of Tobacco in the United States.* Washington, DC: GPO, 1884.

Kilpatrick, Dr. "Historical and Statistical Collections of Louisiana: The Parish of Catahoula, Part 2." *De Bow's Review* 2, no. 6 (1852): 631–46.

King, Wilford I. "The Gasoline Engine and the Farmer's Incomes." *Journal of Farm Economics* 11, no. 1 (1929): 64–73.

Kingsnorth, Paul. "Bovine Growth Hormones." *The Ecologist* 28, no. 5 (1998): 266–9.

Kinsman, C. D. "An Appraisal of Power Used on Farms in the United States." *U.S. Dept. of Agriculture Bulletin*, no. 1348 (1925).

KINZE Manufacturing. *History of Planting: Seeding Milestones 1800–1997* [Web page]. Williamsburg, IA: KINZE, c2007 [cited 3 July 2007]. Available from http://www.kinzemfg.com/company/plantrev/timeline.html.

Kirkland, Edward C. *History of American Life*. New York: Appleton–Century–Crofts, 1951.

Klippart, John H. *The Wheat Plant: Its Origin, Culture, Growth, Development, Composition, Varieties, Diseases, Etc.* New York: A. O. Moore & Co., 1860.

Kloppenburg, Jack Ralph, Jr. *First the Seed: The Political Economy of Plant Biotechnology, 1492–2000*. New York: Cambridge University Press, 1988.

Klose, Nelson. *America's Crop Heritage: The History of Foreign Plant Introduction by the Federal Government.* Ames, IA: Iowa State College Press, 1950.

Knopf, Henry A. "Changes in Wheat Production in the United States." PhD Diss., Cornell University, 1967.

Krugman, Paul R. "Increasing Returns, Monopolistic Competitions, and International Trade." *Journal of International Economics* 9, no. 4 (1979): 460–79.

Kupperman, Karen Ordahl. "The Puzzle of the American Climate in the Early Colonial Period." *American Historical Review* 87, no. 5 (1982): 1262–89.

Labate, Joanne A., Kendall R. Lamkey, Sharon E. Mitchell, Stephen Kresovich, Hillary Sullivan, and John S. C. Smith. "Molecular and Historical Aspects of Corn Belt Dent Diversity." *Crop Science* 43, no. 1 (2003): 89–91.

Lakwete, Angele. *Inventing the Cotton Gin: Machine and Myth in Antebellum America*. Baltimore, MD: Johns Hopkins University Press, 2003.

Lamb, Robert Byron. *The Mule in Southern Agriculture*, University of California Publication in Geography, vol. 15. Berkeley, CA: University of California Press, 1963.

Lamborn, Leebert Lloyd. *Cottonseed Products: A Manual of the Treatment of Cottonseed for its Products and their Utilization in the Arts*. New York: Van Nostrand, 1904.

Lampard, Eric E. *The Rise of the Dairy Industry in Wisconsin: A Study in Agricultural Change, 1820–1920*. Madison, WI: State Historical Society of Wisconsin, 1963.

Landes, David S. *Unbound Prometheus: Technological Change and Industrial Development in Western Europe from 1750 to Present*, 2nd ed. Cambridge, U.K.; New York: Cambridge University Press, 2003.

Lang, Brian, and Mike White. "Canada Thistle, Management." *Iowa State University Extension Fact Sheet*, no. BL–31 (2001).

Lange, Fabian, Alan L. Olmstead, and Paul W. Rhode. "The Impact of the Boll Weevil, 1892–1940." *IGA Working Paper*, no. 2007:06 (2007).

Large, E. C. *The Advance of Fungi*. New York: Henry Holt, 1940.

Larson, C. W., L. M. Davis, C. A. Juve, O. C. Stine, A. E. Wight, A. J. Pistor, and C. E. Langworthy. "The Dairy Industry." In *U.S. Dept. of Agriculture Yearbook 1922*, 281–394. Washington, DC: GPO, 1923.

Lauer, Joe. "Management Needs for Specialty Corn Hybrids."*Agronomy Advice* (December 1998).

Lawton, Harry W. "A History of Citrus in Southern California." In *A History of Citrus in the Riverside Area*, edited by Esther H. Klotz, Harry W. Lawton, and Joan H. Hall, 6–11. Riverside, CA: Riverside Museum Press, 1969.

Leavitt, Charles T. "Attempts to Improve Cattle Breeds in the United States, 1790–1860." *Agricultural History* 7, no. 2 (1933): 51–67.

Leavitt, Charles T. "The Meat and Dairy Livestock Industry, 1819–1860." PhD Diss., University of Chicago, 1931.

Lebergott, Stanley. *The Americans: An Economic Record*. New York: Norton, 1984.

Lebergott, Stanley. "The Demand for Land: The United States, 1820–1860." *Journal of Economic History* 45, no. 2 (1985): 181–212.

Leding, A. R. "Community Production of Acala Cotton in New Mexico." *U.S. Dept. of Agriculture Circular*, no. 314 (1934).

Leigh, G. J. *World's Greatest Fix: A History of Nitrogen and Fertilizer*. New York: Oxford University Press, 2004.

Leithead, Horace L., Lewis L. Yarlett, and Thomas N. Shiflet. "100 Native Forage Grasses in 11 Southern States." *U.S. Dept. of Agriculture Handbook*, no. 389 (1971).

Lemmer, George E. "The Spread of Improved Cattle Breeds in the United States, 1790–1860." *Agricultural History* 21, no. 2 (1947): 79–93.

Leonard, B. R., J. B. Graves, and P. C. Ellsworth. "Insect and Mite Pests of Cotton." In *Cotton: Origin, History, Technology, and Production*, edited by C. Wayne Smith and J. Tom Cothren, 489–551. New York: John Wiley, 1999.

Libecap, Gary D., and Zeynep Kocabiyik Hansen. "'Rain Follows the Plow' and Dryfarming Doctrine: The Climate Information Problem and Homestead Failure in the Upper Great Plains, 1890–1925." *Journal of Economic History* 62, no. 1 (2002): 86–120.

Lindert, Peter H. "Long-Run Trends in American Farmland Values." *Agricultural History* 62, no. 3 (1988): 45–85.

Livingston, Robert R. *Essay on Sheep: Their Varieties: Account of the Merinoes of Spain, France, Etc.*, Early American Imprints, Series 2, no. 28966. Concord, NH: Daniel Cooledge, 1813.

Lloyd, William A. *J. S. Leaming and his Corn.* Wooster, OH: Ohio Agricultural Extension Station, 1912.

Loegering, W. Q., C. O. Johnston, and J. W. Hendrix. "Wheat Rusts." In *Wheat Improvement*, edited by Karl S. Quisenberry and L. P. Reitz, 307–35. Madison, WI: American Society of Agronomy, 1967.

Loehr, Rodney C. "The Influence of English Agriculture on American Agriculture, 1775–1825." *Agricultural History* 11, no. 1 (1937): 3–15.

Loomis, Ralph A., and Glen T. Barton. "Productivity of Agriculture: United States, 1870–1958." *U.S. Dept. of Agriculture Technical Bulletin*, no. 1238 (1961).

Lorain, John. *Nature and Reason Harmonized in the Practice of Husbandry.* Philadelphia: H. C. Carey & I. Lea, 1825.

Los Angeles Times [Los Angeles, CA], Various issues and dates.

Louisville Courier–Journal [Louisville, KY], Various issues and dates.

Lowery, J. C. "Cotton Improvement." *Alabama Extension Service Circular*, no. 144 (1938).

Lucas, George B. *Diseases of Tobacco*, 2nd ed. New York: Scarecrow Press, 1965.

Lush, Jay L. *Animal Breeding Plans*, 3rd ed. Ames, IA: Collegiate Press, 1945.

Lush, Jay L., and A. L. Anderson. "A Genetic History of Poland–China Swine: Early Breed History: The 'Hot Blood' Versus the 'Big Type'." *Journal of Heredity* 30, no. 4 (1939): 149–56.

Lush, Jay L., and A. L. Anderson. "A Genetic History of Poland–China Swine: Founders of the Breed, Prominent Individuals, Length of Generation." *Journal of Heredity* 30, no. 5 (1939): 219–24.

Lusteck, Robert. "The Migrations of Maize into the Southeastern United States." In *Histories of Maize: Multidisciplinary Approaches to the Prehistory, Linquistics, Biogeography, Domestication, and Evolution of Maize*, edited by John E. Staller, Robert H. Tykot, and Bruce F. Benz, 521–8. Amsterdam: Elsevier, 2006.

Lyman, Joseph B. *Cotton Culture.* New York: Orange Judd, 1868.

Malin, Donald F. *The Evolution of Breeds.* Des Moines, IA: Wallace Publishing, 1923.

Malin, James C. *Winter Wheat in the Golden Belt of Kansas.* Lawrence, KS: University of Kansas Press, 1944.

Mangelsdorf, Paul C. *Corn: Its Origin, Evolution, and Improvement.* Cambridge, MA: Belknap Press of Harvard University Press, 1974.

Mann, Charles C. *1491: New Revelations of the Americas Before Columbus.* New York: Knopf, 2005.

Manners, Ian R. "The Persistent Problem of the Boll Weevil: Pest Control in Principle and in Practice." *Geographical Review* 69, no. 1 (1979): 25–42.

Markley, Klare S., and Warren H. Goss. *Soybean Chemistry and Technology.* Ithaca, NY: State College of Agriculture at Cornell University, 1944.

Marlatt, C. L. "The Annual Losses Occasioned by Destructive Insects in the United States." In *U.S. Dept. of Agriculture Yearbook 1904*, 461–74. Washington, DC: GPO, 1905.
Marlatt, C. L. "Fifty Years of Entomological Progress, Part 1, 1889–1899." *Journal of Economic Entomology* 33, no. 1 (1940): 8–15.
Marlatt, C. L. "Losses Due to Insects, Part 1, Insects as a Check on Agricultural Production and as a Source of Waste to Accumulated Supplies." *Report of the National Conservation Commission* 3, 60th Cong., 2nd sess., S. Doc. 676 (1909): 301–9.
Marlatt, C. L. "The Principal Insect Enemies of Growing Wheat." *U.S. Dept. of Agriculture Farmers' Bulletin*, no. 132 (1901).
Marlatt, C. L. "The San Jose Scale: Its Native Home and Natural Enemy." In *U.S. Dept. of Agriculture Yearbook 1902*. Washington, DC: GPO, 1902.
Marshall, F. R. *Breeding Farm Animals*. Chicago: Breeders' Gazette, 1912.
Marx, Karl. *Capital*, 1st English ed., Vol. 1. London: Sonnenschein, 1887.
Mason, A. Freeman. *Spraying, Dusting and Fumigating of Plants*. New York: Macmillan, 1928.
McClain, C. "A New Look at Eliza Tibbets." *California Citrograph* 61, no. 12 (1976): 449–53.
McClelland, Peter D. *Sowing Modernity: America's First Agricultural Revolution*. Ithaca, NY: Cornell University Press, 1997.
McColloch, James W. "The Hessian Fly in Kansas." *Kansas Agricultural Experiment Station Technical Bulletin*, no. 11 (1923).
McDonald, Forrest, and Grady McWhiney. "The South from Self-Sufficiency to Peonage: An Interpretation." *American Historical Review* 85, no. 5 (1980): 1095–118.
McGaw, Judith A. "Specialization and American Agricultural Innovation in the Early Industrial Era: John Hare Powel and Livestock Breeding." *Business and Economic History*, 2nd ser. 13 (1984): 134–49.
McGinty, Brian. *Strong Wine: The Life and Legend of Agoston Haraszthy*. Stanford, CA: Stanford University Press, 1998.
McGregor, S. E. "Insect Pollination of Cultivated Crops." *U.S. Dept. of Agriculture Handbook*, no. 496 (1976).
McInnes, Jean, and Raymond Fogelman. "Wheat Scab in Minnesota." *University of Minnesota Agricultural Experiment Station Technical Bulletin*, no. 18 (1923): 1–43.
McKeever, H. G. "Community Production of Acala Cotton in the Coachella Valley of California." *U.S. Dept. of Agriculture Bulletin*, no. 1467 (1927).
McKelvey, Blake. "The Flower City: Center for Nurseries and Fruit Orchards." *Rochester Historical Society Publications* 18 (1940): 121–69.
McKibben, Eugene G., and R. Austin Griffin. *Changes in Farm Power and Equipment: Tractors, Trucks, and Automobiles*, Works Progress Administration. National Research Project. Studies of Changing Techniques and Employment in Agriculture Report, no. A–9. Philadelphia: WPA, 1938.
McMillen, Wheeler. *Too Many Farmers*. New York: William Morrow, 1929.

McMullen, Marcia, Roger Jones, and Dale Gallenberg. "Scab of Wheat and Barley: A Re-Emerging Disease of Devastating Impact." *Plant Disease* 81 (1997): 1340–8.

McWhorter, Clyde C. "Introduction and Spread of Johnsongrass in the United States." *Weed Science* 19, no. 5 (1971): 496–500.

Meeks, Harold A. *Time and Change in Vermont.* Chester, CT: Globe Pequot Press, 1986.

Menard, Russell R. "The Tobacco Industry in the Chesapeake Colonies, 1617–1703: An Interpretation." *Research in Economic History* 5, (1980): 109–77.

Mendenhall, Marjorie Stratford. "A History of Agriculture in South Carolina 1790 to 1860: An Economic and Social Study." PhD Diss., University of North Carolina, 1940.

Merchant, Carolyn. *Ecological Revolutions: Nature, Gender, and Science in New England.* Chapel Hill, NC: University of North Carolina Press, 1989.

Meredith, William R., Jr., and Robert R. Bridge. "Genetic Contribution to Yield Changes in Upland Cotton." In *Genetic Contributions to Yield Gains of Five Major Crop Plants*, edited by W. R. Fehr, 75–87. Madison, WI: CSSA, 1984.

Merrill, Gilbert R., Alfred R. Macormac, and Herbert R. Mausersberger. *American Cotton Handbook*, 2nd rev. ed. New York: Textile Book Publishers, 1949.

Metcalf, Clell Lee, and Wesley Pillsbury Flint. *Destructive and Useful Insects: Their Habits and Control*, 4th ed. New York: McGraw–Hill, 1962.

Middlesex Gazette [Middletown, CT], Various issues and dates.

Middleton, Arthur Pierce. *Tobacco Coast: A Maritime History of the Chesapeake Bay in the Colonial Era*. Baltimore, MD: Johns Hopkins University Press, 1953.

Midwestern Regional Climate Center. Midwest Climate Watch. *Historical Climate Summaries for Wooster_Exp_Stn, OH* [Web site]. Champaign, IL: The Center, 2007 [updated regularly]. Available from http://mcc.sws.uiuc.edu/cliwatch/watch.htm.

Millardet, Alexis. *The Discovery of Bordeaux Mixture*. Ithaca, NY: American Phytopathological Society, 1933.

Miller, J. D., R. M. Hosford, R. W. Stack, and G. D. Statler. "Diseases of Durum Wheat." In *Durum Wheat: Chemistry and Technology*, edited by Giuseppe Fabriani and Claudia Lintas, 69–92. St. Paul, MN: American Association of Cereal Chemists, 1988.

Miller, Robert D., and Donald J. Fowlkes. "Dark Fire-Cured Tobacco." In *Tobacco: Production, Chemistry, and Technology*, edited by D. Layten Davis and Mark T. Nielsen, 5D, 164–81. Oxford: Blackwell Science, 1999.

Millis, Dale, Morrie Bryant, and Clive Holland. "White Corn Production and Uses." In *Crop Insights (Pioneer Hi–Bred International, Inc.)*, 1–4, 1998.

Mills, Minnie Tibbets. "Luther Calvin Tibbets: Founder of the Navel Orange Industry of California." *Quarterly Publication of the Historical Society of Southern California* 25, no. 4 (1943): 127–61.

Minnesota Agricultural Experiment Station. *Annual Report 1894*. St. Paul, MN: Pioneer Press, 1895.

Mischka, Joseph. *The Percheron Horse in America.* Cedar Rapids, IA: Heart Prairie Press, 1991.

Mississippi Department of Archives and History. Online Archives. Archives & Library, Jackson, MS.

Mitchell, Edward B. "Animal Diseases and Our Food Supply." In *U.S. Dept. of Agriculture Yearbook 1915*, 159–72. Washington, DC: GPO, 1916.

Mitich, Larry W. "Colonel Johnson's Grass: Johnsongrass." *Weed Technology* 1 (1987): 112–3.

Moloney, John F. "Cottonseed." In *Cotton: Production, Marketing, and Utilization*, edited by W. B. Andrews, 431–63. State College, MS: W. B. Andrews, 1950.

Montgomery, E. G. *The Corn Crops: A Discussion of Maize, Kafirs, and Sorghums as Grown in the United States and Canada.* 1916 reprint ed. New York: Macmillan, 1913.

Montgomery, E. G. "Wheat Breeding Experiments." *Nebraska Agricultural Experiment Station Bulletin*, no. 125 (1912).

Moore, F., and B. Moore. "When Was the Navel Orange Imported?" *California Citrograph* 36 (1951): 219.

Moore, John Hebron. *Agriculture in Ante-Bellum Mississippi.* New York: Bookman Associates, 1958.

Moore, John Hebron. "Cotton Breeding in the Old South." *Agricultural History* 30, no. 3 (July 1956): 95–104.

Moore, John Hebron. *The Emergence of the Cotton Kingdom in the Old Southwest, Mississippi, 1770–1860.* Baton Rouge, LA: Louisiana State University Press, 1988.

Moore, John Hebron, and Margaret D. Moore. *Cotton Culture on the South Carolina Frontier: Journal of John Baxter Fraser, 1804–1807*, 1997.

Morilla Critz, José, Alan L. Olmstead, and Paul W. Rhode. "'Horn of Plenty': The Globalization of Mediterranean Horticulture and the Economic Development of Southern Europe, 1880–1930." *Journal of Economic History* 59, no. 2 (1999): 316–52.

Morilla Critz, José, Alan L. Olmstead, and Paul W. Rhode. "International Competition and the Development of the Dried Fruit Industry, 1880–1930." In *The Mediterranean Response to Globalization Before 1950*, edited by S. Pamuk and J. Williamson, 199–232. London: Routledge, 2000.

Morrison, F. B. *Feeds and Feeding: A Handbook for the Student and Stockman.* 20th ed. Ithaca, NY: Morrison Publishing, 1936.

Morrison, William P., and Frank B. Pearis. "Response Model Concept and Economic Impact." In *Response Model for an Introduced Pest – The Russian Wheat Aphid*, edited by Shannon S. Quisenberry and Frank B. Pearis, 1–11. Lanham, MD: Entomological Society of America, 1998.

Morrow, G. E., Frank D. Gardner, and E. H. Farrington. "Corn and Oats Experiments 1893." *University of Illinois Agricultural Experiment Station Bulletin*, no. 31 (1894): 333–88.

Morse, Jedidiah. *The American Geography.* London: John Stockdale, 1794.

Morse, William J., and J. L. Cartter. "Improvement in Soybeans." In *U.S. Dept. of Agriculture Yearbook 1937*, 1154–89. Washington, DC: GPO, 1937.

Morton, Lucie T. *Winegrowing in Eastern America.* Ithaca, NY: Cornell University Press, 1985.

Moses, H. Vincent. "G. Harold Powell and the Corporate Consolidation of the Modern Citrus Enterprise, 1904–1922." *Business History Review* 69, no. 2 (1995): 119–55.

Mosiman, Elizabeth A. *Philadelphia Society for Promoting Agriculture Library and Archive at the Penn Library: Overview* [Web page]. The Society, 1999. Available from http://www.pspaonline.com/collection2.html.

Mumford, Frederick B. *The Breeding of Animals*. New York: Macmillan, 1917.

Mumford, Frederick B. "Some of the Principles of Animal-Breeding." In *Cyclopedia of American Agriculture: A Popular Survey of Agricultural Conditions*, edited by L. H. Bailey, 28–53. New York: Macmillan, 1907–1909.

Murphey, R. M., M. C. Torres Penedo, C. Stormont, and C. J. Bahre. "Blood Type Analyses of Creole-Like Cattle: A Comparison with Longhorns and Mixed Controls." *Journal of Heredity* 70, no. 4 (1979): 231–4.

Murray, Stanley N. *The Valley Comes of Age: A History of Agriculture in the Valley of the Red River of the North, 1812–1920*. Fargo, ND: North Dakota Institute for Regional Studies, 1967.

Musoke, Moses S., and Alan L. Olmstead. "The Rise of the Cotton Industry in California: A Comparative Perspective." *Journal of Economic History* 42, no. 2 (1982): 385–412.

Myers, J. Arthur. *Man's Greatest Victory over Tuberculosis*. Springfield, IL: Charles C. Thomas, 1940.

Myers, J. Arthur, and James H. Steele. *Bovine Tuberculosis Control in Man and Animals*. St. Louis, MO: W. H. Green, 1969.

Myrick, Herbert. *The Book of Corn: A Complete Treatise upon the Culture, Marketing, and Uses of Maize in America and Elsewhere*. New York: Orange Judd, 1903.

Myrick, Herbert. "The Improved Silo and Ensilage." In *Reports of the United States Commissioners to the Universal Exposition of 1889 at Paris*, Vol. 5, *Agriculture*, edited by Charles V. Riley, 743–6. Washington, DC: GPO, 1891.

National Research Council. *Growth and Recommended Nutrient Allowances for Cattle*. Washington, DC: National Academy of Sciences, 1924.

National Research Council. *Livestock Disease Eradication: Evaluation of the Cooperative State–Federal Bovine Tuberculosis Eradication Program* [Online book]. Washington, DC: National Academy Press, 1994. Available from http://books.nap.edu/openbook.php?isbn=NI000139.

National Research Council. *Plan for Cooperative Experiments on Protein Requirements for Growth in Cattle*. Washington, DC: National Academy of Sciences, 1917.

National Research Council. *Recommended Nutrient Allowances for Swine*, 1st ed. Recommended Nutrient Allowances for Domestic Animals, no. 2. Washington, DC: The Council, 1944.

Nebraska Department of Agriculture. *Nebraska Agricultural Statistics, 1941–1949*. Lincoln, NE: State–Federal Division of Agricultural Statistics, Various years.

Neely, J. Winston. "Challenges of Cotton Mechanization to the Plant Breeder." In *Proceedings of the Second Annual Beltwide Cotton Mechanization Conference*,

October 14–16, 1948, 13–22. Memphis, TN: National Cotton Council of America, 1948.

Neely, J. Winston. "Cotton Breeders Pave the Way – For Higher Quality and Yield." *Cotton Trade Journal* 38, no. 10 (1958): 74–5.

New England Farmer [Boston, MA], Various issues and dates.

New York Times [New York, NY], Various issues and dates.

New York State Agricultural Society. *Transactions 1850*. Vol. 10. Albany, NY: Printer to the Legislature, 1851.

Newman, C. L. "The Comparative Yield of Corn from Seed of the Same Variety Grown in Different Latitudes." *Arkansas Agricultural Experiment Station Bulletin*, no. 59 (1899).

Niles, G. A., and C. V. Feaster. "Breeding." In *Cotton*, edited by R. J. Kohel and C. F. Lewis, 201–31. Madison, WI: American Society of Agronomy, 1984.

Nimmo, Joseph. *Report in Regard to the Range and Ranch Cattle Business of the United States*, U.S. Department of the Treasury Document, no. 690. Washington, DC: GPO, 1885.

Norrie, K. H. "The Rate of Settlement of the Canadian Prairies, 1870–1911." *Journal of Economic History* 35, no. 2 (1975): 410–27.

Norse, Clifford C. "The Southern Cultivator, 1843–1861." PhD Diss., Florida State University, 1969.

North Dakota Agricultural Experiment Station (Fargo). "Grain and Forage Crops." *Bulletin*, no. 10 (May 1893).

North Dakota Agricultural Experiment Station (Fargo). "Grain and Forage Crops." *Bulletin*, no. 11 (Nov. 1893).

Nowak, Ronald W. *Walker's Mammals of the World*, 6th ed. Baltimore, MD: Johns Hopkins University Press, 1999.

Nystrom, Amer B. "Dairy Cattle Breeds." *U.S. Dept. of Agriculture Farmers' Bulletin*, no. 1443 (1938).

Oakes, Elinor F. "A Ticklish Business: Dairying in New England and Pennsylvania, 1750–1812." *Pennsylvania History* 47, no. 3 (1980): 195–212.

Oerke, Erich-Christain. "Estimated Crop Losses Due to Pathogens, Animal Pests, and Weeds." In *Crop Production and Crop Protection: Estimated Losses in Major Food and Cash Crops*, edited by Erich-Christain Oerke, Heinz-Wilhelm Dehne, Fritz Schönbeck, and A. Weber., 72–741. Amsterdam: Elsevier, 1994.

Ohio State Board of Agriculture. *Annual Report 1857*. Columbus, OH: State Printer, 1858.

O'Kelly, J. F. "Cotton Varieties and Breeding." In *Cotton Production, Marketing, and Utilization*, edited by W. B. Andrews, 19–53. State College, MS: W. B. Andrews, 1950.

Oklahoma State University Department of Animal Science. *Breeds of Livestock* [Web page]. Stillwater, OK: Oklahoma State University, 2005 [cited 20 June 2007]. Available from http://www.ansi.okstate.edu/breeds/.

Olmstead, Alan L. "The Mechanization of Reaping and Mowing in American Agriculture, 1833–1870." *Journal of Economic History* 35, no. 2 (1975): 327–52.

Olmstead, Alan L., and Paul W. Rhode. "The Agricultural Mechanization Controversy, 1919–1945." *Agricultural History* 68, no. 3 (1994): 35–53.

Olmstead, Alan L., and Paul W. Rhode. "Beyond the Threshold: An Analysis of the Characteristics and Behavior of Early Reaper Adopters." *Journal of Economic History* 55, no. 1 (1995): 27–57.

Olmstead, Alan L., and Paul W. Rhode. "Hog-Round Marketing, Seed Quality, and Government Policy: Institutional Change in U.S. Cotton Production, 1920–1960." *Journal of Economic History* 63, no. 2 (2003): 447–88.

Olmstead, Alan L., and Paul W. Rhode. "An Impossible Undertaking: The Eradication of Bovine Tuberculosis in the United States." *Journal of Economic History* 64, no. 3 (2004): 734–72.

Olmstead, Alan L., and Paul W. Rhode. "Induced Innovation in American Agriculture: A Reconsideration." *Journal of Political Economy* 101, no. 1 (1993): 100–18.

Olmstead, Alan L., and Paul W. Rhode. "Induced Innovation in American Agriculture: An Econometric Analysis." *Research in Economic History* 18 (1998): 103–19.

Olmstead, Alan L., and Paul W. Rhode. "An Overview of California Agricultural Mechanization, 1870–1930." *Agricultural History* 62, no. 3 (1988): 86–112.

Olmstead, Alan L., and Paul W. Rhode. "Quantitative Indices of the Early Growth of the California Wine Industry." In *Homenaje al Profesor Tortella*, edited by José Morilla Critz, Juan Hernánez Andréu, José Luis García Ruiz, and José María Ortiz Villajos. Alcalá, Spain: Servicio de Publicaciones de la Universidad de Alcalá, 2008.

Olmstead, Alan L., and Paul W. Rhode. "The Red Queen and the Hard Reds: Productivity Growth in American Wheat, 1800–1940." *Journal of Economic History* 62, no. 4 (2002): 929–66.

Olmstead, Alan L., and Paul W. Rhode. "Reshaping the Landscape: The Impact and Diffusion of the Tractor in American Agriculture, 1910–60." *Journal of Economic History* 61, no. 3 (2001): 663–98.

Olmstead, Alan L., and Paul W. Rhode. "The 'Tuberculous Cattle Trust': Disease Contagion in an Era of Regulatory Uncertainty." *Journal of Economic History* 64, no. 4 (2005): 929–63.

Olmsted, Frederick Law. *A Journey in the Seaboard Slave States: With Remarks on Their Economy*. New York: Dix and Edwards, 1856.

Olson, P. J., H. L. Walster, and T. H. Hopper. "Corn for North Dakota." *North Dakota Agricultural Experiment Station (Fargo) Bulletin*, no. 207 (1927).

"One-Variety Cotton Program." *Textile World* 92, no. 1 (1942): 103–4.

Ordish, George. *Great Wine Blight*. London: Sudwych and Jackson, 1987.

Orland, Barbara. "Cow's Milk and Human Disease: Bovine Tuberculosis and the Difficulties Involved in Combating Animal Diseases." *Food & History* 1, no. 1 (2003): 179–202.

O'Rourke, Kevin H. "The European Grain Invasion, 1870–1913." *Journal of Economic History* 57, no. 4 (1997): 775–801.

Osband, Kent. "The Boll Weevil Versus 'King Cotton'." *Journal of Economic History* 45, no. 3 (1985): 627–43.

Palm, Einar W. "Estimated Crop Losses Without the Use of Fungicides and Nematicides and Without Nonchemical Controls." In *CRC Handbook of Pest*

Management in Agriculture, Vol. 1, edited by David Pimentel, 139–57. Boca Raton, FL: CRC Press, 1981.

Pardey, Philip G., Julian M. Alston, Jason E. Christian, and Shenggen Fan. *Hidden Harvest: U.S. Benefits from International Research Aid*, Food Policy Report. Washington, DC: International Food Policy Research Institute, 1996.

Parker, J. B. "Cow–Testing Associations a Factor in Low–Cost Dairying." In *U.S. Dept. of Agriculture Yearbook 1926*, 280–3. Washington, DC: GPO, 1926.

Parker, William N., and Judith L. V. Klein. "Productivity Growth in Grain Production in the United States, 1840–60 and 1900–10." In *Output, Employment, and Productivity in the United States After 1800*, Vol. 30, 523–80. New York: Columbia University Press for the National Bureau of Economic Research, 1966.

Parkinson, Richard. *A Tour in America in 1798, 1799, and 1800: Exhibiting Sketches of Society and Manners, and a Particular Account of the American System of Agriculture, with Its Recent Improvements*. London: J. Harding, 1805.

Patterson, F. L., and R. E. Allan. "Soft Wheat Breeding in the United States." In *Soft Wheat: Production, Breeding, Milling, and Uses*, edited by W. T. Yamazaki and C. T. Greenwood, 33–98. St. Paul, MN: American Association of Cereal Chemists, 1981.

Patton, P. "The Relationship of Weather to Crops in the Plains Region of Montana." *Montana Agricultural Experiment Station Bulletin*, no. 206 (1927).

Paullin, Charles O. *Atlas of the Historical Geography of the United States*. Washington, DC: Carnegie Institution, 1932.

Pauly, Philip J. "Fighting the Hessian Fly: American and British Responses to Insect Invasion, 1776–1789." *Environmental History* (July 2002): 485–507.

Peedin, G. F. "Flue-Cured Tobacco." In *Tobacco: Production, Chemistry, and Technology*, edited by D. Layten Davis and Mark T. Nielsen, 5A, 104–42. Oxford; Maiden, MA: Blackwell Science, 1999.

Perkins, John H. "Boll Weevil Eradication." *Science* 207, no. 4435 (1980): 1044–50.

Peters, Walter H. *Livestock Production*. New York: McGraw–Hill, 1942.

Peterson, Paul D., and C. Lee Campbell. "Beverly T. Galloway: Visionary Administrator." *Annual Review of Phytopathology* 35 (1997): 29–43.

Peterson, Paul D., C. Lee Campbell, and Clay S. Griffith. "James E. Teshemacher and the Cause and Management of Potato Blight in the United States." *Plant Disease* 76, no. 7 (1992): 754–6.

Peterson, R. F. *Wheat*. London: Leonard Hill Books, 1965.

Phares, D. L. *Farmer's Book of Grasses and Other Forage Plants for the Southern United States*. Starkville, MS: J. C. Hill, 1881.

Philadelphia Gazette and Universal Daily Adverstiser [Philadelphia, PA], Various issues and dates.

Pinney, Thomas. *A History of Wine in America: From the Beginnings to Prohibition*. Berkeley, CA: University of California Press, 1989.

Pioneer Hi-Bred International. *A History of Pioneer's First Ten Years*. Des Moines, IA: Pioneer, 1979.

Piper, Charles V., and Katherine S. Bort. "Early Agricultural History of Timothy." *Journal of the American Society of Agronomy* 7, no. 1 (1915): 1–14.
Piper, Charles V., and William J. Morse. *The Soybean.* New York: McGraw–Hill, 1923.
Pirtle, T. R. *A Handbook of Dairy Statistics.* Washington, DC: GPO, 1922.
Pirtle, T. R. *A Handbook of Dairy Statistics.* Washington, DC: GPO, 1928.
Pirtle, T. R. *History of the Dairy Industry.* Chicago: Mojonnier Bros., 1926.
Pittsfield Sun [Pittsfield, MA], Various issues and dates.
Plant Disease Bulletin 1–6 (1918–23).
Plant Disease Reporter 7–22 (1924–40).
Plumb, Charles S. *Beginnings in Animal Husbandry.* St. Paul, MN: Webb Publishing, 1912.
Plumb, Charles S. *Indian Corn Culture.* Chicago: Breeder's Gazette Print, 1895.
Plumb, Charles S. *Types and Breeds of Farm Animals.* Boston: Ginn, 1906.
Poehlman, John Milton, and David Allen Sleper. *Breeding Field Crops*, 4th ed. Ames, IA: Iowa State University Press, 1995.
Pollan, Michael. *The Omnivore's Dilemma: A Natural History of Four Meals.* New York: Penguin Press, 2006.
Poneleit, C. G. "Breeding White Endosperm Corn." In *Specialty Corns*, edited by A. R. Hallaur. Boca Raton, FL: CRC Press, 1994.
Porcher, Richard Dwight, and Sarah Fick. *The Story of Sea Island Cotton*, 1st ed. Charleston, SC: Wyrick, 2005.
Porter, Horace G. "Toward Standardized Cotton Production." *Agricultural Situation Spring Planting Issue* 29, no. 2 (1945): 21–2.
Post, Lauren C. "The Domestic Animals and Plants of French Louisiana as Mentioned in the Literature with Reference to Sources, Varieties and Uses." *Louisiana Historical Quarterly* 16 (1933): 554–86.
Powell, Fred Wilbur. *Bureau of Plant Industry: Its History, Activities and Organization.* Baltimore, MD: The Johns Hopkins Press, 1927.
Power, Richard Lyle. *Planting Corn Belt Culture: The Impress of the Upland Southerner and Yankee in the Old Northwest*, Indiana Historical Society Publications, no. 17. Indianapolis, IN: Indiana Historical Society, 1953.
Prentice, E. Parmalee, ed. *American Dairy Cattle: Their Past and Present.* New York: Harper and Sons, 1942.
Prescott, Maurice Sheldon, and Frank Tilden Price. *Holstein–Freisian History.* Lacona, NY: Holstein–Freisian World, 1930.
Prescott, Maurice Sheldon, and Melvin Schall. *Holstein–Freisian History.* Diamond Jubilee, ed. Lacona, NY: Holstein–Freisian World, 1960.
Preston, Samuel H., and Michael R. Haines. *Fatal Years: Mortality in Late Nineteenth-Century America*, NBER Series on Long-Term Factors in Economic Development. Princeton, NJ: Princeton University Press, 1991.
Prince, Eldred E., Jr. *The Long Green: The Rise and Fall of Tobacco in South Carolina.* Athens, GA: University of Georgia Press, 2000.
Pritchett, John Perry. *The Red River Valley, 1811–1849: A Regional Study.* New Haven, CT: Yale University Press, 1942.
Purdue University Agricultural Experiment Station. "Indiana Stallion Enrollment." *Circular*, no. 282 (1942).

Quaintance, A. L. "The San Jose Scale and Its Control." *U.S. Dept. of Agriculture Farmers' Bulletin*, no. 650 (1915).

Quisenberry, Karl S., and L. P. Reitz. "Turkey Wheat: The Cornerstone of an Empire." *Agricultural History* 48, no. 1 (1974): 98–110.

Rabaud, Etienne. "Telegony: A Superstition That Dies Hard." *Journal of Heredity* 5, no. 9 (1914): 389–99.

Ragland, Robert L. *On the Cultivation and Curing of Tobacco, and More Particularly of Fine Yellow Tobacco*. Richmond, VA: Clemitt and Jones, 1872.

Ragland, Robert L. *Tobacco, How to Raise It and How to Make It Pay*. Hyco, VA: Ragland Seed Co., 1895.

Rains, J. F. "Cotton Improvement by One-Variety Community Production." *Arkansas Extension Circular*, no. 414 (1938).

Ramsey, Elizabeth. *The History of Tobacco Production in the Connecticut Valley*, Smith College Studies in History, Vol. 15, no. 3–4. Northampton, MA: Smith College, Department of History, 1930.

Rasmussen, Wayne D. "The Impact of Technological Change on American Agriculture, 1862–1962." *Journal of Economic History* 22, no. 4 (1962): 578–91.

Ratner, Sidney, James H. Soltow, and Richard Sylla. *The Evolution of the American Economy*. New York: Macmillan Publishing, 1993.

Reese, H. H. "Breeding Horses for the United States Army." In *U.S. Dept. of Agriculture Yearbook 1917*. Washington, DC: GPO, 1918.

Report of Seed Certified in . . . by State Certifying Agencies, various years.

Republican Star and General Advertiser [Easton, MD], Various issues and dates.

Rhode, Paul W. "Learning, Capital Accumulation, and the Transformation of California Agriculture." *Journal of Economic History* 55, no. 4 (1995): 773–800.

Rice, J. B., H. H. Mitchell, and R. J. Laible. *A Comparison of White and Yellow Corn for Growing and Fattening Swine and for Brood Sows*. Urbana, IL: Illinois Agricultural Experiment Station, 1926.

Rice, Otis K. "Importations of Cattle into Kentucky, 1785–1860." *Register of the Kentucky State Historical Society* 49 (1951).

Rice, Victor A. *Breeding and Improvement of Farm Animals*. New York: McGraw–Hill, 1926.

Richards, Henry I. *Cotton and the AAA*. Washington, DC: Brookings Institution, 1936.

Riley, Charles V. *Fourth Report of the United States Entomological Commission, Being a Revised Edition of Bulletin No. 3 and the Final Report on the Cotton Worm, Together with a Chapter on the Boll Worm*. Washington, DC: GPO, 1885.

Riley, Charles V. "Injurious and Beneficial Insects in the United States: Insecticides and Insecticide Appliances." In *Reports of the United States Commissioners to the Universal Exposition of 1889 at Paris*, Vol. 5, *Agriculture*, edited by Charles V. Riley, 603–17. Washington, DC: GPO, 1891.

Riley, Charles V. *The Locust Plague in the United States*. Chicago: Rand McNally, 1877.

Ritvo, Harriet. *The Animal Estate: The English and Other Creatures in the Victorian Age*. Cambridge, MA: Harvard University Press, 1987.

Roadhouse, Chester L., and James L. Henderson. *The Market Milk Industry*. New York: McGraw–Hill, 1941.

Robert, Joseph Clarke. *The Tobacco Kingdom: Plantation, Market, and Factory in Virginia and North Carolina, 1800–1860*. Durham, NC: Duke University Press, 1938.

Roberts, I. P., M. V. Slingerland, and J. L. Stone. "The Hessian Fly: Its Ravages in New York in 1901." *Cornell University Agricultural Experiment Station Bulletin*, no. 194 (1901): 239–60.

Roeding, George C. "The Fig in California." In *California, Its Products, Resources, Industries, and Attractions: What It Offers the Immigrant, Home-seeker, Investors, and Tourist*, edited by T. G. Daniells, 96–101. Sacramento, CA: California Louisiana Purchase Exposition Commission; State Printer, 1904.

Roelfs, Alan P. "Effects of Barberry Eradication on Stem Rust in the United States." *Plant Disease* 66, no. 2 (1982): 177–81.

Roelfs, Alan P. "Estimated Losses Caused by Rust in Small Grain Cereals in the United States, 1918–76." *U.S. Dept. of Agriculture Miscellaneous Publication*, no. 1363 (1978).

Rogers, James A., and Larry E. Nelson. *Mr. D.R., A Biography of David R. Coker*. Hartsville, SC: Coker College Press, 1994.

Roistacher, C. N. "A History of the Parent Washington Navel Orange Tree." In *Proceedings of the Global Citrus Germplasm Network, 7–8 Dec. 2000*, App. 6. S.l.: GCGN, 2000.

Rommel, George M. "Government Encouragement of Imported Breeds of Horses." In *Bureau of Animal Industry Annual Report 1905*, 147–59. Washington, DC: GPO, 1905.

Roosevelt, Theodore. *State of the Union Address: Theodore Roosevelt to the Senate and House of Representatives, December 6, 1904* [Online document]. Boston: Information Please, Pearson Education, 2007. Available from http://www.infoplease.com/t/hist/state-of-the-union/116.html.

Roper, Daniel C. "Cotton Ginning." In *Twelfth Census of the United States 1900, Vol. IX Manufactures, Pt. III Special Reports on Selected Industries*, edited by U.S. Census Bureau, 331–40. Washington, DC: GPO, 1902.

Rosenau, Milton J. "Pasteurization." In *Milk and Its Relation to the Public Health*, 591–628. Washington, DC: GPO, 1908.

Rosengraten, Theodore. *Tombee: Portrait of a Cotton Planter*. New York: William Morrow, 1986.

Rothenberg, Winifred Barr. *From Market-Places to a Market Economy: The Transformation of Rural Massachusetts, 1750–1850*. Chicago: University of Chicago Press, 1992.

Rouse, John E. *The Criollo: Spanish Cattle in the Americas*. Norman, OK: University of Oklahoma Press, 1977.

Russell, E. Z. "Breeds of Swine." *U.S. Dept. of Agriculture Farmers' Bulletin*, no. 1263 (1922).

Russell, E. Z., S. S. Buckley, O. E. Baker, C. E. Gibbons, R. H. Wilcox, H. W. Hawthorne, S. W. Mendum, O. C. Stine, G. K. Holmes, A. V. Swarthout, W. B. Bell, G. S. Jamieson, C. W. Warburton, and C. F. Langworthy. "Hog

Production and Marketing." In *U.S. Dept. of Agriculture Yearbook 1922*, 181–280. Washington, DC: GPO, 1923.

Russell, Howard S. *A Long, Deep Furrow: Three Centuries of Farming in New England*. Hanover, NH: University Press of New England, 1976.

Russell, Nicholas. *Like Engend'ring Like: Heredity and Animal Breeding in Early Modern England*. Cambridge: Cambridge University Press, 1986.

Russell, W. A. "Genetic Improvement of Maize Yields." *Advances in Agronomy* 46 (1991): 245–98.

Russelle, Michael P. "Alfalfa." *American Scientist* 89, no. 3 (2001): 252–61.

Ryan, Harold J., and University of California (System). Division of Agricultural Sciences. *Plant Quarantines in California*. Berkeley, CA: University of California Agricultural Publications, 1969.

Sackman, Douglas Cazaux. *Orange Empire: California and the Fruits of Eden*. Berkeley, CA: University of California Press, 2005.

Salmon, S. C. "Climate and Small Grain." In *U.S. Dept. of Agriculture Yearbook 1941*, 334–5. Washington, DC: GPO, 1942.

Salmon, S. C. "Developing Better Varieties of Wheat for Kansas." In *Wheat in Kansas*, 210–17. Topeka, KS: Kansas State Board of Agriculture, 1920.

Salmon, S. C., O. R. Mathews, and R. W. Luekel. "A Half Century of Wheat Improvement in the United States." In *Advances in Agronomy*, Vol. 5, edited by A. G. Norman, 1–151. New York: Academic Press, 1953.

Sanders, Alvin H. *A History of the Percheron Horse*. Chicago, IL: Breeders' Gazette Print, 1917.

Sanders, Alvin H. *Short-Horn Cattle: A Series of Historical Sketches, Memoirs and Records of the Breed and Its Development in the United States and Canada*, 2nd ed. Chicago, IL: Sanders, 1909.

Sanders, James Harvey. *The Breeds of Live Stock and the Principles of Heredity*. Chicago, IL: J.H. Sanders, 1887.

Sargen, Nicholas P. *"Tractorization" in the United States and its Relevance for the Developing Countries*. New York: Garland, 1979.

Schafer, E. G., E. F. Gaines, and O. E. Barbee. "Two Important Varieties of Winter Wheat: A Comparison of Red Russian and Hybrid 128." *Washington Agricultural Experiment Station Bulletin*, no. 159 (1921).

Schafer, E. G., E. F. Gaines, and O. E. Barbee. "Wheat Varieties in Washington." *Washington Agricultural Experiment Station Bulletin*, no. 207 (1926).

Schertz, Lyle P., and others. *Another Revolution in U.S. Farming?: A Summary Analysis of the Structure of U.S. Farming*, U.S. Department of Agriculture Agricultural Economic Report no. 441. Washington, DC: U.S. Department of Agriculture, 1979.

Schlebecker, John T., and Andrew W. Hopkins. *A History of Dairy Journalism in the United States, 1810–1950*. Madison, WI: University of Wisconsin Press, 1957.

Schmidt, Hubert G. *Agriculture in New Jersey: A Three-Hundred-Year History*. New Brunswick, NJ: Rutgers University Press, 1973.

Schöpf, Johann D. *Travels in the Confederation, 1783–1784*, Edited and translated by Alfred J. Morrison. Philadelphia, PA: W. J. Campbell, 1911.

Schumaker, Max George. *The Northern Farmer and His Markets During the Late Colonial Period*. New York: Arno Press, 1975.

Schwartz, P. H., and W. Klassen. "Estimate of Losses Caused by Insects and Mites to Agricultural Crops." In *CRC Handbook of Pest Management in Agriculture*, Vol. 1, edited by David Pimentel, 15–77. Boca Raton, FL: CRC Press, 1981.

Sechrist, Robert P. *Basic Geographic and Historic Data for Interfacing ICPSR Data Sets, 1620–1983 [United States]* [Computer file]. Baton Rouge, LA: Louisiana State University [producer], 1984; Ann Arbor, MI: Inter-university Consortium for Political and Social Research [distributor], 2000. Available from http://www.icpsr.umich.edu [Membership required].

Seftel, Howard. "Government Regulation and the Rise of the California Fruit Industry: The Entrepreneurial Attack on Fruit Pests, 1880–1920." *Business History Review* 59, no. 3 (1985): 369–402.

The Shaker Manuscript Collection. Western Reserve Historical Society, Cleveland, OH.

Shamel, A. D. "Citrus Performance Records and Methods of Judging Citrus Fruits and Trees." *U.S. Dept. of Agriculture Farmers' Bulletin*, no. 794 (1917, Rev. 1928).

Shamel, A. D. "The Improvement of Tobacco by Breeding and Selection." In *U.S. Dept. of Agriculture Yearbook 1904*, 435–42. Washington, DC: GPO, 1905.

Shamel, A. D. "Washington Navel Orange: Important California Citrus Fruit Originated in Brazil Nearly a Century Ago, Brought to United States in 1869: Comparison of Culture in California and Brazil: Importance of Bud Mutations." *Journal of Heredity* 6, no. 10 (1915): 434–45.

Shamel, A. D., and W. W. Cobey. "Tobacco Breeding." *U.S. Bureau of Plant Industry Bulletin*, no. 234 (1907).

Shamel, A. D., L. B. Scott, and C. S. Pomeroy. "Citrus-Fruit Improvement: A Study of Bud Variation in the Valencia Orange." *U.S. Dept. of Agriculture Bulletin*, no. 624 (1918).

Shamel, Archibald Dixon. Papers, 1909–52. Special Collections. UC Riverside Libraries, Riverside, CA.

Shaw, George Wright. "How to Increase the Yield of Wheat in California." *California Agricultural Experiment Station Bulletin* (1911).

Shaw, Thomas. *Weeds and How to Eradicate Them*. Toronto, ON: J. E. Bryant, 1893.

Sheffield, John Holroyd. *Observations on the Commerce of the American States with the West Indies*. Philadelphia, PA: Robert Bell, 1783.

Shelford, V. E., and W. P. Flint. "Populations of the Chinch Bug in the Upper Mississippi Valley from 1823 to 1940." *Ecology* 24, no. 4 (1943): 435–55.

Shepard, S. M. *The Hog in America*. Indianapolis, IN: Swine Breeders' Journal, 1886.

Shepherd, Geoffrey S. *Agricultural Price Control*. Ames, IA: Iowa State College Press, 1945.

Sherbakoff, C. D. "An Improved Method of Delinting Cotton Seed with Sulphuric Acid." *University of Tennessee Agricultural Experiment Station Circular*, no. 3 (1926).

Shoemaker, P. B., and H. D. Shew. "Fungal and Bacterial Diseases." In *Tobacco: Production, Chemistry and Technology*, edited by D. Layten Davis and Mark T. Nielsen, 6A, 183–97. Oxford; Maiden, MA: Blackwell Science, 1999.

Short, Sara D. *Characteristics and Production Costs of U.S. Dairy Operations: Electronic Report from the Economic Research Service* [Online document]. Washington, DC: GPO, 2004 [cited 12 June 2007]. Available from http://www.ers.usda.gov/publications/sb974–6/sb974–6.pdf.

Short, Sara D. *Structure, Management, and Performance Characteristics of Specialized Dairy Farm Businesses in the United States* [Online document]. Washington, DC: U.S. Department of Agriculture Resource Economics Division, 2000 [cited 12 June 2007]. Available from http://www.ers.usda.gov/Publications/ah720/ah720.pdf.

Simpson, D. M. "Natural Cross-Pollination in Cotton." *U.S. Dept. of Agriculture Technical Bulletin*, no. 1094 (1954).

Sims, John A., and Leslie E. Johnson. *Animals in the American Economy*. Ames, IA: Iowa State University Press, 1972.

Sinnott, Edmund W., and L. C. Dunn. *Principles of Genetics: An Elementary Text, with Problems*. New York: McGraw–Hill, 1925.

Slafer, Gustavo A., Emilio H. Satorre, and Fernando H. Andrade. "Increases in Grain Yield in Bread Wheat from Breeding and Associated Physiological Changes." In *Genetic Improvement in Field Crops*, edited by Gustavo A. Slafer, 1–68. New York: Marcel Dekker, 1994.

Silicon Valley History Online Collection. 1859–1991. San Jose Public Library, California Room, San Jose, CA.

Smalley, H. E. "Estimated Losses Without Pesticides and Substituting Only Readily Available Nonchemical Controls for Livestock Pests." In *CRC Handbook of Pest Management in Agriculture*, Vol. 1, edited by David Pimentel, 169–71. Boca Raton, FL: CRC Press, 1981.

Smil, Vaclav. *Creating the Twentieth Century: Technical Innovations of 1867–1914 and Their Lasting Impact*. New York: Oxford University Press, 2005.

Smil, Vaclav. *Enriching the Earth: Fritz Haber, Carl Bosch, and the Transformation of World Food Production*. Cambridge, MA: MIT Press, 2001.

Smil, Vaclav. "Horse Power: The Millennium of the Horse Began with a Whimper, But Went out With a Bang." *Nature* 405 (May 11, 2000): 125.

Smith, C. Wayne. *Crop Production: Evolution, History, and Technology*. New York: John Wiley and Sons, 1995.

Smith, C. Wayne, and J. Tom Cothren, eds. *Cotton: Origin, History, Technology, and Production*. New York: John Wiley, 1999.

Smith, C. Wayne, Roy G. Cantrell, Hal S. Moser, and Stephen R. Oakley. "History of Cultivar Development in the United States." In *Cotton: Origin, History, Technology, and Production*, edited by C. Wayne Smith and J. Tom Cothren, 99–171. New York: John Wiley & Sons, 1999.

Smith, E. F. "The Granville Wilt of Tobacco." *U.S. Bureau of Plant Industry Bulletin*, no. 141 (1908).

Smith, H. P. "Cultural Practices." In *Cotton Production, Marketing, and Utilization*, edited by W. B. Andrews, 135–71. State College, MS: W. B. Andrews, 1950.

Smith, John. *The Generall Historie of Virginia, New England & the Summer Isles*. London: Michael Sparks, 1624.

Smith, P. R. "Introductory Remarks." In *Proceedings of the 1970 Beltwide Cotton Production Research Conferences: Cotton Planting Seed Quality Seminar*. Memphis, TN: National Cotton Council, 1970.

Smith, Ralph E., and collaborators. "Protecting Plants from Their Enemies." In *California Agriculture*, edited by Claude B. Hutchison, 239–315. Berkeley, CA: University of California Press, 1946.

Socolofsky, Homer E., and Huber Self. *Historical Atlas of Kansas*. Norman, OK: University of Oklahoma Press, 1972.

Southern Cultivator [Augusta, GA], Various issues and dates.

Soxman, R. C. *Marketing of Cotton in Producers' Local Markets*. Washington, DC: U.S. Department of Agriculture, Production & Marketing Administration, Cotton Branch, 1949.

Spencer, D. A., M. C. Hall, C. D. Marsh, J. S. Cotton, C. E. Gibbons, O. C. Stine, O. E. Baker, V. N. Valgren, W. B. Bell, and Will C. Barnes. "The Sheep Industry." In *U.S. Dept. of Agriculture Yearbook 1922*, 229–310. Washington, DC: GPO, 1924.

Spillman, W. J. "The Hybrid Wheats." *Washington Agricultural Experiment Station Bulletin*, no. 89 (1909).

Stadler, L. J., and C. A. Helms. "Corn in Missouri: Corn Varieties and Their Improvement." *University of Missouri Agricultural Experiment Bulletin*, no. 181 (1921).

Stakman, E. C. "Plant Diseases are Shifty Enemies." *American Scientist* 35 (1947): 321–50.

Stampp, Kenneth M., ed. *Records of Ante-Bellum Southern Plantations from the Revolution Through the Civil War*. [Microfilm]. Series A, *Selections from the South Caroliniana Library, University of South Carolina*, Pt. 1, *The Papers of James Henry Hammond, 1795–1865*. Frederick, MD: University Publications of America, 1985.

Stavinoha, Katie Dickie, and Lorie A. Woodward. "Texas Boll Weevil History." In *Boll Weevil Eradication in the United States Through 1999*, edited by Willard A. Dickerson, A. L. Brashear, J. T. Brumley, F. L. Carter, W. J. Grefenstette, and F. A. Harris, 451–502. Memphis, TN: Cotton Foundation, 2001.

Steckel, Richard. "The Economic Foundations of East–West Migration During the Nineteenth Century." *Explorations in Economic History* 20, no. 1 (1983): 14–36.

Steenbock, Harry. "White Corn vs. Yellow Corn and a Probable Relation Between the Fat-Soluble Vitamine and Yellow Plant Pigments." *Science, NS* 50, no. 1293 (1919): 352–3.

Stephens, S. G. "The Origin of Sea Island Cotton." *Agricultural History* 50, no. 2 (April 1976): 391–9.

Stephens, S. G. "Some Observations of Photoperiodism and the Development of Annual Forms of Domesticated Cottons." *Economic Botany* 30 (1975): 409–18.

Stevens, Frank L., and John G. Hall. *Diseases of Economic Plants*. New York: Macmillan, 1910.

Stewart, Elliot W. *Feeding Animals: A Practical Work upon the Laws of Animal Growth Specially Applied to the Rearing and Feeding of Horses, Cattle, Dairy Cows, Sheep, and Swine*, 1st ed. Lake View, NY: Author, 1883.

Stewart, Elliot W. *Feeding Animals: A Practical Work upon the Laws of Animal Growth Specially Applied to the Rearing and Feeding of Horses, Cattle, Dairy Cows, Sheep, and Swine*, 4th ed. Lake View, NY: Author, 1888.

Stewart, George. *Alfalfa–Growing in the United States and Canada*. New York: Macmillan, 1926.

Stewart, J. B. "The Production of Cigar-Wrapper Tobacco Under Shade in the Connecticut Valley." *U.S. Bureau of Plant Industry Bulletin*, no. 138 (1908): 1–12.

Stoll, Steven. "Insects and Institutions: University Science and the Fruit Business of California." *Agricultural History* 69, no. 2 (1995): 216–39.

Strauss, Frederick, and Louis H. Bean. "Gross Farm Income and Indices of Farm Production and Prices in the United States, 1869–1937." *U.S. Dept. of Agriculture Technical Bulletin*, no. 703 (1940).

Street, James H. *The New Revolution in the Cotton Economy: Mechanization and Its Consequences*. Chapel Hill, NC: University of North Carolina Press, 1957.

Strohm, John. "The Conestoga Horse." In *Report of the Commissioner of Agriculture 1863*, 175–80. Washington, DC: GPO, 1864.

Strutevant, E. L. "Varieties of Corn." *U.S. Office of Experiment Stations Bulletin*, no. 57 (1899).

Stucky, Harley J. A. *Century of Russian Mennonite History in America*. North Newton, KS: Mennonite Press, 1973.

Talley, Steven. "Hessian Fly Genomics Research Will Benefit Wheat Farmers, Others." *Purdue News* (March 2001).

Taylor, Fred. "Relation Between Primary Market Prices and Qualities of Cotton." *U.S. Dept. of Agriculture Bulletin*, no. 457 (1916).

Taylor, Paul, and Thomas Vasey. "Historical Background of California Farm Labor." *Rural Sociology* 1, no. 3 (1936): 281–95.

Texas Agricultural Extension Service. *The New Agriculture: 1949 Annual Report*. College Station, TX: The Service, 1950.

tfX. *What Are Trans Fats?* [Web page]. Oxford, UK: tfX, c2007 [cited 6 July 2007]. Available from http://www.tfx.org.uk/page3.html.

Thompson, James Westfall. *A History of Livestock Raising in the United States, 1607–1860*, Agricultural History Series, no. 5. Washington, DC: U.S. Department of Agriculture, 1942.

Thorpe, T. B. "Cotton and Its Cultivation." *Harper's Magazine* (March 1854): 447–63.

Tilley, Nannie May. *The Bright–Tobacco Industry, 1860–1929*. Chapel Hill, NC: University of North Carolina Press, 1948.

Tobacco Institute. *Virginia & Tobacco: A Chapter in America's Industrial Growth*, Tobacco History Series. Washington, DC: The Institute, 1960.

Towne, Charles W., and Edward N. Wentworth. *Pigs: From Cave to Corn Belt*, 1st ed. Norman, OK: University of Oklahoma Press, 1950.

Towne, Marvin W., and Wayne D. Rasmussen. "Farm Gross Product and Gross Investment in the Nineteenth Century." In *Trends in the American Economy in the Nineteenth Century*, edited by William N. Parker, 255–315. Princeton, NJ: Princeton University Press, 1960.

Tracy, S. M. "Cultivated Varieties of Cotton." In *The Cotton Plant: Its History, Botany, Chemistry, Culture, Enemies, and Uses*, edited by Charles Dabney, 197–224. Washington, DC: GPO, 1896.

Traill, Catherine Parr. *The Backwoods of Canada, Being Letters from the Wife of an Emigrant Officer, Illustrative of the Domestic Economy of British America.* 1971 Reprint ed. London: Charles Knight, 1836.

Tripp, Robert. "The Institutional Conditions for Seed Enterprise Development." *Overseas Development Institute Working Paper*, no. 105 (1997).

Trowbridge, E. A. "Mule Production." In *Farm Knowledge*, Vol. 1, *Farm Animals*, edited by E. L. D. Seymour. Garden City, NY: Doubleday, Page, 1918.

Troyer, A. Forrest. "Background of U.S. Hybrid Corn." *Crop Science* 39, no. 3 (1999): 601–26.

Troyer, A. Forrest. "Persistent and Popular Germplasm in Seventy Centuries of Corn Evolotuion." In *Corn: Origin, History, Technology, and Production*, edited by C. Wayne Smith, Javier Betran, and E. C. A. Runge, 133–231. New York: Wiley, 2004.

Troyer, A. Forrest. "Temperate Corn: Background, Behavior, and Breeding." In *Speciality Corns*, edited by A. R. Hallaur, 393–466. Boca Raton, FL: CRC Press, 1994.

Troyer, A. Forrest, and Lois G. Hendrickson. "Background and Importance of Minnesota 13 Corn." *Crop Science* 47, no. 3 (2007): 905–14.

Tso, T. C. "Seed to Smoke." In *Tobacco: Production, Chemistry, and Technology*, edited by D. Layten Davis and Mark T. Nielsen, 1–31. Oxford; Maiden, MA: Blackwell Science, 1999.

Tufts, Warren P., and collaborators. "The Rich Pattern of California Crops." In *California Agriculture*, edited by Claude B. Hutchison, 113–315. Berkeley, CA: University of California Press, 1946.

Turner, John. *White Gold Comes to California*. Bakersfield, CA: California Planting Cotton Seed Distributors, 1981.

Tyler, Frederick J. "Varieties of American Upland Cotton." *U.S. Bureau of Plant Industry Bulletin*, no. 163 (1910).

Tysdal, H. M., and H. L. Westover. "Alfalfa Improvement." In *U.S. Dept. of Agriculture Yearbook 1937*, 1122–53. Washington, DC: GPO, 1937.

U.S. Agricultural Marketing Service. *Cotton Varieties Planted* (Various issues and dates).

U.S. Agricultural Marketing Service. "Get Your Green Card Mr. Farmer!" *Agricultural Situation* 40, no. 11 (1956).

U.S. Agricultural Marketing Service. *Report of the Chief 1941*. Washington, DC: GPO, 1941.

U.S. Agricultural Research Administration. *Report of the Administrator of Agricultural Research 1947*. Washington, DC: GPO, 1948.

U.S. Agricultural Research Service. *GrainGenes: A Database for Triticeae and Avena* [Online database]. Washington, DC: USDA, 2007 [updated regularly]. Available from http://wheat.pw.usda.gov/GG2/index.shtml.

U.S. Agricultural Research Service. *Summary of DHI Participation, 1997–2007* [Web page]. Baltimore, MD: Animal Improvement Programs Laboratory, 1997–2007 [cited 29 May 2007]. Available from http://aipl.arsusda.gov/publish/dhi/part.html.

U.S. Animal and Plant Health Inspection Service. *Boll Weevil Eradication: Factsheet* [Online document], 2002 [cited 14 May 2007]. Available from http://www.aphis.usda.gov/lpa/pubs/fsheet_faq_notice/faq_phbweevil.pdf.

U.S. Animal and Plant Protection Service. *Boll Weevil Eradication Program, Update* [Online document], Spring 2007 [cited 24 May 2007]. Available from http://www.aphis.usda.gov/plant_health/plant_pest_info/cotton_pests/index.shtml.

U.S. Bureau of Agricultural Economics. "Horses, Mules, and Motor Vehicles: Year Ended March 31, 1924 with Comparable Data for Earlier Years." *U.S. Dept. of Agriculture Statistical Bulletin*, no. 5 (1925).

U.S. Bureau of Agricultural Economics. *Income Parity for Agriculture,* Pt. 3, *Prices Paid by Farmers for Commodities and Services,* Sec. 4, *Prices Paid by Farmers for Farm Machinery and Motor Vehicles, 1910–38*. Washington, DC: GPO, 1939.

U.S. Bureau of Agricultural Economics. *Livestock on Farms, January 1, 1867–1919, Revised Estimates, Number, Value per Head, Total Value, by States and Division.* Washington, DC: U.S. Department of Agriculture, 1938.

U.S. Bureau of Agricultural Economics. "Statistics on Cotton and Related Data." *U.S. Dept. of Agriculture Statistical Bulletin*, no. 99 (1951).

U.S. Bureau of Agricultural Economics. Division of Cotton Marketing. "Grade, Staple Length, and Tenderability of Cotton in the United States, 1928–29 to 1933–34." *U.S. Dept. of Agriculture Statistical Bulletin*, no. 15 (1936).

U.S. Bureau of Animal Industry. "A Plan for the Improvement of American Breeding Stock." In *Annual Report 1903*, 316–25. Washington, DC: GPO, 1904.

U.S. Bureau of Animal Industry. *Special Report on Diseases of Cattle 1912*. Washington, DC: GPO, 1912.

U.S. Bureau of Animal Industry. *Special Report on Diseases of Cattle 1916*. Washington, DC: GPO, 1916.

U.S. Bureau of Crop Estimates. "White, Yellow, and Mixed Corn." *Monthly Crop Reporter*, various issues and years.

U.S. Bureau of Markets. "Marketing Cotton Seed for Planting." *Market Reporter* 1, no. 5 (1920).

U.S. Bureau of Plant Industry. *Better Cottons*. Beltsville, MD: GPO, 1947.

U.S. Census Bureau. *Cotton Production in the United States (1906–1965).*

U.S. Census Bureau. *Historical Statistics of the United States: Colonial Times to 1970*. 2 vols. Washington, DC: GPO, 1975.

U.S. Census Bureau. *Mortality Rates 1910–1920*. Washington, DC: GPO, 1922.

U.S. Census Bureau. *Quantity of Cotton Ginned in U.S.: Crops of 1900 to 1904, Inclusive*, U.S. Census Bureau Bulletin, no. 19. Washington, DC: GPO, 1905.

U.S. Census Bureau. *Statistical Atlas of the United States*. Washington, DC: GPO, 1925.

U.S. Census Bureau. "Tuberculosis in the United States." In *Mortality Statistics 1907*, 516. Washington, DC: GPO, 1909.

U.S. Census Bureau. 7th Census 1850. *The Seventh Census of the United States, 1850. Agriculture: Farms and Implements, Stock Products, Home Manufactures*. Washington, DC: GPO, 1853.

U.S. Census Bureau. 8th Census 1860. *Agriculture of the United States in 1860: Compiled from the Original Returns of the Eighth Census*. Washington, DC: GPO, 1864.

U.S. Census Bureau. 8th Census 1860. *Population Schedules of the Eighth Census of the United States, 1860: Kentucky. Bracken County* [Microform]. Washington, DC: The Bureau, 1860.

U.S. Census Bureau. 8th Census 1860. *Population Schedules of the Eighth Census of the United States, 1860: Ohio. Brown County* [Microform]. Washington, DC: The Bureau, 1860.

U.S. Census Bureau. 9th Census 1870. *The Statistics of the Wealth and Industry of the United States: Compiled from the Original Returns of the Ninth Census (June 1, 1870)*. Washington, DC: GPO, 1872.

U.S. Census Bureau. 10th Census 1880. *Report on the Productions of Agriculture as Returned at the Tenth Census (June 1, 1880)*. Washington, DC: GPO, 1883.

U.S. Census Bureau. 11th Census 1890. *Report on the Statistics of Agriculture in the United States at the Eleventh Census 1890*. Washington, DC: GPO, 1895.

U.S. Census Bureau. 12th Census 1900. *Census Reports: Twelfth Census of the United States, Taken in the Year 1900*. Vols. 5–6, *Agriculture*. Washington, DC: GPO, 1902.

U.S. Census Bureau. 13th Census 1910. *Thirteenth Census of the United States Taken in the Year 1910*. Vols. 5–7, *Agriculture 1909 and 1910*. Washington, DC: GPO, 1913.

U.S. Census Bureau. 14th Census 1920. *Fourteenth Census of the United States Taken in the Year 1920: Reports*. Vols. 5–6, *Agriculture*. Washington, DC: GPO, 1922.

U.S. Census Bureau. 15th Census 1930. *Fifteenth Census of the United States 1930. Agriculture*. 8 vols. Washington, DC: GPO, 1932.

U.S. Census Bureau. 15th Census 1930. *Fifteenth Census of the United States 1930. Census of Agriculture. The Farm Horse*. Washington, DC: GPO, 1933.

U.S. Census Bureau. 16th Census 1940. *Sixteenth Census of the United States: 1940. Agriculture*. 3 vols. Washington, DC: GPO, 1941.

U.S. Census Bureau. 20th Census 1980. *Census of Population. California*. Vol. 1, pt. 6. Washington, DC: GPO, 1983.

U.S. Census Bureau. 21st Census 1990. *1990 Census of Population. Social and Economic Characteristics. California*. Vol. CP-2-6-1. Washington, DC: GPO, 1993.

U.S. Census Bureau. 22nd Census 2000. *Census 2000 Summary File 3: Industry by Sex 2000* [Computer file]. Washington, DC: GPO, 2003. Available from http://factfinder.census.gov (Table QT–P27).

U.S. Census Bureau. Census of Agriculture 1945. *Special Report: Farms and Farm Characteristics by Value of Products*. Washington, DC: GPO, 1948.

U.S. Census Bureau. Census of Agriculture 1950. *United States Census of Agriculture 1950*, Vol. 2. Washington, DC: GPO, 1952.

U.S. Census Bureau. Census of Agriculture 1954. *United States Census of Agriculture: 1954*, Vol. 3, *Special Reports*, pt. 11, *Farmers' Expenditures in 1955*. Washington, DC: GPO, 1956.

U.S. Census Bureau. Census of Agriculture 1959. *U.S. Census of Agriculture: 1959. Final Report.* 5 vols. Washington, DC: GPO, 1961–63.

U.S. Census Bureau. Census of Agriculture 1969. *1969 Census of Agriculture. General Report.* Washington, DC: GPO, 1971.

U.S. Commission to the Paris Exposition 1889. "List of the Exhibit Material. Division 2. Food Substances of Vegetable Origin." In *Reports of the United States Commissioners to the Universal Exposition of 1889 at Paris*, Vol. 5, *Agriculture*, 861–76. Washington, DC: GPO, 1891.

U.S. Commission to the Paris Exposition 1889. "The Number of Exhibitors in the United States Section." In *Reports of the United States Commissioners to the Universal Exposition of 1889 at Paris*, Vol. 1, 342–3. Washington, DC: GPO, 1891.

U.S. Crop Reporting Board. *Prices Paid by Farmers for Commodities and Services, United States 1910–1960*, U.S. Department of Agriculture Statistical Bulletin, no. 319. Washington, DC: U.S. Department of Agriculture, 1962.

U.S. Department of Agriculture. *Agricultural Statistics* (Various years).

U.S. Department of Agriculture. *Crops and Markets* 12, no. 2 (1935): 34.

U.S. Department of Agriculture. *Monthly Report* (May and June 1872).

U.S. Department of Agriculture. *Report of the Administrator of the Production and Marketing Administration* (Various years).

U.S. Department of Agriculture. *Report of the Commissioner of Agriculture* (Various years).

U.S. Department of Agriculture. *Report of the Secretary of Agriculture* (Various years).

U.S. Department of Agriculture. *Yearbook* (Various years).

U.S. Department of Agriculture. *1997 Census of Agriculture.* Vol. 1, *National, State, and County Tables*, pt. 5, *California State-Level Data.* Washington, DC: National Agricultural Statistics Service; GPO, 1999.

U.S. Department of Agriculture. *Agricultural Prices 1950.* Washington, DC: GPO.

U.S. Department of Agriculture. "Cost of Raising Horses 1912." *Crop Reporter* 15, no. 4 (1913).

U.S. Department of Agriculture. "Cotton Seed for Planting Purposes in Fair Demand." *Weather, Crops, and Markets* 1, no. 3 (1922).

U.S. Department of Agriculture. "Economic Indicators of the Farm Sector, Production and Efficiency Statistics." *Statistical Bulletin*, no. 679 (1980).

U.S. Department of Agriculture. "Livestock Breeding at the Crossroads." In *Yearbook 1936*, 831–62. Washington, DC: GPO, 1936.

U.S. Department of Agriculture. "Progress in Horse Breeding." *Farmers' Bulletin*, no. 419 (1910).

U.S. Department of Agriculture. "A Progress Report on the Investigations of the European Corn Borer." *Department Bulletin*, no. 1476 (1927).

U.S. Department of Agriculture. *Report of European Corn-Borer Control Campaign by the United States Department of Agriculture for the Period May 14, 1927–Oct 31, 1927, Inclusive.* Washington, DC: GPO, 1928.

U.S. Department of Agriculture. "Shortage of Work Animals in Sight." In *Crops and Markets*, 108–9. Washington, DC: GPO, 1925.

U.S. Department of Agriculture. "Silos and Ensilage: A Record of Practical Tests in Several States and Canada." *Special Report*, no. 48 (1882).

U.S. Department of Agriculture. "Statistics of Grain Crops 1921: Wheat." In *Yearbook 1921*, 520–38. Washington, DC: GPO, 1922.

U.S. Department of Agriculture. "Wintering Farm-Animals." *U.S. Dept. of Agriculture Monthly Review* (February/March 1875).

U.S. Department of Agriculture. *Yearbook 1943–1947: Science in Farming*. Washington, DC: GPO, 1947.

U.S. Department of Agriculture. Division of Cereal Crops and Diseases. Photograph Collection. U.S. National Agricultural Library Manuscript Collections, Beltsville, MD.

U.S. Department of Health and Human Services. Health Resources and Services Administration. Bureau of Health Professions. *Bureau of Health Professions Area Resource File, 1940–1990: United States* [Computer file]. Rockville, MD: U.S. Department of Health and Human Services, Office of Data Analysis and Management [producer], 1991; Ann Arbor, MI: ICPSR [distributor], 1994. Available from http://www.icpsr.umich.edu/cocoon/ICPSR/STUDY/09075.xml [Membership required].

U.S. Department of State. "Cattle and Dairy Farming in Ontario, Consul Pace, Port Sarnia." In *Reports from the Consuls of the United States on Cattle and Dairy Farming and the Markets for Cattle, Beef, and Dairy Products in Their Several Districts*, 540–6. Washington, DC: GPO, 1887.

U.S. Department of State. "Cattle-Raising in Quebec, Consul Parker, Sherbrooke." In *Reports from the Consuls of the United States on Cattle and Dairy Farming and the Markets for Cattle, Beef, and Dairy Products in Their Several Districts*, 571–4. Washington, DC: GPO, 1887.

U.S. Economic Research Service. *Dairy: Background* [Online document]. Washington, DC: The Service, 2004 [cited 29 May 2007]. Available from http://www.ers.usda.gov/Briefing/Dairy/Background.htm.

U.S. Economic Research Service. "Statistics on Cotton and Related Data, 1920–73." *U.S. Dept. of Agriculture Statistical Bulletin*, no. 535 (1974).

U.S. Farm Security Administration. Office of War Information. FSA/OWI Photograph Collection. Library of Congress, Prints & Photographs Division, Washington, DC.

U.S. Federal Extension Service. *Report of Cooperative Extension Work in Agriculture and Home Economics* (Various issues and years).

U.S. Federal Extension Service. Records. General Summary Reports. National Archives, College Station, MD.

U.S. Federal Trade Commission. *Report of the Federal Trade Commission of Milk and Milk Products, 1914–1918*. Washington, DC: GPO, 1921.

U.S. House Committee on Agriculture. *Cotton: Hearings*. 78th Cong., 2nd sess., December 4–9, 1944.

U.S. House Committee on Agriculture. *Hearings*. 78th Cong., 1st sess., 1943.

U.S. House Committee on Agriculture. *Hearings Before Committee on Agriculture Dealing with Cotton Standards*. 67th Cong., 4th sess., pt. 1, February 5, 9, and 12, 1928.

U.S. House Committee on Agriculture. *Letter from H. A. Wallace dated March 16, 1937, Congressional Record.* 75th Cong., 1st sess., Vol. 81, pt. 3, April 5, 1937.

U.S. House Committee on Agriculture. *Research and Related Services in the U.S. Department of Agriculture.* 81st Cong., 2nd sess., Vol. 1, December 21, 1950.

U.S. House Committee on Agriculture. *Study of Agricultural and Economic Problems of the Cotton Belt: Hearings.* 80th Cong., 1st sess., pt. 2, October 10, 1947.

U.S. House Committee on Appropriations. *Agricultural Department Appropriation Bill.* (Various years).

U.S. House Committee on Ways and Means. *Hearings on Tariff Readjustment, 1929.* 70th Cong., 2nd sess., Vol. 7, 1929.

U.S. House Committee on Ways and Means. *Tariff Hearings.* 66th Cong., 3rd sess., 1921.

U.S. House Committee on Ways and Means. *Tariff Hearings.* 60th Cong., 2nd sess., Vol. 4, Doc. 1505, 1909.

U.S. House Committee on Ways and Means. *Tariff Schedules and Hearings 1913.* 62nd Cong., 3rd sess., Vol. 3, Doc. 1477, 1913.

U.S. Industrial Commission. *Report on the Distribution of Farm Products*, U.S. Industrial Commission Reports, Vol. 6. Washington, DC: GPO, 1901.

U.S. National Agricultural Statistics Service. *Quick Stats* [Online database]. Washington, DC: U.S. Department of Agriculture; GPO [updated frequently] [cited 8 August 2007]. Available from http://www.nass.usda.gov/Data_and_Statistics/Quick_Stats/index.asp.

U.S. National Parks Service. *Lincoln Notebook: The Plant That Killed Nancy Hanks Lincoln* [Web page]. Lincoln City, IN: U.S. NPS Lincoln Boyhood National Memorial, 2007 [cited 29 May 2007]. Available from http://www.nps.gov/archive/libo/white_snakeroot3.htm.

U.S. Office of Marketing Services. *Report of the Director* (Various issues and dates).

U.S. Patent Office. *Report of the Commissioner. Agriculture*, 1845–1855. Washington, DC: GPO, 1846–56.

U.S. Senate Committee on Agriculture and Forestry. *Authorizing the Secretary of Agriculture to Provide for Classification of Cotton, to Furnish Information on Market Supply, Demand, Location, Condition, and Market Prices for Cotton, and Other Purposes: Report (to accompany S. 1500).* 75th Cong., 1st sess., February 24, 1937.

U.S. Senate Committee on Patents. *Plant Patents: Report to Accompany S. 4015.* 71st Cong., 2nd sess., Report no. 315, April 2, 1930.

U.S. Tariff Commission. *The Wool-Growing Industry.* Washington, DC: GPO, 1921.

Ullrey, Duane E. "Landmark and Historic Contributions of NRC's Committee on Animal Nutrition." In *Scientific Advances in Animal Nutrition: Promise for the New Century, Proceedings of a Symposium*, edited by the National Research Council Board on Agriculture and Natural Resources. Washington, DC: National Academy of the Sciences, 2002.

Ullstrup, A. J. "The Impacts of the Southern Corn Leaf Blight Epidemics of 1970–1971." *Annual Review of Phytopathology* 10 (1972): 37–50.

University of California Integrated Pest Management Program. *UC IPM Online* [Web site]. Davis, CA: UC Agriculture and Natural Resources [updated frequently]. Available from http://www.ipm.ucdavis.edu/.

Van Kuren, S. J. "Dairy Machinery and Its Part in the Dairy Industry." In *History of the Dairy Industry*, by T. R. Pirtle, 73–175. Chicago: Mojonner Bros., 1926.

Van Valen, L. "A New Evolutionary Law." *Evolutionary Theory* 1 (1973): 1–30.

Van Willigen, John, and Susan C. Eastwood. *Tobacco Culture: Farming Kentucky's Burley Belt*. Lexington, KY: University Press of Kentucky, 1998.

Vaughan, Henry W. *Breeds of Live Stock in America*. Columbus, OH: College Book Co., 1948.

Vicksburg Sentinel [Vicksburg, MS], Various issues and dates.

Virts, Nancy. "Change in the Plantation System: American South, 1910–1945." *Explorations in Economic History* 43 (2006): 153–76.

Virts, Nancy. "The Efficiency of Southern Tenant Plantations, 1900–1945." *Journal of Economic History* 51, no. 2 (1991): 385–95.

Voelker, D. E. "History of Dairy Recordkeeping." *National Cooperative Dairy Herd Improvement Program Handbook Fact Sheet*, no. A-2 (1985).

Waddle, Billy M., and Rex F. Colwick. "Producing Seeds of Cotton and Other Fiber Crops." In *U.S. Dept. of Agriculture Yearbook 1961*: 188–92. Washington, DC: GPO, 1962.

Wall Street Journal [New York], Various issues and dates.

Wallace, Henry A., and Earl N. Bressman. *Corn and Corn Growing*. Des Moines, IA: Wallace Publishing, 1923.

Wallace, Henry A., and Earl N. Bressman. *Corn and Corn Growing*, 3rd ed. New York: John Wiley, 1928.

Wallace, Henry A., and Earl N. Bressman. *Corn and Corn Growing*, 4th rev. ed., Wiley Farm Series. New York: John Wiley, 1937.

Wallace, Henry A., and Earl N. Bressman. *Corn and Corn Growing*, 5th rev. ed., Wiley Farm Series. New York: John Wiley, 1949.

Wallace, Henry A., and William L. Brown. *Corn and Its Early Fathers*, rev. ed. Ames, IA: Iowa State University Press, 1988.

Wallschlaeger, F. O. "The World's Production and Commerce in Citrus Fruits and their By-Products." *Citrus Protective League of California Bulletin*, no. 11 (1914).

Walsh, Lorena S. "Consumer Behavior, Diet and the Standard of Living in Late Colonial and Early Antebellum America, 1770–1840." In *American Economic Growth and Standards of Living Before the Civil War*, edited by Robert E. Gallman and John J. Wallis, 217–64. Chicago: University of Chicago Press, 1922.

Walsh, Lorena S. *Feeding the Eighteenth-Century Town Folk, or, Whence the Beef?* [Web page]. Williamsburg, VA: Colonial Williamsburg Foundation, 2000 [cited 6 July 2007]. Available from http://research.history.org/Historical_Research/Research_Themes/ThemeRespect/Feeding.cfm.

Walsh, Lorena S. "Plantation Management in the Chesapeake, 1620–1820." *Journal of Economic History* 49, no. 2 (1989): 393–406.

Walsh, Lorena S., Ann Smart Martin, and Joanne Bowen. *Provisioning Early American Towns, The Chesapeake: A Multidisciplinary Case Study: Final Progress Report, September 1997* [Online document]. Williamsburg, VA: Colonial Williamsburg Foundation, 1997 [cited 6 July 2007]. Available from http://research.history.org/Files/HistRes/Provisioning.pdf.

Walton, Gary M., and Hugh Rockoff. *History of the American Economy*. Orlando, FL: Dryden Press, 1998.

Walton, W. R. "The Green-Bug or Spring Grain-Aphis: How to Prevent Its Periodical Outbreaks." *U.S. Dept. of Agriculture Farmers' Bulletin*, no. 1217 (1921).

Ward, Tony. "The Origins of the Canadian Wheat Boom, 1880–1910." *Canadian Journal of Economics* 27, no. 4 (1994): 864–83.

Warder, J. T. "Mule Raising." In *Report of the Commissioner of Agriculture 1863*, 180–90. Washington, DC: GPO, 1864.

Ware, Jacob Osborn. "Origin, Rise and Development of American Upland Cotton Varieties and Their Status at Present." *University of Arkansas College of Agriculture, Agricultural Experiment Station Mimeo* (1950).

Ware, Jacob Osborn. "Plant Breeding and the Cotton Industry." In *U.S. Dept. of Agriculture Yearbook 1936*, 657–744. Washington, DC: GPO, 1936.

Warntz, William. "An Historical Consideration of the Terms 'Corn' and 'Corn Belt' in the United States." *Agricultural History* 31, no. 1 (1957): 40–5.

Warren, George F. *Elements of Agriculture*, 2nd ed. New York: Macmillan, 1910.

Washington, George. *George Washington to John Beale Bordley, Mount Vernon, August 17, 1788* [Web page]. Charlottesville, VA: University of Virginia Press, 2007 [cited 10 April 2007]. Available from http://www.mountvernon.org/learn/explore_mv/index.cfm/pid/575/.

Washington Post [Washington, DC], Various issues and dates.

Wasson, R. A. "One Variety Cotton Improvement Association." *Louisiana State University Division of Agricultural Extension Circular*, no. 182 (1939).

Watkins, James L. *King Cotton: A Historical and Statistical Review, 1790 to 1908*. Reprint of 1908 ed. New York: Negro Universities Press, 1969.

Watt, George. *The Wild and Cultivated Cotton Plants of the World*. London: Longmans, Green, and Co., 1907.

Weatherwax, Paul. *Indian Corn in Old America*. New York: Macmillan, 1954.

Webb, Henry W. "Private Cotton Breeding in the Southeast." In *Proceedings of the Beltwide Cotton Conferences 1998*, Vol. 1, 522–34. Memphis, TN: National Cotton Council, 1998.

Webb, Walter Prescott. *The Great Plains*. New York: Gosset & Dunlap, 1931.

Webber, Herbert J. "The Effect of Research in Genetics on the Art of Breeding, Part 2." *American Breeders Magazine* 3, no. 2 (1912): 125–34.

Webber, Herbert J. "History and Development of the Citrus Industry." In *The Citrus Industry*, Vol. 1, *History, Botany, and Breeding*, edited by Herbert J. Webber and Leon D. Batchelor, 1–40. Berkeley, CA: University of California Press, 1943.

Webber, Herbert J. "History and Development of the Citrus Industry, rev. by Walter Reuther and Harry W. Lawton." In *The Citrus Industry*, Vol. 1, *History,*

World Distribution, Botany, and Varieties, edited by Walter Reuther, Herbert J. Webber, and Leon D. Batchelor, Ch. 1. Riverside, CA: University of California Division of Agricultural Sciences, 1967.

Weber, G. A. *The Plant Quarantine and Control Administration: Its History, Activities and Organization.* Washington, DC: Brookings Institution, 1930.

Webster, F. M. "The Hessian Fly." *Ohio Agricultural Experiment Station Bulletin*, no. 107 (1899): 257–88.

Wedin, Walter. *Improvements* [Web page]. Corvallis, OR: Alfalfa Information System, Oregon State University. Available from http://forages.oregonstate.edu/IS/AIS/enfpmain.cfm?PageID=261.

Wendel, Jonathan F., Curt L. Brubaker, and A. Edward Percival. "Genetic Diversity in Gossypium Hirsutum and the Origin of Upland Cotton." *American Journal of Botany* 79, no. 11 (1992): 1291–313.

Wendel, Jonathan F., and Richard C. Cronn. "Polyploidy and the Evolutionary History of Cotton." *Advances in Agronomy* 78 (2003): 139–86.

Westbrook, E. C. "One-Variety Community Cotton Production." *University of Georgia Agricultural Extension Service Bulletin*, no. 449 (1935).

Westbrook, E. C. "One-Variety Cotton Communities." *Crops & Soils* 8, no. 6 (1956): 16–17.

Westgate, J. M. "Another Explanation of the Hardiness of Grimm Alfalfa." *Science, NS* 30, no. 762 (1909): 184–6.

Whatley, Warren C. "Labor for the Picking: The New Deal in the South." *Journal of Economic History* 43, no. 4 (1983): 905–29.

Wiley, G. F. "Cotton Improvement in Carroll County, Georgia, Through Use of the One Variety Community Plan of Production." In *Proceedings of the 39th Annual Convention of the Association of Southern Agricultural Workers.* Atlanta, GA: The Association, 1938.

Will, George F. *Corn for the Northwest.* St. Paul, MN: Webb Book Publishing, 1930.

Will, George F., and George E. Hyde. *Corn Among the Indians of the Upper Missouri.* Lincoln, NE: University of Nebraska Press, 1964, © 1917.

Williams, J. O., and William Jackson. "Improving Horses and Mules." In *U.S. Dept. of Agriculture Yearbook 1936*, 929–46. Washington, DC: GPO, 1936.

Williams, Robert C. *Fordson, Farmall, and Poppin' Johnny: A History of the Farm Tractor and Its Impact on America.* Urbana, IL: University of Illinois Press, 1987.

Willis, J. W. "One Variety Cotton Community Organization." *Mississippi Agricultural College Extension Circular*, no. 96 (1938).

Windels, C. E. "Economic and Social Impacts of Fusarium Head Blight: Changing Farms and Rural Communities in the Northern Great Plains." *Phytopathology* 90 (2000): 17–21.

Wing, Henry H. "The Dairy Industry of the United States." In *Reports of the United States Commissioners to the Universal Exposition of 1889 at Paris*, Vol. 5, *Agriculture*, edited by Charles V. Riley, 577–92. Washington, DC: GPO, 1891.

Wing, Henry H. *Milk and Its Products: A Treatise upon the Nature and Qualities of Dairy Milk and the Manufacture of Butter and Cheese*. New York: Macmillan, 1897.

Wing, Joseph E. *Sheep Farming in America*. Chicago, IL: Sanders Publishing, 1912.

Winters, Laurence M. *Animal Breeding*. New York: Wiley, 1925.

Wisconsin Department of Agriculture and Markets. *Biennial Report 1935–36*, Bulletin no. 178. Madison, WI: The Department, 1936.

Wisconsin Department of Agriculture and Markets. *Biennial Report 1939–40*, Bulletin no. 219. Madison, WI: The Department, 1940.

Wisconsin Historical Images. Wisconsin Historical Society Library–Archives, Madison, WI.

Witzel, S. A. "Development of Dairy Farm Engineering." *Journal of Dairy Science* 39, no. 6 (1956): 777–82.

Wood, William Price, Jr. "An Economic Study of the Southern Seed Prices." MA Diss., Cornell University, 1927.

Woodward, Carl R. "The Jersey Red Hog." *New Jersey Agriculture* 11, no. 4 (1929): 14.

Woodward, Carl R. *Ploughs and Politicks: Charles Read of New Jersey and His Notes on Agriculture, 1715–1774*. New Brunswick, NJ: Rutgers University Press, 1941.

Woodward, Carl R. "Sheep at $1,000 a Head." *New Jersey Agriculture* 10, no. 10 (1928): 14–15.

Worcester, Donald Emmet. *The Texas Longhorn: Relic of the Past, Asset for the Future*, 1st ed. College Station, TX: Texas A&M University Press, 1987.

Worthen, Edmund L. *Farm Soils: Their Management and Fertilization*. New York: John Wiley, 1927.

Worthley, L. H., and D. J. Caffrey. "Timely Information about the European Corn Borer." *U.S. Dept. of Agriculture Miscellaneous Circular*, no. 70 (1926).

Wrenn, Lynette Boney. *Cinderella of the New South: A History of the Cottonseed Industry, 1855–1955*. Knoxville, TN: University of Tennessee Press, 1995.

Wright, David E. "Alcohol Wrecks a Marriage: The Farm Chemurgic Movement and the USDA in the Alcohol Fuels Campaign in the Spring of 1933." *Agricultural History* 67, no. 1 (1993): 33–66.

Wright, Gavin. *Old South, New South: Revolutions in the Southern Economy since the Civil War*. New York: Basic Books, 1986.

Wright, J. W. *Marketing Practices in Producers' Local Cotton Markets*. Washington, DC: USDA Bureau of Agricultural Economics, 1938.

Young, Edward. *Special Report on Immigration 1869–70*. Washington, DC: U.S. Department of the Treasury, Bureau of Statistics; GPO, 1872.

Index

Note: Figures and tables are indicated with an italicized page number; footnotes are indicated by an n following the page number.

www.ingramcontent.com/pod-product-compliance
Ingram Content Group UK Ltd.
Pitfield, Milton Keynes, MK11 3LW, UK
UKHW050345180125
453697UK00001B/1

9 780521 857116